Suresh C. Pillai, Vignesh Kumaravel (Eds.)
Photocatalysis

Also of Interest

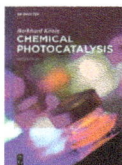

Chemical Photocatalysis
König, 2020
ISBN 978-3-11-057654-2, e-ISBN 978-3-11-057676-4

Quantum Electrodynamics of Photosynthesis.
Mathematical Description of Light, Life and Matter
Braun, 2020
ISBN 978-3-11-062692-6, e-ISBN 978-3-11-062994-1

Host-Guest Chemistry.
Supramolecular Inclusion in Solution
Wagner, 2021
ISBN 978-3-11-056436-5, e-ISBN 978-3-11-056438-9

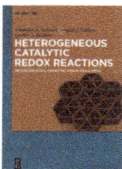

Heterogeneous Catalytic Redox Reactions.
Fundamentals and Applications
Sadykov, Tikho and Isupova, 2019
ISBN 978-3-11-058586-5, e-ISBN 978-3-11-058777-7

Photocatalysis

Edited by
Suresh C. Pillai, Vignesh Kumaravel

DE GRUYTER

Editors
Prof. Suresh C. Pillai
Institute of Technology Sligo
Ash Lane
Sligo F91 YW50
Ireland
pillai.suresh@itsligo.ie

Dr. Vignesh Kumaravel
Institute of Technology Sligo
Ash Lane
Sligo F91 YW50
Ireland
Kumaravel.Vignesh@itsligo.ie

ISBN 978-3-11-066845-2
e-ISBN (PDF) 978-3-11-066848-3
e-ISBN (EPUB) 978-3-11-066858-2

Library of Congress Control Number: 2021934990

Bibliographic information published by the Deutsche Nationalbibliothek
The Deutsche Nationalbibliothek lists this publication in the Deutsche Nationalbibliografie;
detailed bibliographic data are available on the Internet at http://dnb.dnb.de.

© 2021 Walter de Gruyter GmbH, Berlin/Boston
Cover image: Zolnierek / iStock / Getty Images Plus
Typesetting: Integra Software Services Pvt. Ltd.
Printing and binding: CPI books GmbH, Leck

www.degruyter.com

Preface

Infectious diseases, global warming, energy crisis, and pollution are serious threats to human life in the current scenario. Photocatalysis is one of the economically viable and eco-friendly technologies to address these issues in terms of hydrogen production, carbon dioxide conversion, antimicrobial coatings, wastewater treatment, and remediation of persistent organic pollutants with the help of solar energy. Various photocatalysts such as titanium dioxide, zinc oxide, graphitic-carbon nitride, cadmium sulfide, iron oxide, perovskites, and metal halides have been investigated for their potential applications in the past three decades. This book is exclusively drafted for portraying the current trends of photocatalysis. A total of nine chapters have been accepted for publication in this book, which is supposed to furnish the key insights and basic concepts for both experts and early career researchers in the photocatalysis field.

Chapter 1 details the synthesis and characterization of photocatalysts including transition metal oxides, transition metal dichalcogenides, and carbon materials for various applications. The significance of techniques such as sol–gel, precipitation, hydrothermal, solvothermal, wet chemical, electrochemical deposition, and chemical vapor deposition has been discussed.

Chapter 2 describes the fundamentals of the reactor designs for the photocatalytic and photoelectrocatalytic water treatment. The kinetics, mass transport, reactor configurations, scale-up factors, and challenges have been discussed in detail.

Chapter 3 discloses the fundamentals of photocatalytic hydrogen production. The reaction energetics, reactor designs, UV light active photocatalysts, visible light active photocatalysts, photoefficiency of materials, and the commercial challenges have been discussed in this chapter.

Chapter 4 depicts the significance of visible light–assisted photocatalytic degradation of antibiotics using semiconductor heterostructures. The insights of various heterostructures such as type I, type II, type III, Schottky, and p–n junction have been briefed. The surface plasma resonance and Z-scheme photocatalytic mechanisms have also been discussed.

Chapter 5 shows the implications of surface models and experimental advances of photocatalytic self-cleaning surfaces. Various models such as Wenzel, Cassie–Baxter, and Miwa-Hashimoto have been explained in detail. Moreover, the current advances and commercialization of TiO_2-based self-cleaning coatings have also been demonstrated.

Chapter 6 lists the applications of photocatalysts for the degradation of air pollutants. The photocatalysts such as titanium dioxide, zinc oxide, tungsten trioxide, graphene, and MXenes have been examined for the visible light–assisted degradation of various pollutants.

Chapter 7 reveals the importance of artificial photosynthesis (carbon dioxide conversion) using various photocatalytic materials. The photocatalysis mechanism,

https://doi.org/10.1515/9783110668483-202

characterization techniques, and reactor designs have also been discussed. The efficiency of various metal oxides, heterojunctions, quantum dots, ferroelectric materials, perovskites, plasmonic materials, and metal–organic frameworks has been highlighted.

Chapter 8 describes the removal of endocrine disruptors, pesticides, and pharmaceuticals in water/wastewater using various photocatalytic materials. The treatment techniques, effect of operating conditions, and degradation pathway of various pollutants have been discussed in detail. Visible light–assisted photocatalytic applications of various metal-doped semiconductors for water treatment have also been highlighted.

Chapter 9 explores the commercial aspects and future trends of photocatalytic materials with respect to photoreduction, photooxidation, or/and photoinduced superhydrophilicity phenomena. The industrial applications of photocatalysis technology for water/air treatment, hydrogen production, carbon dioxide conversion, sensors, self-cleaning surfaces, antimicrobial coatings, and photovoltaics have been reviewed in detail.

The challenges and prospects of the above chapters clearly testify that numerous attempts should be taken at pilot scale in this coming decade for the successful commercialization of this sustainable technology.

Finally, we would like to express our sincere thanks to all the authors for their substantial contributions in this book. The help and support we have received from the editorial and publishing team of De Gruyter is greatly acknowledged.

Contents

List of Contributing Authors

N. Sandhyarani
Nanoscience Research Laboratory
School of Materials Science and Engineering
National Institute of Technology Calicut
Kerala, India
sandhya@nitc.ac.in

Aruna Vijayan
Nanoscience Research Laboratory
School of Materials Science and Engineering
National Institute of Technology Calicut
Kerala, India

P. Fernandez-Ibañez
School of Engineering
Ulster University
Belfast, UK
p.fernandez@ulster.ac.uk

S. McMichael
School of Engineering
Ulster University
Belfast, UK

A. Tolosana-Moranchel
School of Engineering
Ulster University
Belfast, UK

J. A. Byrne
School of Engineering
Ulster University
Belfast, UK

Laura Clarizia
Department of Chemical Materials and
Industrial Production Engineering
University of Naples "Federico II"
P. le V. Tecchio 80
80125 Naples, Italy
laura.clarizia2@unina.it

Priyanka Ganguly
Nanotechnology and Bio-engineering
Research Group
Department of Environmental Science
Institute of Technology Sligo

Sligo, Ireland
priyanka.ganguly@mail.itsligo.ie

Suyana Panneri
Materials Science and Technology Division
National Institute for Interdisciplinary
Science and Technology (CSIR-NIIST)
Thiruvananthapuram, India
Email: suyansam@gmail.com

U. S. Hareesh
Materials Science and Technology Division
National Institute for Interdisciplinary
Science and Technology (CSIR-NIIST)
Thiruvananthapuram, India
and
Academy of Scientific and Innovative
Research (AcSIR)
New Delhi, India
hareesh@niist.res.in

Saeed Punnoli Ammed
Department of Chemistry
Government Engineering College
Calicut, Kerala
India

Anupama R. Prasad
Department of Chemistry
University of Calicut
Kerala
India

Sanjay Gopal Ullattil
Chemical Engineering Program
Texas A&M University at Qatar
P.O. Box 23874
Doha, Qatar
sanjaygopal.u@gmail.com
sanjay.ullattil@qatar.tamu.edu

A. Joseph Nathanael
Tae Hwan Oh
Department of Chemical Engineering
Yeungnam University
Gyeongsan, South Korea

https://doi.org/10.1515/9783110668483-204

Tae Hwan Oh
Department of Chemical Engineering
Yeungnam University
Gyeongsan, South Korea

James A. Sullivan
UCD School of Chemistry
Belfield, Ireland

Raphaël Abolivier
UCD School of Chemistry
Belfield, Ireland

Beatriz Villajos
Instituto de Catálisis y Petroleoquímica
IPC-CSIC, C/Marie Curie 2
28049 Madrid, Spain

Sara Mesa-Medina
Instituto de Catálisis y Petroleoquímica
IPC-CSIC, C/Marie Curie 2
28049 Madrid, Spain

Marisol Faraldos
Instituto de Catálisis y Petroleoquímica
IPC-CSIC, C/Marie Curie 2
28049 Madrid, Spain

Ciara Byrne
Department of Inorganic Chemistry and
Technology
National Institute of Chemistry
Hajdrihova 19
SI-1001 Ljubljana, Slovenia

Ana Bahamonde
Instituto de Catálisis y
Petroleoquímica
IPC-CSIC, C/Marie Curie 2
28049 Madrid, Spain

Antonio Gascó
Department of Forest and Environmental
Engineering and Management
Universidad Politécnica de
Madrid
Escuela Técnica Superior de Ingeniería de
Montes
Forestal y del Medio Natural
C/José Antonio Novais 10
28040 Madrid, Spain

Daphne Hermosilla
Department of Forest and Environmental
Engineering and Management
Universidad Politécnica de
Madrid
Escuela Técnica Superior de Ingeniería de
Montes
Forestal y del Medio Natural
C/José Antonio Novais 10
28040 Madrid, Spain

Jesna Louis
Department of Polymer Science and Rubber
Technology
Cochin University of Science and Technology
Kerala, India

Nisha T. Padmanabhan
Department of Polymer Science and Rubber
Technology
Cochin University of Science and Technology
Kerala, India

Honey John
Department of Polymer Science and Rubber
Technology
Cochin University of Science and Technology
Kerala, India
honey@cusat.ac.in

Aruna Vijayan, N. Sandhyarani

Chapter 1
Synthesis and Characterization
of Photocatalytic Materials

1.1 Introduction

Solar energy is an ultimate reliable renewable source due to its vast availability. Solar energy puts considerable challenges in terms of harvesting, storage, and utilization. The innovative and efficient processes for solar energy harvesting and its utilization are of great prominence. Photocatalysis is an important method for harvesting solar energy, which is inspired by natural photosynthesis. Photocatalysis can contribute to a major part of our growing energy demands in an efficient and cost-effective manner. International Union of Pure and Applied Chemistry defined photocatalyst as a "catalyst able to produce, upon absorption of light, chemical transformations of the reaction partners. The excited state of the photocatalysts repeatedly interacts with the reaction partners forming reaction intermediates and regenerates itself after each cycle of such interactions" [1]. Although the origin of photoelectrochemical (PEC) cells may date back to Edmond Becquerel, who discovered the photovoltaic effect in 1839, its golden era started with the discovery of PEC water splitting using TiO_2 by Honda and Fujishima in the 1970s, and this, in turn, leads to many other photocatalytic/photoelectrocatalytic applications like hydrogen evolution, CO_2 reduction, removal of pollutants, and disinfection.

1.2 Photocatalysis: Mechanism and Application

The photocatalytic reaction mechanism involves four steps.
1. Electron–hole pair generation by light absorption
2. Excitation of electrons and separation of charges

Acknowledgments: The authors gratefully acknowledge the Science and Engineering Research Board, Department of Science and Technology, Government of India, for the SERB POWER fellowship granted to NS (Grant No. SPF/2021/000008)

Aruna Vijayan, Nanoscience Research Laboratory, School of Materials Science and Engineering, National Institute of Technology Calicut, Kerala, India
N. Sandhyarani, Nanoscience Research Laboratory, School of Materials Science and Engineering, National Institute of Technology Calicut, Kerala, India, e-mail: sandhya@nitc.ac.in

https://doi.org/10.1515/9783110668483-001

3. Transfer of holes and electrons to the photocatalytic surface
4. Redox reactions by surface-adsorbed charges

One of the major drawbacks of photocatalysts is the fast recombination of electron–hole pairs, which led to the dissipation of harvested energy in terms of heat or light emission. Creation of photogenerated charges with longer life period, which in turn depends on the donor or acceptor properties of the surface-adsorbed species, is one of the key challenges in photocatalysis. Figure 1.1 illustrates the general mechanism of semiconductor photocatalysis.

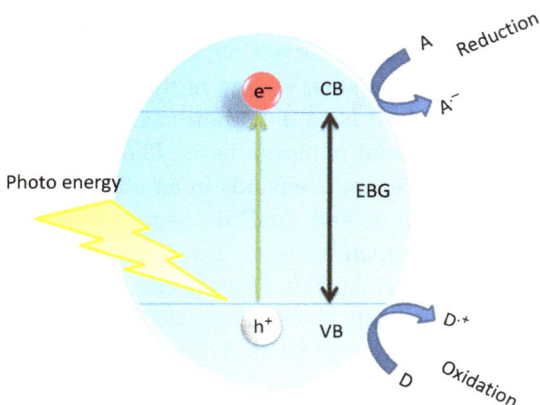

Figure 1.1: Schematic illustration of semiconductor photocatalysis. EBG is electronic bandgap, CB is conduction band, and VB is valence band.

1.2.1 Photocatalytic Hydrogen Evolution

Solar water splitting (SWS) holds a pivotal position in solar energy transformation. This research area witnesses increasing attention because it produces hydrogen, a clean energy source, in an environmentally friendly manner. In water splitting, the conduction band electrons reduce water molecules into hydrogen, and holes from the valence band oxidize water to oxygen [2]. For effective water splitting, a more negative conduction band edge than the hydrogen reduction potential ($H^+/H_2 = -0.41$ V vs normal hydrogen electrode (NHE) at pH 7), and a more positive valence band edge than the oxidation potential of water ($H_2O/O_2 = 0.82$ V vs. NHE at pH 7) are desired [3]. The ideal bandgap for overall water splitting is 1.23 eV, which falls in the infrared region of the solar spectrum. In order to absorb sunlight, the bandgap of the material should lie between 2 and 3 eV [4]. From the kinetic point of view, the two half-reactions must have comparable reaction rates. Hydrogen evolution reaction (HER) involves two electrons and protons, whereas oxygen evolution reaction (OER) involves four electrons and protons [1].

The mechanism of water splitting is as follows:

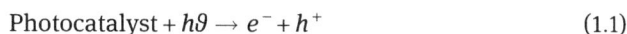

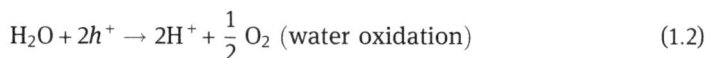

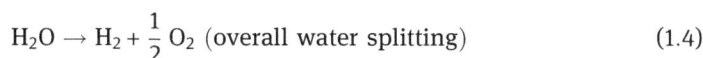

$$\text{Photocatalyst} + h\vartheta \rightarrow e^- + h^+ \tag{1.1}$$

$$H_2O + 2h^+ \rightarrow 2H^+ + \frac{1}{2}O_2 \text{ (water oxidation)} \tag{1.2}$$

$$2H^+ + 2e^- \rightarrow H_2 \text{ (water reduction)} \tag{1.3}$$

$$H_2O \rightarrow H_2 + \frac{1}{2}O_2 \text{ (overall water splitting)} \tag{1.4}$$

1.2.2 Photocatalytic Removal of Pollutants

Compared to conventional pollutant degradation techniques, photocatalytic removal of pollutants gains increasing attention because of its eco-friendliness and lack of toxic by-products. In photocatalytic removal of pollutants, the electrons, holes, and reactive radicals such as $HO^\cdot$, $HO_2^\cdot$, and $O_2^{\cdot-}$ react with the surface-adsorbed pollutants and decompose them. The efficiency of photocatalyst depends on the lifetime of charge carriers, which results in the formation of reactive free radicals.

For organic pollutant degradation, the excited electrons in the conduction band of the photocatalyst react with the oxygen and form superoxide radical anion ($O_2^{\cdot-}$), and it can oxidize organic molecules. Superoxide anion radicals can combine with H^+ to produce hydrogen peroxide (H_2O_2). Apart from this, photoinduced holes react with water molecules and form hydroxyl radicals ($\cdot OH$). These hydroxyl radicals, hydrogen peroxide, as well as superoxide anion radicals oxidize the pollutants [5]. The mechanism is summarized as follows:

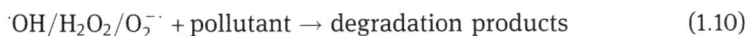

$$\text{Photocatalyst} + h\vartheta \rightarrow e^- + h^+ \tag{1.5}$$

$$H^+ + H_2O/OH \rightarrow H^+ + {}^\cdot OH \tag{1.6}$$

$$O_2 + e^- \rightarrow O_2^{-\cdot} \tag{1.7}$$

$$O_2^{-\cdot} + H^+ \rightarrow HO_2^\cdot \tag{1.8}$$

$$2HO_2^\cdot + 2H^+ \rightarrow H_2O_2 + 2{}^\cdot OH \tag{1.9}$$

$${}^\cdot OH/H_2O_2/O_2^{-\cdot} + \text{pollutant} \rightarrow \text{degradation products} \tag{1.10}$$

1.2.3 Photocatalytic CO$_2$ Reduction

Photoreduction of CO_2 to hydrocarbons is one of the solutions to solve the global energy shortage, which in turn reduces the greenhouse effect. Here, CO_2 is converted into CH_4 and other intermediates such as CO, HCOOH, HCHO, and CH_3OH. It can be considered as a carbon-neutral cycle since no auxiliary carbon source is used other than CO_2. Generally, the photocatalytic CO_2 reduction reaction can be divided into four main steps: (1) CO_2 adsorption, (2) generation of electron–hole pair in photocatalyst by absorbing incident photon energy, (3) electron–hole pair migration to the photocatalyst surface, and (4) reduction of CO_2.

Chemical reduction of CO_2 mainly involves two different reaction pathways:
1. $CO_2 \rightarrow HCOOH \rightarrow HCHO \rightarrow CH_3OH \rightarrow CH_4$
2. $CO_2 \rightarrow CO \rightarrow C\cdot \rightarrow CH_2 \rightarrow CH_4$

In both reactions, the final product is methane and reaction intermediates are different [6].

1.2.4 Photocatalytic Disinfection

The photocatalytic disinfection process is a nontoxic, efficient, and stable method in comparison with chlorination and UV disinfection. Disinfection property of photocatalytic processes makes it promising for medical applications. It will be an effective technique to fight against bioterror, preventing the spread of airborne biological threats. Photocatalytic disinfection methods are used for laboratory and hospital applications. Photocatalytic disinfection by surfaces coated with titanium dioxide is an alternative technique to traditional surface disinfection methods. It has been used in the treatment of industrial nonbiodegradable water contaminants and the treatment of effluents from the resin factor [7].

Photocatalytic disinfection proceeds through the formation of reactive oxygen species. They can cause the breakdown of cell membranes, and this will promote the internalization of the semiconductor photocatalyst and lead to the death of the cell. Improving the separation efficacy of charge carriers increases the photocatalytic disinfection efficiency [8].

1.3 Selection of Photocatalytic Material for Photocatalytic Application

Selection of a photocatalyst depends on several factors such as bandgap, structure and morphology, crystallinity, surface area and particle size, adsorption efficacy, light absorption capacity, composition, mobility, charge separation, the lifetime of exciton, and stability. Metal oxides, transition metal dichalcogenides (TMD), double hydroxides, boron nitride, and graphitic carbon nitride (g-C_3N_4) are some of the promising photocatalysts. Synthesis of a visible light active, nontoxic, inexpensive, metal-free photocatalyst is a challenge (Figure 1.2).

Apart from the positions of the valence and conduction bands, the overall photocatalytic performance of the catalyst depends on the efficiency in charge separation and charge transport. Several strategies can be opted to improve the efficacy of charge separation. For example, the construction of morphologically different nanostructures facilitates better charge transport and promotes charge separation efficiency. Moving from zero-dimensional (0D) to one-dimensional (1D) nanostructure, photocatalytic activity increases due to better charge mobility and reduced charge recombination [10]. Cocatalyst loading on semiconductor and formation of "heterojunction" photocatalysts is found to improve the photocatalytic activity [11]. Semiconductor heterostructures can be classified into three types based on their bandgaps and the electronic affinity: type I, type II, and type III. Energy band modulation using chemical doping, introducing defects in nanostructures, and so on also result in photocatalytic activity enhancement [8, 12].

Photocatalytic materials with differing morphology (powders, fibers, film, etc.) can be obtained using the appropriate synthetic procedures such as hydrothermal/solvothermal, sol–gel process, microwave, sonochemical, electrodeposition, chemical vapor deposition (CVD), and physical vapor deposition (PVD) [13]. Hydrothermal reactions are heterogeneous reactions carried out under high temperature and pressure usually above 100 °C and 1 bar in the presence of aqueous solvents. Whereas in the solvothermal process, the reaction is carried out in nonaqueous solvents. Hydrothermal method exploits the better solubility of inorganic compounds and the hastening of heterogeneous reactions in hot water. The reactions are generally done under isothermal conditions in steel autoclave with Teflon lining. Teflon lining is used to avert the corrosion of the steel vessel. Since the reactor is not stirred during the reaction, diffusive mass transfer dominates and crystallization occurs under stagnant conditions. Solvothermal synthesis generally exhibits better control over shape, size, and crystallinity of nanoparticle than hydrothermal synthesis [14].

The sol–gel method is generally used for the synthesis of colloidal dispersion of inorganic and organic–inorganic hybrid materials. Molecular-level homogeneity and low processing temperature are the major advantages of this method [13]. Commonly, metal alkoxides are used as precursors for the sol–gel synthesis, taking advantage of the formation of homogeneous solution in a range of solvents and their

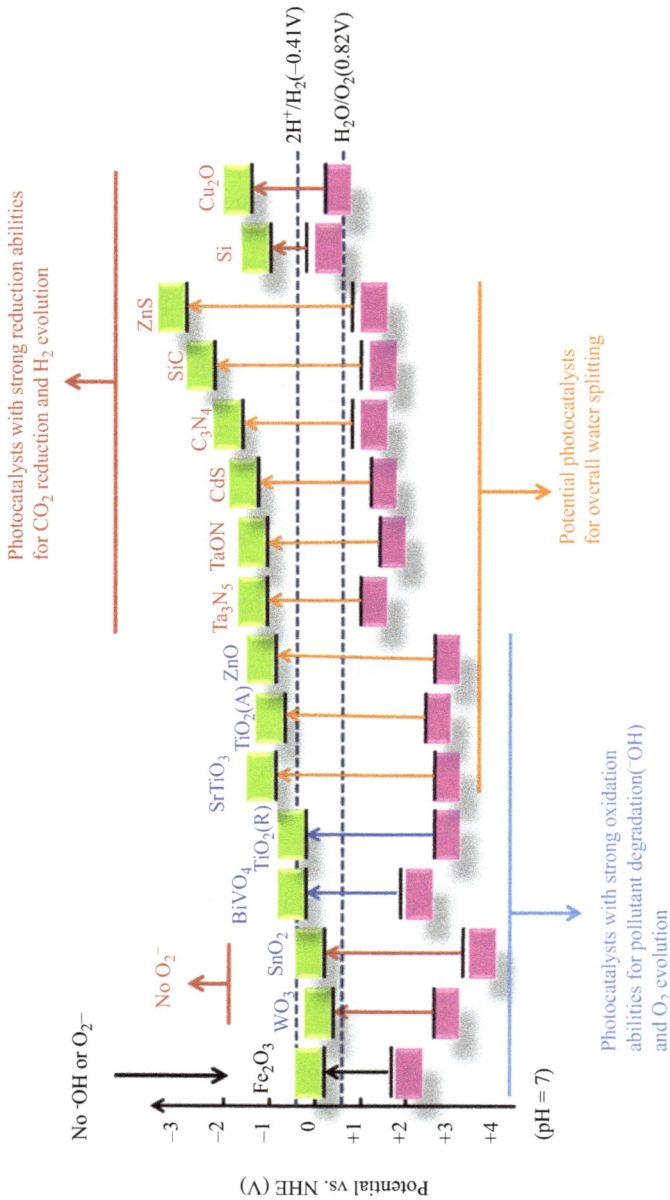

Figure 1.2: Band positions and applications of some important photocatalyst at pH 7 [9].
Reprinted in a modified form from Xin Li and coworkers [9]. Copyright (2018), with permission from
Elsevier.

high reactivity to nucleophiles. The mechanism involves the hydrolysis and condensation reactions. Nature of the final product depends on the reaction conditions, mechanism of reaction, and the heat treatment. Based on the drying process of gel, one can synthesize either dense powder or porous materials.

In the sonochemical method, ultrasound waves are used for the synthesis of nanomaterials. Main event in sonochemistry is the creation, growth, and collapse of a bubble that is formed in the liquid. Collapse of the bubble leads to the formation of reactive radicals and the reaction proceeds. Short reaction time and comparatively mild reaction conditions are the major advantages of this method. The size and morphology of the nanomaterial are determined by the reaction conditions such as the power of ultrasonication, sonication time, and the temperature.

The electrodeposition method is generally used for the synthesis of advanced thin films. Here the deposition is carried out under a given potential. Electrophoretic deposition is also widely used for the deposition of oxide thin films. Here, a potential is applied and depends on the charge of the nanoparticle it moves toward the corresponding electrodes and gets deposited.

CVD is another popular technique used for the synthesis of thin films, composite materials, mesoporous materials, and so on. CVD generally involves the chemical reaction of a volatile compound with other gases to produce a nonvolatile solid. Depending upon the precursors used and the reaction conditions, the different chemical reactions are classified into reduction, oxidation, pyrolysis, disproportionation and reversible transfer, and compound formation. It provides materials with high purity and better crystallinity. Reliant upon the types of precursors, the reaction conditions, and the forms of energy supplied to the system to activate the reaction, a variety of CVD reactors and CVD methods are being developed. High process temperature, complexity in the processes, and evolution of toxic and corrosive gases are some of the disadvantages of CVD. PVD differs from CVD that does not involve the chemical transition of the precursor. PVD is a method of transporting growth species from a source and depositing them on a substrate. The synthetic conditions and precursors of specific photocatalytic nanomaterials are discussed in Section 1.5.

1.4 Characterization Techniques for Photocatalytic Property Measurement

1.4.1 Linear Sweep Voltammetry (LSV)

In linear sweep voltammetry (LSV), the potential is swept from an initial potential to the final potential at a particular scan rate. The LSV plot can be obtained by plotting potential versus current density. It can be performed under the dark and light conditions to investigate the photocatalytic performance of the material. Higher current density in the presence of light indicates the photocatalytic activity of the material.

1.4.2 Chronoamperometric Analysis

The chronoamperometric analysis is performed at a constant potential under the dark and light conditions in a regular interval of time. The chronoamperometric plot can be obtained by plotting time against current density. Higher photocurrent density confirms increased absorption of light and greater separation of charge. A more efficient electron–hole separation provides more electrons into the circuit to trigger a higher photocurrent.

1.4.3 Electrochemical Impedance Spectroscopy (EIS)

Charge separation and its transport in a photocatalytic material can be investigated using electrochemical impedance spectroscopy (EIS). Semicircle arc diameter denotes the charge-transfer resistance (R_{ct}) of the electrode. Smaller the semicircle diameter, lower will be the charge transfer resistance indicating higher conductance and hence the photocatalytic activity [15].

1.4.4 Mott–Schottky Analysis

The Mott–Schottky equation derived from Poisson's equation is given by

$$\frac{1}{C^2} = \frac{2}{\varepsilon\varepsilon_\circ A^2 e N_D} \left(V - V_{fb} - \frac{k_B T}{e} \right)$$ (1.11)

where C is the interfacial capacitance, A is the interfacial area, N_D represents the number of donors, V is the applied voltage, k_B is Boltzmann's constant, T is the absolute temperature, and e is the electronic charge.

A plot between the reciprocal of the square of capacitance against the potential difference between the electrolyte and the semiconductor gives a straight line known as the Mott–Schottky plot. The Y-intercept of this line will give the flat band potential (V_{fb}). For a known value of ε and A, the number of donors (N_D) can also found from the slope [16].

1.4.5 Ultraviolet–Visible Diffuse Reflectance Spectroscopy (UV–Vis DRS)

The optical property of photocatalyst can be investigated with ultraviolet–visible diffuse reflectance spectroscopy (UV–vis DRS). From the absorption spectra, respective bandgap plots of $(\alpha h\nu)^2$ versus E ($=h\nu$) according to Tauc's equation can be plotted. Tauc's equation is given by

$$\alpha h\vartheta = k\left(h\vartheta - E_g\right)^{\frac{1}{2}} \tag{1.12}$$

where $\alpha = -\ln T$, h is Planck's constant, v is the frequency, E_g is the bandgap.

Smaller bandgap of the order of visible light energy indicates enhanced visible light absorption, which leads to increased production of electron–hole pairs which in turn augments the photocatalytic activity [15].

1.4.6 Photoluminescence (PL) Study

Photoluminescence (PL) study is usually employed to identify the recombination and separation of photogenerated electron–hole pairs in semiconductors. PL intensity is a direct indication of the electron–hole recombination rate. An intense peak suggests fast recombination. As the intensity decreases, recombination rate decreases [15].

Figure 1.3 represents different PEC studies performed on CuS/TiO$_2$ nanocomposite. Figure 1.3a represents transient photocurrent response for multiple on–off cycles. The CuS/TiO$_2$ nanocomposite exhibits 3.6 times higher photocurrent than TiO$_2$ nanospindles. EIS and LSV plots (Figure 1.3b and e) also support this observation, which confirms the greater absorption of light and efficient separation of charges in the composite than pristine TiO$_2$ nanospindles. The reduced diameter of the semicircle in the presence of light confirms the efficient charge generation and charge transfer in the presence of light. Further to confirm the major charge carriers, the authors used the Mott–Schottky analysis, and it is represented in Figure 1.3c and d. From the measured slope value, the n-type and p-type behavior of TiO$_2$ and CuS was confirmed. PL studies were performed to analyze the recombination rate of the photoexcited electron–hole pair. PL emission with much lower intensity for composite suggests lower recombination and efficient charge separation of the composite [15].

Optical properties of the as-synthesized catalyst was studied by means of UV–vis DRS study (Figure 1.3g). The fundamental absorption edge for TiO$_2$ was obtained at $\sim$400 nm. Absorption band edge shift toward higher wavelength upon incorporation of CuS indicating the successful formation of the composite. Corresponding bandgaps were calculated from the Tauc plots as 3.15 and 2.59 eV; smaller bandgap of the composite indicates increased visible light absorption and enhanced charge carrier formation.

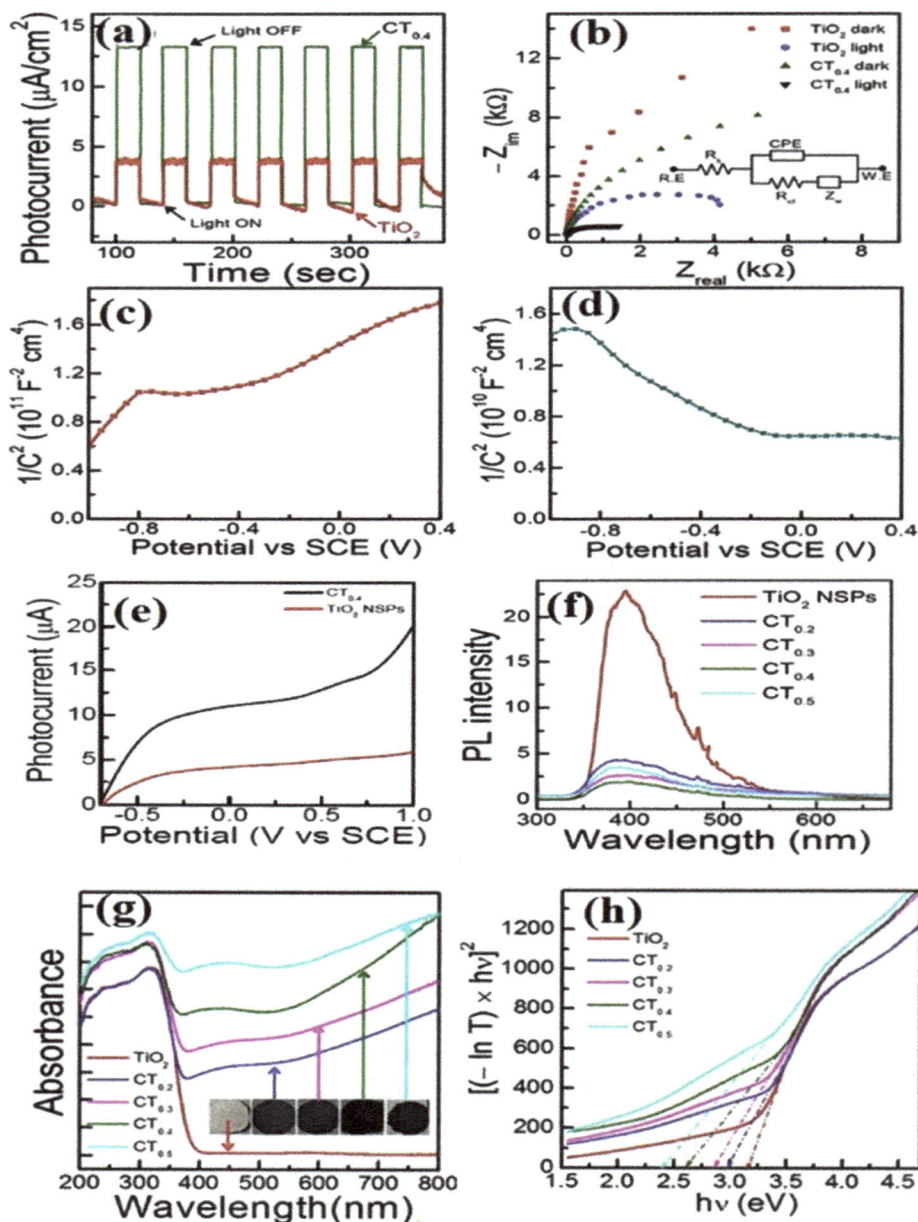

Figure 1.3: Photoelectrochemical properties of TiO_2 and CuS/TiO_2 nanocomposite in the presence and absence of light. (a) Transient photocurrent response, (b) Nyquist plots Mott–Schottky plots of (c) TiO_2 and (d) CuS, (e) linear sweep voltammograms, (f) photoluminescence spectra, (g) UV–vis DRS absorption spectra, and (h) the corresponding Tauc plot [15]. Reprinted with permission from Moumita Chandra et al. [15]. Copyright (2018) American Chemical Society.

1.5 Examples of Photocatalysts

1.5.1 Transition Metal Oxides

Transition metal oxides are one of the most popular categories of semiconductor photocatalysts. Low cost, chemical stability, favorable thermodynamics, good electron transport properties, strong oxidizing power, nontoxicity, and others make transition metal oxides as promising photocatalysts. They find their application in many areas such as water splitting, photocatalytic CO_2 reduction, pollutant degradation, and microbial disinfection. But their large bandgap, low charge carrier mobility, fast recombination rate, and others are impeding the photocatalytic activity [17, 18]. The present challenges in this area of research are to solve these issues.

1.5.1.1 Titanium Dioxide

Titanium dioxide (TiO_2) is a widely explored photocatalyst in numerous applications due to its strong oxidizing capacities toward the decomposition of organic pollutants, chemical stability, hydrophilicity, low toxicity, low cost, and long durability [19]. Because of the presence of oxygen vacancy, TiO_2 usually exists as an n-type semiconductor. Photocatalytic activity of TiO_2 depends on various components such as surface area, particle size, type of polymorphs, type of dopant, defect concentration, and synthesis method [20]. One of the major drawbacks of TiO_2 is its wide bandgap and hence absorbs only in the ultraviolet region (up to 380 nm). Much effort has been taken to narrow the bandgap through bandgap engineering. One another major drawback is the rapid recombination of photogenerated excitons. Many methods, including heterostructure formation, doping with nonmetals and metals, and reduction of crystal size, were employed to reduce the recombination rate.

Crystalline TiO_2 exhibits superior photocatalytic activity than amorphous TiO_2, reason being the minimization of the photoexcited electron–hole recombination owing to the crystalline nature [20]. There are four main types of TiO_2 structures: anatase, rutile, brookite, and TiO_2(B) with a layered titanate structure. The stability of each phase depends on the particle size. The rutile phase is found to be the most stable phase if the particle size is above 35 nm. If the particle size is below 11 nm, the anatase phase is found to be more stable. Particles with a size between 11 and 35 nm brookite phase are found to be more stable [10, 20, 21]. By annealing other three metastable phases at elevated temperature, rutile phase can be synthesized. Among the different TiO_2 polymorphs, anatase and rutile phases exhibit good photocatalytic activity (Figure 1.4). However, anatase shows more photocatalytic activity when compared to rutile due to the higher surface area, enhanced adsorption characteristics, increased number of active reaction sites, and more oxygen vacancy, leading to efficient charge separation. Anatase and rutile phase have different density and band structure, and hence

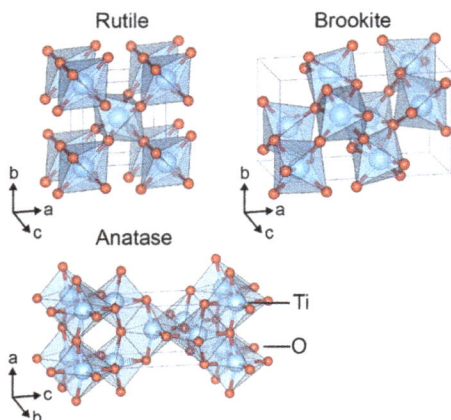

Figure 1.4: Crystalline TiO_2 in different phases: rutile, brookite, and anatase. Adapted with permission from http://creativecommons.org/licenses/by/4.0/. [23].

different bandgaps. Anatase has a bandgap of 3.20 eV equivalent to 384 nm, and rutile has a bandgap of 3.02 eV equivalent to 410 nm [10, 22].

Normally, anatase titania is produced by sol–gel route, where the titania sols are prepared in the presence of acid or base catalysts using alkoxide precursors. The sols are then allowed to age for certain time to obtain the gel. Depending on the drying process of gel, either porous or nonporous structures are obtained. Drying under normal temperature and pressure normally yields dense structures, whereas supercritical drying yields porous nanomaterials with enhanced surface area. On calcination to high temperature, these anatase titania are converted to rutile.

It is shown that the photocatalytic activity also depends on the exposed facet. Degradation of organic pollutant studies shows that as the percentage of exposed {001} facet increases, the photocatalytic activity of the anatase nanocrystal increases [24]. Ratio of {001} to {101} facets has a substantial role on the photocatalytic activity of anatase TiO_2 toward CO_2 reduction. Jiaguo Yu et al. in their studies show that the optimal ratio of {001} to {101} facets was about 58% to 42% for CO_2 reduction [25]. Aiyun Meng et al. synthesized anatase TiO_2 through hydrothermal synthesis using tetrabutyltitanate under highly acidic condition. It is further calcined at 550 °C. They selectively deposited Co_3O_4 nanoparticles as a cocatalyst for water oxidation on {001} and Pt nanoparticles for water reduction on {101} facets, respectively, by a two-step photodeposition strategy, and the catalyst exhibited enhanced photocatalytic hydrogen production [26]. Olga A. Krysiak et al. studied the photocatalytic activity of rutile and anatase TiO_2 electrodes decorated with Au and Ag nanoparticles. To synthesize rutile TiO_2, titanium sheets were etched using HCl and annealed in an oxygen environment either at 600 or 850 °C. In anatase-phase synthesis, etched titanium nanosheets are treated with alkoxide in a nitrogen environment. Subsequent hydrolysis and annealing in air chamber furnace at 450 °C resulted in rutile TiO_2 [27].

It is reported that biphasic TiO_2 exhibits superior photocatalytic activity than rutile TiO_2 and anatase TiO_2. This results from the efficient electron transfer from the conduction band of anatase to those of rutile TiO_2. Huihui Li et al. synthesized g-C_3N_4/rutile–brookite $TiO_{2-x}N_y$ through a facile solvothermal approach. They investigated the photocatalytic activity of the Z-scheme g-C_3N_4/rutile–brookite $TiO_{2-x}N_y$ catalyst for degradation of NO gas under UV–visible irradiation. For the synthesis of composite, $TiCl_3$, hexamethylenetetramine, along with g-C_3N_4, is autoclaved at 190 °C for 2 h. The composite material exhibited higher photocatalytic activity in terms of photocatalytic degradation of NO than pure g-C_3N_4 and similar to nitrogen-doped brookite-phase TiO_2 [28, 29].

As explained earlier, morphology of the photocatalyst has a crucial role in its activity. TiO_2 nano- or microstructures with varying morphologies have been developed. So far, TiO_2 spheres, tubes, fibers, nanorods, sheets, interconnected architectures, etc. have been synthesized. Generally, TiO_2 spheres have high specific surface area, pore size, and pore volume than other structures. This, in turn, increases the rate of mass transfer. Also, the spherical structure favors the light-harvesting capability of TiO_2 [19]. Generally, TiO_2 nanospheres are synthesized using polystyrene spheres as sacrificial templates. After the synthesis, polystyrene spheres can be removed from the catalyst material by calcination. Jiabai Cai et al. synthesized TiO_2–CeO_2 double-shell structure decorated with Au nanoparticle using the abovementioned "template, sol–gel, calcination" method. They tested the photocatalytic dye degradation of methyl orange and confirmed the greater photocatalytic activity of TiO_2–Au–CeO_2 composite [30]. In another work, Yin and coworkers demonstrated the synthesis of self-doped TiO_2 hollow spheres on carbon sphere templates. Initially, using glucose as the precursor, hydrothermal synthesis of carbon spheres was done. Carbon spheres, tetrabutyltitanate, and $NaBH_4$ in ethanol were made into a gel form by continuous stirring. Subsequent drying and calcination give TiO_2 hollow spheres. On this TiO_2 hollow sphere, Ag/AgCl was deposited by a precipitation photoreduction method. The prepared samples showed enhanced photocatalytic activities in terms of degradation of rhodamine B [31].

TiO_2, 1D structures possess remarkable photocatalytic activity owing to its higher surface-to-volume ratio. This aids in the reduction of electron–hole recombination rate and increase in the interfacial charge transfer rate, and both these favor photocatalytic reactions [19]. The specificity of the embedded site is another advantage of 1D TiO_2 structures [32]. TiO_2 nanotubes' preparation methods include solvothermal, hydrothermal, template-assisted method, and electrochemical anodization. In typical hydrothermal synthesis, reactants are allowed to react under controlled pressure and temperature in a stainless steel autoclave vessel. Morphology and size of the resulting tube depend on the concentration of reactants, temperature, and pH [33]. Ailan Qu et al. reported TiO_2 nanotubes decorated with graphene quantum dot (QD) by simple hydrothermal method using TiO_2 powder (Degussa P25) as the precursor. Studies reveal that graphene QD decorated TiO_2 nanotubes exhibit boosted visible light absorption properties and hence the photocatalytic activity [34].

In the template method, morphologically known and characterized templates are used for the synthesis of nanostructures [33]. Jiankang Zhang et al. synthesized porous TiO_2 nanotubes using a facile template-assisted atomic-layer deposition technique using carbon nanocoils as sacrificial templates. In their work, they decorated the inner surface and outer surface of porous TiO_2 nanotubes separately with Pt and CoO_x nanoclusters. The prepared catalyst shows remarkable hydrogen production (275.9 mmol/h) rate [35]. Size-controlled 1D TiO_2 nanotubes can be synthesized using electrochemical anodization. Such a nanotube array exhibits high stability, nontoxicity, recyclability, and also it is cost-effective [33]. Martin Motola et al. synthesized Ti^{3+}-doped TiO_2 nanotubes for enhanced photocatalytic performance. They prepared TiO_2 nanotubes through electrochemical anodization, using titanium foil as the starting material. To improve the photocatalytic activity pre-annealing followed by etching was also performed on the synthesized nanotube. Annealing in the H_2/Ar atmosphere was also done on this. Photocatalytic dye degradation of organic dyes shows the better performance of the catalyst [36].

Two-dimensional (2D) nanosheets exhibit photocatalytic properties like photocatalytic degradation of organic pollutants and superhydrophilicity, under irradiation of UV light. Superior photocatalytic property and highly smooth surface make TiO_2 nanosheet a promising candidate for self-cleaning coatings [19]. TiO_2 nanosheets are mainly preparing via the hydrothermal method. Liu et al. synthesized layered TiO_2 nanosheets–layered black phosphorous composite. TiO_2 nanosheets were synthesized using the hydrothermal method, using tetrabutyltitanate, as a starting material. TiO_2 nanosheets–layered black phosphorous composite was prepared by mechanical stirring. Photocatalytic degradation of rhodamine B shows that composite exhibits greatly enhanced activity [37]. Ma et al. described the fabrication of CdSe QDs/graphene/TiO_2 ternary composite and elaborated their photocatalytic activity toward decomposition and disinfection under irradiation of visible light. Graphene-modified TiO_2 nanosheets were prepared hydrothermally using $Ti(OBu)_4$ as a starting material, which was used as a supporting matrix. CdSe QDs were anchored to the TiO_2/graphene nanosheet via mercaptopropionic acid. The ternary composite under visible light generates active species, such as $\cdot OH$, $O_2^{-}\cdot$, and holes, which contributes to the higher photocatalytic antibacterial activity against *Escherichia coli* [38].

Three-dimensional (3D) porous hierarchical TiO_2 interconnected structures possess large surface-to-volume ratios, which make them as promising photocatalysts. This porous framework supports efficient diffusion as well as superior carrier mobility [19]. Dong et al. prepared 3D interconnected mesoporous anatase TiO_2–silica nanocomposite with enhanced photocatalytic activity. The interconnected mesoporous structure was prepared according to the "extracting silica" approach with titanium isopropoxide, tetraethyl orthosilicate, etc. as the starting material. Characterizations reveal that TiO_2 is in complete anatase crystalline form. The photocatalytic degradation rates of acid red and microcystin-LR on this mesoporous interconnected structure were found to be much higher than TiO_2 powder (P25) [39].

The photocatalytic performance of catalysts can be improved by the use of photonic crystals, which are spatially ordered inverse opals. In our research group, we prepared gold-loaded fluorinated titania inverse opal photocatalysts, which exhibited a remarkably higher hydrogen production rate. These inverse opal structures were made with the opal templates of polystyrene spheres. We found that these inverse opals exhibit a fivefold hydrogen evolution rate than nanocrystalline titania photocatalyst. The enhancement in activity is because of the slow photon effect in the inverse opal structure and oxygen vacancies, which in turn increases the charge transfer and photocatalytic activity [40]. Here, Au nanoparticle acts as the cocatalyst. Figure 1.5 represents scanning electron microscopic images and transmission electron microscopic image of Au/F TiO_2 inverse opal structures fabricated with 460 nm polystyrene opal structures and the dependence of photocurrent on the structure

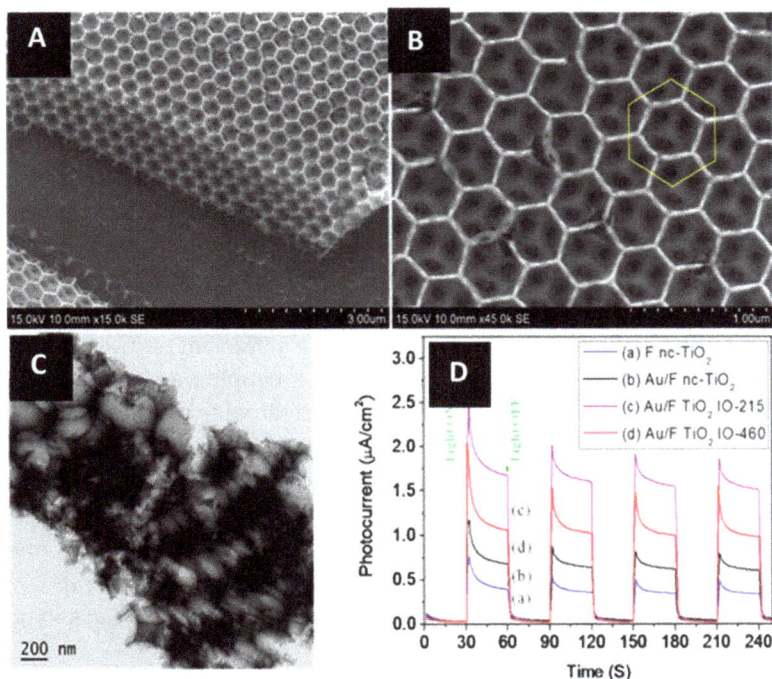

Figure 1.5: Scanning electron microscopic images (A,B) and transmission electron microscopic image (C) of Au/F TiO_2 IO-460. Figure D represents the photocurrent measured using fluorine-doped nanocrystalline titania with and without Au and the photocurrent obtained using Au/F TiO_2 inverse opal structures made with 215 and 460 nm opal structures [40].

of photocatalysts. The increase in photocurrent is evident on going from the nanocrystalline to inverse opal structure.

1.5.1.2 Zinc Oxide

Zinc oxide (ZnO) is a wide bandgap metal oxide semiconductor material which is being used as a promising material for photocatalysis because of its features like enhanced carrier mobility and favorable optical characteristics. ZnO is an n-type semiconductor like TiO_2, and it is considered as an alternative to TiO_2 photocatalyst as it possesses the similar bandgap energy. The advantage of ZnO over TiO_2 is that it absorbs a greater fraction of the solar spectrum when compared with TiO_2 [41]. One of the major obstacles of ZnO is its wider bandgap that greatly restricts the effective utilization of solar light. Proper tuning of the bandgap is required to increase the visible light photocatalytic activity of ZnO. Based on the morphology, ZnO can be divided into 0D, 1D (nanorods, nanofibers, nanowires, nanotubes, and nanoneedles), 2D (nanosheets), and 3D (nanoflower) nanostructures [41].

One-dimensional ZnO nanoarchitectures like nanorods and nanoneedles can be synthesized using various techniques including CVD, template-based micelle, and reverse micelle hydrothermal synthesis [42]. Xiong et al. reported the preparation of 1D multiarmed zinc oxide nanocrystal for water purification through solvothermal strategy in a hexalene glycol–H_2O reaction system. They used zinc nitrate hexahydrate as the precursor for the synthesis. Suppression of recombination of charge carriers in multiarmed zinc oxide nanorod successfully confirmed using PL studies. The multiarmed ZnO nanorods showed superior photocatalytic performance in terms of degradation of rhodamine B and ciprofloxacin [43]. Hekmatara et al. synthesized zinc oxide nanorod with and without the presence of a biocompatible copolymer. First, by simple co-precipitation method without adding a surfactant and second via adding poly(citric acid)-grafted polyethylene glycol as a capping agent. Both the methods used zinc acetate as the precursor and sodium hydroxide as the precipitating agent. It was found that the polymer-capped rod had a higher aspect ratio. Photocatalytic degradation of methylene blue and methyl orange dyes was explored to investigate the activity of both the catalysts. It was found that the polymer-capped zinc nanorod exhibited enhanced degradation rate than the uncapped one [44]. This experiment suggests that though the surface is protected with a capping agent, the increased aspect ratio of the nanorod could enhance the photocatalytic activity.

Two-dimensional ZnO nanosheets are promising candidates for photocatalysis due to its high specific surface area and polar faces [41]. Koutavarapu et al. synthesized ZnO nanosheets anchored on $NaBiS_2$ nanoribbons. $NaBiS_2$ nanoribbons were synthesized hydrothermally using $Bi(NO_3)_3 \cdot 5H_2O$ and $Na_2S \cdot 9H_2O$ as the starting material. The low-temperature solution process using zinc acetate was used for the synthesis of ZnO nanosheets. Further, $ZnO-NaBiS_2$ was fabricated through the hydrothermal method. PL and photocatalytic dye degradation studies suggest the successful fabrication of $NaBiS_2/ZnO$ nanocomposites for visible light photocatalysis. It can be effectively used for organic pollutant degradation [45].

Sol–gel synthesis, precipitation method, hydrothermal, solvothermal, electrochemical deposition process, microwave, microemulsion, wet chemical, electrospinning, and flux methods are some of the solution-based approaches used for the synthesis of ZnO nanostructures [41]. Simplicity of the process is the major advantage of sol–gel synthesis. Apart from this low production cost, good repeatability, facile synthesis condition, etc. make sol–gel synthesis more attractive [41]. Mahdavi and coworkers synthesized Al-doped ZnO through sol–gel synthesis. With varying concentrations of aluminum, morphological changes in ZnO were observed. As the concentration of aluminum increases, morphology changed from spherical to a rod-like structure. From the dye degradation study, they investigated that the presence of an optimal concentration of Al prevents the recombination of charge carriers and increases the hydroxyl groups on catalyst surface [46]. Annealing temperature plays a vital role in determining the photocatalytic activity of ZnO. Pudukudy et al. synthesized ZnO particle by precipitation method and studied the effect of annealing temperature on morphological and photocatalytic properties. ZnO nanoparticles were precipitated from zinc nitrate solution using NaOH solution. This precipitate was annealed over a range of temperature from 300 to 600 °C. They observed that with an increase in annealing temperature, a quasi-spherical morphology was changed to nanocapsules. PL study and photocatalytic degradation of methylene blue suggest increasing photocatalytic activity with annealing temperature [47].

Some vapor-phase techniques like PVD, pulsed laser deposition (PLD), thermal evaporation, molecular beam epitaxy (MBE), CVD, metal–organic CVD, and plasma-enhanced CVD are used for the synthesis of size-controlled ZnO nanostructures [41]. Nevertheless, solution-phase method is more popular because of its simplicity and yield [22]. The lifetime of the photogenerated exciton in ZnO is improved by incorporating noble metal with it (Table 1.1). Zhang et al. synthesized flower-like Ag/ZnO by a low cost, simple method. Ag-deposited ZnO flower was synthesized hydrothermally using $Zn(CH_3COO)_2 \cdot 2H_2O$, along with $C_6H_8O_7 \cdot H_2O$ and $AgNO_3$ [48]. It was found that the deposition of the optimal amount of Ag on ZnO increases the photocatalytic activity. As the amount of Ag increases, surface plasmon absorption increases, and PL decreases. This trend is confirmed through the photocatalytic degradation of methylene blue.

Table 1.1: Effect of fabrication method in the morphology of ZnO.

Fabrication	Starting material	Morphology	Application	Reference
ZnO nanorod in stainless steel by chemical vapor deposition	High-purity Zn powder	ZnO nanorod	Photocatalytic dye degradation	[49]
Graphene–ZnO hybrid by cathodic electrochemical deposition	ITO-coated GO and zinc nitrate electrolyte	ZnO film	Photocatalytic dye degradation	[50]
ZnO–CdS composite by wet chemical method	$Zn(NO_3)_2$, NaOH, CdS nanoparticles	ZnO–CdS flower	Photocatalytic dye degradation	[51]
ZnO nanostructure by sol–gel electrophoretic deposition	Zinc acetate, triethylamine, methanol	Urchin-like ZnO	Photocatalytic dye degradation	[52]
ZnO nanostructure by pulse electrochemical method	Zinc electrodes	Flower-like array of ZnO nanoparticle and nanorod	Photocatalytic dye degradation	[53]

GO, graphene oxide; ITO, Indium Tin Oxide.

1.5.1.3 Copper Oxide

The two stable forms of copper oxide are cuprous oxide (Cu_2O) and cupric oxide (CuO), in which the former one is a direct bandgap material with a cubic structure having a bandgap range of 2.1–2.6 eV and the latter one is a monoclinic structure with a bandgap range of 1.2–1.9 eV. Both Cu_2O and CuO are p-type semiconductors [54]. Cu_2O is a promising photocatalytic material for water splitting and organic pollutant decomposition owing to its low toxicity and cost. However, the rapid recombination of the charge carriers causes a considerable reduction in the photocatalytic efficiency of Cu_2O. Various strategies have been reported for the preparation of Cu_2O and CuO thin films, like reactive sputtering, PLD, spray pyrolysis, MBE, successive ionic layer adsorption and reaction, sol–gel synthesis, hydrothermal synthesis, and CVD [54, 55]. Copper oxide can be synthesized in morphologically different nanostructures such as nanorods, nanospheres, nanotubes, microflowers, nanoneedles, and nanosheets [55–57].

Srinivasa et al. reported the influence of oxygen partial pressure on the formation of the copper oxide film. They have synthesized the films through radio frequency magnetron sputtering on a glass substrate under argon and O_2 atmosphere. They report that with the increase in partial pressure of O_2, Cu_2O with cubic structure changes to monoclinic CuO. X-ray diffraction analysis and Raman analysis confirmed this observation. Dye degradation of methyl orange, methylene blue, and ciprofloxacin under sunlight irradiation indicated that Cu_2O thin films exhibited enhanced photocatalytic activity than CuO films [54].

Tezcan and coworkers optimized the deposition of the Cu_2O layer on ZnO through a two-step electrodeposition method at varying temperatures. Electrodeposition was performed in a conventional three-electrode cell using zinc nitrate tetrahydrate and copper sulfate pentahydrate as the starting material. They observed that Cu_2O deposited at 40 °C exhibits the highest photocurrent density. The flat band potential and bandgap were calculated from the Mott–Schottky measurements and Tauc plot, respectively [58]. Vivek et al. prepared flower-like copper oxide microstructure through hydrothermal treatment (Table 1.2). The hydrothermal reaction was carried out at 120 °C using $CuCl_2 \cdot 2H_2O$ under alkaline conditions. Efficient photocatalytic activity is observed in terms of degradation of methylene blue and crystal violet under UV irradiation [55].

Table 1.2: Different fabrication method for copper oxide-based photocatalysts.

Fabrication	Starting material	Morphology	Application	Reference
Hydrothermal synthesis of CuO quantum dot deposited on TiO_2 nanosheet	$CuCl_2$ along with hydrothermally synthesized TiO_2	CuO quantum dot on TiO_2 nanosheet	Photocatalytic water splitting	[59]
Nanostructured CuO thin film grown on Cu foil using wet chemical method	Cu foil, NaOH aqueous solution, $(NH_4)_2S_2O_8$	CuO thin film	Photocatalytic water splitting	[60]
Electrodeposition of a thin uniform Cu_2O/CuO layer over the silver structure	Electrochemical synthesis of Ag dendrites over Al foil using $AgNO_3$ Cu_2O/CuO was deposited over Ag using copper citrate electrolyte containing $CuSO_4 \cdot 5H_2O$ and $C_6H_5Na_3O_7 \cdot 2H_2O$	Cu_2O/CuO layer on Ag nanodendrites	Photocatalytic CO_2 reduction	[61]
Copper oxide nanocluster grafted over niobate nanosheets by wet chemical method	CuO clusters on Ti metal sheet by an impregnation method using $CuCl_2$ solution. Niobate nanosheets by delamination of bulk niobate with tetrabutylammonium hydroxide	Cu nanocluster on Nb_3O_8 nanosheet	Photocatalytic CO_2 reduction	[62]

Table 1.2 (continued)

Fabrication	Starting material	Morphology	Application	Reference
Reduced graphene oxide–Cu_2O composite by wet chemical synthesis	Reduction of copper sulfate supported on graphene oxide using ascorbic acid in the presence of polyethylene glycol and sodium hydroxide	Cu_2O particles deposited on the reduced graphene oxide layer	Photocatalytic disinfection	[63]

1.5.1.4 Iron Oxide

Iron oxide (Fe_2O_3) exists in different crystalline structures, such as wüstite (FeO), hematite (α-Fe_2O_3), maghemite (γ-Fe_2O_3), and magnetite (Fe_3O_4). Among these phases, the hematite phase is found to be more stable at ambient conditions, which is an n-type semiconductor with a bandgap of 2.1 eV. The lower valence band position than the water oxidation potential makes hematite a favorable material for PEC water splitting. The photocatalytic activity of materials has a strong dependence on the surface area. For attaining maximum surface area, several approaches have been adopted to synthesize Fe_2O_3 in varying morphologies including hydrothermal methods, hydrolysis, ionic liquid-assisted synthesis, co-precipitation, solvothermal methods, thermal decomposition, a combination reflex condensation and hydrothermal method, and combustion methods [64]. Helal et al. synthesized heterostructured Fe_2O_3/Bi_2S_3 nanorods through one-step hydrothermal synthesis. $Bi(NO_3)_3 \cdot 5H_2O$, $Fe(NO_3)_3 \cdot 9H_2O$, and $(NH_2)_2S$ were used as raw materials for the synthesis of Fe_2O_3/ Bi_2S_3 nanorods. The optical behavior and bandgap were calculated using UV–visible DRS, and the calculated bandgap was 2.13 eV. Photodegradation studies using methylene blue and phenol prove that Fe_2O_3/Bi_2S_3 nanorods can be used as a suitable photocatalyst for pollutant degradation [65].

X. She et al. synthesized α-Fe_2O_3/g-C_3N_4 catalyst for effective Z-scheme water splitting. The existence of α-Fe_2O_3 helps to produce ultrathin 2D g-C_3N_4 nanosheets. The solvothermal method was adopted for the synthesis of α-Fe_2O_3 where $FeCl_3$ and CH_3COONa served as precursors. The Z-scheme structure provides a tight interface between α-Fe_2O_3 and g-C_3N_4, which in turn suppresses charge carrier recombination by transferring electrons from the conduction band of α-Fe_2O_3 to the valence band of 2D g-C_3N_4. The decreased intensity of emission in the PL spectra of α-Fe_2O_3/g-C_3N_4 hybrid is direct evidence of lower recombination rate [66].

Fe_2O_3 microflowers can be prepared by polyol-based precursor and annealing method using $FeCl_3$ as the starting material. Incorporation of this structure into metal oxides such as TiO_2 will enhance the photocatalytic activity. Flower-like core–shell

TiO_2/Fe_2O_3 nanostructures that exhibit efficient photocatalytic degradation of aqueous solution of paracetamol has been reported [67].

1.5.2 Carbon Nanomaterials

Carbon is an abundant element in nature. Carbon materials possess structural diversity, good electrical and thermal conductivity and remarkable mechanical properties. Widely used carbonaceous materials in photocatalysis include g-C_3N_4, graphene, and carbon nanotubes (CNTs). Carbonaceous materials alone generally do not display significant HER and OER activities. Creating defects, doping with heteroatoms, forming composites among them are some of the strategies to improve the catalytic activity of this material. Generally, the carbonaceous nanomaterials are used as catalytic supports.

1.5.2.1 Carbon Nitride

Carbon nitride is a polymeric semiconductor possessing extensive π conjugation with imide-linked tri-s-triazine (known as melem) as the basic building unit. This bright yellow powder, which contains sp^2-bonded C and N atoms, is usually prepared by thermal condensation of precursors like cyanamide, melamine, urea, dicyandiamide, or their mixtures that are nitrogen-rich [68]. The most stable allotrope among different carbon nitrides is g-C_3N_4 with a calculated bandgap of about 2.7 eV. The conduction band and valence band edges are positioned at energetically favorable positions, and hence it is a visible light active catalyst. Density functional theory calculations suggest nitrogen and carbon atoms as favored sites for water oxidation and proton reduction, respectively. g-C_3N_4 was first used for visible light photocatalytic water splitting by Wang et al. [69]. The activity of g-C_3N_4 described so far is lower than other photocatalysts. The main limitation of g-C_3N_4 as photocatalyst is the fast recombination of photogenerated excitons due to low charge mobility and the covalent bonding in g-C_3N_4 which leads to low proton reduction and H_2O oxidation efficiencies [68]. Porous C_3N_4 provides large surface area, good charge separation and migration, and easy mass diffusion.

Owing to the stacked layers, bulk C_3N_4 exhibits a lower specific surface area. Mechanical exfoliation like scotch-tape-assisted exfoliation similar to graphene synthesis, thermal exfoliation, and sonication-assisted liquid exfoliation can be utilized for the exfoliation of pristine C_3N_4 which can lead to an improvement in the photocatalytic activity. Y. Liu et al. reported the synthesis of K-doped g-C_3N_4 nanosheet photocatalyst, by the combination of K atom doping and thermal exfoliation. This process resulted in nanosheets with large surface area, shortened migration length of charge to surface, and effective electron–hole separation. This new photocatalyst showed 8.3 times higher photocatalytic activity in terms of visible light photocatalytic degradation of RhB than pure g-C_3N_4 [70]. Individual g-C_3N_4 can be assembled into hierarchical nanostructures with abundant porous network in one of the ways to increase the photocatalytic activity. Fu et al. invented a process to synthesize hierarchical g-C$_3$N$_4$ nanotubes through high-temperature oxidation exfoliation in the presence of airflow followed by curling condensation of g-C$_3$N$_4$. The high-temperature oxidation process will facilitate O doping. When compared to bulk g-C$_3$N$_4$, the O-doped g-C$_3$N$_4$ nanotubes exhibit higher photocatalytic activity in terms of photocatalytic reduction of CO$_2$ [71].

Zhu and group synthesized g-C$_3$N$_4$/Ag$_2$WO$_4$ Z-scheme photocatalyst through in situ precipitation method. In this work, g-C$_3$N$_4$ was prepared by heating melamine at 550 °C, which acts as a support for Ag$_2$WO$_4$. They suggested that g-C$_3$N$_4$ nanosheets control the phase structure of Ag$_2$WO$_4$ nanoparticles. Here Ag$^+$ adsorbed on the g-C$_3$N$_4$ surface served as the active site for nucleation and further growth of β-Ag$_2$WO$_4$ and prevents β to α conversion. The results suggest that the composite exhibits superior photocatalytic efficiency in terms of photodegradation of methyl orange [72]. Wet chemical method is a facile and inexpensive technique used for the synthesis of various heterostructures. Xia et al. synthesized g-C$_3$N$_4$/MnO$_2$ nanocomposite, which exhibits enhanced Z-scheme photocatalytic dye degradation. They prepared heterostructured photocatalyst through wet chemical method in which MnO$_2$ layers were grown on the g-C$_3$N$_4$ layer. g-C$_3$N$_4$ layer was synthesized by the thermal exfoliation of urea. The band structures were determined using the Mott–Schottky analysis and X-ray photoelectron spectroscopy (XPS) valence band spectroscopy. n-Type semiconductivity of g-C$_3$N$_4$ and MnO$_2$ were confirmed using the Mott–Schottky analysis. Flat band potential is approximately considered as the conduction band potential (−1.61 and 1.22 V), and XPS valence band spectroscopy gave valence band potential (1.81 and 3.26 V). The band values suggest a Z-scheme charge transfer in the composite, which facilitates improved utilization of charge carriers [73].

1.5.2.2 Graphene

Graphene is a single layer of sp^2-hybridized carbon atoms that are bound in a hexagonal lattice. Graphene and its derivatives can be synthesized from naturally occurring earth-abundant materials such as natural graphite, biomass waste, and polymers. Graphene can be either synthesized by bottom-up methodologies like CVD or wet chemical synthesis or by a top-down approach like thermal or liquid-phase exfoliation methods. Pristine graphene is a zero bandgap semiconductor and it shows poor catalytic activity. Graphene has to be modified by chemical modification, heteroatom doping, etc. to improve its catalytic activity. Size and morphology tuning also help to improve the activity. Graphene owing to its zero bandgap shows weak photocatalytic activity, whereas graphene oxide (GO) exhibits appreciable activity. Graphene greatly enhances the charge transfer and thereby the separation of charge carriers in semiconductor photocatalysts such as TiO$_2$ due to the excellent electron mobility, high conductivity, and high specific surface area. Hence, graphene and its derivatives are widely used as a photocatalyst support.

R. Atchudan et al. synthesized GO-grafted TiO_2 for photocatalytic degradation. GO and TiO_2 were synthesized using modified Hummers method and sol–gel synthesis, respectively. GO was grafted over TiO_2 using a solvothermal method using tetrahydrofuran as solvent. Photocatalytic degradation studies on methylene blue and methyl orange suggest the improved photocatalytic activity of composite compared to bare TiO_2. Introduction of TiO_2 into the graphene layer increases the activity through interfacial interaction between two semiconductors [74]. It is reported that 2D–2D interfaces effectively harvest photons and enhance exciton generation and reduce the charge carrier recombination, thereby promoting photocatalytic activity. Rajesh Bera et al. grafted CdS nanoparticle, CdS nanorod, CdS nanosheets on reduced GO (rGO) by surface modification method (Table 1.3). It was found that CdS nanosheet–rGO composite, which is having 2D–2D nanoarchitecture, enhances the

Table 1.3: Some of the synthetic methods of graphene-based photocatalysts.

Fabrication	Starting material	Morphology	Application	Reference
Chemically synthesized Au/graphene-wrapped ZnO hollow spheres	ZnO hollow spheres, zinc acetate dehydrate, trisodium citrate, hexamethylenetetramine rGO-modified Hummers method AuNPs/rGO/ZnO–HAuCl₄ Tetra-*n*-butyl ammonium bromide, rGO, ZnO	ZnO hollow sphere wrapped with Au-deposited graphene layer	Photodegradation of methylene blue dye	[76]
PRGO/Ru/SrTiO₃:Rh Z-scheme photocatalyst	PRGO with BiVO₄ by photocatalytic reduction of GO on BiVO₄ PRGO/Ru/SrTiO₃:Rh by photocatalytic reduction of GO on Ru/SrTiO₃:Rh	BiVO₄ and Ru/SrTiO₃:Rh nanoparticle integrated with photocatalytically reduced rGO	Photocatalytic water splitting	[77]
rGO–Ag composite for photocatalytic degradation	rGO-modified Hummers method Silica-coated Ag nanoparticle. Precipitation method using CH₃COOAg as precursor rGO-Ag-sol-gel–like process	Ag nanoparticle attached on rGO layer	Photocatalytic degradation of colorless organic pollutants	[78]

GO, graphene oxide; PRGO, photoreduced graphene oxide; rGO, reduced graphene oxide; AuNPs, gold nanoparticles.

photocatalytic activity in terms of dye degradation than the other two composites. Modified Hummers method was used to synthesize GO. Surface-functionalized CdS nanostructures were incorporated into GO through a hydrothermal method, where hydrazine hydrate was added for reducing GO to rGO [75].

1.5.2.3 Carbon Quantum Dot

Carbon QD (CQD) gains growing research attention owing to its optical, electronic, and physicochemical properties. It is widely used in photocatalytic applications because of its low toxicity, low cost, biocompatibility, easiness of synthesis, etc. Generally, CQDs are synthesized by electrochemical oxidation, laser ablation of graphite, microwave synthesis, electrochemical soaking of CNTs, and bottom-up methods. Quenching of PL and up-conversion PL of CQD makes it a promising candidate for photocatalysis [79].

Atkin et al. synthesized tungsten disulfide (WS_2)/carbon dot hybrid for efficient photocatalysis. WS_2 nanoflakes were synthesized by exfoliating bulk WS_2. WS_2/carbon dot hybrid was synthesized using microwave irradiation of nanoflakes in citric acid solution, where citric acid was used as a carbon source. They observed that at an optimal concentration, carbon dots get nucleated on the surface of WS_2. Presence of carbon dot makes the hybrid to photocatalyze Congo Red more efficiently than bare WS_2 [80].

CQDs were also used for photocatalytic inactivation of microbes. Han et al. synthesized N-doped CQDs and incorporated into molybdenum disulfide (MoS_2) sheets to improve photocatalytic and photothermal properties of MoS_2. Figure 1.6 represents the synthesis procedure adopted for the preparation of MoS_2/CQDs hybrid. The exhibited higher biocompatibility suggests that MoS_2/CQDs hybrid is environmentally friendly. Further, this MoS_2/CQDs coated on a Ti basement through spin coating was proved to be as an antibacterial coating [81].

Single-atom catalysts get increasing attention in the heterogeneous catalysis field since they offer the maximum possible efficacy per metal atom. Wang and coworkers synthesized CQDs with anchoring sites for single cobalt atom in $Co-N_4$ structures. Cobalt atoms were anchored on carbon dot catalyst through facile pyrolysis of vitamin B_{12}. Here Co atoms enhance charge separation and charge transport, and CQDs act as support for anchored Co atoms. Also, it works as a light-harvesting antenna. Increased photocatalytic activity of the hybrid is confirmed through photodegradation of an organic dye, oxidative coupling of aromatic amines, and water oxidation [82].

1.5.3 Transition Metal Dichalcogenides

TMDs are layered materials similar to graphite. They have a general formula MX_2, where M and X denote transition metal element and chalcogen, respectively. In TMDs,

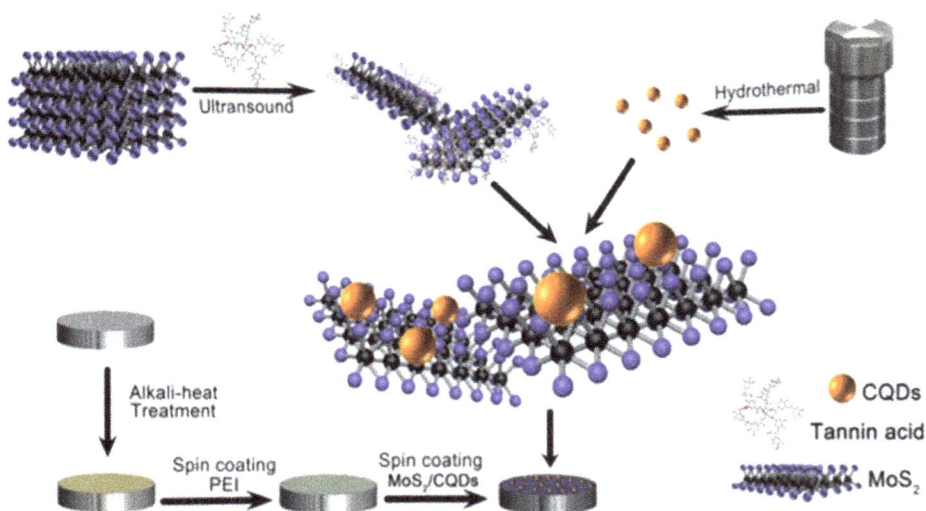

Figure 1.6: Schematic illustration of the synthesis procedure of polyethylenimine-MoS$_2$/CQDs hybrids on Ti plates [81]. Reprinted with permission from Donglin Han et al. [81]. Copyright (2020) American Chemical Society.

monolayers are stacked together via weak van der Waals forces, and in each layer, the transition metal is placed in between two chalcogen layers. TMDs are mainly synthesizing in three ways, namely, (a) substrate growth, (b) colloidal synthesis, and (c) exfoliation. Exfoliation can be performed either by a physical method such as scotch-tape method or by sonication in an appropriate solvent or through intercalation [83, 84]. Exfoliation generally results in polydispersed monolayer without much control in size and shape, whereas the substrate growth method yields monolayers with high crystallinity. Colloidal synthesis can be used for the synthesis of bulk amount of TMD monolayers with good crystallinity [85, 86].

1.5.3.1 Molybdenum Disulfide

Bulk MoS$_2$ has a stacked-layer structure held by weak van der Waals forces. In each monolayer, the plane of Mo atom is inserted between the two layers of sulfur atoms. Based on the crystal structure, MoS$_2$ is classified into 1H, 1T, 2H, and 3R. Among these, layered hexagonal structure is the most common phase, and it can be transformed into the 1T phase. Bulk MoS$_2$ is an indirect bandgap material having a bandgap of 1.3 eV. Upon exfoliation, the crystal structure changes to the 2D phase with a bandgap of 1.9 eV and becomes a direct bandgap material. MoS$_2$ with lower dimensions is considered as a promising visible light active photocatalyst. Sulfur atoms at the edge sites are reported as the catalytically active sites in MoS$_2$ than the basal plane. The number of layers is vital in determining the HER activity of MoS_2. As the number of layers decreases, more sulfur edges get exposed causing an increase in the charge carrier separation and the catalytic activity [8]. Numerous methods can be adopted for the synthesis of MoS_2. It can be prepared either through the top-down method or through the bottom-up methodology. Top-down approach includes mechanical cleavage, laser thinning techniques, chemical intercalation, and liquid-phase exfoliation by sonication. The bottom-up approach includes physical vapor transport, CVD, and wet chemical approaches [8, 87, 88].

One-dimensional $CdS-MoS_2$ nanowires with core–shell structure were developed through a simple hydrothermal strategy by taking CdS nanowires as nanobuilding blocks. Thioacetamide and sodium molybdate along with as-prepared CdS were used as starting materials for synthesis. Owing to their morphology, a strong coaxial contact occurs on the interface of MoS_2 thin shell and CdS core, thus facilitating the electron transfer from CdS core to MoS_2 shell. Improved separation and mobility of charge carriers were confirmed through optical and PEC studies. Photocatalytic H_2 measurement was carried out using lactic acid as a sacrificial agent [89].

Xu and coworkers synthesized metallic $1T-MoS_2/O-g-C_3N_4$ by an in situ growth process. In this process, $MoCl_5$ and thioacetamide along with $O-g-C_3N_4$ in DMF were heated at 200 °C. After washing, the product was freeze-dried to get $1T-MoS_2/O-g-C_3N_4$. PL and UV–vis DRS studies were used to measure the optical properties of the prepared sample. Apart from exposed edge sites, intimate layer interaction between $1T-MoS_2$ and $O-g-C_3N_4$ also helps to enhance the photocatalytic activity. This will help to transfer electrons from conduction band of $O-g-C_3N_4$ to the surface of MoS_2, thereby reducing the charge recombination, and the electrons would rapidly participate in photocatalytic reduction of hydrogen [90]. L. Ye et al. synthesized MoS_2/S-doped $g-C_3N_4$ film which shows superior PEC performance. S-doped $g-C_3N_4$ films were prepared on an ITO conducting glass plate by CVD using thiourea and melamine as precursors. Subsequent deposition of MoS_2 film was done hydrothermally using sodium molybdate and thiourea (Table 1.4). Efficient light absorption and enhanced carrier concentration were confirmed through the UV–vis DRS spectrum and Mott–Schottky plots. High catalytic performance was analyzed in terms of H_2 evolution [91].

MoS_2, because of its unique layered structure, can combine with other semiconductor materials, which enhance the electron–hole pair separation resulting in the promoted photocatalytic disinfection [8].

1.5.3.2 Tungsten Disulfide

WS_2 is a semiconducting material, which has a bandgap of 1.35 eV. WS_2 nanosheet is a promising solar spectrum photocatalyst because of its broad spectral range, biocompatibility, strong W–S bond, high biodegradability, ease of synthesis, and tunable morphology [98]. Several methods, such as exfoliation, CVD, laser ablation,

Table 1.4: Few synthetic techniques for the controlled morphology of MoS_2.

Fabrication	Starting material	Morphology	Application	Reference
Solvothermal synthesis of MoS_2 clusters/ CdS nanorod heterojunction	Solvothermal synthesis using Na_2MoO_4, $Cd(NO_3)_2 \cdot 4H_2O$, and NH_2CSNH_2 in ethylene glycol	MoS_2 clusters on CdS nanorod	Photocatalytic hydrogen production	[92]
Hydrothermal synthesis of few layer MoS_2 flakes deposited on ZnO/FTO nanorods	ZnO nanorod array through the hydrothermal route using zinc nitrate and hexamethylenetetramine. MoS_2 flake deposition using MOCVD system using Mo $(CO)_6$ and H_2S	Few layer MoS_2 flakes on ZnO rod	Photoelectrochemical study	[93]
Solvothermal synthesis of MoS_2/CdS nanodots-on-nanorod heterostructures	Solvothermal synthesis using $Cd(NO_3)_2 \cdot 4H_2O$, $Na_2MoO_4 \cdot 2H_2O$, and CS $(NH_2)_2$ in ethylene diamine	MoS_2 nanodots on CdS nanorod	Photocatalytic pollutant degradation	[94]
Hydrothermal synthesis of magnetically recyclable MoS_2/ Fe_3O_4 hybrid composite	Hydrothermal synthesis using $Na_2MoO_4 \cdot 2H_2O$, thiourea, and Fe_3O_4	Fe_3O_4 nanoparticles deposited on MoS_2 flower	Photocatalytic dye degradation	[95]
MoS_2 quantum dots growth at the S vacancy in $ZnIn_2S_4$	Li-intercalated $ZnIn_2S_4$– $InCl_3 \cdot 4H_2O$, $Zn(Ac)_2 \cdot 2H_2O$, and thioacetamide Hydrothermal synthesis of MoS_2 QDs-thiourea and $Na_2MoO_4 \cdot 2H_2O$ along with Li-intercalated $ZnIn_2S_4$	MoS_2 QD and $ZnIn_2S_4$ nanosheets	Photocatalytic hydrogen production	[96]
Plasma-enhanced CVD method for edge-rich MoS_2 nano array on three-dimensional graphene glass	Three-dimensional graphene prepared using ethanol and H_2O. MoS_2 synthesized using sulfurization of MoO_3	Edge-rich MoS_2 array and edge-oriented three-dimensional graphene	Photocatalytic hydrogen production	[97]

GO, graphene oxide; ITO.MOCVD, metal–organic chemical vapor deposition; CVD, chemical vapor deposition; QD, quantum dots; FTO, Fluorine doped tin oxide.

sputtering, and thermal decomposition, have been used for the synthesis of WS_2 nanomaterials. Like MoS_2, in WS_2 also edge sites are more active than basal planes for HER. Due to its higher intrinsic electrical conductivity, layered WS_2 can be considered as an apt cocatalyst than MoS_2.

Xiang and coworkers developed CdS nanorods anchored on WS_2/graphene layer as an efficient catalyst for photocatalytic HER. Layered WS_2/graphene hybrid was synthesized hydrothermally, and incorporation of CdS nanorods was done using the ethylene-diamine-assisted solvothermal method [11]. Mahler et al. reported the synthesis of stable WS_2 monolayers by colloidal synthesis method. Surprisingly, they obtained octahedral 1T-WS_2 structure instead of expected prismatic 2H-WS_2 structure. Reactivity of the tungsten precursor has a crucial role in determining the structure of WS_2 in the colloidal synthetic method. Less reactive tungsten precursor leads to 1T structure; however, a highly reactive precursor leads to 2H structure. In their work, they investigated the photoactivity of the obtained polymorphs and found the high H_2 evolution rates for distorted 1T-WS_2 nanostructure than the regular 2H-WS_2 form [99].

Dai and coworkers developed WS_2–Bi_2S_3 nanotube using WS_2 QDs as seeds for Bi_2S_3 crystal growth. Zero-dimensional QDs own exceptional electronic and optical properties because of edge effects and quantum confinement. "Seed-mediated" synthesis enables smoother and smaller size WS_2–Bi_2S_3 nanotube preparation. The bottom-up hydrothermal method was used for the preparation of QDs, where sodium tungstate and L-cysteine served as precursors. The WS_2–Bi_2S_3 nanotubes were synthesized solvothermally. Superior PEC properties show a high yield of electron–hole pair during photocatalysis [100].

Li et al. synthesized hollow core–shell β-Bi_2O_3–WS_2 p–n junction as a catalyst for photocatalytic degradation. Using theoretical calculation, they found that photogenerated electron from WS_2 transfers to the β-Bi_2O_3 conduction band, while the photogenerated hole from β-Bi_2O_3 transfers to the WS_2 valence band, resulting in an efficient separation of the electron–hole pair. Longer life and mobility of photogenerated carriers were experimentally verified using photocurrent measurements, PL spectroscopy, EIS, and UV–vis DRS. Photocatalytic degradation of fluoroquinolones ofloxacin, ciprofloxacin, and norfloxacin by WS_2-coated β-Bi_2O_3 photocatalyst was found to be greater than those of β-Bi_2O_3 under the same conditions. In this work, β-Bi_2O_3 hollow microspheres were synthesized by a simple solvothermal-calcining method. The core–shell β-Bi_2O_3–WS_2 heterojunction was prepared through a self-assembly process [101].

1.5.3.3 Rhenium Disulfide

In rhenium disulfide (ReS_2), bulk structure behaves as electronically decoupled monolayers because of the weak interlayer van der Waals interactions. This helps ReS_2 to produce trions in a multilayer nanosheet [102]. It is reported that interlayer decoupling, resulting from the Peierls distortion of the ReS_2 1T structure, will help to prevent ordered stacking and also reduces the wave function overlap [103]. ReS_2 monolayers have an experimental bandgap of 1.55 eV. Conduction band (–6.15 eV) and valence band (–4.26 eV) positions suggest that ReS_2 is suitable as a photocatalyst. ReS_2 naturally exists in 1T phase, and it is highly appreciable for applications like photocatalytic water splitting [104].

Q. Zhang et al. reported ReS_2 nanowalls as a competent catalyst for light-induced HER. In their study, they synthesized vertical and uniform ReS_2 nanowalls via the CVD method. The corresponding hydrogen production activity in Na_2S–Na_2SO_3 and lactic acid solutions was found to be 13 mmol/g h. Efficient photocatalytic activity through two electron transfer mechanisms is confirmed with UV-DRS and PL studies. Q. Jing and coworkers synthesized pure ReS_2 nanosheets through ultrasonic liquid exfoliation. Bulk layered crystals were dispersed in *N*-methyl-pyrrolidone and sonicated at a certain time and purified through centrifugation. The TiO_2–ReS_2 heterostructure was prepared through a self-assembly process. After combining this with TiO_2 nanoparticles, the efficient photocatalytic activity of TiO_2–ReS_2 nanocomposites was measured in terms of rhodamine B degradation under sunlight irradiation. The boosted photocatalytic performance was ascribed to effective electron–hole separation, and it is confirmed using PEC measurements such as transient photocurrent response and PL. A higher photocurrent density was observed for TiO_2–ReS_2 than TiO_2 nanoparticles and ReS_2 nanosheets individually. Faster electron transfer in composite was also confirmed with EIS data [105].

1.6 Summary

This chapter provided an overview of the diverse synthesis methods for some commonly used photocatalysts. Photocatalysis used for various applications such as hydrogen evolution, removal of pollutants, CO_2 reduction, and disinfection is reviewed. The formation of heterojunction, doping, etc. has a decisive role in the photocatalytic activity of the catalyst. Additionally, the morphology of the catalyst is significant in determining the photocatalytic activity. Depending on the application, various synthetic procedures such as sol–gel synthesis, precipitation method, hydrothermal, solvothermal, wet chemical method, electrochemical deposition process, and CVD can be used.

References

[1]   Yang X, Wang D. Photocatalysis: from fundamental principles to materials and applications, ACS Applied Energy Materials, 2018, 1, 6657–6693.

[2]   Navarro Yerga RM, Consuelo Álvarez Galván M, Del Valle F, Villoria De La Mano JA, Fierro JLG. Water splitting on semiconductor catalysts under visible light irradiation, ChemSusChem, 2009, 2, 471–485.

[3]   Zou X, Zhang Y. Noble metal-free hydrogen evolution catalysts for water splitting, Chemical Society Reviews, 2015, 44, 5148–5180.

[4]   Chen X, Shen S, Guo L, Mao SS. Semiconductor-based photocatalytic hydrogen generation, Chemical Reviews, 2010, 110, 6503–6570.

[5]   Kabra K, Chaudhary R, Sawhney RL. Treatment of hazardous organic and inorganic compounds through aqueous-phase photocatalysis: a review, Industrial & Engineering Chemistry Research, 2004, 43, 7683–7696.

[6]   Dey GR. Chemical reduction of CO_2 to different products during photo catalytic reaction on TiO_2 under diverse conditions: an overview, Journal of Natural Gas Chemistry, 2007, 16, 217–226.

[7]   Gamage J, Zhang Z. Applications of photocatalytic disinfection, International Journal of Photoenergy, 2010, 2010, 1–11.

[8]   Li Z, Meng X, Zhang Z. Recent development on MoS_2-based photocatalysis: a review, Journal of Photochemistry and Photobiology C: Photochemistry Reviews, 2018, 35, 39–55.

[9]   Li X, Shen R, Ma S, Chen X, Xie J. Graphene-based heterojunction photocatalysts, Applied Surface Science, 2018, 430, 53–107.

[10]   Ma Y, Wang X, Jia Y, Chen X, Han H, Li C. Titanium dioxide-based nanomaterials for photocatalytic fuel generations, Chemical Reviews, 2014, 114, 9987–10043.

[11]   Xiang Q, Cheng F, Lang D. Hierarchical layered WS_2/graphene-modified CdS nanorods for efficient photocatalytic hydrogen evolution, ChemSusChem, 2016, 9, 996–1002.

[12]   Kumar S, Kumar A, Navakoteswara Rao V, Kumar A, Shankar MV, Krishnan V. Defect-rich MoS_2 ultrathin nanosheets-coated nitrogen-doped ZnO nanorod heterostructures: an insight into in-situ-generated ZnS for enhanced photocatalytic hydrogen evolution, ACS Applied Energy Materials, 2019, 2, 5622–5634.

[13]   Cao G. Nanostructures and Nanomaterials, Published by Imperial College Press and Distributed by World Scientific Publishing co., 2004, 448.

[14]   Hernández-Ramírez A, Medina-Ramírez I. Photocatalytic Semiconductors, Springer International Publishing, Cham, 2015, 1–289.

[15]   Chandra M, Bhunia K, Pradhan D. Controlled synthesis of CuS/TiO_2 heterostructured nanocomposites for enhanced photocatalytic hydrogen generation through water splitting, Inorganic Chemistry, 2018, 57, 4524–4533.

[16]   Gelderman K, Lee L, Donne SW. Flat-band potential of a semiconductor: using the Mott-Schottky equation, Journal of Chemical Education, 2007, 84, 685–688.

[17]   Haque F, Daeneke T, Kalantar-zadeh K, Ou JZ. Two-dimensional transition metal oxide and chalcogenide-based photocatalysts, Nano-Micro Letters, 2018, 10, 23.

[18]   Pozan GS, Isleyen M, Gokcen S. Transition metal coated TiO_2 nanoparticles: synthesis, characterization and their photocatalytic activity, Applied Catalysis B: Environmental, 2013, 140–141, 537–545.

[19]   Nakata K, Fujishima A. TiO_2 photocatalysis: design and applications, Journal of Photochemistry and Photobiology C: Photochemistry Reviews, 2012, 13, 169–189.

[20] Etacheri V, Di Valentin C, Schneider J, Bahnemann D, Pillai SC. Visible-light activation of TiO_2 photocatalysts: Advances in theory and experiments, Journal of Photochemistry and Photobiology C: Photochemistry Reviews, 2015, 25, 1–29.

[21] Fujishima A, Zhang X, Tryk DA. TiO_2 photocatalysis and related surface phenomena, Surface Science Reports, 2008, 63, 515–582.

[22] Ni M, Leung MKH, Leung DYC, Sumathy K. A review and recent developments in photocatalytic water-splitting using TiO_2 for hydrogen production, Renewable and Sustainable Energy Reviews, 2007, 11, 401–425.

[23] Haggerty JES, Schelhas LT, Kitchaev DA, et al. High-fraction brookite films from amorphous precursors, Scientific Reports, 2017, 7, 1–11.

[24] Liu G, Yang HG, Pan J, Yang YQ, Lu GQM, Cheng HM. Titanium dioxide crystals with tailored facets, Chemical Reviews, 2014, 114, 9559–9612.

[25] Yu J, Low J, Xiao W, Zhou P, Jaroniec M. Enhanced photocatalytic CO_2-reduction activity of enhanced photocatalytic CO_2-reduction activity of anatase TiO_2 by, Journal of the American Chemical Society, 2014, 136, 8839.

[26] Meng A, Zhang J, Xu D, Cheng B, Yu J. Enhanced photocatalytic H_2-production activity of anatase TiO_2 nanosheet by selectively depositing dual-cocatalysts on (101) and (001) facets, Applied Catalysis B: Environmental, 2016, 198, 286–294.

[27] Krysiak OA, Barczuk PJ, Bienkowski K, Wojciechowski T, Augustynski J. The photocatalytic activity of rutile and anatase TiO_2 electrodes modified with plasmonic metal nanoparticles followed by photoelectrochemical measurements, Catalysis Today, 2019, 321–322, 52–58.

[28] Li H, Wu X, Yin S, Katsumata K, Wang Y. Effect of rutile TiO_2 on the photocatalytic performance of g-C_3N_4/brookite-$TiO_{2-x}N_y$ photocatalyst for NO decomposition, Applied Surface Science, 2017, 392, 531–539.

[29] Yin S, Aita Y, Komatsu M, Wang J, Tang Q, Sato T. Synthesis of excellent visible-light responsive $TiO_{2-x}N_y$ photocatalyst by a homogeneous precipitation-solvothermal process, Journal of Materials Chemistry, 2005, 15, 674–682.

[30] Cai J, Wu X, Li S, Zheng F. Controllable location of Au nanoparticles as cocatalyst onto TiO_2@CeO_2 nanocomposite hollow spheres for enhancing photocatalytic activity, Applied Catalysis B: Environmental, 2017, 201, 12–21.

[31] Yin H, Wang X, Wang L, et al. Ag/AgCl modified self-doped TiO_2 hollow sphere with enhanced visible light photocatalytic activity, Journal of Alloys and Compounds, 2016, 657, 44–52.

[32] Zhou X, Liu N, Schmuki P. Photocatalysis with TiO_2 nanotubes: "colorful" reactivity and designing site-specific photocatalytic centers into TiO_2 nanotubes, ACS Catalysis, 2017, 7, 3210–3235.

[33] Ge M, Li Q, Cao C, et al. One-dimensional TiO_2 nanotube photocatalysts for solar water splitting, Advancement of Science, 2017, 4, 1–31.

[34] Qu A, Xie H, Xu X, Zhang Y, Wen S, Cui Y. High quantum yield graphene quantum dots decorated TiO_2 nanotubes for enhancing photocatalytic activity, Applied Surface Science, 2016, 375, 230–241.

[35] Zhang J, Yu Z, Gao Z, et al. Porous TiO_2 nanotubes with spatially separated platinum and CoO_x cocatalysts produced by atomic layer deposition for photocatalytic hydrogen production, Angewandte Chemie International Edition, 2017, 56, 816–820.

[36] Motola M, Čaplovičová M, Krbal M, et al. Ti^{3+} doped anodic single-wall TiO_2 nanotubes as highly efficient photocatalyst, Electrochimica Acta, 2020, 331, 1–8.

[37] Liu Y, Zhou M, Zhang W, et al. Enhanced photocatalytic properties of TiO_2 nanosheets@2D layered black phosphorus composite with high stability under hydro-oxygen environment, Nanoscale, 2019, 11, 5674–5683.

[38] Ma X, Xiang Q, Liao Y, Wen T, Zhang H. Visible-light-driven CdSe quantum dots/graphene/ TiO_2 nanosheets composite with excellent photocatalytic activity for E. coli disinfection and organic pollutant degradation, Applied Surface Science, 2018, 457, 846–855.

[39] Dong W, Yao Y, Li L, et al. Three-dimensional interconnected mesoporous anatase TiO_2 exhibiting unique photocatalytic performances, Applied Catalysis B: Environmental, 2017, 217, 293–302.

[40] Rahul TK, Sandhyarani N. In situ gold-loaded fluorinated titania inverse opal photocatalysts for enhanced solar-light-driven hydrogen production, ChemNanoMat, 2017, 3, 503–510.

[41] Ong CB, Ng LY, Mohammad AW. A review of ZnO nanoparticles as solar photocatalysts: Synthesis, mechanisms and applications, Renewable and Sustainable Energy Reviews, 2018, 81, 536–551.

[42] Baral A, Khanuja M, Islam SS, Sharma R, Mehta BR. Identification and origin of visible transitions in one dimensional (1D) ZnO nanostructures: excitation wavelength and morphology dependence study, Journal of Luminescence, 2017, 183, 383–390.

[43] Xiong J, Gan Y, Zhu J, et al. Insights into the structure-induced catalysis dependence of simply engineered one-dimensional zinc oxide nanocrystals towards photocatalytic water purification, Inorganic Chemistry Frontiers, 2017, 4, 2075–2087.

[44] Hekmatara SH, Mohammadi M, Haghani M. Novel water-soluble, copolymer capped zinc oxide nanorods with high photocatalytic activity for degradation of organic pollutants from water, Chemical Physics Letters, 2019, 730, 345–353.

[45] Koutavarapu R., Lee G., Babu B., Yoo K., Shim J. Visible-light-driven photocatalytic activity of tiny ZnO nanosheets anchored on $NaBiS_2$ nanoribbons via hydrothermal synthesis, Journal of Materials Science: Materials in Electronics, 2019, 30(11), 10900–10911.

[46] Mahdavi R, Talesh SSA. Sol-gel synthesis, structural and enhanced photocatalytic performance of Al doped ZnO nanoparticles, Advanced Powder Technology: The International Journal of the Society of Powder Technology, Japan, 2017, 28, 1418–1425.

[47] Pudukudy M, Hetieqa A, Yaakob Z. Synthesis, characterization and photocatalytic activity of annealing dependent quasi spherical and capsule like ZnO nanostructures, Applied Surface Science, 2014, 319, 221–229.

[48] Zhang X, Wang Y, Hou F, et al. Effects of Ag loading on structural and photocatalytic properties of flower-like ZnO microspheres, Applied Surface Science, 2017, 391, 476–483.

[49] Abd Aziz SNQA, Pung S-Y, Ramli NN, Lockman Z. Growth of ZnO nanorods on stainless steel wire using chemical vapour deposition and their photocatalytic activity, Scientific World Journal, 2014, 2014, 1–9.

[50] Wei A, Xiong L, Sun L, et al. One-step electrochemical synthesis of a graphene – ZnO hybrid for improved photocatalytic activity, Materials Research Bulletin, 2013, 48, 2855–2860.

[51] Zgura I, Preda N, Socol G, et al. Wet chemical synthesis of ZnO-CdS composites and their photocatalytic activity, Materials Research Bulletin, 2018, 99, 174–181.

[52] Taheri M, Abdizadeh H, Reza M. Formation of urchin-like ZnO nanostructures by sol-gel electrophoretic deposition for photocatalytic application, Journal of Alloys and Compounds, 2017, 725, 291–301.

[53] Ulyankina A, Leontyev I, Avramenko M, Zhigunov D, Smirnova N. Large-scale synthesis of ZnO nanostructures by pulse electrochemical method and their photocatalytic properties, Materials Science in Semiconductor Processing, 2018, 76, 7–13.

[54] Sai Guru Srinivasan S, Govardhanan B, Aabel P, Ashok M, Santhosh Kumar MC. Effect of oxygen partial pressure on the tuning of copper oxide thin films by reactive sputtering for solar light driven photocatalysis, Solar Energy, 2019, 187, 368–378.

[55] Vivek E, Senthilkumar N, Pramothkumar A, Vimalan M, Potheher IV. Synthesis of flower-like copper oxide microstructure and its photocatalytic property, Physical Review B: Condensed Matter, 2019, 566, 96–102.

[56] Liu Q, Liang Y, Liu H, Hong J, Xu Z. Solution phase synthesis of CuO nanorods, Materials Chemistry and Physics, 2006, 98, 519–522.

[57] Xu Y, Chen D, Jiao X. Fabrication of CuO pricky microspheres with tunable size by a simple solution route, The Journal of Physical Chemistry. B, 2005, 109, 13561–13566.

[58] Tezcan F, Mahmood A, Kardaş G. Optimizing copper oxide layer on zinc oxide via two-step electrodeposition for better photocatalytic performance in photoelectrochemical cells, Applied Surface Science, 2019, 479, 1110–1117.

[59] Wang Y, Zhou M, He Y, Zhou Z, Sun Z. In situ loading CuO quantum dots on TiO_2 nanosheets as cocatalyst for improved photocatalytic water splitting, Journal of Alloys and Compounds, 2020, 813, 152184.

[60] Momeni MM, Nazari Z. Hydrogen evolution from solar water splitting on nanostructured copper oxide photocathodes, Materials Research Innovations, 2017, 21, 15–20.

[61] Landaeta E, Masitas RA, Clarke TB, et al. Copper-oxide-coated silver nanodendrites for photoelectrocatalytic CO_2 reduction to acetate at low overpotential, ACS Applied Nano Materials, 2020, 3, 3478–3486.

[62] Yin G, Nishikawa M, Nosaka Y, et al. Photocatalytic carbon dioxide reduction by copper oxide nanocluster-grafted niobate nanosheets, ACS Nano, 2015, 9, 2111–2119.

[63] Yang Z, Hao X, Chen S, et al. Long-term antibacterial stable reduced graphene oxide nanocomposites loaded with cuprous oxide nanoparticles, Journal of Colloid and Interface Science, 2019, 533, 13–23.

[64] Mishra M, Chun DM. α-Fe_2O_3 as a photocatalytic material: a review, Applied Catalysis. A, General, 2015, 498, 126–141.

[65] Helal A, Harraz FA, Ismail AA, Sami TM, Ibrahim IA. Hydrothermal synthesis of novel heterostructured Fe_2O_3/Bi_2S_3 nanorods with enhanced photocatalytic activity under visible light, Applied Catalysis B: Environmental, 2017, 213, 18–27.

[66] Jiang Z, Wan W, Li H, Yuan S, Zhao H, Wong PK. A hierarchical Z-scheme α-Fe_2O_3/g-C_3N_4 hybrid for enhanced photocatalytic CO_2 reduction, Advanced Materials, 2018, 30, 1706108.

[67] O TF, Abdel-Wahab A, Al-Shirbini A, Mohamed O, Nasr O. Photocatalytic degradation of paracetamol over magnetic flower-like, Journal Photochem Photobiol A Chem, 2017, 347, 186–198.

[68] Xu Y, Kraft M, Xu R. Metal-free carbonaceous electrocatalysts and photocatalysts for water splitting, Chemical Society Reviews, 2016, 45, 3039–3052.

[69] Wang X, Maeda K, Thomas A, et al. A metal-free polymeric photocatalyst for hydrogen production from water under visible light, Nature Materials, 2009, 8, 76–80.

[70] Liu Y, Tian J, Wang Q, Wei L, Wang C, Yang C. Enhanced visible light photocatalytic activity of g-C_3N_4 via the synergistic effect of K atom bridging doping and nanosheets formed by thermal exfoliation, Optical Materials, 2020, 99, 109594.

[71] Fu J, Zhu B, Jiang C, Cheng B, You W, Yu J. Hierarchical porous O-doped g-C_3N_4 with enhanced photocatalytic CO_2 reduction activity, Small 2017, 1603938, 1–9.

[72] Zhu B, Xia P, Li Y, Ho W, Yu J. Fabrication and photocatalytic activity enhanced mechanism of direct Z-scheme g-C_3N_4/Ag_2WO_4 photocatalyst, Applied Surface Science, 2017, 391, 175–183.

[73] Xia P, Zhu B, Cheng B, Yu J, Xu J. 2D/2D g-C_3N_4/MnO_2 nanocomposite as a direct Z-scheme photocatalyst for enhanced photocatalytic activity, ACS Sustainable Chemistry & Engineering, 2018, 6, 965–973.

[74] Atchudan R, Jebakumar Immanuel Edison TN, Perumal S, Karthikeyan D, Yr Lee. Effective photocatalytic degradation of anthropogenic dyes using graphene oxide grafting titanium

dioxide nanoparticles under UV-light irradiation, Journal of Photochemistry and Photobiology. A, Chemistry, 2017, 333, 92–104.

[75] Bera R, Kundu S, Patra A. 2D hybrid nanostructure of reduced graphene oxide–CdS nanosheet for enhanced photocatalysis, ACS Applied Materials & Interfaces, 2015, 7, 13251–13259.

[76] Khoa NT, Kim SW, Yoo DH, Cho S, Kim EJ, Hahn SH. Fabrication of Au/graphene-wrapped ZnO-nanoparticle-assembled hollow spheres with effective photoinduced charge transfer for photocatalysis, ACS Applied Materials & Interfaces, 2015, 7, 3524–3531.

[77] Iwase A, Ng YH, Ishiguro Y, Kudo A, Amal R. Reduced graphene oxide as a solid-state electron mediator in Z-scheme photocatalytic water splitting under visible light, Journal of the American Chemical Society, 2011, 133, 11054–11057.

[78] Bhunia SK, Jana NR. Reduced graphene oxide-silver nanoparticle composite as visible light photocatalyst for degradation of colorless endocrine disruptors, ACS Applied Materials & Interfaces, 2014, 6, 20085–20092.

[79] Kumar S, Terashima C, Fujishima A. Photocatalytic Degradation of Organic Pollutants in Water Using Graphene Oxide Composite, Springer International Publishing, 2019, 413–438.

[80] Atkin P, Daeneke T, Wang Y, et al. 2D WS_2/carbon dot hybrids with enhanced photocatalytic activity, Journal of Materials Chemistry A, 2016, 4, 13563–13571.

[81] Han D, Ma M, Han Y, et al. Eco-friendly hybrids of carbon quantum dots modified MoS_2 for rapid microbial inactivation by strengthened photocatalysis, ACS Sustainable Chemistry & Engineering, 2020, 8, 534–542.

[82] Wang Q, Li J, Tu X, et al. Single atomically anchored cobalt on carbon quantum dots as efficient photocatalysts for visible light-promoted oxidation reactions, Chemistry of Materials: A Publication of the American Chemical Society, 2020, 32, 734–743.

[83] Coleman JN, Khan U, Young K, et al. Two-dimensional nanosheets produced by liquid exfoliation of layered materials, Science 80-, 2011, 331, 568–571.

[84] Miremadi BK, Cowan T, Morrison SR. New structures from exfoliated MoS_2, Journal of Applied Physics, 1991, 69, 6373–6379.

[85] Song JG, Park J, Lee W, et al. Layer-controlled, wafer-scale, and conformal synthesis of tungsten disulfide nanosheets using atomic layer deposition, ACS Nano, 2013, 7, 11333–11340.

[86] Mahler B, Hoepfner V, Liao K, Ozin GA. Colloidal synthesis of 1T-WS_2 and 2H-WS_2 nanosheets: applications for photocatalytic hydrogen evolution, Journal of the American Chemical Society, 2014, 136, 14121–14127.

[87] Zhang G, Liu H, Qu J, Li J. Two-dimensional layered MoS_2: rational design, properties and electrochemical applications, Energy & Environmental Science, 2016, 9, 1190–1209.

[88] Joensen P, Frindt RF, Morrison SR. Single-layer MoS_2, Materials Research Bulletin, 1986, 21, 457–461.

[89] Han B, Liu S, Zhang N, Xu YJ, Tang ZR. One-dimensional CdS@MoS_2 core-shell nanowires for boosted photocatalytic hydrogen evolution under visible light, Applied Catalysis B: Environmental, 2017, 202, 298–304.

[90] Xu H, Yi J, She X, et al. 2D heterostructure comprised of metallic 1T-MoS_2/monolayer O-g-C_3N_4 towards efficient photocatalytic hydrogen evolution, Applied Catalysis B: Environmental, 2018, 220, 379–385.

[91] Ye L, Wang D, Chen S. Fabrication and enhanced photoelectrochemical performance of MoS_2/S-doped g-C_3N_4 heterojunction film, ACS Applied Materials & Interfaces, 2016, 8, 5280–5289.

[92] Feng C, Chen Z, Hou J, et al. Effectively enhanced photocatalytic hydrogen production performance of one-pot synthesized MoS_2 clusters/CdS nanorod heterojunction material under visible light, Chemical Engineering Journal, 2018, 345, 404–413.

[93]   Nguyen TD, Man MT, Nguyen MH, Seo D, Kim E. Effect of few-layer MoS$_2$ flakes deposited ZnO/FTO nanorods on photoelectrochemical characteristic, Materials Research Express, 2019, 6, 085070.

[94]   Li L, Yin X, Sun Y. Facile synthesized low-cost MoS$_2$/CdS nanodots-on-nanorods heterostructures for highly efficient pollution degradation under visible-light irradiation, Separation and Purification Technology, 2019, 212, 135–141.

[95]   Lin X, Wang X, Zhou Q, et al. Magnetically recyclable MoS$_2$/Fe$_3$O$_4$ hybrid composite as visible light responsive photocatalyst with enhanced photocatalytic performance, ACS Sustainable Chemistry & Engineering, 2019, 7, 1673–1682.

[96]   Zhang S, Liu X, Liu C, et al. MoS2 quantum dot growth induced by S vacancies in a ZnIn$_2$S$_4$ monolayer: atomic-level heterostructure for photocatalytic hydrogen production, ACS Nano, 2018, 12, 751–758.

[97]   Li X, Guo S, Li W, et al. Edge-rich MoS$_2$ grown on edge-oriented three-dimensional graphene glass for high-performance hydrogen evolution, Nano Energy, 2019, 57, 388–397.

[98]   Ashraf W, Fatima T, Srivastava K, Khanuja M. Superior photocatalytic activity of tungsten disulfide nanostructures: role of morphology and defects, Applied Nanoscience, 2019, 9, 1515–1529.

[99]   Mahler B, Hoepfner V, Liao K, Ozin A. Colloidal synthesis of 1T-WS$_2$ and 2H-WS$_2$ nanosheets: Applications for Photocatalytic Hydrogen Evolution. 2014,

[100] Dai W, Yu J, Luo S, et al. WS2 quantum dots seeding in Bi2S3 nanotubes: A novel Vis-NIR light sensitive photocatalyst with low-resistance junction interface for CO2 reduction, Chemical Engineering Journal, 2020, 389, 123430.

[101] Li L, Yan Y, Liu H, et al. Hollow core/shell β-Bi$_2$O$_3$@WS$_2$ p–n heterojunction for efficient photocatalytic degradation of fluoroquinolones: a theoretical and experimental study, Inorganic Chemistry Frontiers, 2020, 7, 1374–1385.

[102] Zhang Q, Wang W, Zhang J, et al. Highly efficient photocatalytic hydrogen evolution by ReS$_2$ via a two-electron catalytic reaction, Advanced Materials, 2018, 30, 1–7.

[103] Tongay S, Sahin H, Ko C, et al. Monolayer behaviour in bulk ReS$_2$ due to electronic and vibrational decoupling, Nature Communications, 2014, 5, 3252.

[104] Rahman M, Davey K, Qiao S-Z. Advent of 2D rhenium disulfide (ReS$_2$): fundamentals to applications, Advanced Functional Materials, 2017, 27, 1606129.

[105] Jing Q, Zhang H, Huang H, et al. Ultrasonic exfoliated ReS$_2$ nanosheets: fabrication and use as co-catalyst for enhancing photocatalytic efficiency of TiO$_2$ nanoparticles under sunlight, Nanotechnology, 2019, 30, 184001.

P. Fernandez-Ibañez, S. McMichael, A. Tolosana-Moranchel,
J. A. Byrne

Chapter 2
Photocatalytic Reactors Design for Water Applications

2.1 Introduction

The World Health Organization has estimated that 2.2 billion people do not have access to safely managed water sources [1], and these people have to use contaminated water sources that increase their risk to contract infections and waterborne diseases; for example, diarrhea, typhoid, and cholera. Furthermore, water demands and pollution continue to increase rapidly as the population and industrialization rates grow. Consequently, over 2 billion people live in countries experiencing high water stress [2], which is aggravated by the actual climate change scenario, thus making that the arid areas of the world become drier and have to rely more and more on rainfalls and river basins to meet their basic needs. There is a need for new approaches and technologies that can remediate contaminated water in a sustainable and environmentally friendly manner.

Photocatalysis is a clean technology that can utilize UV and visible light (solar or lamps) and oxygen from the atmosphere to inactivate microorganisms and degrade organic pollutants in water. If solar energy can be used, then it is the ultimate green technology for water treatment; however, its main limitation is the capacity of recombination of charge carriers in the photocatalyst, photoactivated by radiation absorption, reducing the photocatalytic efficiency. To remediate this, photoelectrocatalysis (PEC) utilizes a photoelectrochemical cell (can be configured in different ways, e.g., cathode–anode–anode and anode–cathode–anode) with an applied bias that prevents the recombination of carriers in the photoanode and favors the oxidation of chemicals and microorganisms by photogenerated reactive oxygen species (ROS). The key research areas include the development of novel photoelectrochemical cells, the selection of materials that can better utilize the UV and visible part of

Acknowledgments: The authors wish to acknowledge the Department for Economy (DfE) Northern Ireland for funding Stuart McMichael, the Global Challenges Research Fund (GCRF) UK Research and Innovation for funding SAFEWATER (grant reference EP/P032427/1), and the funding from the European Union's Horizon 2020 research and innovation program under grant agreement no. 820718 (PANIWATER).

P. Fernandez-Ibañez, School of Engineering, Ulster University, UK,
e-mail: p.fernandez@ulster.ac.uk
S. McMichael, A. Tolosana-Moranchel, J. A. Byrne, School of Engineering, Ulster University, UK

https://doi.org/10.1515/9783110668483-002

the solar spectrum, the stability of these materials, the design of the cells for an optimal mass transfer and a radiation absorption, scaling up of the process, and the economic implications of using an electrical bias to reduce substantially the treatment time in comparison with other advanced oxidation processes. Understanding the fundamental mechanism of the process, especially when organic and biological pollutants are targets simultaneously, will permit enhancement and optimization of the PEC systems for water purification.

2.2 Reactors for Photocatalysis

The development of an efficient photocatalytic reactor is a challenging task, as there are many aspects and parameters that need to be optimized to facilitate commercialization or implantation into an industrial application. This has resulted in a variety of lab-scale and pilot-scale reactor designs. Regardless of the design and operation, photocatalytic reactors can be categorized into two distinct groups: suspension, where the photocatalyst remains mixed with the solution, and immobilized, where the photocatalyst is bonded to a supporting substrate. Both systems have their advantages and disadvantages. In order to optimize either system it is important to analyze the following: illumination and optical properties, efficiency, kinetics, and mass transfer. Each of these areas will be discussed as well as a short overview of existing reactor setups.

2.2.1 Preliminary Design Aspects

As with any design, it is important to develop a product design specification to establish and define many design parameters, and some of which have been summarized in Figure 2.1. Overall a water treatment reactor should have a high throughput with acceptable performance, and energy efficient while also being cost-effective. It should be noted that cost-effectiveness does not necessarily mean a system with a low capital cost, but rather the effective cost per unit of water treated (L or m^3) for the lifetime of the system is economical.

The purpose of the reactor will have drastic influence on the overall form of the reactor, whether it is laboratory device for testing new photocatalytic materials or for a pilot-scale device; despite this, there are many overlapping parameters that must be analyzed. Before upscaling, it is essential to conduct laboratory-scale experiments and simulations to predict the performance of large-scale devices. The simulations or modeling should consider how the reactor is illuminated, the subsequent intrinsic kinetic model accounting for photon absorption, mass balance (fluid dynamics), and finally how the model correlates with the experimental data.

Reactor Design Considerations						
Purpose	Reactor Type	Illumination	Flow	Materials	Contaminant	Geometry
Lab Scale Testing	Suspension	Solar	Batch	Photocatalyst	Microorganism	Scale / Volume
New Materials	Catalyst removal	Real	Continuous	Chemical Resistance	Chemical	Photon Reflectors
Kinetics	Immobilised	Simulated	Flow Regime	Window		Unique Features
Pilot Scale Device		Artificial		UV Transparent		
		Lamps				
		LEDs				

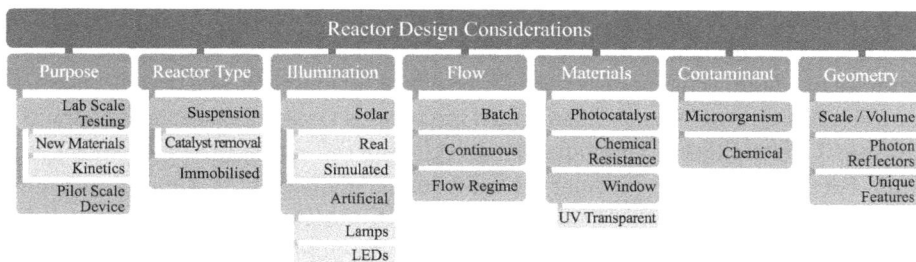

Figure 2.1: Design considerations for a photocatalytic reactor for water treatment.

2.2.2 Illumination and Optical Properties

2.2.2.1 Illumination

The illumination of the reactor is a key parameter that must be analyzed to ensure that the reactor is fully irradiated and to allow for adequate photon absorption with the key wavelengths for the photocatalyst throughout the reactor. For initial laboratory experiments, solar simulators can be utilized, and the experimental setup aligned to match the irradiance standard (G-173) set by the American Society for Testing and Materials is commonly referenced to air mass (AM) 1.5, which has global UV irradiation of 46.6 W/m^2. When using solar simulators, there is an uneven irradiation profile with higher intensities at the center of the beam compared to the edges, which needs to be accounted. The high-powered lamps emit significant amount of heat, which can have an influence on the results; however, this can be minimized by using an infrared water filter. When moving from laboratory-based experiment to real solar irradiation, the difficulty arrives with changing irradiance with the time of day/year that needs also consideration. Alternatively, the use of artificial irradiation in the form of UV fluorescent lamps or highly efficient light-emitting diodes (LEDs) can be used as the source of irradiation. This has the advantage of being able to run at any time with specific intensity of irradiation ensuring that the reaction is not photon limited; however, there is of course an associated cost both capital and running. Depending on the type of system (suspension or immobilized), how the radiation is transferred, absorbed, or scattered is going to be different.

2.2.2.2 Illumination in Suspension Systems

One of the key parameters for a suspension reactor commonly referred to as slurry reactors is the photocatalyst loading (g/L) and the corresponding optical penetration depth which allows for adequate photon absorption (90–95%). The optimal light penetration through the water layers of the reactor has to be calculated using the classic laws of light scattering and absorption of suspended photocatalyst nano-particles in a fluid – commonly water, which dissolved inorganic and organic matter – for the range of wavelengths under which the process is taking place, usually the UV range and depending on the catalyst some regions of the visible part of the solar spectrum. For example, if we consider the most commonly used photocatalyst, titanium dioxide, it is well known that suspensions of TiO_2 absorb radiation at wavelengths below 387 nm but there is a significant fraction of the light extinction due to the scattering of TiO_2 nanoparticles. Both effects – which are dependent on the wavelength of the incident radiation – must be accounted for the estimation (theoretically or empirically) of the optimal catalyst load for a certain reactor path length.

A Lambert–Beer law-type equation can be used in this case, where the light absorbance in the photoreactor by the catalyst suspension is proportional to the load of catalyst, the reactor path length, and the light extinction (absorption and scattering) properties of the suspended catalyst at each wavelength. The reasonable hypothesis on the ratio of incoming light must be absorbed by the photocatalyst in the photoreactor (e.g., from 95% to 99%) will lead to an estimate of the ideal path length of the reactor, which usually is equal to the radius or diameter of the reactor tube if we are considering a tubular reactor with or without a mirror; or equal to the water-layer thickness in a flat-plane reactor without any radiation concentration system [3].

Equation (2.1) takes into consideration scattering, where L is the thickness of the solution (m), τ is the optical thickness (dimensionless), σ is the scattering coefficient (m^2/kg), κ is the absorption coefficient (m^2/kg) for a certain wavelength, and C_{cat} is the catalyst loading (kg/m^3). The optical thickness (τ) is dependent on albedo effect (ω), which can be calculated to less than 5% of the asymptotic value at infinite using equation (2.2). In the albedo effect (ω), a value of 0 equals only absorption and 1 equals only scattering, and this can be found in literature or through experimentation and calculated using equation (2.3), where β is the extinction coefficient. For example, in the literature TiO_2 (P25) is reported to have an albedo value between 0.8 and 1 depending on the wavelength [4]. Results of Motegh et al. [3] indicate that the optimal thickness should be at least 3.5 for low photonic flux and 6.5 for high photonic flux:

$$L = \frac{\tau}{(\sigma + \kappa)C_{cat}} \tag{2.1}$$

$$\tau = \cosh^{-1}\left\{[\omega(760 + 761\omega)]^{-1} \times \left[\omega\left(14{,}839 + 7{,}620\omega - 7{,}220\sqrt{1-\omega^2}\right)\right.\right.$$
$$\left.\left. - 7{,}220\left(-1 + \sqrt{1-\omega^2}\right)\right]\right\} \tag{2.2}$$

$$\omega = \frac{\sigma}{\beta} \tag{2.3}$$

2.2.2.3 Illumination in Immobilized Systems

The photocatalyst can be immobilized on a support such as a flat plate. In doing so, there are two ways of irradiating the photocatalyst as shown in Figure 2.2: front face and back face. The optimal loading is different for either orientation.

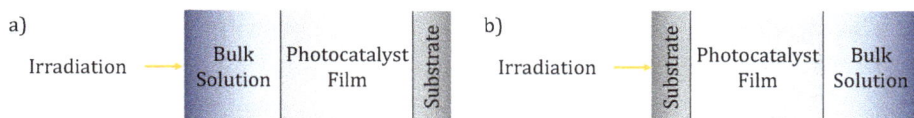

Figure 2.2: (a) Front-face irradiated and (b) back-face irradiated.

For front-face irradiation, the rate of the reaction will increase with film thickness until a point of saturation at which increasing the film thickness will not affect the rate of the reaction. The irradiation flux also affects the optimal size; under low radiation, a thinner film is required to achieve the same normalized reaction rate, compared to under high irradiation [5]. Given the asymptotic nature of front-face irradiation, it is best to have thicker film (4–10 μm) as there no inherent disadvantages to the rate of the reaction, provided there is adequate adhesion to the substrate and photocatalyst is inexpensive. However, as the irradiation must pass through the bulk solution, the intensity will decrease with distance as the photons are absorbed in the solution. This results in less photons being absorbed by the photocatalyst, which in turn results in slow rates compared to back-face irradiation for the degradation of chemicals [6]. This is not necessarily the case for microorganisms as the UV irradiation alone contributes to inactivation [7]. For back-faced irradiation the loading is more critical as lower than the optimal film thickness produces inadequate photon absorption which results in slow reaction rate as there is less ROS produced. Higher than the optimal thickness results in high photon absorption but the longer diffusion length of the material results in the undesirable recombination of the electron–hole pairs [5]. The optimal thickness based on the normalized reaction rate also changes with irradiation, with high irradiation the optimal thickness is less than with low irradiation due to the exponential decay of photons by absorption.

2.2.3 Efficiency

The efficiency of the reactor can be calculated in a variety of ways for chemical compounds as defined by the International Union of Pure and Applied Chemistry [8]:

- Quantum yield – the events occurring per photon absorbed for a photoreaction equation (2.4) is used. This term, however, is only for monochromatic excitation:

$$\Phi_\lambda = \frac{\text{amount of reactant consumed or product formed}}{\text{photons absorbed } (\lambda)} \qquad (2.4)$$

- Quantum efficiency – for polychromic excitation, the term quantum efficiency is more appropriate, using averaged values of photons for the interval of wavelengths considered:

$$\eta(\lambda 1 \rightarrow \lambda 2) = \frac{\text{amount of reactant consumed or product formed}}{\text{average photons absorbed } (\lambda 1 \rightarrow \lambda 2)} \qquad (2.5)$$

- Photonic efficiency – this is the efficiency based on all incident photons (polychromic) and not just absorbed photons (equation (2.6)). This is also reported in the literature as the apparent quantum efficiency:

$$\xi = \frac{\text{amount of reactant consumed or product formed}}{\text{incident photons (spectral range)}} \qquad (2.6)$$

- Photonic yield – this is identical to photonic efficiency but for monochromatic light.

The quantum efficiency/yield requires the measurement of the absorbed photons by the photocatalysts. For heterogeneous photocatalysis, this can be difficult to measure due to scattering and photon absorption by aqueous solution. The photonic efficiency/yield offers a much simpler method to experimentally calculate the efficiency by neglecting the absorption component and measuring the incident photons using a spectrometer or an actinometer. This also gives greater evaluation of the reactor design as it is affected by the configuration and operational conditions including the photoreactor geometry, the position of the lamp, the catalyst loading, the optical path length, while the quantum efficiency is an evaluation of the intrinsic property of the photocatalytic material.

If the incident photonic flux has a major effect on efficiency with low intensity, then there is a higher quantum efficiency as the reaction is limited by the number of photons. In the case of high photonic flux, the reaction is limited by the intrinsic reaction kinetics as all reactive sites will be active, which may indicate the need for surface modification to improve the kinetics of the reaction [9]. When using solar radiation, a reduction in quantum efficiency due to high photonic flux is not necessarily a major concern as the input energy is free. However, when using artificial irradiation in the form of lamps or LEDs, the quantum efficiency will be more important as there is an associated cost, as electrical energy must be used. It is then essential to perform a cost analysis to evaluate potential savings by running the reactor with the optimal photonic flux, that is, the highest quantum efficiency.

2.2.4 Kinetics

Kinetic studies examine the rate of a reaction, and this is important as it establishes the required time for a reaction to take place. The kinetics of a photocatalytic reaction has the additional complexity of light absorption, as the reaction rate is influenced by the rate of photon absorption. Thus, an intrinsic kinetic model is used based on the reaction steps: photon absorption and electron–hole pair generation, recombination, electron trapping, hole trapping, ROS attack on the target contained, and finally termination or scavenging. A general expression that has been derived by Marugán et al. [10] (equations (2.7)–(2.9)) for the reaction rate (r) of compound (A) at a concentration (C) is as follows:

$$r_A = -\alpha_1 f(C)g(e^a) \tag{2.7}$$

where α_1 is a constant that intergrades kinetic parameters and concentrations of species when constant, $f(C)$ is the function of the reaction rate based on the concentration, and $g(e^a)$ is a function for the dependence on photon absorption and electron–hole generation. If the concentration and photoabsorption are linear, then the equation takes the following form:

$$r_A = -\alpha_1[A]e^a \tag{2.8}$$

If the concentration (A) follows Langmuir–Hinshelwood expression, then equation (2.9) can be used, where the rate dependence on photon flux changes linearly with low flux ($a = 1$), but with high photon flux the reaction rate no longer scales linearly and the exponent a can be changed to 0.5 to account for the change in flux; experimentally this change happens at 25 mW/cm^2 [11]:

$$r_A = -\alpha_1 \frac{[A]}{1+\alpha_2[A]}\left(-1+\sqrt{1+\alpha_3 e^a}\right) \tag{2.9}$$

The kinetic parameters α_1, α_2, and α_3 can be estimated using experimental data and a Levenberg–Marquardt nonlinear regression algorithm to compare with the r_A value.

2.2.5 Mass Transport

Mass transport is the movement/diffusion of species from one position to another. The mass transport can be a limiting factor in photocatalysts and thus must be understood and accounted for in a photocatalytic reactor. There are three types of transport:
- Internal mass transport – the movement of reactants and the resulting products within the porous photocatalytic structure.
- External mass transport – the movement/diffusion of reactants or products at the external surface of the photocatalyst and the bulk solution.
- Convective transport – the movement of fluids within the reactor also known as fluid dynamics.

Internal mass transport can occur if nanoparticles agglomerate in a slurry system or in a porous immobilized film. This is only applicable for chemical compounds; for microorganisms, their size will prevent diffusion into the porous material. Visan et al. [12] have derived a mathematical model for the mass transport; firstly, the internal diffusion and reaction are described in equation (2.7), where D_{eff} is the effective diffusion, c is the concentration, and r is the reaction rate. For a flat photocatalyst geometry, when $y = -\delta$ (material thickness), $D_{eff}(\partial/\partial y) = 0$ and when $y = 0$, $c = c_s$ (surface concentration at the particle–liquid interface). The effective diffusion can be calculated using equation (2.11), where D is the molecular diffusion coefficient, ε is porosity, and τ is tortuosity:

$$D_{eff}\frac{\partial^2 c}{\partial y^2} - r = 0 \tag{2.10}$$

$$D_{eff} = D\frac{\epsilon}{\tau} \tag{2.11}$$

When the equation is solved for a first-order reaction, it yields the following equations for concentration profile (equation (2.12), where Φ is the Thiele modulus) and the net reaction rate (equation (2.13)):

$$c(y) = c_s \frac{\cosh\left(\Phi\left(1 + \frac{y}{\delta}\right)\right)}{\cosh(\Phi)} \tag{2.12}$$

$$N_{y=0} = D_{eff}\frac{\partial c}{\partial y} = c_s \frac{D_{eff}}{\delta}\Phi\tanh\Phi \tag{2.13}$$

The Thiele module is the ratio between the reaction and the diffusion time; a high value will indicate that internal diffusion is limiting the reaction and small value indicates the surface reaction is usually rate-limiting. The Thiele modules can be calculated for first-order reactions using equation (2.14) for flat surface and equation (2.15) for a spherical surface:

$$\Phi = \sqrt{\frac{k}{D_{eff}}}\delta \tag{2.14}$$

$$\Phi = \sqrt{\frac{k}{D_{eff}}}R_p \tag{2.15}$$

The internal effectiveness is another ratio (ranging from 0 to 1) that can be used to evaluate the mechanism, in this instance, the ratio between the net reaction rate and the surface reaction rate or in other words the fraction of the catalyst that is being utilized. The first-order reaction can be clouted using equation (2.16) for a flat surface and equation (2.17) for a spherical surface:

$$\eta = \frac{\tanh\Phi}{\Phi} \tag{2.16}$$

$$\eta = \frac{3}{\Phi^2}(\Phi\coth \Phi - 1) \tag{2.17}$$

The external mass transport is normally expressed using a boundary-layer model at the photocatalyst surface (equation (2.18)), where k_m is the mass transfer coefficient and c_b/c_s is the bulk and surface concentration, respectively):

$$N_{y=0} = k_m(c_b - c_s) \tag{2.18}$$

The mass transfer coefficient (k_m) depends on the flow regime; with higher turbulence there is higher coefficient, which reduces the external mass transport limitation. The supply of oxygen can be a rate-limiting step and may be necessary to evaluate the external mass transfer of oxygen [13].

In both suspension and immobilized, it is important to understand how the fluid moves and to establish if there is adequate mixing and the residence time within the reactor for flow systems. For batch systems, a stirred tank reactor can be used to ensure sufficient mixing. However, when scaling up to flow systems, it becomes more complex. For tubular reactors, a plug flow model is an ideal model where the residence time is same at all points of the reactor; in reality, this does not happen as the velocity at the center of the tube is larger than at the surface of the tube due to the viscosity of the fluid (no-slip condition). Under laminar flow, a Poiseuille velocity profile is observed but as the turbulence increases, predicated by the Reynolds number, the peak velocity at the center will decrease [14]. There is also very little mixing in the axial direction (perpendicular to the flow). This is where Multiphysics simulation software can be used to test and visualize the flow for a variety of conceptual designs and scenarios, followed by the incorporate of the reaction rate into the simulation.

2.2.6 Reactor Design and Setups

A number of different design configurations and setup have been examined by researchers, some are depicted in Figure 2.3. The designs can be divided into experimental reactors for initial testing of photocatalytic materials and scaled-up systems examining the design parameters.

2.2.6.1 Experimental Reactor

When testing new photocatalytic materials, a common laboratory setup involves a batch process using a stirred tank reactor, which can be used for suspension or incorporate an immobilized film, and the reactor should be jacketed to regulate temperature, a lamp source with a fixed known irradiation, and an inlet for oxygen/air

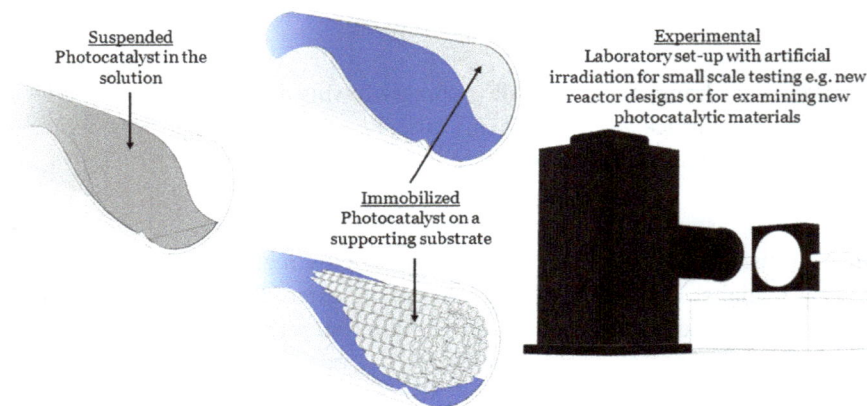

Figure 2.3: Diagram of experimental setup for tubular photocatalytic reactors with suspended (left) or immobilized catalyst (center) and solar simulator (right).

purging. Stirring ensures that the photocatalyst is held in suspension and limits any potential mass transport problems. Regulation of temperature is important as the high energy from the lamp can heat the solution this can increase the rate of the reaction for chemicals and microorganisms [15]. A constant light source ensures that the each photocatalyst particle in the photoreactor is subjected to the same irradiation exposure to ensure photonic flux does not affect the experiment. Oxygen or air purging ensures that there is adequate oxygen to form ROS ensuring that the reductive pathway of the photocatalytic process is not inhibited. All of this ensures that the photocatalyst is tested accurately; however, when comparing different materials in a suspension batch system the optimal path length will change, as discussed in Section 2.2.2.2; therefore the optimal catalyst loading will vary if the volume is fixed.

2.2.6.2 Suspension Reactors

The most fundamental scaled-up reactor is a tubular reactor. These reactors commonly have a recirculating system passing the contaminated water with the photocatalyst through glass tubes (borosilicate or quartz due to their high transparency in the UV range) where it can be irradiated with solar energy or lamps, then into a reservoir tank to be then recirculated. According to Vargas et al. [16], a tubular reactor was made of quartz tubes shaped into a coil with a volume of 7 L connected to a reservoir which gives a total working volume of 15 L for photodegradation of p-nitrophenol, naphthalene, and dibenzothiophene using TiO_2 (0.1 g/L). Tubular reactors are not limited to just solar irradiation [17], Marugan et al. [18] used a 36 W UVA (ultraviolet A) lamp in the center of the tube for the oxidation of cyanide using TiO_2 (0.1–1 g/L).

To enhance tubular reactors, solar concentrators have been used to increase the photon flux. Mehos et al. [19] used a parabolic trough which had a concentration ratio of 20 (reflector width/half-circumference of the pipe) in conjunction with a heat exchanger to reduce the temperature for the treatment of groundwater at a rate of 15 L/ min that was contaminated with trichloroethylene and using TiO_2 (1 g/L). However, the disadvantages of using concentrators are: they require a tracking system, they can only use direct illumination, high temperatures will decrease the solubility of oxygen, and, finally, Mehos et al. showed that one-sun concentrator had a higher efficiency. Consequently, compound parabolic concentrator (CPC) has been examined by other researchers, as CPCs do not require tracking, can use both direct and diffused irradiation, and ensure that all the circumferences of the tubular reactor are irradiated with one-sun concentrator. Polo-López et al. [20] demonstrated the use of a scaled-up CPC system to treat 60 L using TiO_2 (0.1 g/L) for different *Fusarium* spores.

The flat plate reactor as the name suggests uses flat transparent plate instead of a tube. This was used as an attempt to optimize the UV irradiation profile ensuring that the length of the reactor is evenly irradiated for solar applications. Plantard An example is the design of Plantard et al. [21], in which they examined the degradation of pyrimethanil as a contaminant with a working volume of 15 L using TiO_2 (1.5 g/L).

A slurry bubble column reactor introduces the use of air sparging into the system. This has two major advantages: firstly, it ensures that the photocatalyst is held in suspension, and secondly, it supplies oxygen to the solution, which can be reduced to form ROS, thereby enhancing overall results. The addition of a sieve plate reduces back-mixing and has been shown to improve the rate of degradation of phenol [22].

Lastly, an important aspect is the inclusion of a filter to the reactor system to separate the photocatalyst from the solution, and some have reported these reactors as a photocatalytic membrane reactor. These reactors can come in two forms: firstly, where the membrane filter is integrated into the reactor, and secondly, when the filter is separated from the reactor design [23]. The inclusion of the filter introduces additional problems that need to be investigated such as membrane fouling, backwashing, photocatalyst recovery, and cost analysts.

2.2.6.3 Immobilized Reactors

Alrousan et al. [7] examined a variety of reactor setups: front face, back face, and combined by using one tube inside another for the inactivation of E. coli in water. By using a tubular configuration, it was possible to use a CPC to increase the irradiance. Their results show that the most effective configuration was dual tube with only the internal tube coated.

Even with the optimal catalyst loading, there is still a low efficiency for fixed film reactors compared to slurry reactors when using the same photocatalyst load. To further improve the efficiency, it is necessary to increase the surface area to volume ratio of the photocatalyst. This can be achieved through new reactor designs; the simplest is by using a corrugated substrate. This was investigated by Zhang et al. [24], who used a zigzag corrugation of various angles to increase the surface area. The results showed that as the surface area increased, there was an improvement in the rate of degradation compared to flat plate reactor. The corrugation also increases the Reynolds number under a flow regime improving the mass transport in the bulk solution due to increased turbulence, highlighting the importance of adequate mass transport from the bulk solution to the surface of the photocatalyst.

Further methods to increase the surface area are to immobilize the catalyst onto spheres or Raschig rings which are in the millimeter scale. The coated substrates can then be added to the reactor producing what is called a packed bed reactor, as the size of the spheres or Raschig rings ensures easy separation of the solution while increasing the geometric surface area of photocatalyst to over 20 times compared to fixed film reactor [25]. The next aspect that needs to be considered for a packed bed reactor is the optimal depth that ensures adequate photon absorption. This can be done experimentally using various depths of the coated substrates and measuring till 95% of light absorption. Mathematically, Manassero et al. [25] used one-dimensional Monte Carlo simulations to account for photon path length before interactions, scattering, refraction, transmission, and absorption. It showed a uniform distribution of irradiation in the reactor and improved efficiency compared to a fixed film when tested experimentally. However, the slurry reactor still had higher efficiency.

A growing research area is photocatalytic membrane reactors. In this instance, the photocatalyst can be immobilized onto a membrane and combined with UV lamps to activate the photocatalyst. This has the advantage of being able to break down compounds that the filter cannot stop, enables sufficient contact between photocatalysts and the contaminant, that is, reduced any mass transport limitation, and reduces membrane fouling (Figure 2.3) [26].

To conclude the main aspects of illumination, kinetics, efficiency, and mass transport have been discussed for purposes for optimization and modeling. A short overview of some of the different configurations to give an indication of what has been done to further enhance the rector designs and its ability to be scaled up. Even with reactor optimization, finding a photocatalysis that can absorb more of the solar spectrum will undoubtedly improve photocatalytic water treatment because such photocatalytic reactor is worth, but a very challenging task.

2.2.6.4 Disinfection Reactors

Research in photocatalytic disinfection has evolved from very basic laboratory studies to real water samples and larger scale applications in the last decades. Hundreds of publications demonstrating the strong bactericidal effect of photocatalytic processes over a wide range of bacteria can be found [27]. The scope of these studies investigates a broad range of variables that affect substantially the efficiency of the disinfection process, including the chemical and microbiological composition of water, the optical properties of water to be treated, the light intensity, the scale of the problem (from few milliliters to hundreds of liters per day) and the water quality required (drinking, agriculture reclaim, industrial reuse, etc.). Nevertheless, in relation to characteristics and design criteria of photoreactors, there are a number of parameters and microbiological considerations to take into account and have not been considered above, which referred to the application of removal of chemical pollutants from water. These are specific for disinfection applications and are related to the behavior of microorganisms in water, which may contain a consortium of other microorganisms, dissolved organic and inorganic compounds, and suspended matter. They can be either associated with their interaction with the radiation (photobiological response), or the physical–chemical interaction with the photocatalyst, or materials of the reactor. If these are misled or not adequately evaluated, the photocatalytic treatment could lead to undesired effects like the presence of residual communities of microorganisms (resistant microorganisms) after the treatment, the formation of biofilms in the inner walls of the reactor, and poor disinfection efficiency. Therefore, an extensive knowledge and deep understanding of the previously mentioned interactions is necessary to make the correct decisions when designing a photoreactor for photocatalytic disinfection.

i) *Light source.* Besides the well-known effect that different wavelengths have in the light absorption by the photocatalyst – as described before – the radiation spectral characteristics are also critical for microorganisms. Regardless of the use of either sunlight or lamps, it is very important to know very well the spectral distribution (frequency of each wavelength) and the spectral irradiance (surface intensity per wavelength, mW/cm^2 nm). Depending on the wavelengths present in the photoreactor, different effects can be observed in microorganisms.

UltravioletC(UVC) is the most energetic and lethal cell due to the maximum absorption of DNA molecules at close to 260 nm. Cells radiated by UVC may damage dimers of the different bases (pyrimidine and purine) and pyrimidine adducts in DNA that eventually end in cell death or reproduction inhibition. UVC also damages proteins as they absorb photons of 190 nm, tailing up to 220 nm, due to the peptide bond [–C(O)–NH–]. However, UVC does not reach the Earth's surface because it is mostly absorbed in the atmosphere by ozone. A lamp emitting on the UVC leads to an extremely fast bactericidal rate; it takes seconds to kill all present bacteria, from 4 to 6 logarithmic units [28], even in the absence of a catalyst.

The spectral composition of the solar radiation, mainly in the UVA and UVB (ultraviolet B) regions, has a special significance, as the effects of these two ranges are quite different on the microorganisms. The shortest wavelengths of UVB are also absorbed in the atmosphere by ozone, reaching only a very small percentage of this radiation on the Earth's surface. This is strongly dependent on cloud cover and atmospheric conditions. The UVB overlaps with the tail of DNA absorption, so that the UVB radiation causes DNA alterations [29]. The main UVB-induced DNA photoproducts are cyclobutane–pyrimidine dimers, pyrimidine–pyrimidone (six to four) dimers, and the reaction with other pyrimidine bases. The adsorption of UVA and UVB can also inhibit DNA replication via Dewar valence isomers. Adenine or guanine may also absorb UVB radiation generating Dewar adducts. Several proteins and amino acids absorb in the UVB as well, including tryptophan, tyrosine, phenylalanine, histidine, and cysteine. Enterobactin is an iron-chelating agent, whose peak absorption occurs at 316 nm resulting in an iron concentration increase under UVB light [30].

UVA can also damage directly DNA although the mechanism differs from UVB. UVA is involved in the formation of CPD and Dewar valance isomers. The wavelengths that can induce the CPD formation tail up to 365 nm with simultaneous formation of six to four photoproducts [29]. In addition, UVA absorption is done by chromophores and photosensitizers, including porphyrins, flavins, quinones, and NADH/NADPH. These induce oxidation reactions where DNA bases (mainly guanine) are electron donors and form ROS by the generation of superoxide radicals, that is, the promoter of hydrogen peroxide and hydroxyl radicals. Other reactions generate singlet oxygen and the transformation of DNA bases to unstable stereoisomers. In the next section, a detailed description of the formation of ROS is presented. Accumulated damages induced by ROS are considered as a main agent of damage produced by solar radiation. These species have been proven to induce lipid peroxidation, protein oxidation, DNA damages by formation of pyrimidine dimers, or generation of single-strand breaks [31]. Additionally, this part of the spectrum can also be absorbed by natural exogenous photosensitizers present in waters (i.e., humic acids and chlorophylls) and produce ROS by the reaction with dissolved oxygen.

UVA light could also alter the functionality of diverse intracellular compounds with critical effects on cell viability. The main function of catalase (CAT) is to decompound the hydrogen peroxide and to maintain it below the lethal doses to cells. It is also sensitive to UVA radiation. The vital and photosensitive dihydroxy acid dehydratase is an iron–sulfur molecule that could also be inactivated by UVA light. Further details on the importance of CAT and iron–sulfur cluster in cells are explained in the following sections [32].

ii) *Illumination mode.* Besides the spectral distribution, the irradiance (e.g., the radiant energy per unit of time and cross section) is critical for successful disinfection results. Therefore, the optical photoreactor path length has a strong influence on this matter as well as the optical properties of water to be treated. In other words, there must be a compromise between the reactor path-length dimension and the optical absorption of water in the range of action of the photocatalytic process, so that a significant amount of (solar or artificial) radiation penetrates the photoreactor wall and the water layer to eventually reach the photocatalyst and produce efficient generation of ROS and the oxidation of microbes. This effect is strongly dependent on the wavelength also, as each material responds differently for every wavelength.

The continuity of the light exposure, as opposed to an interrupted mode, makes a huge difference in the germicidal effects of the process, being much more efficient the continuous than intermittent illumination [33]. This has been traditionally attributed to the induction of bacterial repair mechanisms at genetic level when they are exposed to sublethal radiation doses that make them more resilient to the treatment and generate residual populations of resistant bacteria at the end of the treatment [56]. The implications of this effect on the reactor design are clear; firstly, the dark spots and dark regions in a reactor are mostly undesirable, which makes really challenging the design of the perfectly illuminated photoreactor. Secondly, in the case of not avoiding dark regions in the photoreactor, the flow-rate determines the grade of intermittence of the light exposure of the water, making hard to manage the ideal flow rate to avoid bacterial recovery during the treatment. These undesired effects that produce residual communities of resilient bacteria can be avoided by simply applying a strongly oxidative process that prevent bacterial repair and recovery and assure total bacteria killing in short times (next section).

iii) *Type of photocatalytic disinfection process.* There are various photooxidation processes that can be used for water disinfection, including solar water disinfection or photo-inactivation by UVA and UVB radiation, UVC disinfection (not covered in this chapter), UV photoinactivation, or solar disinfection improved by adding other oxidizing agents like hydrogen peroxide [34, 35] and persulfate [36], peracetic acid [37], photo-Fenton [38], Fenton-like and heterogeneous photocatalysis using TiO_2 or visible light active materials (VLA materials); for example, graphene–TiO_2 composites [39, 40]. The last four processes are considered photocatalytic processes but the potential effects of the first mentioned processes have to be also considered as all or some of them may co-occur during the photocatalytic disinfection process; therefore, they have to be considered also for the reactor design.

When the photocatalytic reactor utilizes heterogeneous photocatalysis (suspended or immobilized TiO_2 or VLA materials), the previously explained considerations about illumination, light scattering, and absorption performance also apply, as well as their implications in the reactor design and path-length selection. If the reactor applies an immobilized photocatalyst over a support in either the reactor walls or other surfaces, the polluted water exposed to illumination will experience at some degree the direct effects of photon absorption and therefore photo-inactivation of microorganisms, which obeys to different mechanisms of inactivation and different rates as compared with the photocatalytic process. Therefore, some reactors with a noneffective photocatalytic immobilized material may show poor disinfection performance due to the photo-inactivation due to the UVA&UVB exposure more than the photocatalytic action of the photocatalytic material. When the photocatalytic process is enhanced by adding oxidant agents like hydrogen peroxide, persulfate, or peroximonosulfate, the kinetics of disinfection of these photocatalysts or oxidants with UVA or solar radiation must be considered [34–40].

iv) *Temperature.* For any type of photocatalytic process that involves the destruction or inactivation of microorganisms, the temperature of water plays an important role. If we are thinking about the use of solar light, the UVB can be absorbed directly by microorganisms inducing, among others, DNA damages that interfere with the DNA replication or produce mutations [30]. UVA also affects the microbial reproduction by an indirect route. UV radiation also affects the interaction enzymes like catalase (CAT) and superoxide dismutase that reduce or inhibit the action of ROS to prevent oxidative damages [32]. These photobiological processes can be strongly affected by the temperature. Several authors have demonstrated that the effect of temperature and solar radiation is not simply additive but is synergistic in its effect [41–43], finding that the amount of UV energy required to kill certain amount of microorganisms in water is much smaller at temperatures above 48–50 °C (depending on the type). Recently, Castro et al. developed a solar disinfection model for drinking water disinfection that considered the effect of temperature over the microorganism activity and the synergistic effect between radiation and water temperature. This model described the solar disinfection process of *E. coli* at temperatures between 10 and 55 °C at solar UV fluence between 30 and 50 W/m^2 [32].

Besides the effect of temperature over microorganisms in water in the presence of UV radiation, the use of photocatalyst in the process adds a new factor to be considered. The photonic activation of the photocatalyst does not require heating and operates at room temperature. The true activation energy is nil, whereas the apparent activation energy is often very small (a few kJ/mol) in the medium temperature range (between 20 and 80 °C). The unnecessary activation by heating [44] is very interesting for photocatalytic reactors in water applications, where the solubility of oxygen in water must be considered, as it decreases with increasing temperature, as this is a key factor for the correct performing of the systems.

Few researches have been carried out to investigate the effects of temperature in a photocatalytic disinfection process. Garcia-Fernandez et al. [45] demonstrated that the use of heterogeneous photocatalysis using TiO$_2$ suspensions in a solar pilot reactor for wastewater disinfection can be enhanced by the increasing of temperature from 15 to 45 °C for the reduction of bacteria (*E. coli*) and fungi (*Fusarium solani*). Ortega-Gomez et al. [46] similarly demonstrated that a higher water disinfection rate using *Enterococcus faecalis* was attained by increasing temperature according to the Arrhenius equation in the photo-Fenton process. This factor may be worth of consideration for the design of photoelectrochemical cells for water disinfection, as small changes in water temperature can benefit with significant treatment time reductions and therefore increasing the treatment capacity and reducing overall treatment costs.

2.3 Photoelectrocatalytic Cells

2.3.1 Fundamentals of Photoelectrocatalysis

Photocatalysis has attracted a lot of attention during the past decades and has been regarded as a promising technology for air and water purification. However, photocatalytic processes have some disadvantages:
- High electron–hole recombination rates that lead to low efficiencies of absorbed radiation (quantum efficiencies below 10% since over 90% of the photogenerated electrons and holes recombine) [47].
- Most studies have focused on the use of suspended nanoparticles as photocatalysts with high surface area to achieve faster treatment processes. However, in order to recover these nanoparticles, more units are required, increasing the cost of the photocatalytic process [48, 49].

Electrochemically assisted photocatalysis or PEC allows solving these drawbacks [50]. PEC processes consist on the immobilization of a photocatalyst on an electrode that will act either as a photocathode or as a photoanode. When the wavelength of the incident radiation is equal or higher than the bandgap energy $hv \geq E_{bg}$ an electron is promoted from the valence band (VB) to the conduction band (CB) of the semiconductor, giving rise to the generation of electron–hole pairs (e^--h^+) (reaction (2.19)). An external anode potential (E_{anod}) or cell potential (E_{cell}) or a constant current density (j) is applied by using a power supply. This allows controlling the Fermi level of a semiconductor, and therefore ban bending, leading to an efficient separation of e^--h^+ pairs and reducing their recombination (reaction (2.20)) [51]. In the case of an n-type semiconductor, if the potential is greater than the flat band potential (E_{FB}), holes migrate to reach the photocatalyst surface where they will react directly to oxidize organic compounds (reaction (2.21)) or will oxidize water to form hydroxyl radicals (HO˙) that will further oxidize the organic matter (reactions (2.22) and (2.23)). On the other hand, electrons will migrate to the bulk and it will be extracted and transferred to the cathode through the external electric circuit. Electrons will react to carry out reduction reactions. For example, oxygen reduction reactions that will promote the generation of different ROS, such as superoxide radical $(O_2^{˙-}$, reaction (2.24)), hydroperoxyl radical (HO$_2$˙, reaction (2.25)), and hydrogen peroxide (H$_2$O$_2$, reaction (2.26)). Further, HO˙ can be generated from these ROS (reactions (2.27), (2.28), and (2.29)) [50, 52]:

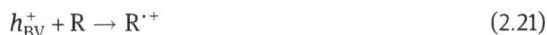

$$TiO_2 + hv \rightarrow e_{BC}^- + h_{BV}^+ \tag{2.19}$$

$$e_{BC}^- + h_{BV}^+ \rightarrow \text{Energía(calor)} \tag{2.20}$$

$$h_{BV}^+ + R \rightarrow R^{˙+} \tag{2.21}$$

$$h_{BV}^+ + H_2O \rightarrow HO^\cdot + H^+ \tag{2.22}$$

$$h_{BV}^+ + HO^- \rightarrow HO^\cdot \tag{2.23}$$

$$e_{BC}^- + O_2 \rightarrow O_2^{\cdot -} \tag{2.24}$$

$$O_2^{\cdot -} + H^+ \rightarrow HO_2^\cdot \tag{2.25}$$

$$2HO_2^\cdot \rightarrow H_2O_2 + O_2 \tag{2.26}$$

$$H_2O_2 + O_2^{\cdot -} \rightarrow HO^\cdot + O_2 + HO^- \tag{2.27}$$

$$H_2O_2 + e_{Cb}^- \rightarrow HO^\cdot + HO^- \tag{2.28}$$

$$2O_2^{\cdot -} + 2H_2O \rightarrow 2HO^\cdot + O_2 + 2HO^- \tag{2.29}$$

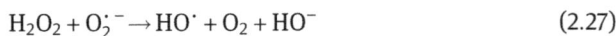

In the case of a p-type semiconductor, electrons will move to the photocatalyst surface while holes will migrate to the back-contact of the electrode. Thus, reduction reactions will take place on the photocatalyst (photocathode in this case) surface while the oxidation reaction will occur on the counter-electrode.

The main advantages and disadvantages of PEC reactors compared to slurry photoreactors are summarized in Table 2.1.

Table 2.1: Advantages and disadvantages of PEC cells compared to slurry photoreactors.

Advantages	Disadvantages
Slower electron–hole recombination rates	The cost of the applied potential must be considered
Higher photogeneration of h^+ and HO·	Lower efficiency due to loss of active surface area
The photocatalyst is immobilized so no separation units are necessary to recover it	Mass transfer limitations are more likely to occur
Mass transfer rates of bacteria are enhanced by electromigration	
When the photocatalyst is irradiated backside, loss of radiation due to absorption by the solution is prevented	

Several articles have been published reporting the application of PEC cells with different configurations to achieve faster degradation of pollutants and inactivation of bacteria in different water matrixes and under different operational conditions. In Table 2.2, the removal efficiency of some of these PEC reactors is summarized with the different types of configurations and the reactor conditions.

Table 2.2: Degradation efficiency of various pollutants and bacterial inactivation using different types of PEC reactors.

Reference	Type of reactor	Anode	Cathode	Pollutant	Operating conditions	Degradation efficiency
Tang et al. [53]	50 mL continuous-flow PEC	g-C$_3$N$_4$/TiO$_2$ nanotube arrays	Pt plate	TC in real lake water	$C_0 = 10$ mg/L 1.0 V vs. Ag/AgCl 0.1 M Na$_2$SO$_4$ simulated solar light irradiation $Q = 0.84$ mL/min	80% of TC removal efficiency 73% of TOC removal efficiency
Kim et al. [54]	1 L annular bubble reactor	TiO$_2$ nanotube arrays	DSA sheet	BP-3 in real swimming pool water	10 mg/L BP-3 no electrolyte addition UV-B 36 W lamp. Applied potential = 2.0 V O$_3$ input rate of 1.25×10^{-4} mol/min	95% and 100% of BP-3 degraded in 30 and 45 min
Changanaqui et al. [55]	Undivided three-electrode cell	ZnO/TiO$_2$/Ag$_2$Se on fluorine-doped tin oxide	Pt wire	Naproxen	5 mg/L of naproxen 50 mM Na$_2$SO$_4$ 1.0 V/Ag|AgC 173 W/m^2 at $\lambda_{max} = 450$–460 nm	100% degradation in 210 min
Domínguez-Espíndola et al. [56]	150 mL cylindrical three-electrode cell	Ag-TiO$_2$ on ITO	Stainless steel spiral	*Pseudomonas aeruginosa Bacillus atrophaeus*	$C_0 = 10^6$ CFU/mL 25 mM Na$_2$SO$_4$ 125 W UVA lamp $E_{bias} = 1.70$ V vs. Ag|AgCl (3 M KCl)	100% inactivation of *P. aeruginosa* and *B. atrophaeus* after 5 and 15 min, respectively

(continued)

Table 2.2 (continued)

Reference	Type of reactor	Anode	Cathode	Pollutant	Operating conditions	Degradation efficiency
Pablos et al. [57]	Annular reactor	TiO_2 on ITO	Nickel mesh	E. coli in simulated wastewater treatment plant effluent	Electric potential: +1.0 V No additional electrolyte E. coli initial concentration = 10^3 CFU/mL 0.47 W/m^2 in UVA	99% inactivation in 210 min
José Martín de Vidales et al. [58]	Single-compartment rectangular flat plate PEC cell	TiO_2 nanotubes	Ti plate with a Ti mesh with ruthenium mixed metal oxides	Methyl orange in distilled water	0.25×10^{-3} mol/L of MO 0.1 mol/L Na_2SO_4 1.50 V vs. Ag/AgCl Xenon power supply flow rate = 100 L/h	50% degradation of MO in 130 min
Zhang et al. [59]	Wedge structured rotating disk reactor	TiO_2/Ti wedge disk	Ti sheet	Textile wastewater	Initial COD = 380 mg/L Applied potential = 1.5 V No additional electrolyte 6 mW/cm^2 (main emission wavelength 254 nm)	80% COD reduction after 3 h
Bai et al. [60]	Thin-layer PEC reactor	Double-faced TiO_2 nanotubes arrays	Pt electrode	TC in distilled water	C_0 = 200 mg/L Applied potential = 2.0 V 0.1 M Na_2SO_4 Flow rate = 10 mL/min 4 W UV lamp (10 mW/cm at 254 nm)	90% conversion in 36 min

PEC, photoelectrocatalysis; TC, tetracycline; BP-3, benzophenone-3; ITO, indium-doped tin oxide; UVA, ultraviolet A; MO, CFU, TOC, COD, DSA.

2.3.2 Photoelectrochemical Reactor Design

Even though several reviews concerning photoelectrochemical cells and reactors have been published, most of them only deal with H_2 production [61–64]. The following sections will be focused mainly in the design of photoelectrochemical cells for photodegradation of organic compounds in wastewater or disinfection. For these applications, the PEC cells should meet the following criteria to improve their catalytic performance: (1) electrodes must be stable, (2) large surface area per unit reactor volume [60, 65], (3) high flow rates to achieve a homogeneous solution and avoid mass transfer limitations [65–67], (4) the configuration of the PEC reactor should minimize ohmic losses [68, 69], and (5) high radiation absorption. In the following sections, these criteria will be discussed in more detail.

2.3.3 Photoanode

Different semiconductors have been used in photoelectrocatalytic processes, such as TiO_2, WO_3, ZnO, CdS, Fe_2O_3, and SnO_2 [51, 70]. Among them TiO_2 has been the most studied photocatalyst due to their optimal properties: cost-efficient, environmentally friendly, nontoxic, with compatible energy positions of the VB and CB, high catalytic activity, and good thermal and chemical stability [49]. However, due to the fact that TiO_2 is only able to absorb radiation with a wavelength below 380 nm, different strategies have been developed to obtain materials that facilitate the chemical reaction, with higher radiation absorption, higher photoactivity under visible light, and lower $e^- - h^+$ pair recombination rates. Among them, it is important to mention doping the semiconductor, heterostructured photoelectrodes (which includes Z-scheme photocatalytic systems, dye-sensitized semiconductors, and plasmonic metal coupling), synthesis of nanostructured photoelectrodes with various morphologies from zero- to three-dimensional nanostructures or faceted photoelectrodes [70, 71].

To prepare photoelectrodes, various techniques have been reported to synthesize photocatalysts and coat the substrates. Physical vapor deposition, molecular beam epitaxy, chemical vapor deposition, electrodeposition, sol–gel methods, spin coating, spray coating, and spray painting technologies [50, 62, 72]. The simplest photoelectrodes, nanoparticle films, have proven to reach low conversions and deactivation of bacteria as their surface area is low and the electron–hole separation is not efficient [51]. However, TiO_2 nanotubes (TNTs) have attracted a great interest as they exhibit excellent stability, large internal surface area, and excellent electron percolation pathways for charge transfer because of their ordered structure. The preparation involves the connection of a cleaned Ti plate to the positive pole (anode) of a DC power supply and a counter-electrode (usually Pt) that is connected to the negative pole (cathode). The electrodes are placed in an electrolytic solution containing fluoride ions, and when a potential is applied in the system, the metal reacts with oxygen ions from the electrolyte growing in the oxide film on the surface (growing on the oxide film surface) [51, 54, 60, 66, 68]. Recently, Marugan et al. reported the fabrication of tubular porous membranes made of TNTs by potentiostatic and galvanostatic anodization under different operating conditions [73]. They found that a water content below 5 wt% is necessary to achieve long nanotubes. They concluded that only by applying potentials over 45 V TNTs were formed and they were better designed when low concentrations of NH_4F were used. This was attributed to the tubular porous Ti surfaces, which required greater potentials and shorter TNT were obtained. In fact, when the TNTs were scaled up using different potentiostatic conditions, no formation of TNT was observed, which indicates the complexity of scaling up these structures.

2.3.4 Cathode

Different cathodes have been used and studied in electrochemically assisted photocatalytic processes for wastewater treatment, among them stainless steel [74], nickel [75], Pt [53, 76], DSA [54], graphite, vitreous carbon, carbon felt [77], and oxygen diffusion cathodes [50, 51]. However, even though the cathode has only been regarded as a counter-electrode, it can play a big role and enhance the PEC process by generating different ROS, such as H_2O_2, which will act as an electron scavenger according to reaction (2.28). Hence, its decomposition not only increases the lifetime of holes but leads to higher production of $HO^{\cdot}$ radicals [51].

Among the different cathodes, graphite, vitreous carbon, carbon felt, and oxygen diffusion cathodes, such as carbon–polytetrafluoroethylene air-diffusion cathodes, stand out because of the high efficiency to produce H_2O_2 [51, 67, 78–80] as the contact between cathode, oxygen, and water must be maximized [67]. It was reported by Mousset et al. [77] that bubbling O_2 in the reactor led to higher amounts of produced H_2O_2. They ascribed this result to the fact that adsorption of gaseous O_2 on the cathode surface is favored compared to dissolved O_2.

Furthermore, several works have shown the possibility of degrading organic compounds from water; at the same time, hydrogen is produced [81, 82] or metals are recovered [74], which is an attractive application because of the double environmental advantage.

2.3.5 Separators and Membranes

In PEC cells, separators, such as porous separators or ion exchange membranes, can be used to separate counter-electrodes from working electrodes. This will increase the performance and safety of the cell when it is necessary to avoid product and electrolyte crossover. For example, the use of Nafion is the most frequently used, which is important in photoelectrochemical water splitting to separate compartments and extract H_2 generated in the cathode and O_2 produced in the anode avoiding their mixture [83–85]. Membranes are also used in PEC cells in which oxidation of organic compounds is carried out in the photoanode while hydrogen is generated in the cathode [51]. However, for wastewater treatment, membrane-less PEC cells are usually employed to avoid unnecessary costs and overpotentials [70].

2.3.6 Radiation Absorption

Since radiation absorption is the first step in photocatalytic and photoelectrocatalytic processes, it is crucial to know the amount of radiation effectively absorbed by the photocatalyst. Reaction rate is going to depend on radiation absorption, since the higher the light intensity is, and therefore radiation absorption, the higher the amount of photogenerated electron–hole pairs are, which will lead to greater concentrations of ROS. Similarly to photocatalysis, reaction rate dependence on light intensity varies according to the radiant flux level [49]. While under low radiation intensities, the reaction rate is proportional to radiation intensity; under moderate or high radiant fluxes, the reaction rate is proportional to the square root of radiation intensity [52, 86, 87]. Under very high radiant flux, the reaction rate will not depend on incident radiation. Furthermore, it has been proven that it is essential that the photoelectrode is uniformly irradiated [52]. Uniform radiation intensity distribution is essential to increase reaction rates and reduce power consumption [88].

   Different artificial radiation sources can be used in PEC systems, such as mercury, xenon, tungsten, fluorescent lamps, and LEDs, especially in lab-scale reactors since they are necessary to keep the PEC cell irradiated with a constant spectral radiation intensity [70]. Furthermore, by using filters, radiation with a wavelength within a certain range can be cut off, for example, to block infrared radiation to prevent the system from heating, and by using monochromators wavelength can be selected. Xenon lamps are one of the commonest radiation sources since they can act as solar simulators using AM filters [78, 81, 89]. AM 1.5G is usually used to mimic the natural solar standard spectrum. That way results obtained by different research groups can be compared [62, 70]. It is also important to mention LEDs, as radiation wavelength and intensity can be selected providing high radiant fluxes within a short range of wavelengths. Using LEDs also allows designing reactors with different geometries and achieving more uniform radiation intensity distribution. Furthermore, they have long lifetime, and the cost is continuously dropping [70, 86].

   For practical and large-scale application solar radiation must be used in order to make the process more environmentally friendly and reduce operating costs and capital [90]. However, since solar radiation is only composed of around 4% of UV radiation, solar light can be concentrated using parabolic mirrors or Fresnel lenses to increase the amount of radiation arriving at the surface of the photocatalyst [91, 92].

When the photocatalyst is supported on non-transparent substrates only a front-side configuration can be used, for example, irradiating the photocatalysts through a mesh used as a cathode [58]. Therefore, radiation will have to go through the solution, so its intensity will be decreased along the radiation path due to its absorption by the electrolyte and other chemical compounds according to Beer's law:

$$A = -\log_{10}\left(\frac{I}{I_0}\right) = \varepsilon bc \tag{2.30}$$

where A is the absorbance, I_0 is the intensity arriving at the solution, I is the intensity after traveling a distance b, ε is the molar absorptivity, b represents the path length of the sample, and c is the concentration of the solution. In the case of wastewater with high absorption coefficients, that is, with compounds able to absorb radiation within the same range than the photocatalyst, the design of thin layer or rotating disk PEC reactors has been reported, both of them showing high performance [59, 60, 65].

One solution to avoid shading effects and get high irradiated surface areas is to use transparent conductive oxides (TCO), such as fluorine-doped tin oxide (FTO) or indium-doped tin oxide-coated glass [64]. TCOs can reduce the active surface area losses at the expense of the introduction of larger ohmic losses by the electron transport as TCO shows typically larger resistances compared to metals [64]. However, they have larger area available for electron transport, have relative high conductivity (compared to the semiconductor), and allow transmitting a considerable amount of the incident radiation. These substrates allow irradiating the photocatalyst either front-side or backside (Figure 2.4). Upon using front-side illumination, most of the radiation will be absorbed by the photocatalyst layers closer to semiconductor–liquid interface. This entails that higher concentrations of electron–hole pair will be photogenerated near the surface of the electrode compared to the back-contact of the electrode. Therefore, the diffusion length of holes will not need to be very long as most of them will be photogenerated near the semiconductor–electrolyte interface. On the hand, photogenerated electrons will have to travel longer distances to reach the back-contact of the electrode, where they will be transferred to the cathode through the external electric circuit. On the contrary, when backside illumination is used, a greater amount of charge carriers will be photogenerated close to the back-contact of the electrode compared to the semiconductor–electrolyte interface. Thus, in contrast to photogenerated electrons, holes will have to diffuse long distances through the bulk of the semiconductor to get to the semiconductor–electrolyte interface. This points out the importance of finding the optimal semiconductor thickness to strike a balance between maximum radiation absorption and transport of photogenerated electron–hole pairs. The optimal thickness depends on the electron and hole mobility and lifetime, which is different for each material [68, 75].

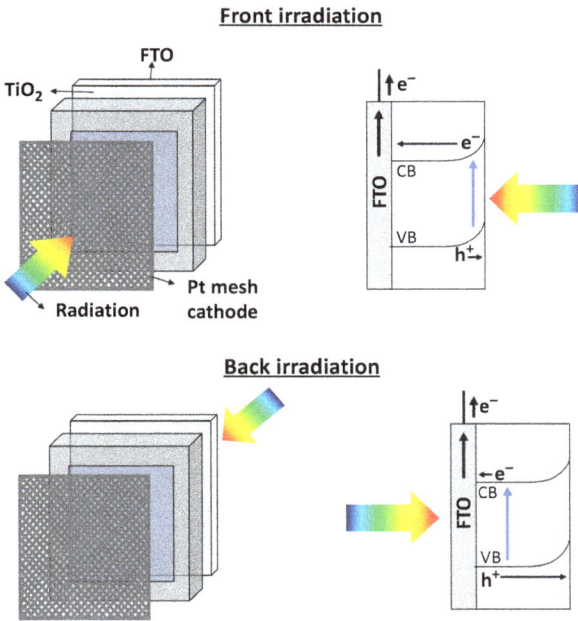

Figure 2.4: Effect of front-side irradiation (top) and backside irradiation (bottom) of the PEC cell on the electron–hole pair generation.

Several research articles studied the effect of backside and front-side illumination of the photoanode. Similar transmittance values were reported for both configurations, either for TNTs grown on TCO or on quartz [93, 94]. Regarding photocurrent measurements, greater values were observed when the photoanode was irradiated through the FTO than through the electrolyte. However, no differences were observed in phenol degradation rate when the PEC removal of phenol was evaluated [94]. On the other hand, it has been reported that backside illumination is detrimental for photoelectrocatalytic inactivation of bacteria, since most of the holes are photogenerated near the back-contact of the electrode, whereas bacteria can only react with holes available at the semiconductor–electrolyte interface [57].

The amount of radiation effectively absorbed by the photocatalyst is named local surface rate of photon absorption (LSRPA, $e_\lambda^{a,s}$) [95]. In order to obtain this value, it is necessary to determine first the spectral optical properties of the photocatalyst immobilized on the photoelectrode and the spectral incident radiation that arrives at the photocatalyst surface. The incident radiation can be obtained from emission models, actinometric measurements, or by using a spectroradiometer. The following radiation balance for each position of the photocatalyst allows calculating the LSRPA [95]:

$$e_\lambda^{a,s}(x) = q_{\lambda,\text{in}}(x) - q_{\lambda,\text{tr}}(x) - q_{\lambda,\text{rf}}(x) \tag{2.31}$$

where $q_{\lambda,in}(x)$ is the local radiative flux that arrives at the photocatalyst surface, $q_{\lambda,tr}(x)$ is the local radiative flux transmitted through the semiconductor, and $q_{\lambda,rf}(x)$ is the local radiative flux reflected by the semiconductor surface. In the event that the system is not uniformly irradiated, it is necessary to take into account the radiation profiles over the irradiated coating. Using average values of the local radiative fluxes over the photocatalytic surface and rearranging equation (2.31) the average value of LSRPA can be obtained according to the following equation:

$$e_\lambda^{a,s} A_{cat} = q_{\lambda,in} A_{cat} \left(1 - \frac{q_{\lambda,tr} A_{cat}}{q_{\lambda,in} A_{cat}} - \frac{q_{\lambda,rf} A_{cat}}{q_{\lambda,in} A_{cat}} \right) = q_{\lambda,in} A_{cat} \alpha_\lambda \tag{2.32}$$

where α_λ is the fraction of the radiation absorbed by the photocatalyst at a given wavelength. Thus, for polychromatic radiation the sum of the photons absorbed at the different wavelengths of the lamp emission spectrum must be calculated:

$$e_\lambda^{a,s} A_{cat} = q_{\lambda,in} A_{cat} \sum_\lambda f_\lambda \alpha_\lambda \tag{2.33}$$

where f_λ is the spectral emission of the lamp. The fraction of the radiation absorbed can be obtained from the following expression:

$$\alpha_\lambda = 1 - T_{f,\lambda} - R_{f,\lambda} \tag{2.34}$$

where $T_{f,\lambda}$ and $R_{f,\lambda}$ represent the ratio of radiation that is transmitted and reflected by the film (f), respectively, which can be estimated from diffuse transmittance and reflectance measurements of the bare (b) and coated TCO (c) by solving the following equations [95]:

$$R_{f,\lambda} = \frac{R_{c,\lambda} T_{b,\lambda}^2 - R_{b,\lambda} T_{c,\lambda}^2}{T_{b,\lambda}^2 - R_{b,\lambda}^2 T_{c,\lambda}^2} \tag{2.35}$$

$$T_{f,\lambda} = \frac{T_{c,\lambda}}{T_{b,\lambda}} (1 - R_{f,\lambda} R_{b,\lambda}) \tag{2.36}$$

2.3.7 Ohmic Losses

Ohmic losses (η_{ohmic}) are functions of the resistances, that is, ionic (R_{ionic}), electronic (R_{elec}), and contact resistances ($R_{contact}$), which appear in the electrodes, electrolyte, and current collectors and contacts due to the fact that every material has inherent resistance to charge flow (i) [96]:

$$\eta_{ohmic} = iR_{ohmic} = i(R_{ionic} + R_{elec} + R_{contact}) \tag{2.37}$$

Ohmic losses in electrical systems can be considered negligible compared to losses caused by the ionic transport in the electrolyte. Nevertheless, it is necessary to fabricate effective electrical contacts for current collection that minimizes losses [64]. The conductivity of the electrolyte must be high to achieve good ion transfer, since relatively low conductivities of the electrolyte can lead to larger resistance, and therefore ohmic losses, which involve higher overpotentials to carry out the degradation reactions [52, 68]. That is the reason why the composition and concentration of the electrolyte is crucial to enhance ionic transport in the PEC cell. However, even though the addition of an electrolyte enhances the efficiency of the process, it would be recommended taking advantage of the composition of some water matrices [53] as some of them are conductive enough to carry out PEC treatments. To reduce ohmic losses in the solution, the anode must be as close as possible from the cathode, since large interelectrode distances lead to an increase in the ohmic potential losses in the electrolyte [64]. Even though for wastewater treatment is uncommon the use of membranes, it is important to remark that they give rise to higher ohmic losses. Depending on the application, a compromise between porosity and thickness of the membrane must be searched [83].

Furthermore, it is important to minimize current density distribution in the PEC reactor, which will depend on the potential distribution and the concentration of electroactive species, so that utilization of the electrocatalysts is optimized [63, 90], for example, by placing both electrodes symmetrically opposed. When the photoanode faces away from the cathode, such as in front-side illumination configuration, severe current distribution problems are observed [97]. Secondary current distribution, which takes into account the kinetics at the electrodes and assumes good mixing of the electrolyte and negligible concentration gradients [90], can be easily determined by using Multiphysics Software [63, 83, 85, 97] and will be summarized briefly.

Assuming dilute-solution theory, the ionic current in the electrolyte is given by the Nernst–Planck equation (equation (2.38)), where the flux of charged electroactive solution species (i) is described by the sum of their transport by migration in an electric field ($\nabla\phi$), diffusion in a concentration gradient (∇c_i), and convection due to the local fluid velocity (v) [83, 97]:

$$N_i \cong -\underbrace{\frac{z_i F D_i}{RT} c_i \nabla\phi}_{\text{migration}} - \underbrace{D_i \nabla c_i}_{\text{diffusion}} + \underbrace{c_i v}_{\text{convection}} \qquad (2.38)$$

where z_i represents the charge of the species, F is the Faraday constant, D_i its diffusion coefficient, R is the gas constant, and T is the temperature. Following Faraday's law, the current density of charged solute species (j_{ionic}) is

$$j_{\text{ionic}} = F \sum_i z_i N_i \cong -F^2 \nabla\phi \sum_i \frac{z_i^2 D_i}{RT} c_i - F \sum_i z_i D_i c_i + Fv \sum_i c_i z_i \qquad (2.39)$$

Applying the electroneutrality condition (equation (2.40)) and assuming negligible concentration gradients (equation (2.41)), the result is Laplace's equation (equation (2.42)):

$$\sum_i c_i\, z_i = 0 \tag{2.40}$$

$$\nabla c_i = 0 \tag{2.41}$$

$$\nabla^2 \phi = 0 \tag{2.42}$$

Since the condition of dilute solutions was considered, ionic conductivities (κ) are linearly proportional to the species concentration:

$$\kappa = F^2 \sum_i |z_i|^2 \frac{D_i}{RT} c_i \tag{2.43}$$

Since the transfer current depends on the electrochemical reaction kinetics, the Butler–Volmer or the Tafel equations have been used to describe the cathodic and anodic dark currents [83, 85, 97]. To model the photocurrent density produced in the photoanode, good results were obtained by using a modified Gärtner–Butler equation (equation (2.44)) [85, 97–99] to take into account polychromatic radiation. However, this model should not be applied when nanoparticles are used as a photocatalyst since in this case there is no band bending:

$$j_{phot,1} \approx \left(\frac{2e\left(\sum_\lambda I_{0,\lambda}\alpha_\lambda\right)^2 \varepsilon_0 \varepsilon_R}{n_0} \right)^{\frac{1}{2}} (U_{applied} - U_{FB})^{\frac{1}{2}} \tag{2.44}$$

where e is the electronic charge, $\sum_\lambda I_{0,\lambda}\alpha_\lambda$ represents the radiation absorbed by the photocatalyst, ε_0 is the permittivity of free space, ε_R is the relative permittivity of the photocatalyst, n_0 is the charge carrier density, $U_{applied}$ is the applied potential, and U_{FB} is the flat band potential. Other corrections were made to account for the limited current density by the absorbed photon flux, the bulk recombination (φ_{bulk}), and the interfacial charge transfer efficiency ($\varphi_{surface}$):

$$j_{phot} \approx \varphi_{bulk}\, \varphi_{surface}\, \frac{j_{phot,1}}{1 + \dfrac{j_{phot,1}}{e\sum_\lambda I_{0,\lambda}}} \tag{2.45}$$

The bulk recombination and the interfacial charge transfer efficiency can be estimated from impedance spectroscopy measurements, and the experimental photocurrent density is obtained in the presence of a fast hole scavenger [85, 97].

In the photoanode, subsequent to electron photogeneration, they migrate to the back-contact of the electrode where they are transferred to the substrate. Then they pass through the substrate layer until they are collected to be removed from the photoanode. However, because of the sheet resistance of the substrate, the flow of current through the conductor layer will give rise to ohmic losses and potential drops on the substrate. It is important to quantify this effect, especially when only one current collector or substrates with relatively low conductivities are used, such as TCOs. The electric current through the conductive substrate can be modeled by the Ohm's law (equation (2.46)) and the continuity of charge (equation (2.47)) [83, 100]:

$$j_s = \lambda \sigma \nabla \phi \tag{2.46}$$

$$\nabla \cdot j_s = - \nabla \cdot j_l = A_0 j_R \tag{2.47}$$

where j_s and j_l are the current density in the solid and liquid, respectively, $\nabla \phi$ is the electric field, σ is the substrate conductivity, A_0 is its specific surface area, and j_R is the transfer current density between the ionic and electronic phases. Several works have shown that large electrode sizes can lead to potential drops along the surface of the electrode and substantial potential distributions, which could entail that a high percentage of the photoelectrode area is working at a voltage below the minimum required to carry out the reaction [63, 100]. It was observed that electrodes with larger areas than 5×10^{-4} m^2 cannot be used in large PEC reactor. Otherwise, large electrodes areas, for example, in scaled-up systems, can limit current collection because of the substrate ohmic losses, in particular, under high current densities or when solar radiation is concentrated. To minimize substrate current losses, small electrode sizes and low currents are recommended to be used. Other solution would be to improve the conductivity of the substrates or use substrates with lower resistivity, such as metals.

2.3.8 Mass Transfer

To enhance the performance of the cell, it is crucial to determine the flow inside the cell. Simulations can be performed to reach a good mixture of reagents and avoid concentration gradients, flow distributions, bubble accumulation, product build-up, dead volumes, and other nonidealities [64, 69, 101]. High mass transfer rates of reactants to the photoelectrode surface must be ensured to increase degradation rates and prevent the process from being mass transfer limited [65, 90, 102]. Higher turbulence will not only promote mass transfer but it will also reduce the fouling of the reactor. To get some insights about the mass transport inside the reactor, dimensionless groups can be calculated, such as the Sherwood number (Sh) or the Schmidt number (Sc), which describe the rate of mass transport and compare the rate of transport by convection to that by diffusion. From empirical dimensionless group, correlations allow estimating the mass transport coefficient related to the flow conditions [90]. Mass transfer can be improved with a proper design of the PEC cell. To increase turbulence in the system, different approaches can be found, such as using mesh electrodes and recycling a part of the treated water with a high flow rate or bubble air or oxygen inside the PEC reactor [52]. Furthermore, by including some strategically located baffles, the electrolyte might be forced to adopt a serpentine flow pattern, which maximizes the pathway the solution has to flow inside the reactor, increasing its effective contact time and favoring uniform electrolyte distributions [64]. The design of certain PEC reactors, such as thin-layer or rotating disk photoelectrocatalytic reactors, has proven to achieve better mass transfer due to high ratios of electrode area to solution volume [60, 65]. It is important to mention that applying the right potential in the PEC reactor can lead to faster bacterial inactivation rates since good electrostatic interactions of catalyst–bacteria are essential. In the PEC process, bacteria, which are negatively charged, moves toward the positively charged photoanode, so mass transfer rates of bacteria are enhanced by electromigration [57, 103].

2.3.9 Considerations for Up-Scaling

Scaling up a PEC reactor is not straightforward, and few studies have been carried out regarding large-scale applications [75, 104, 105]. A helpful parameter for scale-up analysis is the Wagner number, a dimensionless group that describes the ratio of a polarization resistance per unit area (due to the faradaic reaction) to the electrolyte resistance per unit area [90]:

$$\text{Wa} = \frac{R_a}{R_e} = \frac{\left(\frac{d\eta}{dj}\right)}{\left(\frac{h}{\kappa_e}\right)} \tag{2.48}$$

where $d\eta/dj$ is the reciprocal slope of a j–η curve, h is a characteristic length, κ_e is the electrolyte conductivity. In the case the conductivity of the substrate is relatively low, the denominator would be the sum of the electrolyte resistance and the sheet resistance ($L/\kappa_{substrate}$), where L is the farthest distance between the substrate and the current collector. The higher the Wa number, the more uniform the secondary current distribution. Moreover, PEC reactors with similar Wa numbers will show similar current distributions [90]. It must be pointed out that the Wa number does not take into account light absorption, so it is important to calculate the amount of radiation absorbed by the photocatalyst, as previously explained in Section 2.3.6, to try to resemble the same irradiation conditions.

Only few studies have been published in which the performance of a pilot plant PEC system was used to decontaminate water [104, 106]. Zhao et al. [106] studied the photoelectrochemical treatment of landfill leachate in a 6.5-L continuous-flow reactor. A DSA was used as the anode at the center of the reactor and stainless steel as the cathode. Experiments under different current densities with and without UV radiation were carried out (Figure 2.5). At a current density of 67.1 mA/cm^2 and 2.5 h

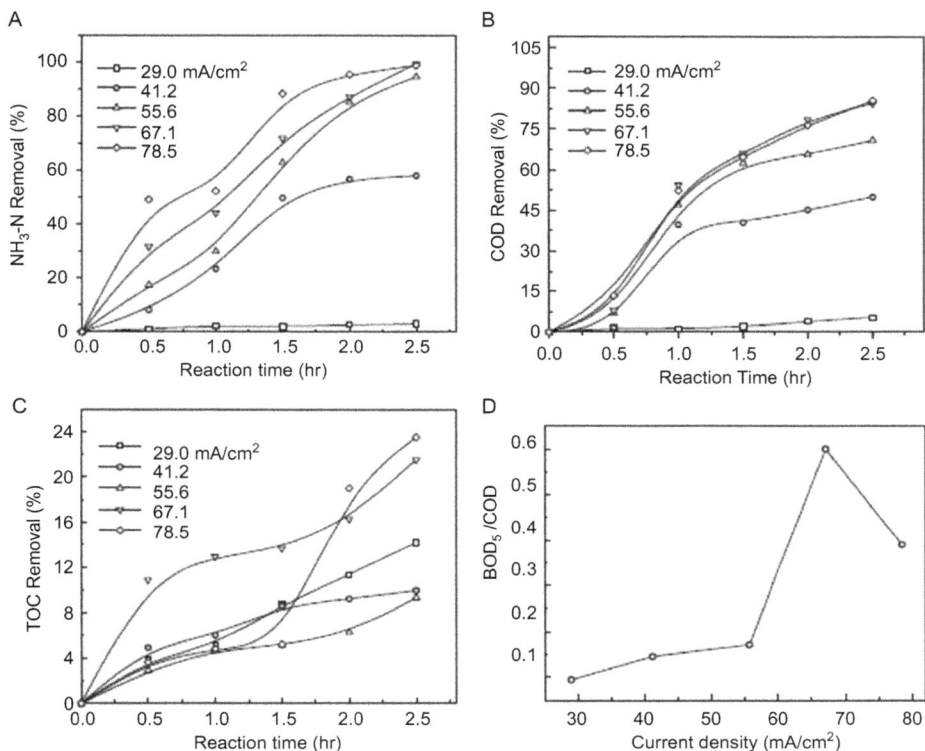

Figure 2.5: Effect of current densities on PEC degradation of landfill leachate. (A) NH_3-N; (B) COD removal; (C) TOC removal; and (D) BOD_5/COD. Reprinted from Zhao et al. [106], Copyright © (2009), with permission from Elsevier.

reaction time, they observed conversions of 74.1% for COD, 41.6% for TOC, and 94.5% for ammonium in the electrolysis process with UV light irradiation.

Figures of merit concerning electric energy consumption and solar collector area are suggested not only to compare and evaluate the electric or solar energy efficiency of a PEC system but also to scale up processes [107]. These figures of merit allow performing a simple analysis of economics; since these types of systems are electric-energy-intensive, energy consumption can be regarded as the main operating cost. The main advantage of using these figures of merit is that they consider the effect of the pollutant concentration. The electric energy per order, E_{EO}, (equation (2.49)), is the electric energy in kilowatt hours (kWh) required to degrade a contaminant C by one order of magnitude, for example, from 10 to 1 mg/L, in a unit volume (e.g., 1 m^3) of contaminated water [54, 107]:

$$E_{EO} = \frac{P \times t \times 1{,}000}{V \times \log(C_i/C_f)} \tag{2.49}$$

where P stands for the rated power (kW) of the treatment system, t being the treatment time (h), V the treated volume (m^3), and C_i and C_f the initial and final concentrations of the pollutant after the t treatment time.

Regarding solar-energy-driven systems, figures of merit based on solar collector area are necessary, since the capital cost can be related to the needed area of the solar collector and there are no irradiation sources that consume electric energy. The collector area per order A_{CO}, is defined as the collector area required to reduce the concentration of a contaminant C in polluted water in a unit volume (e.g., 1 m^3) by one order of magnitude in a time t_o (1 h) when the incident solar irradiance ($E_s{}^o$) is 1,000 W/m^2. This figure of merit can be used in the low concentration range, which is typical for contaminants of emerging concern:

$$A_{CO} = \frac{A \times \overline{E_s} \times t}{V \times \log(C_i/C_f)} \tag{2.50}$$

where A is the actual collector area (m^2), $\overline{E_s}$ (W/m^2) is the average solar irradiance over period t of the treatment, V the treated volume (m^3), and C_i and C_f the initial and final concentrations of the pollutant after the t treatment time. More figures of merit, including some for flow-through operation and in the high concentration range, can be found in [107].

2.4 Challenges and Prospects

Promising results concerning pollutant degradation and bacteria inactivation have been obtained using PEC systems at lab scale up to now. Nevertheless, next steps on the assessment of the viability of these processes including design, development, and testing new pilot PEC cells to treat real wastewater or other water matrixes are still required to find out suitable applications of this technology. For this reason, it is crucial for the development of models that allow estimating intrinsic kinetic constants and optimizing the geometry of the cell to maximize absorption of solar radiation. To do so, the development of new visible active materials that are able to increase organic compound photodegradation and disinfection rates is crucial. Finally, detailed cost analysis is key to compare the feasibility of this technology to other oxidation processes.

2.5 Concluding Remarks

PEC for the disinfection of water, degradation of persistent organic contaminants, and other applications remains an exciting and complex area for research. Even if there is a large number of articles investigating the materials that can be used as photo-anode or -cathode, and the fundamentals of the mechanisms of the reactive species generated during the degradation or inactivation process, there are still some gaps in key aspects. More complex and complete models are necessary to design photoreactors and PEC cells and estimate intrinsic kinetic parameters using computational fluid dynamics, since several side reactions will take place in the process, especially when using real water matrixes. As long as the model used is suitable, the results obtained from these simulations will allow optimization of the reactor by minimizing potential losses, improving fluid dynamics, and selecting the right materials. Regarding the water matrix, research must focus on the degradation of complex mixtures of pollutants (chemical and biological) in real conditions, that is, using real concentrations of pollutants in wastewater or surface water. Finally, proper economical and energetic assessment of the process has to be carried out to put it in perspective with the rest of the water treatment processes currently used for wastewater remediation and recovery.

References

[1]   "WHO. 1 in 3 people globally do not have access to safe drinking water UNICEF, WHO., https://www.who.int/news-room., 2020 (accessed 12. 08.20).".
[2]   "UN, 2018, SDG 6 Synthesis Report 2018 on Water and Sanitation. https://www.Unwater.org (accessed 12. 08.2020).".
[3]   Motegh M, Cen J, Appel PW, Van Ommen JR, Kreutzer MT. Photocatalytic-reactor efficiencies and simplified expressions to assess their relevance in kinetic experiments, Chemical Engineering Journal, 2012, 207–208, 607–615. Doi: 10.1016/j.cej.2012.07.023.
[4]   Cabrera MI, Alfano OM, Cassano AE. Absorption and scattering coefficients of titanium dioxide participate suspensions in water, The Journal of Physical Chemistry, 1996, 100(51), 20043–20050. Doi: 10.1021/jp962095q.
[5]   Padoin N, Soares C. An explicit correlation for optimal TiO2 film thickness in immobilized photocatalytic reaction systems, Chemical Engineering Journal, 2017, 310, 381–388. Doi: 10.1016/j.cej.2016.06.013.
[6]   Vezzoli M, Farrell T, Baker A, Psaltis S, Martens WN, Bell JM. Optimal catalyst thickness in titanium dioxide fixed film reactors: Mathematical modelling and experimental validation, Chemical Engineering Journal, 2013, 234, 57–65. Doi: 10.1016/j.cej.2013.08.049.
[7]   Alrousan DMA, Polo-López MI, Dunlop PSM, Fernández-Ibáñez P, Byrne JA. Solar photocatalytic disinfection of water with immobilised titanium dioxide in re-circulating flow CPC reactors, Applied Catalysis B: Environmental, 2012, 128, 126–134. Doi: 10.1016/j.apcatb.2012.07.038.

[8]   McNaught AD, Wilkinson A. I. U. of P. and A. Chemistry., Compendium of Chemical Terminology: IUPAC Recommendations, Blackwell Science, Oxford [England]; Malden, MA, USA, 1997.

[9]   Bloh JZ. A holistic approach to model the kinetics of photocatalytic reactions, Frontiers in Chemistry, 2019, 7, no. MAR, 1–13. Doi: 10.3389/fchem.2019.00128.

[10]   Marugán J, Van Grieken R, Cassano AE, Alfano OM. CHAPTER 15 Photocatalytic Reactor Design, In: Photocatalysis: Fundamentals and Perspectives, The Royal Society of Chemistry, 2016, 367–387.

[11]   Herrmann J-M. Heterogeneous photocatalysis: fundamentals and applications to the removal of various types of aqueous pollutants, Catalysis Today, 1999, 53(1), 115–129. Doi: https://doi.org/10.1016%2FS0920-5861(99)00107-8.

[12]   Visan A, Van Ommen JR, Kreutzer MT, Lammertink RGH. Photocatalytic reactor design: guidelines for kinetic investigation, Industrial & Engineering Chemistry Research, 2019, 58(14), 5349–5357. Doi: 10.1021/acs.iecr.9b00381.

[13]   Bouchy M, Zahraa O. Photocatalytic reactors, International Journal of Photoenergy, 2003, 5(3), 191–197. Doi: 10.1155/S1110662X03000321.

[14]   Kurup GK, Basu AS. Field-free particle focusing in microfluidic plugs, Biomicrofluidics, 2012, 6(2), 1–10. Doi: 10.1063/1.3700120.

[15]   Hu Q, Liu B, Zhang Z, Song M, Zhao X. Temperature effect on the photocatalytic degradation of methyl orange under UV-vis light irradiation, Journal of Wuhan University of Technology. Sci. Ed., 2010, 25(2), 210–213. Doi: 10.1007/s11595-010-2210-5.

[16]   Vargas R, Núñez O. Photocatalytic degradation of oil industry hydrocarbons models at laboratory and at pilot-plant scale, Solar Energy, 2010, 84(2), 345–351. Doi: 10.1016/j. solener.2009.12.005.

[17]   Spasiano D, Marotta R, Malato S, Fernandez-Ibañez P, Di Somma I. Solar photocatalysis: Materials, reactors, some commercial, and pre-industrialized applications. A comprehensive approach, Applied Catalysis B: Environmental, 2015, 170–171, 90–123. Doi: 10.1016/j. apcatb.2014.12.050.

[18]   Marugán J, Van Grieken R, Cassano AE, Alfano OM. Scaling-up of slurry reactors for the photocatalytic oxidation of cyanide with TiO2 and silica-supported TiO2 suspensions, Catalysis Today, 2009, 144(1–2), 87–93. Doi: 10.1016/j.cattod.2008.12.026.

[19]   Mehos MS, Turchi CS. Field testing solar photocatalytic detoxification on TCE-contaminated groundwater, Environmental Progress, Aug 1993, 12(3), 194–199. Doi: 10.1002/ ep.670120308.

[20]   Polo-López MI, Fernández-Ibáñez P, García-Fernández I, Oller I, Salgado-Tránsito I, Sichel C. Resistance of Fusarium sp spores to solar TiO2 photocatalysis: Influence of spore type and water (scaling-up results), Journal of Chemical Technology and Biotechnology, 2010, 85(8), 1038–1048. Doi: 10.1002/jctb.2397.

[21]   Plantard G, Janin T, Goetz V, Brosillon S. Solar photocatalysis treatment of phytosanitary refuses: Efficiency of industrial photocatalysts, Applied Catalysis B: Environmental, 2012, 115–116, 38–44. Doi: 10.1016/j.apcatb.2011.11.034.

[22]   Kamble SP, Sawant SB, Pangarkar VG. Novel solar-based photocatalytic reactor for degradation of refractory pollutants, AIChE Journal, 2004, 50(7), 1647–1650. Doi: 10.1002/ aic.10125.

[23]   Janssens R, Mandal MK, Dubey KK, Luis P. Slurry photocatalytic membrane reactor technology for removal of pharmaceutical compounds from wastewater: Towards cytostatic drug elimination, The Science of the Total Environment, 2017, 599–600, 612–626. Doi: 10.1016/j.scitotenv.2017.03.253.

[24] Zhang Z, Anderson WA, Moo-Young M. Experimental analysis of a corrugated plate photocatalytic reactor, Chemical Engineering Journal, 2004, 99(2), 145–152. Doi: 10.1016/j.cej.2004.01.001.

[25] Manassero A, Satuf ML, Alfano OM. Photocatalytic reactors with suspended and immobilized TiO2: Comparative efficiency evaluation, Chemical Engineering Journal, 2017, 326, 29–36. Doi: 10.1016/j.cej.2017.05.087.

[26] Zheng X, Shen ZP, Shi L, Cheng R, Yuan DH. Photocatalytic membrane reactors (PMRs) in water treatment: Configurations and influencing factors, Catalysts, 2017, 7(8). Doi: 10.3390/catal7080224.

[27] Robertson PKJ, Robertson JMC, Bahnemann DW. Removal of microorganisms and their chemical metabolites from water using semiconductor photocatalysis, Journal of Hazardous Materials, 2012, 211–212, 161–171. Doi: 10.1016/j.jhazmat.2011.11.058.

[28] Hijnen WAM, Beerendonk EF, Medema GJ. Inactivation credit of UV radiation for viruses, bacteria and protozoan (oo)cysts in water: A review, Water Research, 2006, 40(1), 3–22. Doi: 10.1016/j.watres.2005.10.030.

[29] Giannakis S, Polo López MI, Spuhler D, Sánchez Pérez JA, Fernández Ibáñez P, Pulgarin C. Solar disinfection is an augmentable, in situ-generated photo-Fenton reaction – Part 1: A review of the mechanisms and the fundamental aspects of the process, Applied Catalysis B: Environmental, 2016, 199, 199–223. Doi: 10.1016/j.apcatb.2016.06.009.

[30] Nelson KL, et al. Sunlight-mediated inactivation of health-relevant microorganisms in water: a review of mechanisms and modeling approaches, Environmental Science. Processes & Impacts, 2018, 20(8), 1089–1122. Doi: 10.1039/c8em00047f.

[31] Goodsell DS. The molecular perspective: ultraviolet light and pyrimidine dimers, Oncologist, 2001, 6(3), 298–299. Doi: 10.1634/theoncologist.6-3-298.

[32] Castro-Alférez M, Polo-López MI, Marugán J, Fernández-Ibáñez P. Mechanistic modeling of UV and mild-heat synergistic effect on solar water disinfection, Chemical Engineering Journal, 2017, 316, 111–120. Doi: 10.1016/j.cej.2017.01.026.

[33] Ubomba-Jaswa E, Navntoft C, Polo-López MI, Fernandez-Ibáñez P, McGuigan KG. Solar disinfection of drinking water (SODIS): an investigation of the effect of UV-A dose on inactivation efficiency, Photochemical and Photobiological Sciences, 2009, 8(5), 587–595. Doi: 10.1039/B816593A.

[34] Polo-López MI, Castro-Alférez M, Oller I, Fernández-Ibáñez P. Assessment of solar photo-Fenton, photocatalysis, and H2O2 for removal of phytopathogen fungi spores in synthetic and real effluents of urban wastewater, Chemical Engineering Journal, 2014, 257, 122–130. Doi: 10.1016/j.cej.2014.07.016.

[35] Ferro G, Polo-López MI, Martínez-Piernas AB, Fernández-Ibáñez P, Agüera A, Rizzo L. Cross-contamination of residual emerging contaminants and antibiotic resistant bacteria in lettuce crops and soil irrigated with wastewater treated by sunlight/H2O2, Environmental Science & Technology, 2015, 49(18), 11096–11104. Doi: 10.1021/acs.est.5b02613.

[36] Bianco A, Polo–López MI, Fernández–Ibáñez P, Brigante M, Mailhot G. Disinfection of water inoculated with Enterococcus faecalis using solar/Fe(III)EDDS-H2O2 or S2O82– process, Water Research, 2017, 118, 249–260. Doi: 10.1016/j.watres.2017.03.061.

[37] Rizzo L, Agovino T, Nahim-Granados S, Castro-Alférez M, Fernández-Ibáñez P, Polo-López MI. Tertiary treatment of urban wastewater by solar and UV-C driven advanced oxidation with peracetic acid: Effect on contaminants of emerging concern and antibiotic resistance, Water Research, 2019, 149, 272–281. Doi: 10.1016/j.watres.2018.11.031.

[38] Ortega-Gómez E, Ballesteros Martín MM, Carratalà A, Fernández Ibañez P, Sánchez Pérez JA, Pulgarín C. Principal parameters affecting virus inactivation by the solar photo-Fenton

process at neutral pH and µM concentrations of H2O2 and Fe2+/3+, Applied Catalysis B: Environmental, 2015, 174–175, 395–402. Doi: 10.1016/j.apcatb.2015.03.016.

[39] Cruz-Ortiz BR, et al. Mechanism of photocatalytic disinfection using titania-graphene composites under UV and visible irradiation, Chemical Engineering Journal, 2017, 316, 179–186. Doi: 10.1016/j.cej.2017.01.094.

[40] Fernández-Ibáñez P, et al. Solar photocatalytic disinfection of water using titanium dioxide graphene composites, Chemical Engineering Journal, 2015, 261, 36–44. Doi: 10.1016/j.cej.2014.06.089.

[41] Šolić M, Krstulović N. Separate and combined effects of solar radiation, temperature, salinity, and pH on the survival of faecal coliforms in seawater, Marine Pollution Bulletin, 1992, 24(8), 411–416. Doi: https://doi.org/10.1016%2F0025-326X(92)90503-X.

[42] Wegelin M, Canonica S, Mechsner K, Fleischmann T, Pesaro F, Metzler A. Solar water disinfection: Scope of the process and analysis of radiation experiments, Aqua Journal of Water Supply: Research and Technology, 1994, 43(4), 154–169.

[43] Berney M, Weilenmann HU, Simonetti A, Egli T. Efficacy of solar disinfection of Escherichia coli, Shigella flexneri, Salmonella Typhimurium and Vibrio cholerae, Journal of Applied Microbiology, 2006, 101(4), 828–836. Doi: 10.1111/j.1365-2672.2006.02983.x.

[44] Herrmann JM. Heterogeneous photocatalysis: State of the art and present applications, Topics in Catalysis, 2005, 34(1–4), 49–65. Doi: 10.1007/s11244-005-3788-2.

[45] García-Fernández I, Fernández-Calderero I, Inmaculada Polo-López M, Fernández-Ibáñez P. Disinfection of urban effluents using solar TiO2 photocatalysis: A study of significance of dissolved oxygen, temperature, type of microorganism and water matrix, Catalysis Today, 2015, 240, no. PA, 30–38. Doi: 10.1016/j.cattod.2014.03.026.

[46] Ortega-Gómez E, Fernández-Ibáñez P, Ballesteros Martín MM, Polo-López MI, Esteban García B, Sánchez Pérez JA. Water disinfection using photo-Fenton: Effect of temperature on Enterococcus faecalis survival, Water Research, 2012, 46(18), 6154–6162. Doi: 10.1016/j.watres.2012.09.007.

[47] Schneider J, et al. Understanding TiO2 photocatalysis: Mechanisms and materials, Chemical Reviews, 2014, 114(19), 9919–9986. Doi: 10.1021/cr5001892.

[48] Moreira NFF, et al. Solar treatment (H2O2, TiO2-P25 and GO-TiO2 photocatalysis, photo-Fenton) of organic micropollutants, human pathogen indicators, antibiotic resistant bacteria and related genes in urban wastewater, Water Research, 2018, 135, 195–206. Doi: 10.1016/j.watres.2018.01.064.

[49] Malato S, Fernández-Ibáñez P, Maldonado MI, Blanco J, Gernjak W. Decontamination and disinfection of water by solar photocatalysis: Recent overview and trends, Catalysis Today, Sep 2009, 147(1), 1–59. Doi: 10.1016/J.CATTOD.2009.06.018.

[50] Garcia-Segura S, Brillas E. Applied photoelectrocatalysis on the degradation of organic pollutants in wastewaters, Journal of Photochemistry and Photobiology C: Photochemistry Reviews, 2017, 31, 1–35. Doi: 10.1016/j.jphotochemrev.2017.01.005.

[51] Bessegato GG, Guaraldo TT, De Brito JF, Brugnera MF, Zanoni MVB. Achievements and Trends in Photoelectrocatalysis: from Environmental to Energy Applications, Electrocatalysis, Sep 2015, 6(5), 415–441. Doi: 10.1007/s12678-015-0259-9.

[52] Daghrir R, Drogui P, Robert D. Photoelectrocatalytic technologies for environmental applications, Journal of Photochemistry and Photobiology A: Chemistry, Jun 2012, 238, 41–52. Doi: 10.1016/j.jphotochem.2012.04.009.

[53] Tang H, et al. Static and continuous flow photoelectrocatalytic treatment of antibiotic wastewater over mesh of TiO2 nanotubes implanted with g-C3N4 nanosheets, Journal of Hazardous Materials, Feb 2020, 384, 121248. Doi: 10.1016/J.JHAZMAT.2019.121248.

[54]   Kim JYU, Bessegato GG, De Souza BC, Da Silva JJ, Zanoni MVB. Efficient treatment of swimming pool water by photoelectrocatalytic ozonation: Inactivation of Candida parapsilosis and mineralization of Benzophenone-3 and urea, Chemical Engineering Journal, Dec 2019, 378, 122094. Doi: 10.1016/j.cej.2019.122094.

[55]   Changanaqui K, Alarcón H, Brillas E, Sirés I. Blue LED light-driven photoelectrocatalytic removal of naproxen from water: Kinetics and primary by-products, Journal of Electroanalytical Chemistry, 2020, 867, 114192. Doi: 10.1016/j.jelechem.2020.114192.

[56]   Domínguez-Espíndola RB, Bruguera-Casamada C, Silva-Martínez S, Araujo RM, Brillas E, Sirés I. Photoelectrocatalytic inactivation of Pseudomonas aeruginosa using an Ag-decorated TiO2 photoanode, Separation and Purification Technology, 2019, 208, no. May 2018, 83–91. Doi: 10.1016/j.seppur.2018.05.005.

[57]   Pablos C, Marugán J, Adán C, Osuna M, Van Grieken R. Performance of TiO2 photoanodes toward oxidation of methanol and E. coli inactivation in water in a scaled-up photoelectrocatalytic reactor, Electrochimica Acta, 2017. Doi: 10.1016/j.electacta.2017.11.103.

[58]   José Martín De Vidales M, et al. Photoelectrocatalytic Oxidation of Methyl Orange on a TiO2 Nanotubular Anode Using a Flow Cell, Chemical Engineering & Technology, Jan 2016, 39(1), 135–141. Doi: 10.1002/ceat.201500085.

[59]   Li K, et al. Novel wedge structured rotating disk photocatalytic reactor for post-treatment of actual textile wastewater, Chemical Engineering Journal, May 2015, 268, 10–20. Doi: 10.1016/j.cej.2015.01.039.

[60]   Bai J, Liu Y, Li J, Zhou B, Zheng Q, Cai W. A novel thin-layer photoelectrocatalytic (PEC) reactor with double-faced titania nanotube arrays electrode for effective degradation of tetracycline, Applied Catalysis B: Environmental, Aug 2010, 98(3–4), 154–160. Doi: 10.1016/J.APCATB.2010.05.024.

[61]   Minggu LJ, Wan Daud WR, Kassim MB. An overview of photocells and photoreactors for photoelectrochemical water splitting, International Journal of Hydrogen Energy, Jun 2010, 35(11), 5233–5244. Doi: 10.1016/J.IJHYDENE.2010.02.133.

[62]   Chen Z, et al. Photoelectrochemical Water Splitting, 2013.

[63]   Carver C, Ulissi Z, Ong CK, Dennison S, Kelsall GH, Hellgardt K. Modelling and development of photoelectrochemical reactor for H2 production, International Journal of Hydrogen Energy, Feb 2012, 37(3), 2911–2923. Doi: 10.1016/J.IJHYDENE.2011.07.012.

[64]   Bosserez T, Rongé J, Van Humbeeck J, Haussener S, Martens J. Design of compact photoelectrochemical cells for water splitting, Oil & Gas Science and Technology – Revue d'IFP Energies Nouvelles, 2015, 70(5), 877–889. Doi: 10.2516/ogst/2015015.

[65]   Xu Y, He Y, Cao X, Zhong D, Jia J. TiO2/Ti rotating disk photoelectrocatalytic (PEC) reactor: A combination of highly effective thin-film PEC and conventional PEC processes on a single electrode, Environmental Science & Technology, Mar 2008, 42(7), 2612–2617. Doi: 10.1021/es702921h.

[66]   Cho K, et al. Effects of reactive oxidants generation and capacitance on photoelectrochemical water disinfection with self-doped titanium dioxide nanotube arrays, Applied Catalysis B: Environmental, Nov 2019, 257, 117910. Doi: 10.1016/j.apcatb.2019.117910.

[67]   Moreira FC, Boaventura RAR, Brillas E, Vilar VJP. Electrochemical advanced oxidation processes: A review on their application to synthetic and real wastewaters, Applied Catalysis B: Environmental, 2017, 202, 217–261. Doi: 10.1016/j.apcatb.2016.08.037.

[68]   Giménez S, Bisquert J. Photoelectrochemical Solar Fuel Production: From Basic Principles to Advanced Devices, 2016.

[69] Vilanova A, Lopes T, Spenke C, Wullenkord M, Mendes A. Optimized photoelectrochemical tandem cell for solar water splitting, Energy Storage Materials, 2018, 13, no. December 2017, 175–188. Doi: 10.1016/j.ensm.2017.12.017.

[70] Wang Y, Zu M, Zhou X, Lin H, Peng F, Zhang S. Designing efficient TiO2-based photoelectrocatalysis systems for chemical engineering and sensing, Chemical Engineering Journal, Feb 2020, 381, 122605. Doi: 10.1016/j.cej.2019.122605.

[71] Zarei E, Ojani R. Fundamentals and some applications of photoelectrocatalysis and effective factors on its efficiency: a review, Journal of Solid State Electrochemistry, 2017, 21(2), 305–336. Doi: 10.1007/s10008-016-3385-2.

[72] Acar C, Dincer I. A review and evaluation of photoelectrode coating materials and methods for photoelectrochemical hydrogen production, International Journal of Hydrogen Energy, May 2016, 41(19), 7950–7959. Doi: 10.1016/j.ijhydene.2015.11.160.

[73] Casado C, Mesones S, Adán C, Marugán J. Comparing potentiostatic and galvanostatic anodization of titanium membranes for hybrid photocatalytic/microfiltration processes, Applied Catalysis A: General, 2019, 40–52, Doi: 10.1016/j.apcata.2019.03.024.

[74] Zhao X, Guo L, Qu J. Photoelectrocatalytic oxidation of Cu-EDTA complex and electrodeposition recovery of Cu in a continuous tubular photoelectrochemical reactor, Chemical Engineering Journal, Mar 2014, 239, 53–59. Doi: 10.1016/J.CEJ.2013.10.088.

[75] Pablos C, Marugán J, Van Grieken R, Adán C, Riquelme A, Palma J. Correlation between photoelectrochemical behaviour and photoelectrocatalytic activity and scaling-up of P25-TiO2 electrodes, Electrochimica Acta, Jun 2014, 130, 261–270. Doi: 10.1016/J.ELECTACTA.2014.03.038.

[76] Orimolade BO, Koiki BA, Peleyeju GM, Arotiba OA. Visible light driven photoelectrocatalysis on a FTO/BiVO4/BiOI anode for water treatment involving emerging pharmaceutical pollutants, Electrochimica Acta, Jun 2019, 307, 285–292. Doi: 10.1016/j.electacta.2019.03.217.

[77] Mousset E, et al. A new 3D-printed photoelectrocatalytic reactor combining the benefits of a transparent electrode and the Fenton reaction for advanced wastewater treatment, Journal of Materials Chemistry A, 2017, 5(47), 24951–24964. Doi: 10.1039/c7ta08182k.

[78] Olvera-Rodríguez I, et al. TiO2/Au/TiO2 multilayer thin-film photoanodes synthesized by pulsed laser deposition for photoelectrochemical degradation of organic pollutants, Separation and Purification Technology, Oct 2019, 224, 189–198. Doi: 10.1016/j.seppur.2019.05.020.

[79] Oriol R, Sirés I, Brillas E, De Andrade AR. A hybrid photoelectrocatalytic/photoelectro-Fenton treatment of Indigo Carmine in acidic aqueous solution using TiO2 nanotube arrays as photoanode, Journal of Electroanalytical Chemistry, Aug 2019, 847, 113088. Doi: 10.1016/j.jelechem.2019.04.048.

[80] Salmerón I, et al. Optimization of electrocatalytic H_2O_2 production at pilot plant scale for solar-assisted water treatment, Applied Catalysis B: Environmental, 2019, 242, no. June 2018, 327–336. Doi: 10.1016/j.apcatb.2018.09.045.

[81] Wang D, et al. Dye-sensitized photoelectrochemical cell on plasmonic Ag/AgCl @ chiral TiO2 nanofibers for treatment of urban wastewater effluents, with simultaneous production of hydrogen and electricity, Applied Catalysis B: Environmental, Jun 2015, 168–169, 25–32. Doi: 10.1016/J.APCATB.2014.11.012.

[82] Lianos P. Production of electricity and hydrogen by photocatalytic degradation of organic wastes in a photoelectrochemical cell: The concept of the photofuel cell: A review of a re-emerging research field, Journal of Hazardous Materials, Jan 2011, 185(2–3), 575–590. Doi: 10.1016/J.JHAZMAT.2010.10.083.

[83] Haussener S, Xiang C, Spurgeon JM, Ardo S, Lewis NS, Weber AZ. Modeling, simulation, and design criteria for photoelectrochemical water-splitting systems, Energy & Environmental Science, 2012, 5(12), 9922–9935. Doi: 10.1039/C2EE23187E.

[84] Becker J-P, et al. A modular device for large area integrated photoelectrochemical water-splitting as a versatile tool to evaluate photoabsorbers and catalysts, Journal of Materials Chemistry A, 2017, 5(10), 4818–4826. Doi: 10.1039/C6TA10688A.

[85] Bedoya-Lora FE, Hankin A, Kelsall GH. En route to a unified model for photo-electrochemical reactor optimisation. I – Photocurrent and H2 yield predictions, Journal of Materials Chemistry A, 2017, 5(43), 22683–22696. Doi: 10.1039/c7ta05125e.

[86] Casado C, et al. Design and validation of a LED-based high intensity photocatalytic reactor for quantifying activity measurements, Chemical Engineering Journal, Nov 2017, 327, 1043–1055. Doi: 10.1016/J.CEJ.2017.06.167.

[87] Li XZ, Li FB, Fan CM, Sun YP. Photoelectrocatalytic degradation of humic acid in aqueous solution using a Ti/TiO2 mesh photoelectrode, Water Research, May 2002, 36(9), 2215–2224. Doi: 10.1016/S0043-1354(01)00440-7.

[88] Martín-Sómer M, Pablos C, Van Grieken R, Marugán J. Influence of light distribution on the performance of photocatalytic reactors: LED vs mercury lamps, Applied Catalysis B: Environmental, Oct 2017, 215, 1–7. Doi: 10.1016/J.APCATB.2017.05.048.

[89] Wang H, Liang W, Zhang W, Zhou D. Photoelectric performance of TiO2 nanotube array film of highly transparent vs nontransparent on FTO glass, Materials Science in Semiconductor Processing, Nov 2017, 71, 50–53. Doi: 10.1016/j.mssp.2017.07.005.

[90] Pletcher FC, Walsh D. Industrial Electrochemistry, 1993.

[91] Aroutiounian VM, Arakelyan VM, Shahnazaryan GE. Metal oxide photoelectrodes for hydrogen generation using solar radiation-driven water splitting, Solar Energy, May 2005, 78(5), 581–592. Doi: 10.1016/J.SOLENER.2004.02.002.

[92] Bicer Y, Dincer I. Experimental investigation of a PV-Coupled photoelectrochemical hydrogen production system, International Journal of Hydrogen Energy, Jan 2017, 42(4), 2512–2521. Doi: 10.1016/J.IJHYDENE.2016.02.098.

[93] Yoo JE, et al. Anodic TiO2 nanotube arrays directly grown on quartz glass used in front- and back-side irradiation configuration for photocatalytic H2 generation, Physica Status Solidi, 2016, 213(10), 2733–2740. Doi: 10.1002/pssa.201600140.

[94] Jeong HW, Park KJ, Park Y, Han DS, Park H. Exploring the photoelectrocatalytic behavior of free-standing TiO2 nanotube arrays on transparent conductive oxide electrodes: Irradiation direction vs. alignment direction, Catalysis Today, Sep 2019, 335, 319–325. Doi: 10.1016/j.cattod.2018.12.014.

[95] Manassero A, Zacarías SM, Satuf ML, Alfano OM. Intrinsic kinetics of clofibric acid photocatalytic degradation in a fixed-film reactor, Chemical Engineering Journal, Jan 2016, 283, 1384–1391. Doi: 10.1016/J.CEJ.2015.08.060.

[96] Gharehpetian GB, Mousavi Agah SM, Abdi H, Rasouli Nezhad R, Salehimaleh M. Fuel Cells, Distributed Generation Systems, Jan 2017, 221–300. Doi: 10.1016/B978-0-12-804208-3.00005-4.

[97] Hankin A, Bedoya-Lora FE, Ong CK, Alexander JC, Petter F, Kelsall GH. From millimetres to metres: the critical role of current density distributions in photo-electrochemical reactor design, Energy & Environmental Science, 2017, 10(1), 346–360. Doi: 10.1039/C6EE03036J.

[98] Gärtner WW. Depletion-layer photoeffects in semiconductors, Physical Review, Oct 1959, 116(1), 84–87. Doi: 10.1103/PhysRev.116.84.

[99] Butler MA, Ginley DS. Principles of photoelectrochemical, solar energy conversion, Journal of Materials Science, Jan 1980, 15(1), 1–19. Doi: 10.1007/BF00552421.

[100] Holmes-Gentle I, Agarwal H, Alhersh F, Hellgardt K. Assessing the scalability of low conductivity substrates for photo-electrodes via modelling of resistive losses, Physical Chemistry Chemical Physics, 2018, 20(18), 12422–12429. Doi: 10.1039/C8CP01337C.

[101] Farivar F. CFD simulation and development of an improved photoelectrochemical reactor for H2 production, International Journal of Hydrogen Energy, Jan 2016, 41(2), 882–888. Doi: 10.1016/J.IJHYDENE.2015.11.045.

[102] Martín De Vidales MJ, et al. Scale-up of electrolytic and photoelectrolytic processes for water reclaiming: a preliminary study, Environmental Science and Pollution Research, 2016, 23(19), 19713–19722. Doi: 10.1007/s11356-016-7189-9.

[103] An T, Zhao HJ, Wong PK. Advances in Photocatalytic Disinfection, 2017.

[104] Fernandez-Ibañez P, Malato S, Enea O. Photoelectrochemical reactors for the solar decontamination of water, Catalysis Today, Dec 1999, 54(2–3), 329–339. Doi: 10.1016/S0920-5861(99)00194-7.

[105] Meng X, Zhang Z, Li X. Synergetic photoelectrocatalytic reactors for environmental remediation: A review, Journal of Photochemistry and Photobiology C: Photochemistry Reviews, Sep 2015, 24, 83–101. Doi: 10.1016/J.JPHOTOCHEMREV.2015.07.003.

[106] Zhao X, et al. Photoelectrochemical treatment of landfill leachate in a continuous flow reactor, Bioresource Technology, 2010, 101(3), 865–869. Doi: 10.1016/j.biortech.2009.08.098.

[107] Bolton JR, Bircher KG, Tumas W, Tolman CA. Figures-of-merit for the technical development and application of advanced oxidation technologies for both electric- and solar-driven systems, Pure and Applied Chemistry. Chimie Pure Et Appliquee, 2001, 73(4), 627–637. Doi: 10.1351/pac200173040627.

Laura Clarizia, Priyanka Ganguly

Chapter 3
Fundamentals of Photocatalytic Hydrogen Production

3.1 Introduction

At present, hydrogen is the key energy carrier for the green economy due to its lower heating value (LHV, ≈ 120 MJ/kg) greater than prevalent fuels (i.e., $LHV_{Methane} = 50$ MJ/kg, $LHV_{Gasoline} = 44.5$ MJ/kg) and its clean combustion product (H_2O) avoiding toxic and greenhouse responsible emissions.

The main current sources for hydrogen generation are reported in Figure 3.1 [1]. The use of fossil fuels for H_2 generation is undesirable due to CO_2 emissions responsible for global warming and the limited supply of these raw materials.

In recent years, energy production based on clean and renewable sunlight irradiation has attracted growing attention among researchers and industrial stakeholders [2]. In particular, photochemical, photoelectrochemical (PEC) [3], and thermochemical processes [4] have been employed for solar-to-hydrogen (STH) energy conversion.

In this scenario, the possibility to employ low-cost photocatalytic processes activated by sunlight irradiation for H_2 production well matches the green targets of a future circular economy. Solar-driven photocatalytic H_2 evolution may be achieved by splitting water [5] or reforming sacrificial organic species [6]. Despite major efforts in developing materials with significant photoactivity have been recently reported in the literature review, the STH energy conversion efficiencies are still far from ensuring minimal requirements for large-scale implementations based on sunlight capture [7]. For both photosplitting and photoreforming processes, effective semiconductor photocatalysts capable of (i) absorbing visible light irradiation, (ii) promptly using photon energy to generate charge carriers without heat loss, and (iii) promoting photogenerated electron–hole separation can be adopted [8].

To this aim, metal oxides of transition elements with d^0 or d^{10} configuration cations (i.e., Ti^{4+}, Zr^{4+}, Nb^{5+}, Ta^{5+}, W^{6+}, Ga^{3+}, Ge^{4+}, In^{3+}, Sn^{4+}, Sb^{5+}, etc.) have been so far employed in the literature [9]. Owing to its chemical stability, extensive availability, and low cost, TiO_2 was extensively adopted as in its pristine or appropriately

Laura Clarizia, Department of Chemical, Materials and Industrial Production Engineering, University of Naples "Federico II", P. le V. Tecchio 80, 80125 Naples, Italy, e-mail: laura.clarizia2@unina.it

Priyanka Ganguly, Nanotechnology and Bio-engineering Research Group, Department of Environmental Science, Institute of Technology Sligo, Sligo, Ireland, e-mail: priyanka.ganguly@mail.itsligo.ie

https://doi.org/10.1515/9783110668483-003

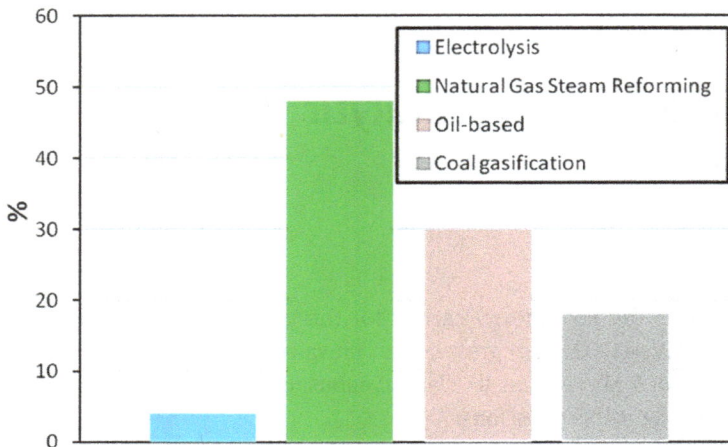

Figure 3.1: Main sources for hydrogen generation (data source: Voldsund et al. [1]).

doped in its different crystalline phases [10]. However, the electronic band structure of TiO_2 allows radiation absorption only in the UV range with a higher likelihood of photogenerated charge carrier recombination, thus significantly lowering the overall process photoefficiency. These aspects were previously examined in detail in excellent review articles in the literature [9, 11, 12].

The basic tenets of photocatalytic water splitting and organic reforming are discussed. The survey on recent advances in semiconductor-based UV-active and visible-light-responsive photocatalysts and a brief overview of standard reactor configurations for hydrogen evolution are herein discussed. A particular focus was given to recent developments for enhancing the photoefficiency of H_2 production for the semiconductor-based photocatalysts.

3.2 Reaction Energetics

3.2.1 Photocatalytic Water Splitting

Photocatalytic water splitting and organic reforming involve the use of semiconductor photocatalysts, which are excited by incident photons to produce photogenerated electron–hole pairs. The high likelihood of photogenerated charge recombination significantly affects the overall process photoefficiency.

Photocatalytic water splitting process is based on the reduction and oxidation of water by reaction with photogenerated electrons and positive holes, respectively. The overall reaction is water decomposition, with the formation of hydrogen and oxygen gases (Figure 3.2).

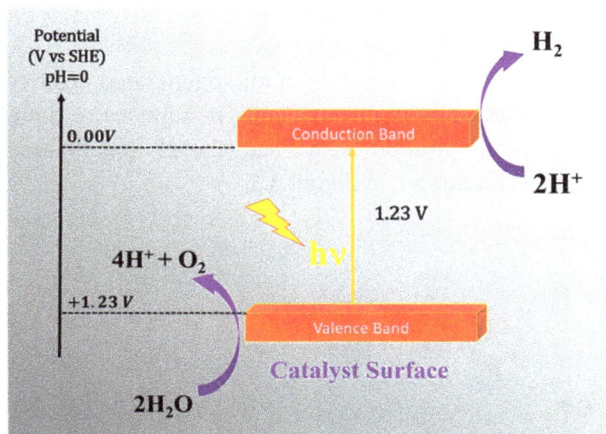

Figure 3.2: Water photosplitting process: water oxidation by photogenerated positive holes at the valence band (VB) and proton reduction by photogenerated electrons at the conduction band (CB).

Concerning the reaction energetics, water splitting is an endergonic reaction with a free Gibbs energy change equal to 238 kJ/mol (equation (3.1)). As the water-splitting reaction is not spontaneous in nature, external energy to trigger the reaction needs to be provided:

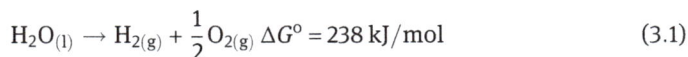

$$H_2O_{(1)} \rightarrow H_{2(g)} + \frac{1}{2}O_{2(g)} \; \Delta G° = 238 \, kJ/mol \tag{3.1}$$

The change in free Gibbs energy of water splitting corresponds to an electrical potential of 1.23 eV or an applied voltage of +1.23 V, according to the following equation:

$$\Delta G° = -n \cdot F \cdot \Delta E° \tag{3.2}$$

where n represents the number of electrons transferred in the reaction, F is Faraday's constant, and $\Delta E°$ represents the voltage difference.

By considering the energy of a single photon, this change in free Gibbs energy accounts for around 1,000 nm of wavelength for a photon, according to the following equation:

$$E \, (eV) = \frac{1,240 \, eV \, nm}{\lambda \, nm} \tag{3.3}$$

Thus, an electrical potential of 1.23 eV or light radiation with wavelength lower than around 1,000 nm is needed to trigger water splitting into electrochemical cells or photochemical devices, respectively.

3.2.2 Photocatalytic Reforming of Organics

Photocatalytic reforming of organics relies on the ability of photogenerated positive holes to oxidize selected organic species (i.e., sacrificial agents) in aqueous solution with proton release. The arising protons promptly react with the photogenerated electrons to produce hydrogen gas, as illustrated in Figure 3.3.

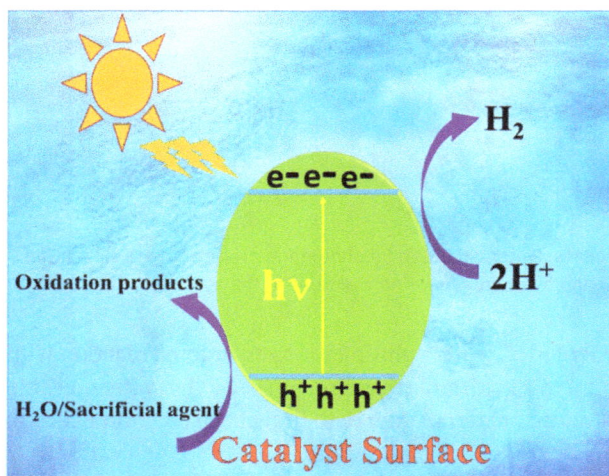

Figure 3.3: Schematic illustration for the photoreforming process. Oxidation of organics (A) by positive holes in the valence band (VB) and reduction of proton ions by photogenerated electrons in the conduction band (CB).

The presence of sacrificial organic species helps prevent O_2/H_2 back reaction, thereby to prevent O_2/H_2 back reaction and affecting traditional water photosplitting. This reduces the photogenerated charge carrier recombination, thus enhancing the overall process photoefficiency for H_2 production. Carboxylic acids, short chain alcohols, and biomass derived substrates have been successfully employed as sacrificial organic species [13].

Theoretically, the use of sacrificial organics in photoreforming processes is an additional cost. Nevertheless, the possibility of employing civil or industrial wastewater streams containing selected organic species may result in a combined strategy of wastewater treatment with simultaneous production of a high value-added product, that is, hydrogen gas.

Figure 3.4 brings out the relevance of the energy band positions in semiconductor-based photocatalysts (i.e., TiO_2). Indeed, photogenerated electrons can reduce protons in aqueous solution only if the conduction band potential is lower than the H^+/H_2 couple. Similarly, the valence band's potential containing photogenerated holes should be higher than the H_2O/O_2 couple for water photosplitting or selected redox couples in organic photoreforming. In general, sacrificial organic species (e.g., methanol) have oxidation potentials lower than water (1.23 V vs SHE).

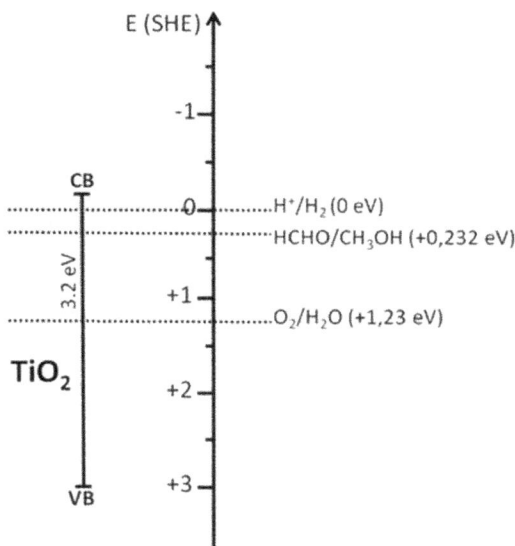

Figure 3.4: Energy levels of selected redox couples and electronic bands of anatase–TiO_2 [14–17].

Among the most frequently adopted sacrificial organic agents, methanol, ethanol, and glycerol should be mentioned. Carboxylic and short-chain organic acids, along with dye molecules, were also often reported as highly effective electron donors in photocatalytic hydrogen production [13]. Species like methanol and ethanol are often employed as fuels in several industrial applications with yields higher than hydrogen evolution reaction (HER). They have been selected in photoreforming processes mainly for a direct understanding of the reaction mechanism related to their oxidation. The possibility of selecting different organic species as effective electron donors, such as waste products or by-products, relies on the purpose of coupling photocatalytic hydrogen generation with decontamination of industrial or urban wastewater. Overall, previous studies dedicated inadequate consideration to the identification of oxidation intermediates and products in the case of complex molecules (i.e., dyes) selected as sacrificial agents, thus raising significant concerns about their environmental fate and exotoxicity after the final discharge.

A combination of photosplitting and photoreforming is possible using selected inorganic electron donors (i.e., I^- and Fe^{2+}) [13]. Their oxidized state can be reduced on a further photocatalyst, enabling water oxidation and O_2 production simultaneously. In the case of iodide, two photocatalysts can be suitably selected and added in the aqueous mixture: the first photocatalyst enables proton (or water) reduction and oxidation of iodide to iodate; on the other hand, the second photocatalyst allows a further reduction of iodate to iodide, with the simultaneous oxidation of water to O_2. According to such an approach, photocatalytic water splitting without sacrificial organic consumption is still the resulting reaction.

3.3 Semiconductor-Based Photocatalysts for H_2 Generation

The importance of band edge values and their corresponding redox potential for hydrogen evolution is well defined in the previous section. There exists a wide array of semiconducting materials potentially used as a photocatalyst. They could be predominantly classified as UV-active photocatalysts and visible-light-active photocatalysts. In this following section, recent reports of both the kinds of catalyst and their improved mechanism are elaborated.

3.3.1 UV-Active Photocatalysts

Discussing UV-active photocatalyst could not be initiated without entailing the pioneering work of Honda and Fujishima to use titania (TiO_2) films for PEC water splitting in 1972 [18]. Since then, a great deal of improvement in catalysis has been observed in terms of hydrogen output with improved heights of the reactor as well as catalyst design. Although recent articles have focused more on improving catalytic performance by enhancing the spectral absorption, UV-induced water splitting reports are rarely reported in the past few years. However, this section details a few of such composite structures and their improved hydrogen generation rate.

Addition of platinum (Pt) as a potential cocatalyst has always been the easiest choice to improve the hydrogen generation efficiency. Kandiel et al. reported the photodeposition of Pt over TiO_2 powders. The resultant samples were studied for their water-splitting ability by irradiation of UV light of wavelength (λ) 365 nm through an light-emitting diode (LED) light source and methanol (0.03 M). Hydrogen generation of up to 5,400 μmol/h was observed for 0.5 wt% Pt concentration [19]. The methanol in the solution acts as a sacrificial agent and transforms or undergoes photooxidation to generate useful products as formaldehyde or formic acid. This process is commonly known as photoreforming; in this method, apart from hydrogen generation, a significant portion of CO_2 is also released, which also converts into useful products [20]. Strataki et al. studied the optimal condition for H_2 generation for Pt-doped titania thin-film sample undergoing photoreforming of alcohol. The study was performed in a reactor with black light tubes with an emission profile in the UV region which is similar to the typical absorption observed for TiO_2. The results displayed were impressive;

it was found in a 50 vol% ethanol–water mixture, the maximum efficiency was observed for titania loading of 40 mg. Lower concentration of titania displayed complete transparency and hence did not displayed much yield. While at higher concentration showed a scattering of light in the thin film, the maximum rate was finally 9 μmol/min [21]. The immobilization of cellulose molecules on platinized TiO_2 powder (0.5 wt% Pt) under UV irradiation results in H_2 and CO_2 production in the process. The presence of a small amount of Pt plays a crucial role in the reduction process and yields H_2 of 195.2 μmol/h [22]. In a recent report, Fang et al. discussed rhodium-doped titania's efficiency in the presence and absence of a sacrificial agent (methanol, in this case). The doped sample displayed enhanced visible light absorption. Still, the hydrogen yield reduced significantly from 7.27 to 0.02 mmol/h/g, when the incident radiation was >420 nm at room temperature. This provides the importance of UV irradiation in the system. Moreover, in a sacrificial-less dispersion system (only with catalyst) but at a high temperature of 260 °C, the yield improved to as high as 42 mmol/h/g under UV irradiation and about 30 mmol/h/g, when the incident radiation was >420 nm. The enhanced activity in the visible region is ascribed to the kinetic energy imparted in the system with elevated temperature, enhancing the oxidation process [23]. Zalas et al. studied the influence of lanthanum-doped elements in titania for hydrogen generation. Moreover, the authors also examined the importance of various sacrificial agents in the water-splitting process. In this process, methanol, ethanol, glycol, or glycerin were applied as the sacrificial agents. The 0.5 mol% of Gd, Eu, Yb or Ho oxides as dopants displayed the best results. Among them, the gadolinium-doped titania exhibited the optimum hydrogen yield with methanol as the sacrificial agent [24]. In another report, the authors studied the effect of nickel doping in mesoporous TiO_2 using methanol (20 vol%) as the sacrificial agent. The hydrogen production increased with the increase in nickel content of up to 1 wt%. On moving more than 1 wt%, the hydrogen amount produced was declined. Hence, 1 wt% was considered as the optimal concentration. The addition of these metal atoms introduces small trap energy levels, which aids in enhancing the charge separation [25]. The co-doping of Sn/Eu in TiO_2 displayed a higher activity for hydrogen generation compared to bare TiO_2 sample, with a quantum efficiency of ~40% with Pd as the cocatalyst under irradiation from a fluorescent lamp [26]. Chae et al. studied the influence of gallium doping in titania for the hydrogen generation. The 2 mol% doped sample displayed greater efficiency compared to undoped titania samples [27]. Binary combination of different semiconductors is also reported to display enhanced hydrogen generation. Naggar et al. reported the combination of ZnO and CdS at a 1:1 ratio. The resultant composite structure in the presence of methanol as a sacrificial agent displayed enhanced H_2 rate under UV light illumination (5.76 mmol/h) [28]. Perovskite material such as strontium titanate ($SrTiO_3$) has been widely researched for photocatalytic water splitting. This material's band edge potential is comparable to the potential redox values for hydrogen and oxygen formation. In this study, the authors have utilized Cu and Pt as a bimetallic cocatalyst to the system, added via simple photodeposition technique. Methanol is used as a sacrificial agent; the system irradiated with UV light displayed a H_2 generation rate of 369.4 μmol/h. The synergistic effect of the presence of both Cu and Pt in the catalyst structure resulted in the suppression of the charge carrier recombination [29].

3.3.2 Visible-Light-Responsive Photocatalysts

Visible-light-responsive photocatalysis has dominated overall for different photocatalytic applications [30]. Accounting for much of the solar spectrum share makes it an ideal region to be explored [31]. Ni-silicate nanosheet onto CdS nanorods was fabricated via a hydrothermal process. Ni^{2+} behaved as a cocatalyst, which improved the charge carrier separation and resulted in enhancement of H_2 generation up to 401.7 μmol/h in the presence of sodium sulfide (0.35 M) and sodium sulfite (0.25 M) sacrificial agents [32]. A three-combination (TiO_2–Au–CdS) photocatalyst was fabricated utilizing a hardcore template method with surface modification, cation exchange, and sulfidation process. The study focused on finding an optimum diameter for TiO_2 hollow structure, which displayed significant hydrogen generation efficiency. The hollow nanoshells of 185 nm size displayed the maximum photocatalytic hydrogen generation rate. The enhanced activity was attributed to the decent matching of Mie resonance in hollow nanoshells as well as the enhanced absorption activity due to CdS [33]. In a recent study, an n–p heterojunction of TiO_2–NiO is reported with ultrafine core–shell nanoparticles. The nanostructure composite displayed enhanced H_2 generation rate and also showed considerable rate of methane generation along with few fatty acids (palmitic acid and stearic acid) from lignin photoreforming. The NiO nanoclusters along with the crystalline nature of the titania nanoparticles contributed to the synergistic enhancement of the hydrogen generation rate [34]. Organic molecules with inorganic oxide compounds have been a promising technique to improve spectral absorption. In a recent report, an efficient copper-doped ZnO nanorod/5,10,15,20-tetrakis(4-sulfophenyl)porphyrin composite photocatalysts were facilely developed. This multicomponent structure has effectively delayed the carrier recombination and displayed enhanced hydrogen yield in the presence of 16.7 wt% triethanolamine (TEOA) [35]. A highly efficient ternary photocatalytic composite $MoO_2/MoS_2/TiO_2$ was reported. This sandwich heterostructure was fabricated by chemical vapor deposition technique. The composite displayed an enhanced H_2 generation of 110 μmol/h/cm^2 in the presence of 0.5 M ethylene glycol and 0.5 M $NaSO_4$ as hole scavengers (Figure 3.5) [36].

Sulfuration or sulfidation is addition of sulfur atoms to the material at elevated temperature. The introduction of sulfur results in the creation of defects and even vacancies. This results in the creation of mid-level bands that effectively enhance the charge carrier separation. The in situ sulfurization of hydrous WO_3 results in the formation of the layered structure of WS_2/WO_3 heterostructures. In a recent report, a p–n junction of $MoSe_2@Co_3O_4$ illustrates improved charge transfer at the interface. Only

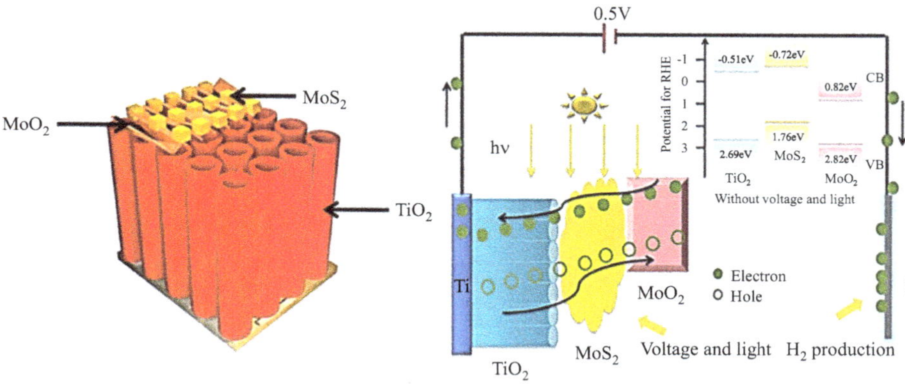

Figure 3.5: Schematic diagram of the sandwich structure of the composite and light harvesting process and the energy band structure of TiO_2, MoS_2, and MoO_2. Reprinted with permission of Wang et al. (2020). Full details are given in the respective publication [36].

15% vol/vol TEAO solution was used as the sacrificial agent. The composite structure displayed 2.34 times higher evolution rate (7,029.2 µmol/h/g) than $MoSe_2$ [37]. In another study, Gong et al. reported the introduction of cobalt nitride in $Mn_{0.2}Cd_{0.8}S$. CoN acts as a potential electron acceptor and an electrostatic self-assembly of these nanoparticle surfaces results in easy flow of electrons. This henceforth results in enhanced efficiency (17.3 times the magnesium cadmium sulfide) with a rate of 14.612 µmol/h/g [38]. The resultant heterostructure displays a Z-scheme mechanism for the enhanced hydrogen generation. The close intimate junctions were created to mitigate the problem of charge carrier recombination. The PEC reduction in the presence of 10 vol% lactic acids displayed hydrogen generation of 680 µmol/h/g [39]. In a recent report, sulfuration treatment of the ZIF-8 molecules was done to introduce specific point defects (Figure 3.6). The introduction of impure band levels aids in enhanced charge carrier separation, which contributes to the improved hydrogen yield of 741.80 µmol/h/g [40].

Apart from oxides/sulfides/selenides of various d-block elements, a broad class of two-dimensional (2D) class of nanomaterials has shown promising results. Enhanced surface area and unique electronic and optical properties are some of the advantages displayed by these classes of materials [8]. Graphitic carbon nitride (g-C_3N_4) has shown the excellent possibility for photocatalytic hydrogen production owing to its medium bandgap of 2.7 eV, metal-free nature, and outstanding resistance [41, 42]. Coupling g-C_3N_4 with nitrogen-doped carbon materials has shown promising efficiency. Increased output of 23.0 µmol/h was observed for nitrogen- and carbon-doped g-C_3N_4 samples with 10 vol% TEOA as the sacrificial reagent and the use of 0.5 wt% platinum (Pt) cocatalyst [43]. While a recent study also shows superior hydrogen generation for nitrogen-deficient carbon nitride sheets. About 10 vol% TEOA

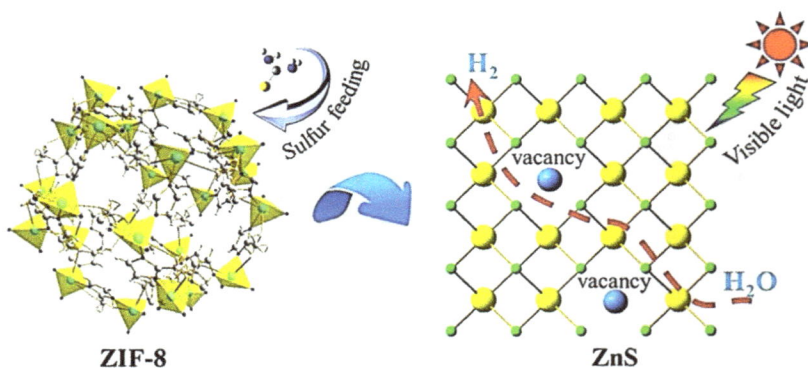

Figure 3.6: Schematic mechanism displaying the sulfuration of ZIF-8. Reprinted with permission of Zhou et al. (2020). Full details are given in the respective publication [40].

as hole scavenger and H_2PtCl_6 solution (3 wt% Pt) were used as sacrificial agents, which resulted in the hydrogen yield of 5,375 µmol/h/g [44]. Zhang et al. reported the enhanced efficacy of hydrogen evolution rate in oxygen-doped crystalline graphitic nitride with K^+ implantation. The doping of the nitride sheets results in the enhanced visible light absorption extending up to the infrared region. The photocatalyst in the presence of H_2PtCl_6 (3 wt% Pt) and TEOA solution (20 vol%) displays an enhanced hydrogen generation rate of 392.2 µmol/h (3.8 times greater to pristine carbon nitride) [45]. Another recent article highlights the importance of the use of metal-free localized plasmon resonance. The one-dimensional (1D) carbon nanotubes/2D ultrathin carbon nitride (1D single-walled carbon nanotube/2D C_3N_4) displays nonresonant plasmonic photocatalyst nature. The catalyst in the presence of 10 vol% TEOA and 3.0 wt% Pt under visible light radiations displayed the yield of 1,346 µmol/h/g. The heterojunction creation increases the spectral absorption in both UV and visible regions [46]. A recent report displays the importance of transition metal phosphosulfides (copper phosphosulfides). It coupled with g-C_3N_4 ($Cu_3P|S/CN$) displays enhanced charge transfer and displays high yield of H_2 gas of 20.22 µmol/h/g with 0.5 wt% Pt. The P–N bond developed in the binary structure enhanced the overall interfacial charge transfer and demonstrated a Z-scheme junction [47]. Similarly MoS_2 nanodot coupled with g-C_3N_4 sheets was recently reported by Wu et al. An eightfold enhanced efficiency (164.0 µmol/h/g) was observed for the heterostructure formed compared to conventional MoS_2–g-C_3N_4 composite in the presence of 10 vol% lactic acid. The trifunctional ammonium tetrathiomolybdate-assisted synthesis aided in formation of in situ quantum dots of MoS_2 [48]. In another study, a ternary heterojunction of $BiVO_4$/graphene quantum dot/g-C_3N_4 was reported. The theoretical understanding validated the importance of graphene quantum dots as an electron mediator in the heterojunction created when compared to reduced graphene oxide (rGO). The PEC hydrogen generation reported the highest output of up to 84.9 mmol/h [49]. MXenes structure-based composite has also demonstrated a promising result. In a recent report, Zhao et al. reported a unique combination of $Ti_3C_2T_x$ MXene coupled with nanostructure of CdS which are developed via one-step gamma-ray radiation-induced reduction under ambient parameters. Consequent alkaline treatment results in the formation of a novel cocatalyst of carbon-supported amorphous TiO_2 nanosheets (TiO_2-C), which is derived from $Ti_3C_2T_x$/CdS and results in TiO_2-C/CdS nanocomposites. The resultant nanocomposite of TiO_2-C/CdS under visible light illumination displays a high H_2 evolution rate of 1.48 mmol/h/g [50]. In another report, 2D Ti_3AlC_2/TiO_2/Ni_2P composite displayed enhanced visible absorption and efficient charge carrier separation. The composite structure exhibited the highest H_2 production of 13,000 µmol/h (Figure 3.7) [51].

A recent report of metal-free photocatalysts with a combination of nitrogen-doped quantum dots of carbon (NCQDs) and covered with PS-PEGCOOH and also with PFTBTA, an organic semiconducting polymer, resulting in a polymer dot heterostructure. The addition of NCQDs into the polymer dot resulted in the fivefold increase of the hydrogen production yield. A white LED ($\lambda > 420$ nm) was used as the light source and ascorbic acid is used as the sacrificial agent. The reported heterostructure resulted in enhanced H_2 rate of 4,490 µmol/h/g [52]. Guo et al. reported the synthesis of a 3D-structured kaolin and red phosphorus composite synthesized through hydrothermal technique. The composites displayed a hydrogen generation of 252 µmol/h/g, an almost eightfold enhancement to the red phosphorus nanomaterials. The presence of kaolin prevents the agglomeration of red phosphorus, thereby declining the distance of charge carrier migration. The three-dimensional structure also enhances the overall increase in the visible light absorption [53]. The ternary chalcogenides are narrow bandgap semiconducting materials which display variable absorption and few extending till near-infrared region. Composites with graphene, MoS_2, and TiO_2 have shown interesting results. Ag-doped $ZnIn_2S_4$ incorporated on rGO is one of the recent studies. Using 10 vol% TEOA solution as a sacrificial agent, the composite shows a hydrogen yield of 6,343.86 µmol/h/g [54]. A 2D–2D MoS_2/Cu–$ZnIn_2S_4$ has also shown improved results of up to 5,463 µmol/h/g in the presence of 0.1 M ascorbic acid [55]. A 2% $MoSe_2$/$ZnIn_2S_4$ photocatalyst shows a high H_2 generation rate of 2,228 µmol/h/g [56]. In a recent report, we have developed ternary chalcogenide composites with titania for water splitting application. Composites of $AgBiS_2$, $AgInS_2$, $AgBiSe_2$, and $AgInSe_2$ with TiO_2, respectively, were synthesized at different weight ratios. These composites in the presence of 10 vol% methanol as a sacrificial agent displayed higher hydrogen production of 1,300 µmol/min, about 300 µmol/min, 180 µmol/min, and 250 µmol/min for $AgBiS_2$–TiO_2, $AgInS_2$–TiO_2, $AgBiSe_2$–TiO_2, and $AgInSe_2$–TiO_2, respectively [31, 57, 58].

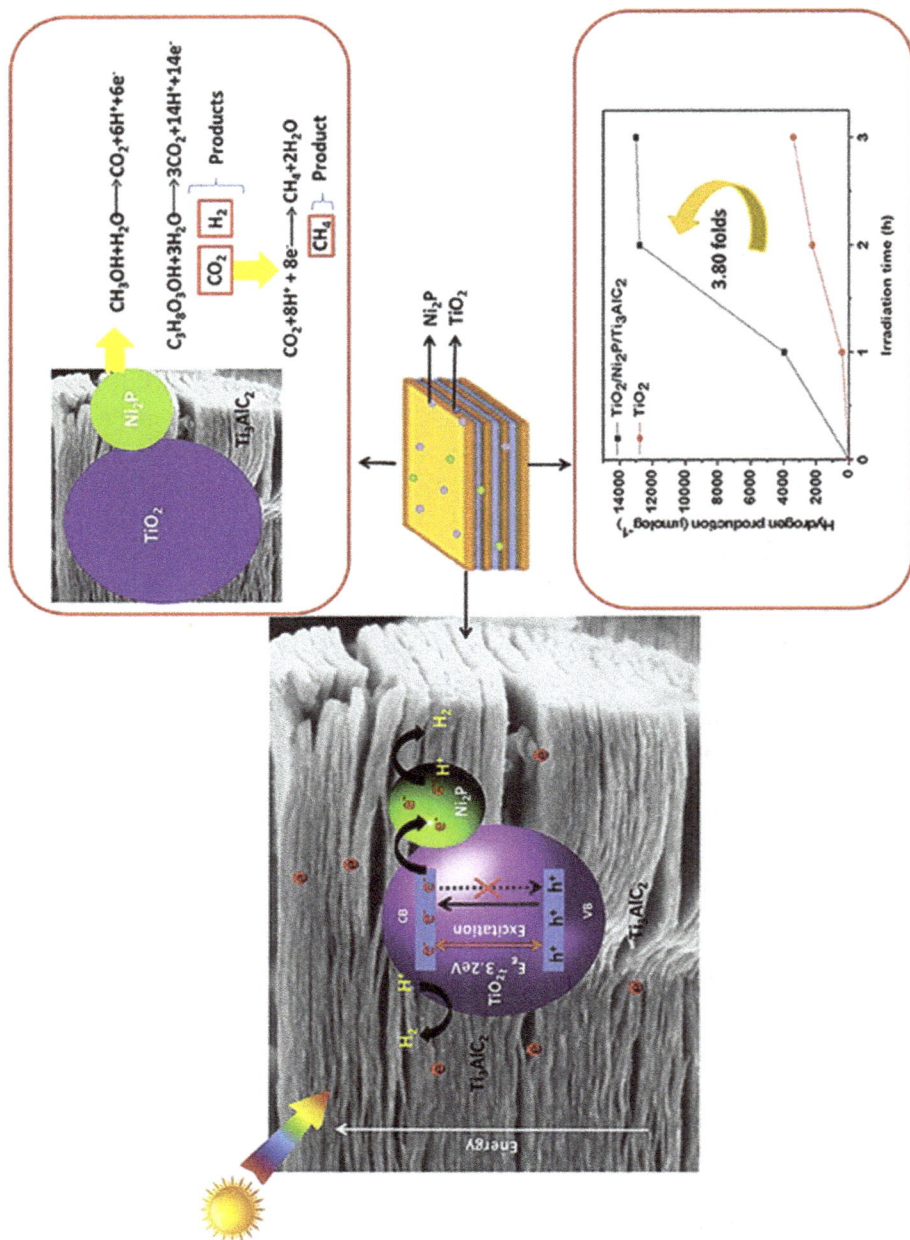

Figure 3.7: Schematic illustration of improved hydrogen generation by 2D Ti$_3$AlC$_2$/TiO$_2$/Ni$_2$P composite. Reprinted with permission of Tasleem et al. (2020). Full details are given in the respective publication [51].

3.4 Photocatalytic Reactors for H$_2$ Generation

Photocatalytic hydrogen production from aqueous solutions can be carried out through different experimental setups. The first possible configuration relies on the use of PEC cells with the catalyst immobilized on a photoanode. In the second type of configuration, the photocatalyst is merely suspended in the solution. An overview of the main features of both systems is provided below.

3.4.1 Immobilized Systems: Photoelectrochemical Cells

In PEC systems, HER induced by water decomposition or organic oxidation requires an external potential difference involving an additional energy supply.

As regards the reaction mechanism in PEC systems, water in photosplitting or organic species in photoreforming is oxidized at the photoanode. Simultaneously, water or protons are reduced at the cathode in a different compartment, with the electrodes electrically connected (Figure 3.8).

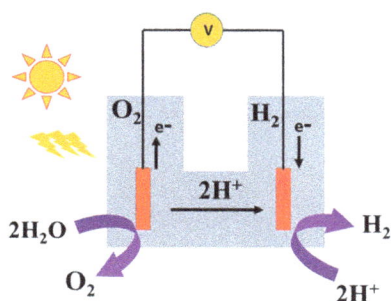

Figure 3.8: Photoelectrochemical cell for water photosplitting.

In general, PEC cells can have four different setups, as follows [59]:
i.   an n-type semiconductor photoanode directly connected with a metal counter-electrode;
ii. an n-type semiconductor and a p-type semiconductor coupled in series;
iii. a single n-type photoanode coupled with a photovoltaic material. In this case, the photovoltaic device provides external bias and is in turn connected with a metal cathode for HER;
iv. connection in a series of two photovoltaic devices.

It is worth mentioning that, in order to provide an extra bias to drive HER, photovoltaic systems can be effectively replaced by dye-sensitized solar cells (DSSCs), also known as Gratzel cells [60]. Concerning their traditional silicon-based counterpart, DSSCs exhibit the capability to work at wider angles and in cloudy/low-light conditions; long life in sunlight over time and mechanical robustness; and the ability to reduce internal operating temperature and avoid electron transfer in the conduction band of semiconductors.

3.4.2 Slurry Systems: Batch-Type Photocatalytic Reactors

Eminent review papers examine the main requirements of reliable photocatalytic reactors for water decontamination but limited literature survey specifically applies to HER [61–63].

Unlike PEC systems requiring an additional energy input, in photocatalytic reactors only light irradiation from a natural or artificial light source is needed to activate HER. In slurry-based photocatalytic reactors, the photocatalyst particles are mixed with the aqueous solution to form heterogeneous or homogeneous suspensions. After absorbing proper photon energy, a three-phase reaction (solid–liquid–gas) occurs leading to hydrogen generation.

Overall, an effective batch-type photocatalytic reactor for HER should be able to maximize incident light absorption by the reacting suspension and minimize photonic losses. Suitable stirring should be provided to the photocatalyst suspension enabling extensive contact at the solid–liquid interface. Moreover, complete sealing is required for photocatalytic reactors to avoid both parasitic reactions between oxygen and photogenerated electrons and hydrogen losses. A schematic illustration of a typical batch reactor for photocatalytic HER is reported in Figure 3.9.

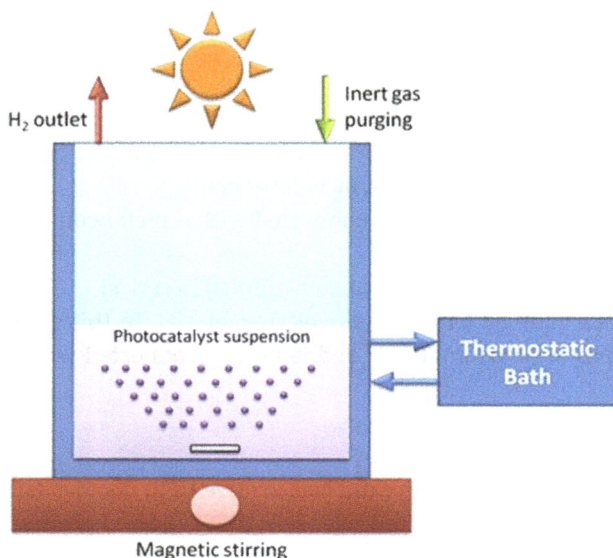

Figure 3.9: Schematic of a batch-type photocatalytic reactor for H_2 evolution from aqueous solutions.

In general, photocatalytic hydrogen generation through water splitting or organic reforming in slurry systems may be affected by the following operating aspects: solution pH; operating temperature; nature and concentration of a sacrificial agent; size, shape, and morphology of a photocatalyst particle; bandgap energy of the semiconductor photocatalyst; and the presence of cocatalysts [9, 64].

3.5 Photoefficiency of HER

A few possible phenomena should be taken into account in order to assess the photoefficiency for HER of semiconductor-based photocatalysts, such as the following [65]:

i.   incident photons possess energy lower than the bandgap value, thus not allowing an effective light capture and activation by the semiconductor material;
ii. incident photons possess energy greater than the bandgap value: in this case, excess energy can be lost as heat or light;
iii. a large percentage of photogenerated charge carriers is not available for redox reactions of interest due to the high likelihood of electron–hole recombination.

Depending on the setup employed for HER, various quantities have been employed in the literature survey to evaluate the overall process photoefficiency. In general, only a few papers report values of photoefficiency for HER, thus preventing a proper comparison among different systems.

As regards PEC systems, the STH conversion efficiency is a commonly used indicator [66]:

$$\text{STH} = \left[\frac{\eta_\text{F} \times \left| J_\text{sc} \left(\frac{\text{mA}}{\text{cm}^2} \right) \right| \times 1.23 (\text{V})}{I \left(\frac{\text{mW}}{\text{cm}^2} \right)} \right]_{\text{AM 1.5 G}}$$

where η_F is the Faradaic efficiency for HER, J_sc is the short-circuit photocurrent density, and P is the power density of incident light irradiation. All quantities mentioned above should be evaluated under standard solar irradiation conditions (AM 1.5 G), and in the absence of any sacrificial species, pH or electrical bias between working and counter electrodes. Among the highest values of STH reported in the literature review on PEC systems, a value of about 3% has been recorded by Pan et al. over a $Cu/Cu_2O/Ga_2O_3/TiO_2/NiMo$ photocathode used along with a $BiVO_4$ photoanode for unassisted solar water splitting [67].

With respect to the batch-type photocatalytic reactors, in the literature survey, only indications on H_2 generation rate have been often reported and compared to primary reference catalysts (e.g., bare TiO_2) [65]. In some studies, the authors estimated the apparent quantum efficiency (AQE) as follows:

$$AQE = \frac{2 \times r_{H_2}}{\text{moles of incident photons/time}} \times 100$$

where r_{H_2} (mol/time) is the hydrogen production rate. A remarkable value of AQE equal to 45.7% has been recorded under visible light irradiation by employing an eosin Y-sensitized $Cu/Cu_2O/Cu/TiO_2$ photocatalyst [68].

Nevertheless, to perform a comparison among different data on photoefficiency for HER available in the literature review, the effective emission power of the light sources employed should be specified, whereas only lamp model and its rated power are commonly reported.

3.6 Commercial Challenges and Future Perspectives

Most photoreactor design and scale-up aspects so far considered in the literature survey focus on water decontamination processes. In particular, solar-driven photocatalytic processes were effectively implemented on a pilot scale for the treatment of industrial and urban wastewater, water disinfection and purification, and fine chemical production.

Regarding reactor design, a major requirement of solar photocatalytic devices is to ensure optimal contact between solar photons and photocatalysts. To this aim, solar photoreactor geometry should be designed so as to maximize radiation capture and uniform distribution inside the device. Indeed, reflection and scattering phenomena are recognized to affect effective radiation absorption by the photocatalytic materials, thus lowering the overall efficiency in case of heterogeneous photocatalytic processes. In this regard, the photon path length inside the radiant field of a solar photocatalytic device should be uniform concerning the photon flux density throughout the reaction.

Moreover, in the case of heterogeneous photocatalytic processes, it is important to note that turbulent conditions should be established to prevent sedimentation and accumulation of photocatalytic materials. On the other hand, pressure losses should be avoided in the view of scale-up and optimization of practical trials [69].

In the literature survey, selected review papers accurately compare the main requirements and challenges in design and engineering of photoreactors for various applications [70, 71]. Overall, different configurations for pilot-scale photoreactors include parabolic trough collectors, compound parabolic collectors (CPCs), inclined plate collectors, double skin sheet photoreactors, and rotating disk reactors.

Although in the last decades several promising studies on photocatalytic hydrogen generation through water photosplitting and organic photoreforming were performed by researchers, they appear to be limited to bench-scale testing. A CPC along with an inner circulated reactor have been carefully designed and employed by Jing et al. [72] to produce H_2 and maximize solar photons capture. In such application, turbulent flow and uniform photocatalyst dispersion were ensured by reaching the Reynolds numbers between 10×10^3 and 50×10^3, thus preventing the deposition of photocatalytic particles. A maximum efficiency of 0.47% by employing a CdS photocatalyst in an 11.4 L aqueous solution containing Na_2SO_3 and Na_2S as sacrificial agents was achieved by Jing et al. [73].

Xing et al. [63] proposed useful standardization criteria that help to directly compare the performances of the different photocatalytic systems so far developed, in the view of their possible scale-up for real applications. For this purpose, the following design and engineering aspects should be considered: illumination, geometric shape of photoreactor, reactor construction material, amount of reactants, and vacuum conditions.

With regard to scale-up of PECs, the following photocell configurations have been so far tested in the literature review: p/n PECs, PECs driven by photovoltaic devices, hybrid PEC and photovoltaic cells, and hybrid PEC and DSSCs. Similar to slurry systems, low efficiency of H_2 production and stability issues should be overcome for future applications. Xing et al. [63] listed the following standardization aspects to be considered in evaluating PEC performances to aim large-scale implementation: illumination, light concentration, photocell buildup, and electrolyte.

3.7 Conclusions

Hydrogen production through solar-based catalytic water photosplitting and organic photoreforming are one of the most appealing technologies for an environmentally and commercially sustainable energy production.

A thorough literature survey revealed that the photoefficiency for HER of metal-based composites recorded both in PECs and in batch-type photoreactors is still far from satisfying minimum standards for commercial implementation. Additionally, the need for highly efficient new generation catalyst material to display photoreforming and photocatalytic hydrogen generation still needs to be found. However, by considering all remarkable results so far collected in photocatalytic H_2 generation since Fujishima and Honda's pioneering research in 1972. The search for practical and low-cost photocatalytic systems for water photosplitting and photoreforming active under visible light irradiation could become readily available in the next future.

Moreover, to promote the use of photocatalytic hydrogen production in large-scale trials, toxicity assessment of the photocatalytic materials should also be performed.

Hence, with a view to green targets of a circular economy, engineering and design of photocatalytic composites with remarkable photoefficiency for HER under natural sunlight irradiation is a crucial challenge for future research.

References

[1]   Voldsund M, Jordal K, Anantharaman R. Hydrogen production with CO2 capture, International Journal of Hydrogen Energy, 2016, 41, 4969–4992.
[2]   Centi G, Van Santen RA. Catalysis for renewables: from feedstock to energy production, John Wiley & Sons, 2008.
[3]   Wang Z, Roberts R, Naterer G, Gabriel K. Comparison of thermochemical, electrolytic, photoelectrolytic and photochemical solar-to-hydrogen production technologies, International Journal of Hydrogen Energy, 2012, 37, 16287–16301.
[4]   Steinfeld A. Solar thermochemical production of hydrogen——a review, Solar Energy, 2005, 78, 603–615.
[5]   Omer AM. Energy, environment and sustainable development, Renewable and Sustainable Energy Reviews, 2008, 12, 2265–2300.
[6]   Puga AV. Photocatalytic production of hydrogen from biomass-derived feedstocks, Coordination Chemistry Reviews, 2016, 315, 1–66.
[7]   Mathew S, Ganguly P, Kumaravel V, Harrison J, Hinder SJ, Bartlett J, Pillai SC. Effect of chalcogens (S, Se, and Te) on the anatase phase stability and photocatalytic antimicrobial activity of TiO2, Materials Today: Proceedings, 2020.
[8]   Ganguly P, Harb M, Cao Z, Cavallo L, Breen A, Dervin S, Dionysiou DD, Pillai SC. 2D nanomaterials for photocatalytic hydrogen production, ACS Energy Letters, 2019.
[9]   Maeda K. Photocatalytic water splitting using semiconductor particles: history and recent developments, Journal of Photochemistry and Photobiology C: Photochemistry Reviews, 2011, 12, 237–268.
[10]   Linic S, Christopher P, Ingram DB. Plasmonic-metal nanostructures for efficient conversion of solar to chemical energy, Nature Materials, 2011, 10, 911–921.
[11]   Abe T, Kaneko M. Reduction catalysis by metal complexes confined in a polymer matrix, Progress in Polymer Science, 2003, 28, 1441–1488.
[12]   Grimes C, Varghese O, Ranjan S. Light, water, hydrogen: the solar generation of hydrogen by water photoelectrolysis, Springer Science & Business Media, 2007.
[13]   Clarizia L, Spasiano D, Di Somma I, Marotta R, Andreozzi R, Dionysiou DD. Copper modified-TiO2 catalysts for hydrogen generation through photoreforming of organics, A Short Review, International Journal of Hydrogen Energy, 2014, 39, 16812–16831.
[14]   Zhang X, Chen YL, Liu R-S, Tsai DP. Plasmonic photocatalysis, Reports on Progress in Physics, 2013, 76, 046401.
[15]   Kotoulas I, Schizodimou A, Kyriacou G. Electrochemical reduction of formic acid on a copper-tin-lead cathode, The Open Electrochemistry Journal, 2013, 5.
[16]   Grätzel M. Photoelectrochemical cells, materials for sustainable energy: a collection of peer-reviewed research and review articles from Nature Publishing Group, World Scientific, 2011, 26–32.
[17]   Gomathisankar P, Hachisuka K, Katsumata H, Suzuki T, Funasaka K, Kaneco S. Enhanced photocatalytic hydrogen production from aqueous methanol solution using ZnO with

simultaneous photodeposition of Cu, International Journal of Hydrogen Energy, 2013, 38, 11840–11846.

[18] Fujishima A, Honda K. Electrochemical photolysis of water at a semiconductor electrode, nature, 1972, 238, 37.

[19] Kandiel TA, Ivanova I, Bahnemann DW. Long-term investigation of the photocatalytic hydrogen production on platinized TiO2: an isotopic study, Energy & Environmental Science, 2014, 7, 1420–1425.

[20] Al-Mazroai LS, Bowker M, Davies P, Dickinson A, Greaves J, James D, Millard L. The photocatalytic reforming of methanol, Catalysis Today, 2007, 122, 46–50.

[21] Strataki N, Lianos P. Optimization of parameters for hydrogen production by photocatalytic alcohol reforming in the presence of Pt/TiO2 nanocrystalline thin films, Journal of Advanced Oxidation Technologies, 2008, 11, 111–115.

[22] Zhang G, Ni C, Huang X, Welgamage A, Lawton LA, Robertson PK, Irvine JT. Simultaneous cellulose conversion and hydrogen production assisted by cellulose decomposition under UV-light photocatalysis, Chemical Communications, 2016, 52, 1673–1676.

[23] Fang S, Liu Y, Sun Z, Lang J, Bao C, Hu YH. Photocatalytic hydrogen production over Rh-loaded TiO2: what is the origin of hydrogen and how to achieve hydrogen production from water?, Applied Catalysis B: Environmental, 2020, 278, 119316.

[24] Zalas M, Laniecki M. Photocatalytic hydrogen generation over lanthanides-doped titania, Solar Energy Materials and Solar Cells, 2005, 89, 287–296.

[25] Jing D, Zhang Y, Guo L. Study on the synthesis of Ni doped mesoporous TiO2 and its photocatalytic activity for hydrogen evolution in aqueous methanol solution, Chemical Physics Letters, 2005, 415, 74–78.

[26] Sasikala R, Sudarsan V, Sudakar C, Naik R, Sakuntala T, Bharadwaj SR. Enhanced photocatalytic hydrogen evolution over nanometer sized Sn and Eu doped titanium oxide, International Journal of Hydrogen Energy, 2008, 33, 4966–4973.

[27] Chae J, Lee J, Jeong JH, Kang M. Hydrogen production from photo splitting of water using the Ga-incorporated TiO2s prepared by a solvothermal method and their characteristics, Bulletin of the Korean Chemical Society, 2009, 30, 302–308.

[28] El Naggar AM, Nassar IM, Gobara HM. Enhanced hydrogen production from water via a photo-catalyzed reaction using chalcogenide d-element nanoparticles induced by UV light, Nanoscale, 2013, 5, 9994–9999.

[29] Qin L, Si G, Li X, Kang S-Z. Synergetic effect of Cu–Pt bimetallic cocatalyst on $SrTiO_3$ for efficient photocatalytic hydrogen production from water, RSC Advances, 2015, 5, 102593–102598.

[30] Padmanabhan N, Ganguly P, Pillai S, John H. Morphology engineered spatial charge separation in superhydrophilic TiO2/graphene hybrids for hydrogen production, Materials Today Energy, 2020, 17, 100447.

[31] Ganguly P, Mathew S, Clarizia L, Kumar SR, Akande A, Hinder S, Breen A, Pillai CS. Theoretical and experimental investigation of visible light responsive $AgBiS_2$-TiO_2 heterojunctions for enhanced photocatalytic applications, Applied Catalysis B: Environmental, 2019.

[32] Wang L, Chen K, Gao Z, Wang Q. Synthesis of Ni-silicate superficially modified CdS and its highly improved photocatalytic hydrogen production, Applied Surface Science, 2020, 529, 147217.

[33] Yao X, Hu X, Zhang W, Gong X, Wang X, Pillai SC, Dionysiou DD, Wang D. Mie resonance in hollow nanoshells of ternary TiO2-Au-CdS and enhanced photocatalytic hydrogen evolution, Applied Catalysis B: Environmental, 2020, 119153.

[34] Etacheri V, Seery MK, Hinder SJ, Pillai SC. Highly visible light active TiO_2–xN_x heterojunction photocatalysts, Chemistry of Materials, 2010, 22, 3843–3853.

[35] Xi M, Wang P, Zhang M, Qin L, Kang S-Z, Li X. ZnO nanorods/sulfophenylporphyrin nanocomposites facilely embedded with special copper for improved photocatalytic hydrogen evolution, Applied Surface Science, 2020, 529, 147200.

[36] Wang J, Zhu H, Tang S, Li M, Zhang Y, Xing W, Xue Q, Yu L. Sandwich structure MoO2/MoS2/TiO2 photocatalyst for superb hydrogen evolution, Journal of Alloys and Compounds, 2020, 155869.

[37] Li H, Hao X, Gong H, Jin Z, Zhao T. Efficient hydrogen production at a rationally designed MoSe2@ Co3O4 pn heterojunction, Journal of Colloid and Interface Science, 2020.

[38] Gong H, Hao X, Li H, Jin Z. A novel materials manganese cadmium sulfide/cobalt nitride for efficiently photocatalytic hydrogen evolution, Journal of Colloid and Interface Science, 2020, 585, 217–228.

[39] Zhang S, Chen S, Liu D, Zhang J, Peng T. Layered WS2/WO3 Z-scheme photocatalyst constructed via an in situ sulfurization of hydrous WO3 nanoplates for efficient H2 generation, Applied Surface Science, 2020, 529, 147013.

[40] Zhou J, Zhao J, Liu R. Defect engineering of zeolite imidazole framework derived ZnS nanosheets towards enhanced visible light driven photocatalytic hydrogen production, Applied Catalysis B: Environmental, 2020, 278, 119265.

[41] Panneri S, Ganguly P, Mohan M, Nair BN, Mohamed AAP, Warrier KG, Hareesh U. Photoregenerable, bifunctional granules of carbon-doped g-C3N4 as adsorptive photocatalyst for the efficient removal of tetracycline antibiotic, ACS Sustainable Chemistry & Engineering, 2017, 5, 1610–1618.

[42] Panneri S, Ganguly P, Nair BN, Mohamed AAP, Warrier KG, Hareesh UN. Copyrolysed C3N4-Ag/ZnO ternary heterostructure systems for enhanced adsorption and photocatalytic degradation of tetracycline, European Journal of Inorganic Chemistry, 2016, 2016, 5068–5076.

[43] Hou X, Cui L, Du H, Gu L, Li Z, Yuan Y. Lowering the Schottky barrier of g-C3N4/Carbon graphite heterostructure by N-doping for increased photocatalytic hydrogen generation, Applied Catalysis B: Environmental, 2020, 278, 119253.

[44] Shen Q, Li N, Bibi R, Richard N, Liu M, Zhou J, Jing D. Incorporating nitrogen defects into novel few-layer carbon nitride nanosheets for enhanced photocatalytic H2 production, Applied Surface Science, 2020, 529, 147104.

[45] Zhang G, Xu Y, He C, Zhang P, Mi H. Oxygen-doped crystalline carbon nitride with greatly extended visible-light-responsive range for photocatalytic H2 generation, Applied Catalysis B: Environmental, 2021, 283, 119636.

[46] Wang S, Chen L, Zhao X, Zhang J, Ao Z, Liu W, Wu H, Shi L, Yin Y, Xu X. Efficient photocatalytic overall water splitting on metal-free 1D SWCNT/2D ultrathin C3N4 heterojunctions via novel non-resonant plasmonic effect, Applied Catalysis B: Environmental, 2020, 278, 119312.

[47] Zhang X, Yan J, Lee LYS. Highly promoted hydrogen production enabled by interfacial PN chemical bonds in copper phosphosulfide Z-scheme composite, Applied Catalysis B: Environmental, 2021, 283, 119624.

[48] Wu X, Zhong W, Ma H, Hong X, Fan J, Yu H. Ultra-small molybdenum sulfide nanodot-coupled graphitic carbon nitride nanosheets: trifunctional ammonium tetrathiomolybdate-assisted synthesis and high photocatalytic hydrogen evolution, Journal of Colloid and Interface Science, 2020, 586, 719–729.

[49] Samsudin MFR, Ullah H, Tahir AA, Li X, Ng YH, Sufian S. Superior photoelectrocatalytic performance of ternary structural BiVO4/GQD/g-C3N4 heterojunction, Journal of Colloid and Interface Science, 2020, 586, 785–796.

[50] Zhao N, Hu Y, Du J, Liu G, Dong B, Yang Y, Peng J, Li J, Zhai M. Ti3C2Tx MXene-derived amorphous TiO2-C nanosheet cocatalysts coupled CdS nanostructures for enhanced photocatalytic hydrogen evolution, Applied Surface Science, 2020, 530, 147247.

[51] Tasleem S, Tahir M, Zakaria ZY. Fabricating structured 2D Ti3AlC2 MAX dispersed TiO2 heterostructure with Ni2P as a cocatalyst for efficient photocatalytic H2 production, Journal of Alloys and Compounds, 2020, 155752.

[52] Elsayed MH, Jayakumar J, Abdellah M, Mansoure TH, Zheng K, Elewa AM, Chang C-L, Ting L-Y, Lin W-C, Yu H-H. Visible-light-driven hydrogen evolution using nitrogen-doped carbon quantum dot-implanted polymer dots as metal-free photocatalysts, Applied Catalysis B: Environmental, 2020, 283, 119659.

[53] Guo C, Du H, Ma Y, Qi K, Zhu E, Su Z, Huojiaaihemaiti M, Wang X. Visible-light photocatalytic activity enhancement of red phosphorus dispersed on the exfoliated kaolin for pollutant degradation and hydrogen evolution, Journal of Colloid and Interface Science, 2020, 585, 167–177.

[54] Gao Y, Xu B, Cherif M, Yu H, Zhang Q, Vidal F, Wang X, Ding F, Sun Y, Ma D. Atomic insights for Ag interstitial/substitutional doping into ZnIn2S4 nanoplates and intimate coupling with reduced graphene oxide for enhanced photocatalytic hydrogen production by water splitting, Applied Catalysis B: Environmental, 2020, 119403.

[55] Yuan Y-J, Chen D, Zhong J, Yang L-X, Wang J, Liu M-J, Tu W-G, Yu Z-T, Zou Z-G. Interface engineering of a noble-metal-free 2D–2D MoS_2/Cu-$ZnIn_2S_4$ photocatalyst for enhanced photocatalytic H_2 production, Journal of Materials Chemistry A, 2017, 5, 15771–15779.

[56] Zeng D, Xiao L, Ong WJ, Wu P, Zheng H, Chen Y, Peng DL. Hierarchical ZnIn2S4/MoSe2 nanoarchitectures for efficient noble-metal-free photocatalytic hydrogen evolution under visible light, ChemSusChem, 2017, 10, 4624–4631.

[57] Ganguly P, Mathew S, Clarizia L, Kumar S, Akande R,A, Hinder SJ, Breen A, Pillai SC. Ternary Metal Chalcogenide Heterostructure (AgInS2–TiO2) Nanocomposites for Visible Light Photocatalytic Applications, ACS Omega, 2019.

[58] Ganguly P, Muscetta M, Padmanabhan NT, Clarizia L, Akande A, Hinder S, Mathew S, John H, Breen A, Pillai SC. New insights into the efficient charge transfer of ternary chalcogenides composites of TiO2, Applied Catalysis B: Environmental, 2020, 119612.

[59] Walter MG, Warren EL, McKone JR, Boettcher SW, Mi Q, Santori EA, Lewis NS. Solar water splitting cells, Chemical Reviews, 2010, 110, 6446–6473.

[60] Grätzel M. Dye-sensitized solar cells, Journal of Photochemistry and Photobiology C: Photochemistry Reviews, 2003, 4, 145–153.

[61] Alfano O, Bahnemann D, Cassano A, Dillert R, Goslich R. Photocatalysis in water environments using artificial and solar light, Catalysis Today, 2000, 58, 199–230.

[62] Bahnemann D. Photocatalytic water treatment: solar energy applications, Solar Energy, 2004, 77, 445–459.

[63] Xing Z, Zong X, Pan J, Wang L. On the engineering part of solar hydrogen production from water splitting: photoreactor design, Chemical Engineering Science, 2013, 104, 125–146.

[64] Ahmad H, Kamarudin S, Minggu L, Kassim M. Hydrogen from photo-catalytic water splitting process: a review, Renewable and Sustainable Energy Reviews, 2015, 43, 599–610.

[65] Muscetta M, Andreozzi R, Clarizia L, Di Somma I, Marotta R. Hydrogen production through photoreforming processes over Cu2O/TiO2 composite materials: a mini-review, International Journal of Hydrogen Energy, 2020.

[66] Dotan H, Mathews N, Hisatomi T, Grätzel M, Rothschild A. On the solar to hydrogen conversion efficiency of photoelectrodes for water splitting, ACS Publications, 2014.

[67] Pan L, Kim JH, Mayer MT, Son M-K, Ummadisingu A, Lee JS, Hagfeldt A, Luo J, Grätzel M. Boosting the performance of Cu 2 O photocathodes for unassisted solar water splitting devices, Nature Catalysis, 2018, 1, 412–420.

[68] Zhen W, Jiao W, Wu Y, Jing H, Lu G. The role of a metallic copper interlayer during visible photocatalytic hydrogen generation over a $Cu/Cu_2O/Cu/TiO_2$ catalyst, Catalysis Science & Technology, 2017, 7, 5028–5037.

[69] Spasiano D, Marotta R, Malato S, Fernandez-Ibañez P, Di Somma I. Solar photocatalysis: materials, reactors, some commercial, and pre-industrialized applications. A comprehensive approach, Applied Catalysis B: Environmental, 2015, 170, 90–123.

[70] Zapata A, Oller I, Bizani E, Sánchez-Pérez J, Maldonado M, Malato S. Evaluation of operational parameters involved in solar photo-Fenton degradation of a commercial pesticide mixture, Catalysis Today, 2009, 144, 94–99.

[71] Braham RJ, Harris AT. Review of major design and scale-up considerations for solar photocatalytic reactors, Industrial & Engineering Chemistry Research, 2009, 48, 8890–8905.

[72] Jing D, Liu H, Zhang X, Zhao L, Guo L. Photocatalytic hydrogen production under direct solar light in a CPC based solar reactor: reactor design and preliminary results, Energy Conversion and Management, 2009, 50, 2919–2926.

[73] Jing D, Guo L, Zhao L, Zhang X, Liu H, Li M, Shen S, Liu G, Hu X, Zhang X. Efficient solar hydrogen production by photocatalytic water splitting: from fundamental study to pilot demonstration, International Journal of Hydrogen Energy, 2010, 35, 7087–7097.

Suyana Panneri, U. S. Hareesh

Chapter 4
Visible Light/Sunlight-Induced Photocatalytic Degradation of Antibiotics by Semiconductor Heterostructures

4.1 Introduction

Availability of clean and safe potable water is a major concern faced by countries worldwide. The widespread use of pharmaceutical products and the improper disposal of unused/outdated drugs in open environment have induced lethal affects not only to humans but also to the local ecosystem [1]. These effects include endocrine disruption and reproductive anomalies, and in most cases the metabolite after drug consumption is more dangerous than parent drug entity [2, 3]. Additionally, the emergence of multidrug-resistant microbes due to the uncontrolled and unethical release of pharmaceutical contaminants into aquatic environment is acknowledged as a fundamental threat to global health and safety [4]. Even with the most optimistic estimations, close to a million people die each year due to drug-resistant infections [5]. Thus, the resistance of microorganism to an antimicrobial drug that was originally effective for its disinfection has led to this threat of antimicrobial resistance. In order to mitigate the harmful effects of these emerging pharmaceutical pollutants, many projects and technologies on water reclamation, recycle, and reuse are vigorously pursued in the recent years [6, 7].

The commonly employed methods for the removal of pharmaceutical wastes range from the conventional techniques of physical adsorption and biological treatments to the advanced methods of membrane adsorption and advanced oxidation processes (AOP) [8]. Most of the conventional methods are not able to provide a comprehensive solution for the pharmaceutical waste disposal due to its high chemical stability and nonbiodegradable nature [9]. AOP is a viable technique that utilizes the generation of reactive oxygen species (ROS) in aqueous medium for the degradation and complete mineralization of the pharmaceutical products [10]. Depending on the mode of production of ROS, AOPs can be classified into chemical, photo/electrochemical, and sonochemical [11]. Out of the various AOPs, heterogeneous photocatalysis employing visible light/sunlight-active semiconductor photocatalysts

Suyana Panneri, Materials Science and Technology Division, National Institute for Interdisciplinary Science and Technology (CSIR-NIIST), Thiruvananthapuram, India, e-mail: suyansam@gmail.com
U. S. Hareesh, Materials Science and Technology Division, National Institute for Interdisciplinary Science and Technology (CSIR-NIIST), Thiruvananthapuram, India, Academy of Scientific and Innovative Research (AcSIR), New Delhi, India, e-mail: hareesh@niist.res.in

https://doi.org/10.1515/9783110668483-004

is a greener, environmentally benign approach to address many of the issues related to sustainability and survival [12, 13].

This chapter provides an insight about photocatalysis, its mechanism, and applications explored in the area of pollutant removal. It is also attempted to analyze the existing shortcomings of photocatalysts and how a viable solution can be obtained through the creation of semiconductor heterostructures. An overview about different types of photocatalytic heterostructures and its advantages is briefly explained.

4.2 Photocatalysis and Mechanism

The acceleration of reaction kinetics upon the absorption of a suitable frequency of light energy by semiconductor materials is termed photocatalysis [14]. If the photocatalytic reaction occurs in more than one medium, it is termed as heterogeneous photocatalytic reactions [14, 15]. When a semiconductor photocatalyst is irradiated with a frequency of light equal to or greater than its bandgap energy, the valence band (VB) electron (e^-) is excited to the conduction band (CB) leaving a hole (h^+) in the VB (Figure 4.1). The generation of electron–hole pair (exciton) constitutes a photoredox couple that will lead to photocatalytic reactions which in turn produce ROS [16]. For organic pollutant (dyes, antibiotics, pesticides, etc.) degradation reactions, the photoexcited electrons in the CB of the semiconductor photocatalyst react with the oxygen to form superoxide anion radicals ($O_2^{-\cdot}$). These superoxide anion radicals are highly reactive and can oxidize organic molecules. Superoxide anion radicals can combine with H^+ to produce hydrogen peroxide (H_2O_2), which is again a strong oxidizing agent. The photoinduced holes can directly oxidize the pollutants but in most cases they react with the water molecules leading to hydroxyl radicals ($OH^\cdot$) that are highly reactive. ROS is highly oxidative in nature and hence decomposes organic molecules (antibiotic, dyes, pesticides, microbes) into smaller nonharmful molecules such as water and CO_2 [17, 18]. The corresponding reactions are shown in the following equations:

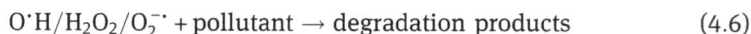

$$Photocatalyst + h\nu \rightarrow e^- + h^+ \tag{4.1}$$

$$h^+ + H_2O/OH \rightarrow H^+ + O^\cdot H \tag{4.2}$$

$$O_2 + e^- \rightarrow O_2^{-\cdot} \tag{4.3}$$

$$O_2^{-\cdot} + H^+ \rightarrow HO_2^\cdot \tag{4.4}$$

$$2HO_2^\cdot + 2H^+ \rightarrow H_2O_2 + 2O^\cdot H \tag{4.5}$$

$$O^\cdot H/H_2O_2/O_2^{-\cdot} + pollutant \rightarrow degradation\ products \tag{4.6}$$

The main four tandem steps (Figure 4.1) in a photocatalytic reaction are (1) light absorption by the semiconductor photocatalyst; (2) photoexcitation and separation of

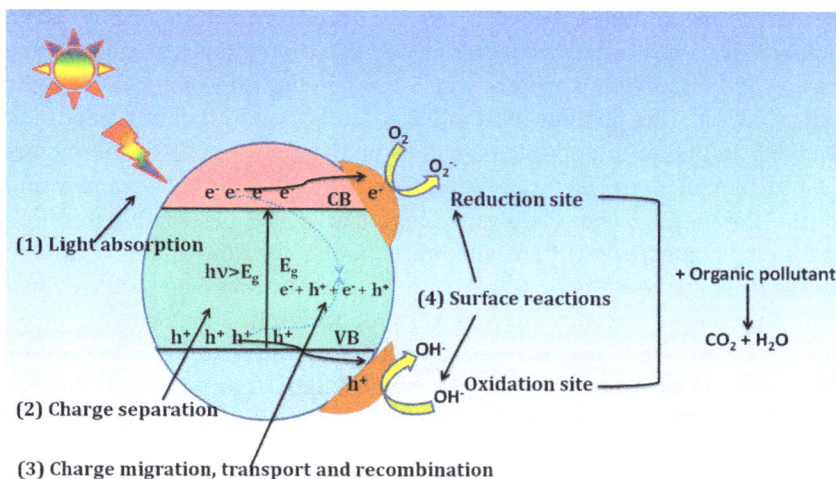

Figure 4.1: Mechanism and different steps involved in photocatalysis [18, 19].

charge carriers, (3) charge carrier movement, transport, and recombination; and (4) effective utilization of the excitons for oxidation and reduction reactions [18]. All these steps are vital in determining the overall efficiency and quantum yield of photocatalytic reactions.

For practical applications, visible light/sunlight photocatalysts are found to be suitable over conventional UV-active candidates since 45% of solar light falls in the visible light (400–750 nm) region and only 5% falls in the UV region [20]. The high electron–hole recombination, low specific surface area, and poor light absorption of many visibly active photocatalysts tamper the activity and reduce the overall quantum efficiency. Many attempts are already practiced to overcome these demerits [21, 22]. Engineering semiconductor heterostructures/heterojunction with compatible materials of appropriate band positions is a promising approach to enhance photocatalytic activity by the effective spatial separation of electron–hole pairs [23].

4.3 Photocatalytic Semiconductor Heterostructures

Interfacial charge transfer and recombinations are the two key factors that determine the quantum efficiency of a photocatalytic reaction. The quantum as well as photocatalytic efficiencies of single-component photocatalysts are normally inadequate to be practically viable. Photocatalytic quantum efficacy can be improved by the formation of heterostructure/heterojunction of semiconductors. By definition, heterojunction is the interface between two semiconductors with uneven band structure that results in band alignment. The creation of heterojunction leads to a band bending at the junction interface, which leads to built-in electric field in the space charge region. This phenomenon induces spatial separation and migration of the electrons and holes in opposite directions minimizing the charge recombination [25, 26]. Heterojunctions can be classified according to their band alignment and charge carrier flow mechanism. Based on charge carrier flow, heterostructures are classified into heterojunctions with (a) unidirectional charge flow (Figure 4.2a) and (b) bidirectional charge flow (Figure 4.2b) [24]. In heterostructures with unidirectional charge flow only one photocatalyst (PS I) will produce exciton pair under a given light radiation, whereas in bidirectional charge flow both photocatalysts (PS I and PS II) are able to produce exciton pairs under the incident light. The widely used classification of heterostructure is in line with the band alignment of the semiconductors and is classified into six types as type I, type II, type III, Schottky, p–n, and Z-scheme heterostructures [23, 25–27].

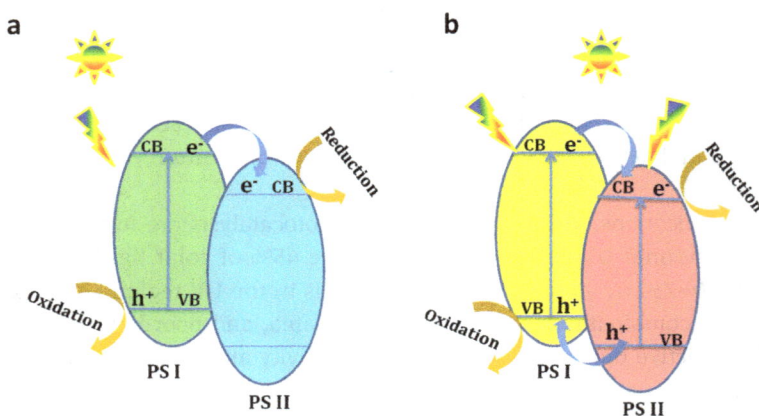

Figure 4.2: Schematic of heterojunctions with (a) unidirectional charge flow and (b) heterojunctions with bidirectional charge flow [24].

4.3.1 Type I Heterostructure

A straddling kind (Figure 4.3a) of band positions by the confinement of VB and CB of the smaller bandgap semiconductor within the large bandgap material is realized in type I heterostructures [28]. Upon light irradiation, the charge carriers are accumulated in one semiconductor and therefore no overall photocatalytic enhancement is observed in type I heterostructure. The typical examples include $V_2O_5/BiVO_4$ [29] and Bi_2S_3/CdS [30].

4.3.2 Type II Heterostructure

The creation of a built-in electric field by the chemical potential difference, when the band positions of semiconductors are aligned in a staggered fashion, results in type II heterojunction (Figure 4.3b) [27, 31]. Charge carriers are migrated to opposite directions, induce spatial separation, and reduce exciton recombination. Moreover, oxidation and reduction half reactions take place in the two semiconductors. In general, type II photocatalytic heterostructures exhibit good wide light-absorption, fast mass transfer, and reduced exciton recombination. The important examples include C_3N_4–ZnS [32], CdS–TiO_2 [33], and g-C_3N_4–$BiPO_4$ [34].

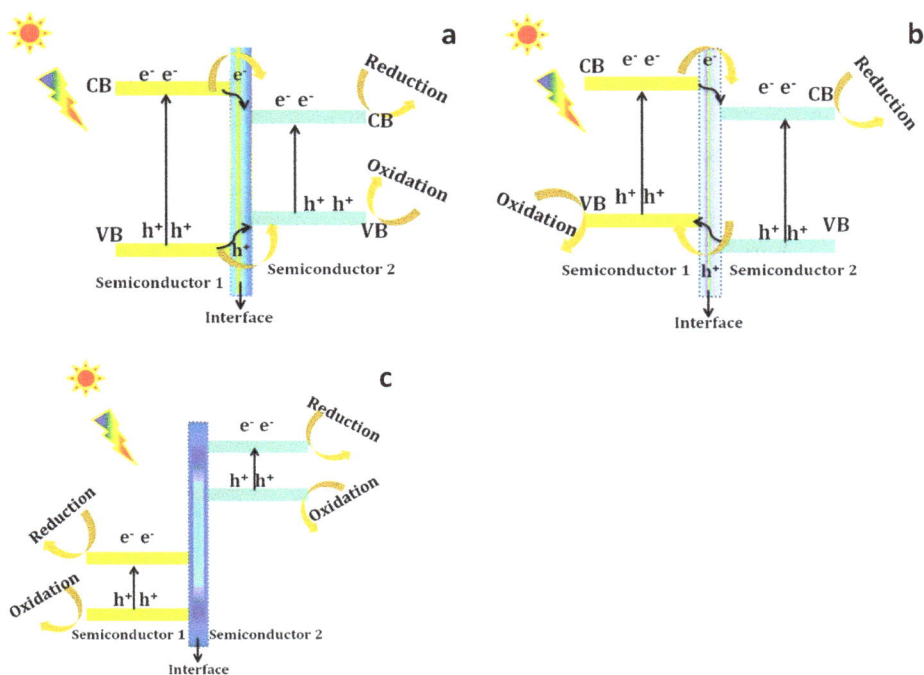

Figure 4.3: Band alignments of (a) type I, and (b) type I and type III heterostructures [27, 31].

4.3.3 Type III Heterostructure

Type III heterostructure (Figure 4.3c) is a broken type since no crossing of the band positions are observed; hence, it is very rare in practical application [31].

4.3.4 Schottky Heterostructure

When a semiconductor is in contact with a metal or noble metal, (Figure 4.4a), Schottky heterostructure is formed due to the difference in work functions of the metal (W_{nm}) and electron affinity of the semiconductor (χ_S) (i.e., energy difference between the CB minimum of semiconductor and the vacuum energy level). Work function is the energy difference between the Fermi level (EF) of the metal and the vacuum energy level. Upon thermodynamic equilibration, an internal electric field is generated by the redistribution of charges that facilitate charge separation by creating a Schottky barrier (Φ_B) [35]. Normally electrons can transfer from a material with a lower work function to a region of higher work function and form Schottky junction [26]. The noble metal behaves differently with different types of semiconductor. For example, for an n-type semiconductor, the W_{nm} is greater than W_s, when the noble metal comes in contact with the semiconductor, the free electrons in the semiconductor transfers to the noble metal until equilibrium Fermi-level EF is reached, which results an upward shifting of the band edges, leading to Schottky barrier formation [24, 35]. On the other hand, for a p-type semiconductor W_{nm} is lesser than W_s, the free electrons from noble metal will shift to the semiconductor leading to a downward shift of band edges. Thus, the Schottky barrier makes the noble metal an "electron sink" and limits the electron back flow to the semiconductor [35]. The noble metal before and after contact with an n- and p-type semiconductors is shown in Figure 4.4b–e. The

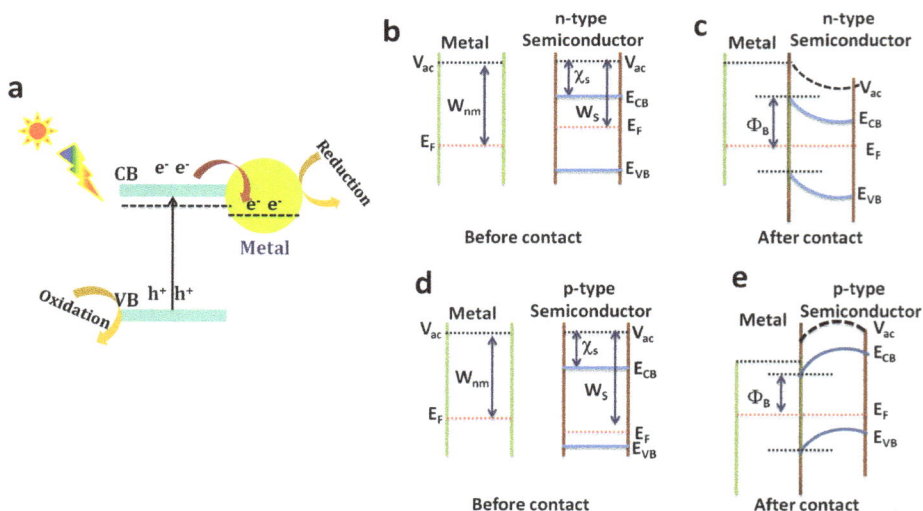

Figure 4.4: (a) Band alignment of Schottky-type heterostructure [39], formation of the Schottky barrier between noble metal and n-type semiconductors; (b) before contact; and (c) after contact; formation of the Schottky barrier between p-type semiconductors and noble metal (d) before contact and (e) after contact [24].

presence of noble metal on the semiconductor photocatalyst leads to surface plasmon resonance (SPR) [36]. In SPR, when the noble metal is irradiated with light, a collective oscillation of the free electrons in the noble metal occurs and maximum oscillation amplitude results when the frequency of both incident light and the oscillating free electrons become alike [35]. Hence, SPR is responsible for strong visible light absorption and creation of more powerful electrons that can enhance the photocatalytic activity. Examples include heterostructures of Au/TiO_2 [36], $Ag–Ag_3PO_4$ [37], and Pt/TiO_2 [38].

4.3.5 p–n Heterojunction

The interface formed between p-type semiconductor with an n-type semiconductor is termed p–n heterojunction (Figure 4.5). The Fermi level of n-type semiconductor is higher than p-type semiconductor due to which the electrons will flow from n-type semiconductor to p-type semiconductor and holes in the opposite direction. This diffusion occurs until Fermi-level equilibrium is reached, which in turn generates an internal electric field comprising a positive n-side and a negative p-side. Band bending occurs at the junction with an upward shift in n-type and downward shift in p-type bands with better separation of excitons for quiet long time [27, 40]. The heterostructures formed in MoS_2/S-doped $g-C_3N_4$ [41], $Co_3O_4–C_3N_4$ [42], NiS/CdS [43] are some significant examples.

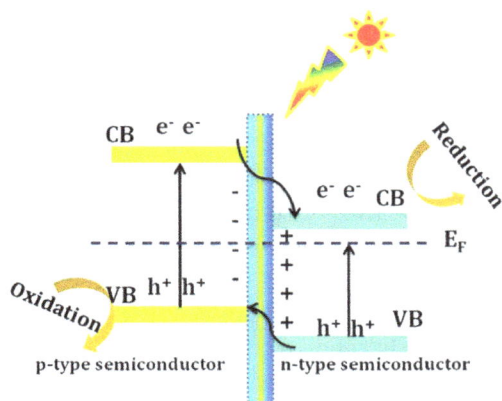

Figure 4.5: Electronic band structure of p–n heterostructure [27].

4.3.6 Z-Scheme Heterostructures

Conventional Z-scheme photocatalysis involves a two-step photoexcitation as in natural photosynthesis and is introduced by Bard et al. [44]. It comprises two photocatalysts and an electron acceptor–donor (A/D) pair. During photocatalysis in Z-scheme heterostructures, strongly reductive electrons are on one photocatalyst, where as the strongly oxidative holes are on the other photocatalyst. A spatial separation of electron–hole pairs is thus attained thereby improving the photocatalytic activity. (Figure 4.6a) [27, 31]. These conventional Z-scheme heterostructures can be created only in the liquid medium, limiting their applicability as redox mediators diminish the light absorption of semiconductor photocatalyst by strongly absorbing the visible light. Ru-loaded $SrTiO_3$:Rh with Fe^{3+}/Fe^{2+} electron mediator [45], Pt/ZrO_2/TaON, Pt/WO_3 with IO_3^-/I^- pairs [46] are some typical examples of conventional Z-scheme photocatalytic heterostructures. In order to overcome the difficulty associated with conventional Z-scheme, Tada et al. proposed an all-solid-state Z-scheme photocatalyst, which use a solid electron mediator (Figure 4.6b) [47]. As in conventional Z-scheme heterostructures, the photocatalytic activity of all-solid-state Z-scheme improves by the effective spatial separation of electron–hole pairs, in addition they can be used in solution, gas, and solid media, thereby broaden their practical applicability. CdS/Au/ZnO [48], AgBr/Ag/Bi_2WO_6 [49] are some established all-solid-state Z-scheme heterostructures. In all these cases, the noble metal particles (Au or Ag) act as the solid-state electron mediator (SSEM). The use of noble metal nanoparticles as electron mediator limits the large-scale applications of all-solid-state Z-scheme pathway. Recently, a Z-scheme heterostructure without electron mediator called direct Z-scheme system has emerged (Figure 4.6c) [50]. In a typical direct Z-scheme, the photogenerated electrons with superior reduction capacity in the CB of semiconductor 1 and the holes with superior oxidizing capacity in the VB of semiconductor 2 is preserved. At the same time, the photogenerated electrons in the CB of semiconductor 2 and holes in the VB of semiconductor 1 having poor redox capacity tends to recombine. In this manner, direct Z-scheme photocatalyst shows outstanding redox capability in photocatalytic reaction by virtue of the spatially separated oxidation and reduction sites [51–55]. Thus, direct Z-scheme heterostructures received much attention due to its effective spatial separation of excitons, preservation strong oxidative and reductive sites, viable charge transfer characteristics, low fabrication cost, optimization of redox potential for specific photocatalytic reaction, the ability to create wide spectrum photocatalysts with extended solar absorption, etc. Typical examples include heterostructures of ZnO/CdS [56], $BiVO_4$–Ru/$SrTiO_3$:Rh [57], SiC–CdS, WO_3–CdS [58], and rutile–anatase TiO_2 [59].

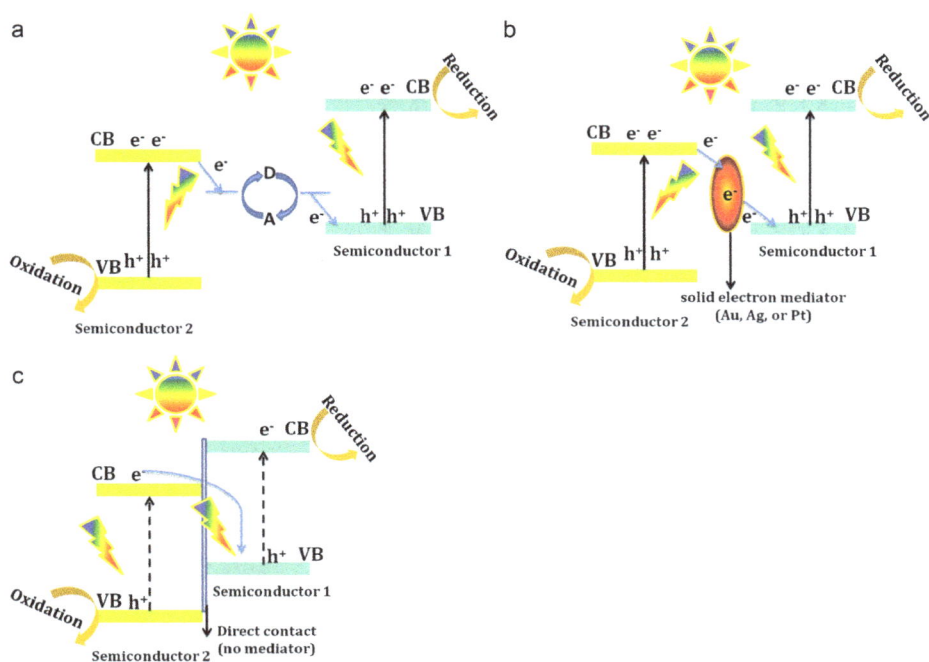

Figure 4.6: Band arrangements of (a) conventional Z-scheme, (b) all-solid-state Z-scheme, and (c) direct Z-scheme [31].

4.4 Photocatalytic Degradation and Mineralization Pathways of Emerging Antibiotic Molecules Under Visible/Sunlight Irradiation

In recent times, the visible/sunlight-active photocatalytic degradation of antibiotic gathered attention due to the adverse effect of pharmaceutical products on environment. Antibiotic molecules are class of medicines that are used for medical application in humans and animals. They are also supplemented in animal food to tackle disease spread as well as to enhance growth. A variety of classifications of antibiotics based on chemical structure, origin, range of activity (broad spectrum or narrow spectrum), mode of action, effects of their activity, and route of administration are available [8, 60, 61]. Based on the physicochemical properties, a variety of antibiotic classes like penicillins, imidazoles, quinolones, sulfonamides, tetracyclines (TC), and trimethoprim are in practice. Main features and examples of some important antibiotics are tabulated in Table 4.1 [8, 60].

Table 4.1: Classification, action, and structural properties of antibiotics.

S. no.	Class	Action	Structural features	Basic chemical structure	Range of activity	Subclass / examples
1	Aminoglycosides	Inhibit protein synthesis of microbes	Consists of two or more amino sugars joined by a glycoside linkage to a hexose nucleus of the drug		Gram-negative and gram-positive bacteria	Streptomycin, neomycin, kanamycin
2	Beta-lactams	Inhibit cell wall synthesis	Contain a beta-lactam ring		Broad spectrum, against gram-positive bacteria	Penicillins (amoxicillin, ampicillin), cephalosporins (cephalexin)
3	Chloramphenicol	Inhibits protein synthesis, inhibiting growth and reproduction of bacteria	It contains a nitrobenzene moiety connected to a propanol group as well as an amino group binding a derivative of dichloroacetic acid		Broad spectrum	

| 4 | Glycopeptides | They are composed of carbohydrate moieties (glycans) covalently attached to the side chains of an amino acid | 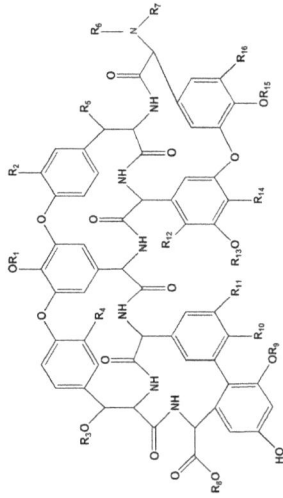 | Narrow spectrum | Vancomycin, teicoplanin, corbomycin |
| 5 | Macrolides | Preventing growth and reproduction by inhibiting protein synthesis of bacteria | Highly substituted monocyclic lactone with one or more saccharides glycosidically attached to hydroxyl groups. The lactone rings are usually 12, 14, or 16 membered | 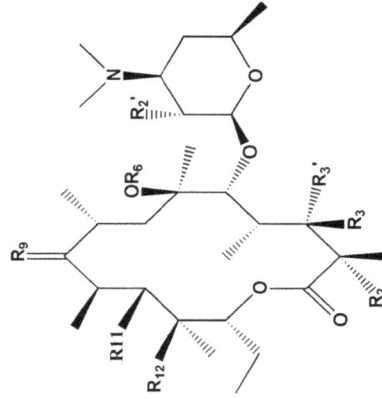 | Gram-positive bacteria, bacteriostatic manner | Erythromycin, clarithromycin, azithromycin |

(continued)

Table 4.1 (continued)

S. no.	Class	Action	Structural features	Basic chemical structure	Range of activity	Subclass / examples
6	Quinolones	Interfere with the replication and transcription of DNA in bacterial cells	Their structure contains two fused rings with a carboxylic acid and a ketone group. If R4=F, then it is a fluoroquinolone compound		Broad spectrum	Ciprofloxacin, norfloxacin, levofloxacin
7	Sulfonamides	Inhibit B vitamin folate synthesis and inhibit growth and reproduction of the bacteria	Sulfonyl group connected to amine group		Broad-spectrum, gram-positive and gram-negative bacteria	Prontosil, sulfanilamide
8	Tetracyclines	Inhibit protein synthesis, inhibiting growth and reproduction of bacteria	These antibiotics contain an octahydronaphthacene ring skeleton, consisting of four fused rings		Broad spectrum against both gram-positive and gram-negative bacteria	Tetracycline, doxycycline, lymecycline, oxytetracycline

The presence of pharmaceutical and personal care products in surface water and groundwater was first identified in the 1960s in the USA and Europe [24]. The first case of antibiotic contamination (macrolides, TC, and sulfonamides) in aquatic sources was reported in England in 1982 [8]. Since then, a vast number of cases of water contamination by antibiotic molecules are noticed. Most of the antibiotic molecules are polar, nonvolatile, water soluble, and have high photo/thermal stability [10, 62]. This has led to their persistence in the environment as a parent molecule or as a stable, more harmful metabolite product. Photocatalysis using visible light/ sunlight -active photocatalysts through the generation of ROS is a promising means to deal with nonbiodegradable or recalcitrant compounds like antibiotics in water bodies [9, 62]. Many reports on photocatalytic degradation of pharmaceutical products are already available. This chapter gives an insight on the recent advances in the degradation and pathways of some of the harmful antibiotic molecules using visible/sunlight-active photocatalytic heterostructure.

4.4.1 p–n Heterostructure

p–n heterostructures have gained much attention recently due to the formation of built-in electric field at the junction interface by the large work function difference between n-type and p-type photocatalysts. Hence, better charge carrier separation and thereby increased photocatalytic efficiencies are displayed by such heterostructures. There are a plenty of reports on p–n heterostructures for antibiotic degradation. P. Suyana et al. reported a one-pot synthetic strategy for the creation of p–n heterostructure comprised of ultrafine p-type Co_3O_4 and n-type C_3N_4 for sunlight-induced degradation of mixture of pollutants: tetracycline hydrochloride (TCH) and methylene blue [42]. The extended visible light absorption and high surface area values along with the intimate interfaces formed in connection with band bending at p–n junction enabled excellent photocatalytic activity in C_3N_4–Co_3O_4 system. A mechanism based on band bending in p–n junction as shown in Figure 4.7 was proposed. Likewise, CoO/g-C_3N_4 p–n heterojunction for the degradation of TC was reported via a solvothermal strategy [63]. The introduction of C_3N_4 reduces the aggregation of CoO nanoparticles, which are beneficial in p–n junction formation and photocatalytic TC degradation. They have also conducted TC degradation in different wavelengths of light and concluded that the TC degradation is dependent on the optical density of the photocatalyst. Bismuth (Bi)-based photocatalyst forms another important class of photocatalysts due to their unique structural and electronic properties [64]. A number of p–n heterostructure containing Bi-based photocatalysts are available for antibiotic degradation. Wen et al. reported the fabrication of n–p heterostructure of $SnO_2/BiOI$ through a chemical bath method for the degradation of oxytetracycline hydrochloride (OTCH) [65]. The n–p heterojunction formation facilitated improved charge transfer pathways, which are confirmed by the optical and electrical characteristics and eventually ended up in 94% degradation of OTCH within 90 min of visible light irradiation. The effect of nitrate ions and ionic strength on the photocatalytic performance was also illustrated. An in situ growth method for the synthesis of p–n junction BiOCl decorated $NaNbO_3$ nanocubes for the degradation of ofloxacin (OFX) was reported [66]. A 2D in-plane CuS/Bi_2WO_6 p–n photo-Fenton heterostructures for the visible-light-driven degradation of TCH was reported by Guo et al. through a hydrothermal method [67]. The formation of p–n junction decisively helped to have enhanced degradation profile. The effects of pH, H_2O_2 concentration, and catalyst dosage were systematically studied. Recently, $CuBi_2O_4/Bi_2MoO_6$ p–n heterojunction with nanosheets on microrod morphology was synthesized and used for the degradation of wide-spectrum antibiotics like TC and quinolones (Figure 4.8) [68]. The optimum performance was observed for a composition having Bi_2MoO_6 sheets with 10% $CuBi_2O_4$ microrods and the degradation efficiencies for TC, OTC, chlortetracycline

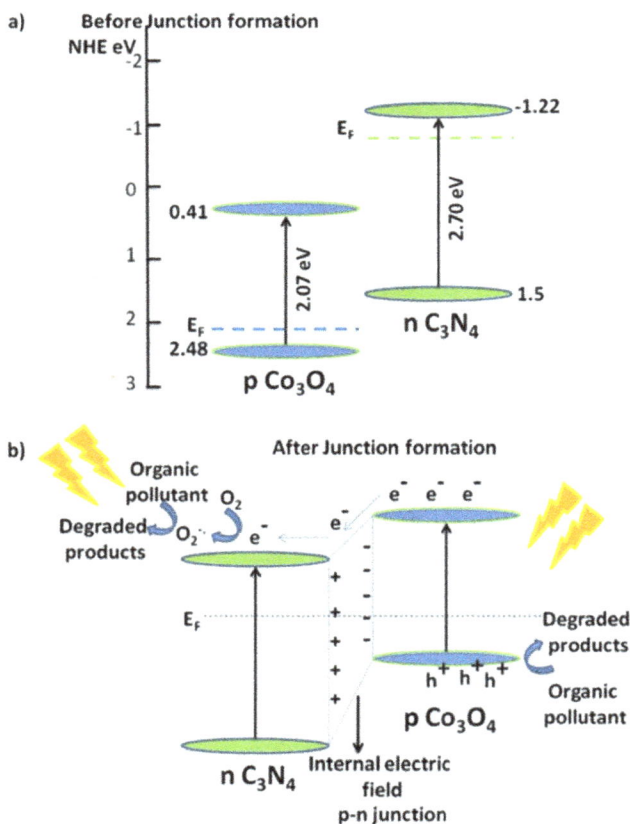

Figure 4.7: Band alignment of p-type Co_3O_4 and n-type C_3N_4 (a) before junction formation and (b) after junction formation [42].
Reprinted with permission of P. Suyana et al. [42].

and ciprofloxacin (CFC) were 72.8%, 74.0%, 74.4%, and 36.7% within 60 min visible light irradiation. Liquid chromatography–mass spectrometry technique was employed to identify the reaction pathway and by-products of TC degradation. Two possible degradation pathways as shown in Figure 4.9 were postulated. The important findings of p–n heterostructures for the antibiotic degradation are summarized in Table 4.2.

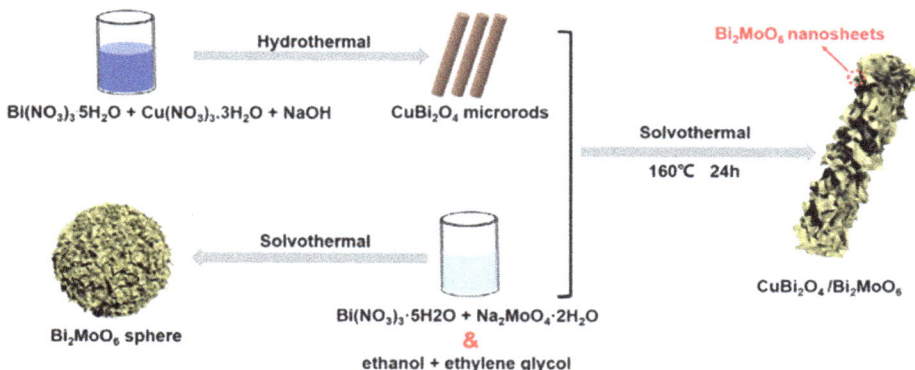

Figure 4.8: Solvothermal route to prepare $CuBi_2O_4/Bi_2MoO_6$ heterojunction photocatalyst [68]. Reprinted with permission of Weilong Shi et al. [68].

4.4.2 Type II

Type II heterostructures are formed by coupling a narrow bandgap semiconductor with wide bandgap semiconductor. It allows optimized light absorption, charge carrier kinetics, and redox capacities for photocatalytic reaction. Although a number of reports on type II heterostructures for different photocatalytic reactions are reported, few are only available on antibiotic degradation. In a remarkable work, type-II $Bi_2W_2O_9/g$-C_3N_4 heterojunction for the degradation of TC under simulated sunlight is reported by Obregon et al. From the band edge potential calculations, they have proposed a type II heterojunction due to which spatial separation of charge carriers was realized leading to remarkable degradation of TC (95% in 90 min irradiation) [69]. Yu et al. reported the fabrication of type II heterojunctions comprised of layered $Bi_2O_2[BO_2(OH)]$ and BiOI via an in situ precipitation for the degradation of TCH [5].

4.4.3 Schottky/SPR Heterojunctions

A variety of visible light/sunlight-active heterostructures with noble metals like Au, Ag, Pt, Pd, and Cu are reported for the antibiotic degradation, making use of the advantages of Schottky barrier and SPR phenomena in them. Xue et al. reported a Au/Pt/g-C_3N_4 photocatalyst via a calcination–photodeposition technique for the TCH

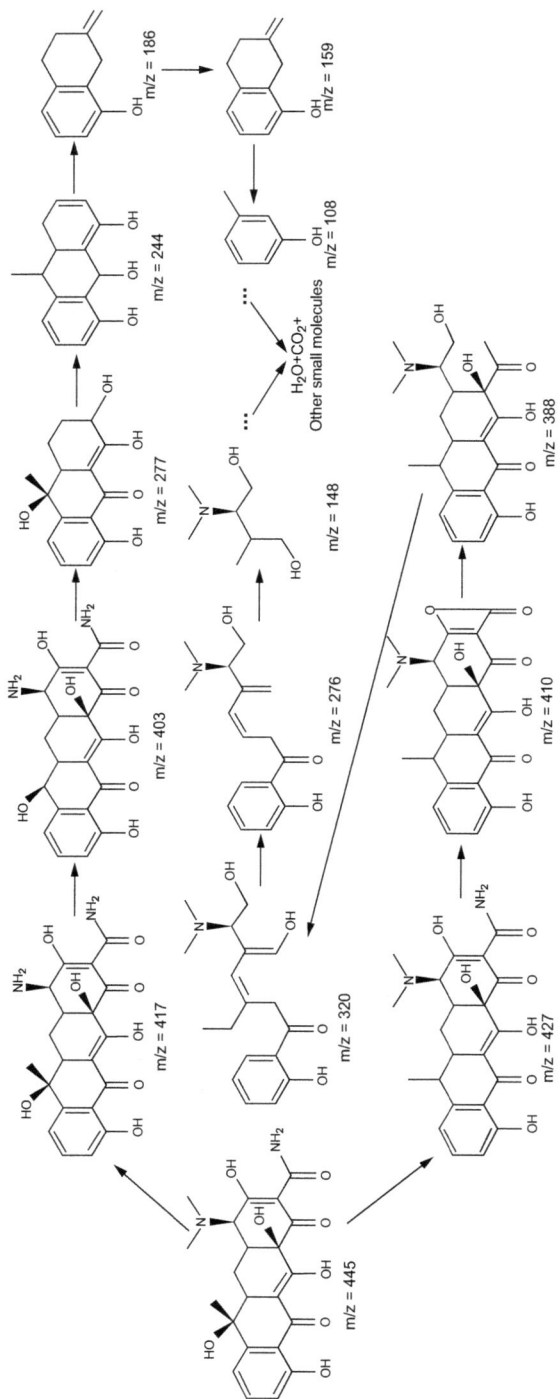

Figure 4.9: Photocatalytic degradation routes of TC with the help of $CuBi_2O_4/Bi_2MoO_6$ p–n heterojunction [68]. Reprinted with permission of Weilong Shi et al. [68].

Table 4.2: Visible light–active photocatalytic p–n heterojunctions, reaction conditions, and efficiency for antibiotic degradation.

Photocatalytic system	Antibiotic	Loading		Degradation efficiency/time in minutes	Reference
		Antibiotic	Photocatalyst		
Co_3O_4–C_3N_4	TCH	10^{-4} M	0.5 g/L	78%/180	[42]
CoO–C3N4	TC	10 mg/L	0.05 g	90%/60	[63]
SnO_2/BiOI	OTCH	10 mg/L	0.05 g	94%/90	[65]
BiOCl–NaNbO3	OFX	5 mg/L	0.25 g	90%/60	[66]
CuS/Bi_2WO_6	TCH	40 mg/l	0.02 g	73%/50	[67]
$CuBi_2O_4$/Bi_2MoO_6	TC			72.8%/60	
	OTC	20 mg/L	0.03 g	74%/60	[68]
	CFX			74.4%/60 36.7%/60	

TCH, tetracycline hydrochloride; TC, tetracycline; OTCH, oxytetracycline hydrochloride; OFX, ofloxacin; OTC, oxytetracycline; CFX, ciprofloxacin.

degradation under visible light irradiation [70]. The Au/Pt-incorporated C_3N_4 exhibited outstanding photocatalytic reaction rate owing to the SPR effects of Au and electron sink function of Pt nanoparticles that enhanced the optical absorption property and photogenerated exciton separation of g-C_3N_4. In a similar approach, Ag-loaded fiber-like carbon nitride Schottky junctions for TC degradation and hydrogen evolution were reported by He et al. [71]. A series of Ag-loaded C_3N_4 was prepared through a combination of solvothermal methods followed by sodium borohydride reduction. The Schottky junction formed between Ag (1.2%) and C_3N_4 reduced exciton recombinations and together with the SPR of Ag-aided enhanced TC degradation efficiency as well as hydrogen evolution rate. More recently, a Schottky junction between Ag_3PO_4/ Ti_3C_2 MXene interface materials was reported [72]. Generally, Schottky junctions are formed at the semiconductor–metal interface but Ti_3C_2 MXene is able to form a Schottky junction with Ag_3PO_4 due to its excellent metallic conductivity. The prepared Schottky junction outperformed the individual components in photocatalytic degradations of 2,4-dinitrophenol and TC, photoreduction of Cr^{6+} owing to the "electron sink" characteristics of Ti_3C_2 due to the built-in electric field. Very recently, Xu et al reported a Ag/porous C_3N_4 plasmonic photocatalyst via thermal exfoliation strategy followed by photoreduction route for the visible light–induced degradation of TC [73]. They have detailed the effect of light energy and hindering effect of inorganic anions on photocatalytic TC degradation. Interestingly, a 92% TC removal efficiency was observed when full-spectrum light ($\lambda > 350$ nm) was used instead of visible light ($\lambda > 420$ nm) which showed only 83% degradation. The hindering effect of inorganic anions on TC degradation is shown as in the order $H_2PO_4^- > CO_3^- > Cl^-$. An interesting plasmonic heterostructure with 0D Bi nanodots/2D Bi_3NbO_7 nanosheets was reported for the degradation of Ciprofloxacin (CFX) [74]. The uniform distribution of semimetallic plasmonic Bi nanodots with sizes of 2–5 nm over Bi_3NbO_7 nanosheets through strong covalent interaction led to enhanced visible light absorption. The viable charge transfer kinetics thus fostered the molecular oxygen activation into superoxide radicals ($\bullet O_2^-$) and singlet oxygen ($_1O^2$). These activated molecular oxygen species were responsible for CFX degradation and using Liquid chromatography–mass spectrometry (LCMS) three possible degradation pathways are postulated as shown in Figure 4.10. Some recent examples of visible/sunlight-active Schottky/SPR heterojunction for antibiotic degradation is tabulated in Table 4.3.

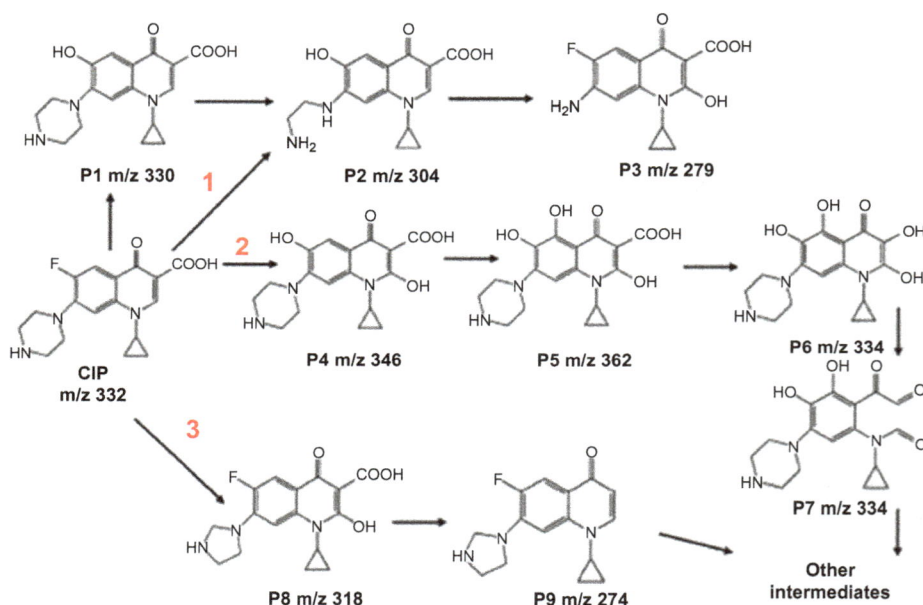

Figure 4.10: Photodegradation pathways of CFX using 0D Bi nanodots/2D Bi_3NbO_7 nanosheets [74]. Reprinted with permission of Kai Wang et al. [74].

4.4.4 Z-Scheme

Owing to the advantages of Z-scheme heterostructures as mentioned above, a wide variety of Z-scheme junctions are accessible for the photocatalytic degradation of antibiotics. The photogenerated charge carriers with superior redox capacity are preserved for prolonged time in Z-scheme heterostructures. The Z-scheme photocatalytic approach for antibiotic removal can be mainly classified according to the presence

Table 4.3: Visible light–active photocatalytic Schottky/SPR heterojunctions, reaction conditions, and efficiency for antibiotic degradation.

Photocatalytic system	Antibiotic	Loading		Degradation efficiency/time in minutes	Reference
		Antibiotic	Photocatalyst		
1D Ag–AgBr/AlOOH	TCH	10 mg/L	1 g	79%/120	[75]
Ag/AgCl/activated carbon	TC	20 mg/L	0.2 g	97.3%/60	[76]
Ag/fiber-like carbon nitride	TC	10 mg/L	0.025 g	72.9%/60	[71]
C_3N_4–Ag/ZnO	TC	20 mg/L	0.04 g	80%/60	[77]
Ag_3PO_4/Ti_3C_2 MXene	TCH TPL CPL	20 mg/L	0.02 g		[72]
Ag/porous graphite carbon nitride	TC	20 mg/L	0.05 g	83%/120	[73]
Ag/Bi_3TaO_7	TC	10 mg/L	0.5 g/L	85.42%/120	[78]
Ag/$AgIn_5S_8$	TCH	10 mg/L	0.03 g	95.3%/120	[79]
Au/Pt/g-C3N4	TCH	20 mg/L	1 g/L	93%/180	[70]
Pt/WO_3 nanosheets	TC	20 mg/L	1 g/L	72.82%/60	[80]
0D Bi nanodots/2D Bi_3NbO_7 nanosheets	CFX	10 mg/L	0.05 g	86%/120	[74]

TCH, tetracycline hydrochloride; TC, tetracycline; CFX, ciprofloxacin; TPL, Thiamphenicol; CPL, Chloramphenicol.

or absence of electron mediators as (a) SSEM Z-scheme and (b) direct Z-scheme. SSEMs such as Ag, Au, and Pt act as electron mediators in SSEM Z-scheme. Li et al. reported an SSEM Z-scheme photocatalyst consisting of BiOCl and CdS with Au as SSEM for the degradation of rhodamine B, phenol, and sulfadiazine antibiotic [81]. The facet engineered deposition of Au nanoparticles on BiOCl and the in situ deposition of CdS on Au via strong S–Au contact led to SSEM Z-scheme heterostructure with improved photophysical properties and enhanced antibiotic degradation under sunlight illumination (Figure 4.11a, b). A reduced graphene oxide (RGO)-mediated SSEM Z-scheme strategy having g-C_3N_4/$BiVO_4$ photocatalysts for TC degradation was synthesized through a hydrothermal method [82]. RGO acts as a promoter for charge carrier separation between g-C_3N_4 and $BiVO_4$ and thereby enhancing the degradation performance. Wen et al. emphasized the formation of CeO_2–Ag/AgBr SSEM Z-scheme for the degradation of CFX [83]. The total organic carbon (TOC) removal efficiency is reported to be 60.79% in 160 min showing the excellent degradation owing to the advantages of Z-scheme. They have studied the effect of other anions on photocatalytic performance and degradation pathways of CFX as well. An iodine vacancy-rich BiOI–Ag–AgI SSEM Z-scheme for the degradation of TC was done by Yang et al. [84]. Defective surface of BiOI was deposited with Ag/AgI nanoparticles and a Z-scheme heterostructure was constructed. The impacts of initial TC dosage, light irradiation conditions, and presence of inorganic ions were investigated. When the photocatalytic reactions were carried out under full-spectrum condition, maximum TC degradation efficiency was achieved. Some of the notable reports on SSEM Z-scheme for antibiotic degradation are listed in Table 4.4. Nevertheless, the use of electron mediator limits the large-scale practical application of SSEM Z-scheme photocatalysts.

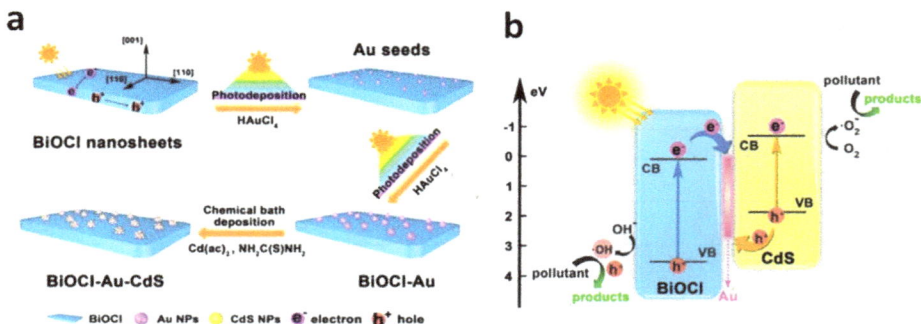

Figure 4.11: (a) Schematic of the formation and (b) proposed photocatalytic mechanism of Z-scheme BiOCl–Au–CdS heterostructures [81].
Reprinted with permission of Qiaoying Li et al. [81].

Direct Z-scheme photocatalysts with outstanding redox capabilities, and spatially separated and preserved redox sites have gained much attention in the field of photocatalytic antibiotic degradation. An adsorptive photocatalyst based on $CuBi_2O_4/BiOBr$ direct Z-scheme heterostructure for the visible light degradation of TC was exploited by Huang et al. by a simple precipitation method [85]. An adsorption efficiency of 75.6% was attained within 120 min, and the effects of pH and reaction temperature on adsorption were drawn systematically. The optimum photocatalytic degradation efficiency was 64.7% within 150 min. A surfactant-free method for the fabrication of direct Z-scheme photocatalytic heterostructures of CdS/Bi_3O_4Cl for the degradation of TC and CFX was reported [85]. High TOC removal and the possible intermediate and pathways of antibiotic degradation were postulated. Wang et al. constructed a 0D/2D Z-scheme heterojunctions of Bi_3TaO_7 quantum dots/g-C_3N_4 nanosheets via an ultrasonication method for the visible light–induced degradation of CFX [88]. Apart from the valuable features of Z-scheme in having spatial separation of charge carriers, the strong coupling between the 0D/2D photocatalysts also benefited the degradation performance. Shao et al. reported a novel double Z-scheme photocatalyst involving $Ag_3PO_4/Bi_2S_3/Bi_2O_3$ for the degradation of sulfamethazine and cloxacillin antibiotics [89]. Interestingly, the band structure and density of states of the individual and composite samples were analyzed by computational means. The effect of different parameters like pH, dosage of antibiotics, surfactant, and other electrolytes in water were assessed clearly and a double Z-scheme mechanism for the catalytic degradation of antibiotics in which Bi_2S_3 acts as a photosensitizer (Figure 4.12) was proposed. A core–shell $Fe_2O_3/CuBi_2O_4$ Z-scheme heterostructure for the degradation of TC through a one-step hydrothermal process was stated by Li et al. [90]. In a noteworthy contribution by Guo et al., TC degradation using AgX (X = Br or I)/$CuBi_2O_4$ was mentioned [91]. Based on energy band structure and reactive radical trapping experiments, it was concluded that a Z-scheme heterostructure was formed when AgBr is used, that is, $AgBr/CuBi_2O_4$, whereas type II heterostructure was obtained when AgI was used, that is, $AgI/CuBi_2O_4$. In both cases, the enhanced photocatalytic activity was due to the effective interfacial charge transfer between photocatalysts. A Z-scheme system of $MoO_3/g-C_3N_4$ was noted for the degradation of OFX through a synergistic role of peroxydisulfate [92]. Several pathways for degradation of OFX were postulated as shown in Figure 4.13. A recent report on all-solid Z-scheme heterojunction photocatalyst of $Bi_7O_9I_3/g-C_3N_4$ dealt with the degradation of doxycycline hydrochloride [93]. The formation of Z-scheme as well as the mild synthesis conditions was feasible for practical application and the degradation byproducts are shown in Figure 4.14. Some of the recent direct Z-scheme photocatalytic structures for the antibiotic degradation are summarized in Table 4.5.

Table 4.4: Visible light–active photocatalytic SSEM Z-scheme heterojunctions, reaction conditions, and efficiency for antibiotic degradation.

SSEM Z-scheme photocatalyst	Antibiotic	Loading		Degradation efficiency/time in minutes	Reference
		Antibiotic	Photocatalyst		
BiOCl–Au–CdS	SD	20 mg/L	0.05 g	100%/240	[81]
C_3N_4/RGO/$BiVO_4$	TC	35 mg/L	0.04 g	72.5%/150	[82]
CeO_2–Ag/AgBr	CFX	10 mg/L	0.05 g	93.05%/120	[83]
BiOI–Ag–AgI	TC	20 mg/L	0.03 g	86.4%/60	[84]
Graphene-bridged Ag_3PO_4/ Ag/$BiVO_4$	TC	10 mg/L	0.5 g/L	94.96%/60	[86]
Nitrogen-doped graphene quantum dots $BiVO_4$/ g-C_3N_4	TC OTC CFX	10 mg/L	0.5 g/L	91.5%/30 66.7%/120 72.4%/120	[87]

SD, sulfadiazine; RGO, reduced graphene oxide; TC, tetracycline; CFX, ciprofloxacin; OTC, oxytetracycline.

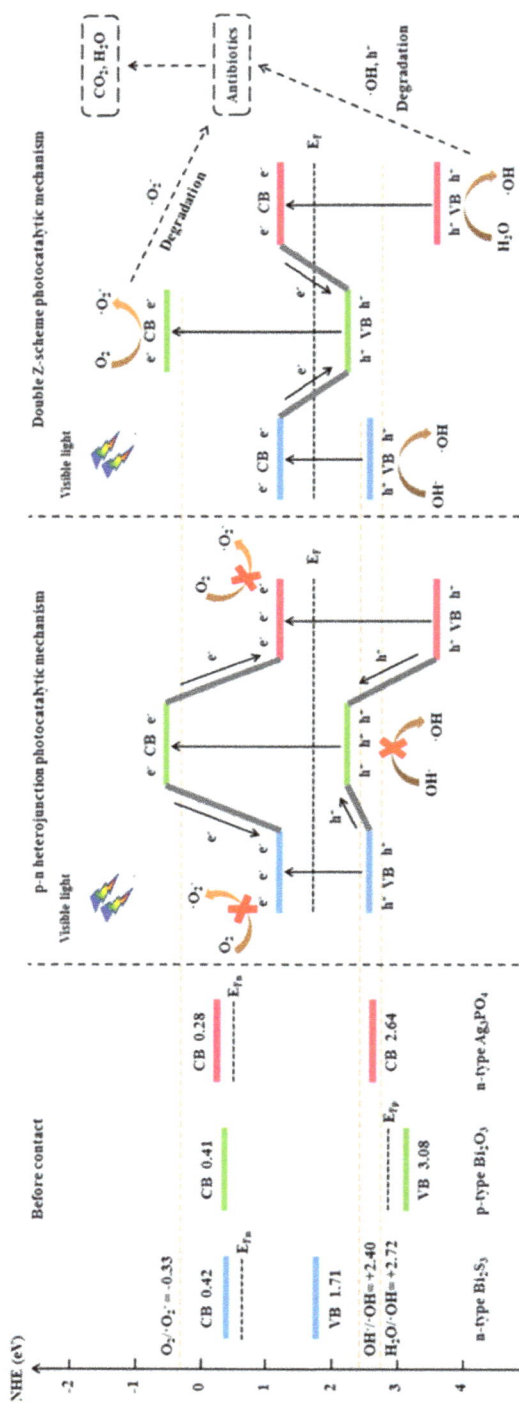

Figure 4.12: Possible charge separation and photocatalytic mechanism of $Ag_3PO_4/Bi_2S_3/Bi_2O_3$ double Z-scheme heterostructures for antibiotic degradation [89].
Reprinted with permission of Binbin Shao et al. [89].

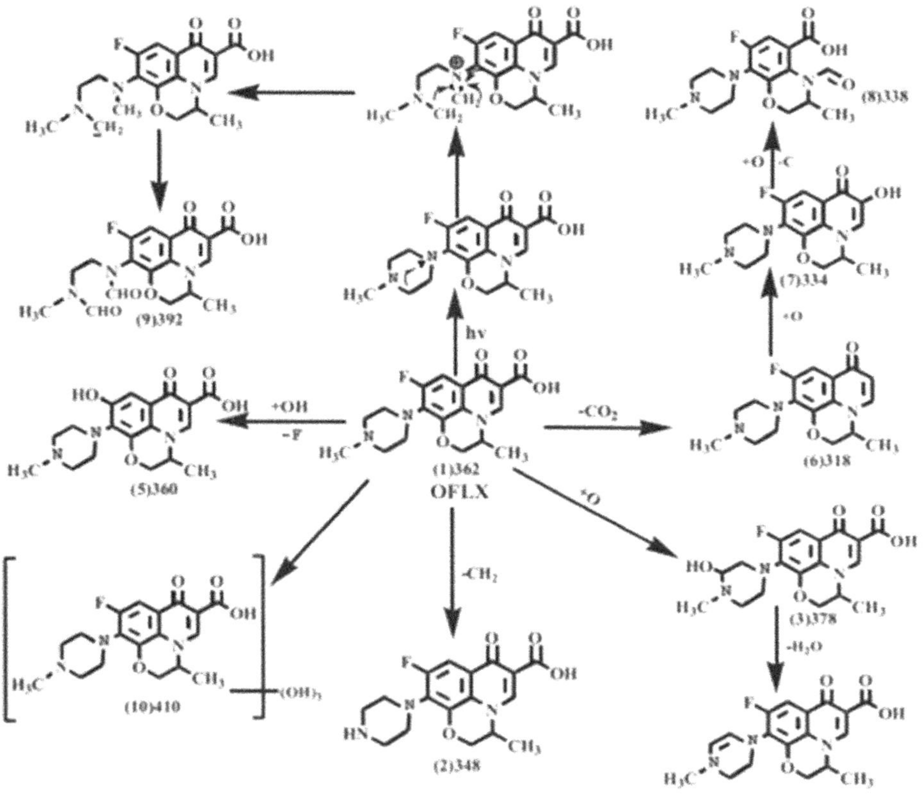

Figure 4.13: Possible transformation pathways for OFX degradation using $MoO_3/g\text{-}C_3N_4$ direct Z-scheme heterostructures [92].
Reprinted with permission of Danni Chen et al. [92].

4.4.5 Adsorptive Photocatalysts for Antibiotic Removal

Adsoprtion is a well-known technique for the removal of organic pollutants from water. However, the inflexible secondary remediation process for the regeneration of the adsorbents and the fate of the adsorbed pollutant render the process nonviable for practical applications [100]. In order to overcome this issue, adsorption combining with advanced oxidation processes like photocatalysis is a fruitful solution for pollutant removal [101]. The so-called bifunctional adsorptive photocatalyst will adsorb as well as simultaneously degrade the pollutant completely without any secondary remediation measures. Recently, antibiotic removal using visible light–active adsorptive photocatalysis has emerged owing to the above-mentioned advantages. Some of the recent studies on the antibiotic removal using adsorptive photocatalysis are discussed below.

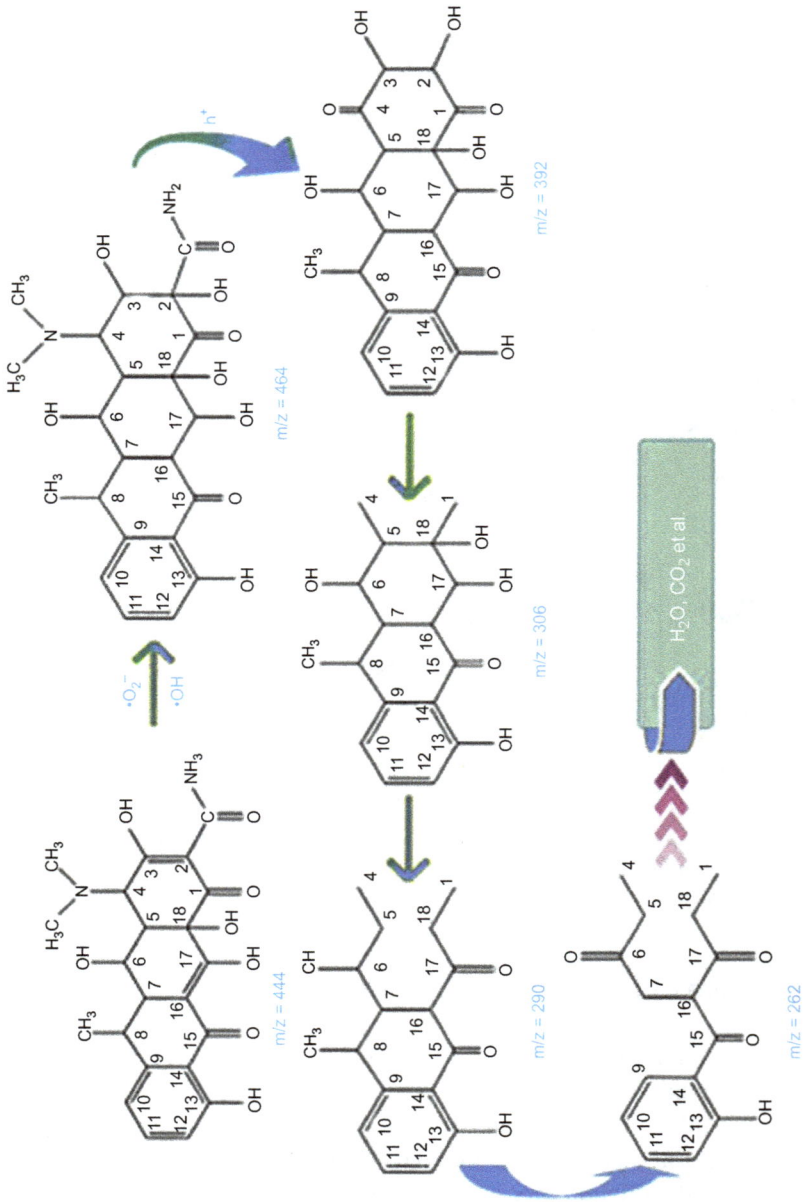

Figure 4.14: Doxycycline hydrochloride degradation pathways by means of $Bi_7O_9I_3/g$-C_3N_4 heterostructures [93]. Reprinted with permission of Zhenzong Zhang et al. [93].

Table 4.5: Visible light–active photocatalytic direct Z-scheme heterojunctions, reaction conditions, and efficiency for antibiotic degradation.

Direct Z-scheme photocatalyst	Antibiotic	Loading		Degradation efficiency/time in minutes	Reference
		Antibiotic	Photocatalyst		
$CuBi_2O_4/BiOBr$	TC	20 mg/L	0.025 g	64.7%/150	[94]
CdS/Bi_3O_4Cl	TC CFX	10 mg/L	0.05 g	~90%/120 84.2%/120	[85]
Bi_3TaO_7 QD/g-C_3N_4	CFX	10 mg/L	0.05 g	91%/120	[88]
$Ag_3PO_4/Bi_2S_3/Bi_2O_3$	SAZ CLX	10 mg/L	0.1 g	98.06%/90 90.26%/90	[89]
$Fe_2O_3/CuBi_2O_4$	TC	10 mg/L	0.05 g	80%/120	[90]
MoO_3/g-C_3N_4/PDS	OFX	10 mg/L	0.06 g	94.4%/60	[92]
$Bi_7O_9I_3/g$-C_3N_4	DXCH	10 mg/L	0.05 g	80%/120	[93]
WO_3-g-C_3N_4	TCH CS	10 mg/L	0.05 g	70%120 82%/120	[95]
$In_2O_3@ZnFe_2O_4$		50 mg/L	–	90%/30	[96]
AgI/WO_3	TC	35 mg/L	0.04 g	75%/60	[97]
$ZnO/AgVO_3$	CFX	10 mg/L	0.02 g	95%/120	[98]
MoS_2/Bi_2O_3	TC	20 mg/L	0.05 g	89.6%/120	[99]

TC, tetracycline; QD, quantum dot; SAZ, sulfamethaxine; CLX, cloxacillin; PDS, peroxydisulfate; DXCH, doxycycline hydrochloride; OFX, Ofloxacin; CFX, ciprofloxacin; CS, Ceftiofur sodium.

P. Suyana et al. reported the use of adsorptive photocatalyst made of spray-dried granules of g-C_3N_4 for the removal of TC [102]. The template-free granulation process led to porous C_3N_4 microspheres with high surface area. Additionally, the in situ carbon doping arising from post-granulation thermal oxidation facilitated improved photophysical properties. The carbon-doped C_3N_4 porous granules adsorbed ~75% TC by adsorption followed by >95% TC degradation under sunlight on repetitive cycles without much reduction in activity. Hao et al. materialized mesoporous BiOI microspheres for the removal of TC by simultaneous adsorption and visible light–induced photocatalysis [103]. The electronic band structure and high-surface to volume ratio enabled the material to be an excellent catalyst for TC removal. Cobalt oxide–loaded graphitic carbon nitride for the adsorptive photocatalytic removal of TC was reported by Niu et al. via an impregnation-calcination method [104]. The NH_4Cl bubble-templated strategy-mediated C_3N_4 having high surface area was successfully impregnated with CoO showed a high adsorption capacity of 391.4 mg/g and photocatalytic removal efficiency of 29.87% with 300 min irradiation. A complexation of Co with TC was established to prove the high adsorption capacity of the material. Recently, a combination of photocatalyst with high surface area sorbent materials like metal organic framework (MOF) is reported for the efficient removal of organic pollutants from water. In a noticeable work, Wang et al. detailed $In_2S_3@MIL$-125(Ti) core–shell microparticle for the removal of TC by integrated adsorption and photocatalysis [105]. An adsorption capacity of 119.2 mg/g was reported owing to the surface complexation, pi–pi interactions, and hydrogen bonding and electrostatic interactions. The degradation of TC was 63.3% on account of its high surface area, and Ti^{3+}–Ti^{4+} intervalence electron transfer. A comparison of TC removal by three Fe-based MOFs like Fe-MIL-101, Fe-MIL-100, and Fe-MIL-53 by the adsorption and photocatalytic degradation was shown by Wang et al. [106]. Owing to the high surface area, excellent pore characteristics, and enhanced visible light absorption, Fe-MIL-101 exhibited higher TC removal rates. Photoregenerable, adsorptive photocatalyst based on C_3N_4 anchored ZIF-8 microcrystals synthesized in an in situ approach were used for TC removal (Figure 4.15a) [107]. A high adsorption capacity of 420 mg/g and subsequent degradation of TC (>95%) under sunlight degradation was achieved owing to the micro-mesoporous architectures, high surface area, etc. The authors postulated the formation of a heterostructure (Figure 4.15b) for the improved photophysical properties and application studies. Notable reports on adsorptive photocatalyst for the removal of antibiotics are shown in Table 4.6.

Table 4.6: Visible light–active adsorptive photocatalyst for antibiotic removal.

Adsorptive photocatalyst	Anti-biotic	Loading			Efficiency		Reference
		Antibiotic	Adsorbent	Photo-catalyst	Adsorption efficiency / capacity	Degradation efficiency/ time in minutes	
Spray-dried C_3N_4 granules	TCH	~10–140 mg/L	0.5 g/L	0.04 g	70 mg/g	>90%/90	[102]
C_3N_4-anchored ZIF-8	TCH	~20–180 mg/L	0.1 g/L	0.1 g/L	420 mg/g	>95%/60	[107]
CoO/g-C_3N_4	TC	10–260 mg/L	20 mg	0.032 g/L	391.4 mg/g	~30%/300	[104]
$In_2S_3@MIL$-125(Ti)	TC	100 mg/L stock solution	–	0.03 g	119.2 mg/g	63.3%/60	[105]

Table 4.6 (continued)

Adsorptive photocatalyst	Antibiotic	Loading			Efficiency		Reference
		Antibiotic	Adsorbent	Photocatalyst	Adsorption efficiency / capacity	Degradation efficiency/ time in minutes	
MIL-101(Cr)-loaded nano zero-valent iron	TC	200 mg/L	–	10 mg	625.0 mg/g	90%/120	[108]
Fe-based MOFs	TC	50 mg/L	50 mg	50 mg	–	88%/180	[106]

TCH, tetracycline hydrochloride; TC, tetracycline; MOF, metal organic framework.

4.4.6 Issues and Challenges

Sunlight/visible light–assisted photocatalysis is viewed upon as a possible solution – for the removal of pharmaceutical pollutants. However, several issues that require critical care need to be solved before the systems are exploited for practical applications. Thus, future prospects of research include

- The extremely poor surface area of the base semiconductor systems shall be improved through synthetic modifications that are easy to adopt. The introduction of fugitive inclusions such as bubble gas template has significantly altered the pore characteristics of base systems and this approach is a subject of intense research as has been seen recently [109, 110]. The very recent reports on single atom catalysis where isolated metal atoms with catalytically active surfaces are linked to porous supports for accelerating chemical reactions shall provide breakthrough achievements enabling wider research interests in this field [111–113].
- Extensive studies are reported for the photocatalytic degradation of antibiotics in aqueous environments. The study shall be further extended to toxic organic pollutants like pesticides and other harmful organic additives and will be of significant societal impact.
- The chemical engineering aspects of translating lab-level achievements to prototype reactors capable of handling pharmaceutical effluents will be of great significance in the road map to commercialization.

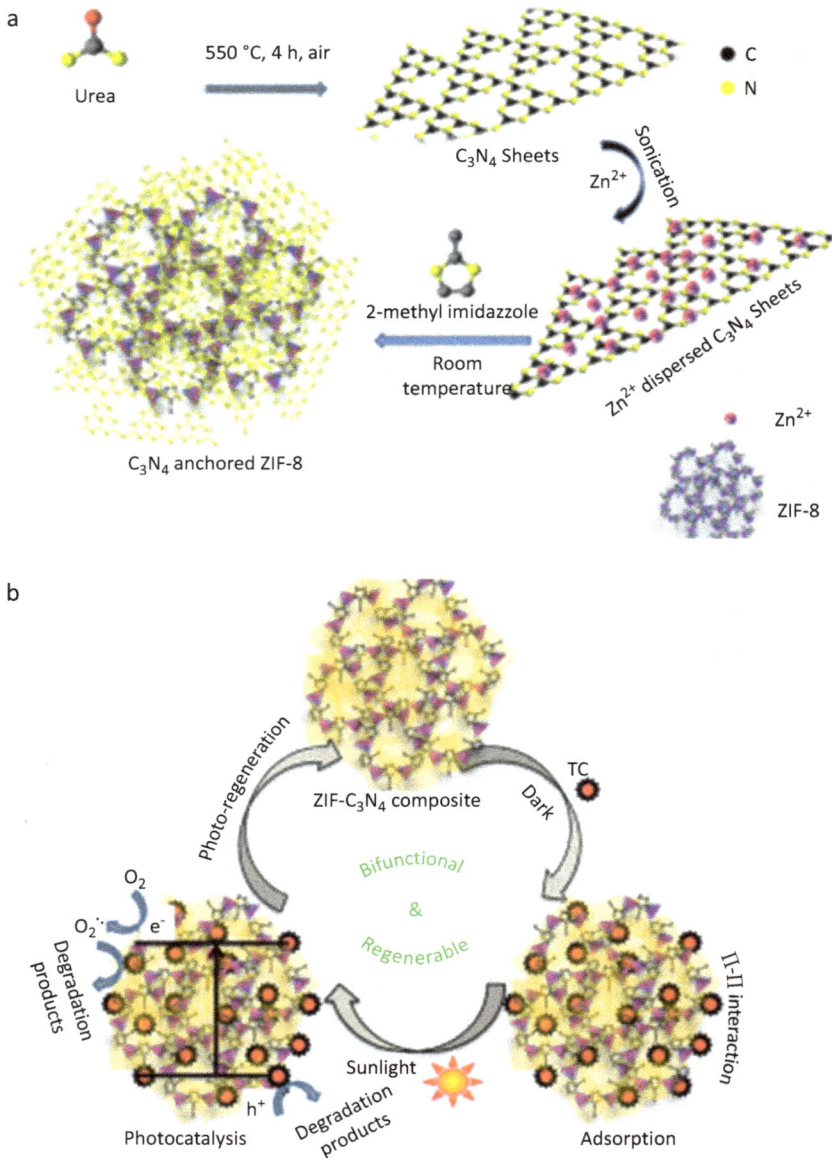

Figure 4.15: (a)Mechanism of formation and (b) adsorption and photocatalytic degradation mechanisms of the C_3N_4–ZIF-8 composite [107].
Reprinted with permission of Suyana Panneri et al. [107].

4.4.7 Conclusions

The past few years have seen the emergence of sunlight/visible light–active photocatalytic systems for a wide range of pollutant degradation applications. The fundamental drawbacks associated with surface area and charge recombinations are effectively addressed through the approaches of pore structure modifications through hierarchical architectures and band engineering through intimate interfaces of heterostructures. Henceforth, a variety of photocatalytic nanocomposites with compatible semiconductors are effectively demonstrated for antibiotic degradation and the mechanistic pathways are elucidated. The extensive research in this area has also led to new research prospects in approaches like single atom catalysis which shall further provide new avenues of application like photocatalytic conversion of carbon dioxide to value-added products.

References

[1]   Fanourakis SK, Peña-Bahamonde J, Bandara PC, Rodrigues DF. Nano-based adsorbent and photocatalyst use for pharmaceutical contaminant removal during indirect potable water reuse, NPJ Clean Water, 2020, 3, 1.

[2]   Lopez B, Ollivier P, Togola A, Baran N, Ghestem JP. Screening of French groundwater for regulated and emerging contaminants, The Science of the Total Environment, 2015, 518–519, 562–573.

[3]   Oller I, Malato S, Sánchez-Pérez JA. Combination of Advanced Oxidation Processes and biological treatments for wastewater decontamination – a review, Science of the Total Environment, 2011, 409, 4141–4166.

[4]   Vortmann M, Balsari S, Holman SR, Greenough PG. Water, sanitation, and hygiene at the world's largest mass gathering, Current Infectious Disease Reports, 2015, 17, 461.

[5]   Yu W, Ji N, Tian N, Bai L, Ou H, Huang H. BiOI/Bi2O2[BO2(OH)] heterojunction with boosted photocatalytic degradation performance for diverse pollutants under visible light irradiation, Colloids and Surfaces. A, Physicochemical and Engineering Aspects, 2020, 603, 125184.

[6]   Sorenson SB, Morssink C, Campos PA. Safe access to safe water in low income countries: water fetching in current times, Social Science & Medicine, 2011, 72, 1522–1526.

[7]   Wilcox J, Nasiri F, Bell S, Rahaman MS. Urban water reuse: a triple bottom line assessment framework and review, Sustainable Cities and Society, 2016, 27, 448–456.

[8]   Homem V, Santos L. Degradation and removal methods of antibiotics from aqueous matrices – A review, Journal of Environmental Management, 2011, 92, 2304–2347.

[9]   Bagheri S, Termeh Yousefi A, Do T-O. Photocatalytic pathway toward degradation of environmental pharmaceutical pollutants: structure, kinetics and mechanism approach, Catalysis Science & Technology, 2017, 7, 4548–4569.

[10] Dalrymple OK, Yeh DH, Trotz MA. Removing pharmaceuticals and endocrine-disrupting compounds from wastewater by photocatalysis, Journal of Chemical Technology and Biotechnology, 2007, 82, 121–134.

[11]   Wang JL, Xu LJ. Advanced oxidation processes for wastewater treatment: formation of hydroxyl radical and application, Critical Reviews in Environmental Science and Technology, 2012, 42, 251–325.

[12]   Ibhadon AO, Fitzpatrick P. Heterogeneous photocatalysis: recent advances and applications, Catalysts, 2013, 3, 189–218.

[13]   Nosaka Y, Nosaka AY. Generation and detection of reactive oxygen species in photocatalysis, Chemical Reviews, 2017, 117, 11302–11336.

[14]   Zhu S, Wang D. Photocatalysis: basic principles, diverse forms of implementations and emerging scientific opportunities, Advanced Energy Materials, 2017, 7, 1700841.

[15]   Pelaez M, Nolan NT, Pillai SC, et al. A review on the visible light active titanium dioxide photocatalysts for environmental applications, Applied Catalysis B: Environmental, 2012, 125, 331–349.

[16]   Kabra K, Chaudhary R, Sawhney RL. Treatment of hazardous organic and inorganic compounds through aqueous-phase photocatalysis: a review, Industrial & Engineering Chemistry Research, 2004, 43, 7683–7696.

[17]   Fox MA, Dulay MT. Heterogeneous photocatalysis, Chemical Reviews, 1993, 93, 341–357.

[18]   Li X, Yu J, Jaroniec M. Hierarchical photocatalysts, Chemical Society Reviews, 2016, 45, 2603–2636.

[19]   Kudo A, Miseki Y. Heterogeneous photocatalyst materials for water splitting, Chemical Society Reviews, 2009, 38, 253–278.

[20]   Miyauchi M, Irie H, Liu M, et al. Visible-light-sensitive photocatalysts: nanocluster-grafted titanium dioxide for indoor environmental remediation, The Journal of Physical Chemistry Letters, 2016, 7, 75–84.

[21]   Djurišić AB, Leung YH, Ching Ng AM. Strategies for improving the efficiency of semiconductor metal oxide photocatalysis, Materials Horizons, 2014, 1, 400–410.

[22]   Wang X, Wang F, Sang Y, Liu H. Full-spectrum solar-light-activated photocatalysts for light–chemical energy conversion, Advanced Energy Materials, 2017, 7, 1700473.

[23]   Moniz SJA, Shevlin SA, Martin DJ, Guo Z-X, Tang J. Visible-light driven heterojunction photocatalysts for water splitting – a critical review, Energy & Environmental Science, 2015, 8, 731–759.

[24]   Majumdar A, Pal A. Recent advancements in visible-light-assisted photocatalytic removal of aqueous pharmaceutical pollutants, Clean Technologies and Environmental Policy, 2020, 22, 11–42.

[25]   Li X, Yu J, Low J, Fang Y, Xiao J, Chen X. Engineering heterogeneous semiconductors for solar water splitting, Journal of Materials Chemistry A, 2015, 3, 2485–2534.

[26]   Wang H, Zhang L, Chen Z, et al. Semiconductor heterojunction photocatalysts: design, construction, and photocatalytic performances, Chemical Society Reviews, 2014, 43, 5234–5244.

[27]   Li H, Zhou Y, Tu W, Ye J, Zou Z. State-of-the-art progress in diverse heterostructured photocatalysts toward promoting photocatalytic performance, Advanced Functional Materials, 2015, 25, 998–1013.

[28]   Zhou H, Qu Y, Zeid T, Duan X. Towards highly efficient photocatalysts using semiconductor nanoarchitectures, Energy & Environmental Science, 2012, 5, 6732–6743.

[29]   Su J, Zou -X-X, Li G-D, et al. Macroporous V2O5–BiVO4 Composites: effect of Heterojunction on the Behavior of Photogenerated Charges, The Journal of Physical Chemistry C, 2011, 115, 8064–8071.

[30]   Li X, Chen J, Li H, et al. Photoreduction of CO2 to methanol over Bi2S3/CdS photocatalyst under visible light irradiation, Journal of Natural Gas Chemistry, 2011, 20, 413–417.

[31] Low J, Yu J, Jaroniec M, Wageh S, Al-Ghamdi AA. Heterojunction photocatalysts, Advanced Materials, 2017, 29, 1601694.

[32] Suyana P, Sneha KR, Nair BN, et al. A facile one pot synthetic approach for C3N4–ZnS composite interfaces as heterojunctions for sunlight-induced multifunctional photocatalytic applications, RSC Advances, 2016, 6, 17800–17809.

[33] Li C, Yuan J, Han B, Jiang L, Shangguan W. TiO2 nanotubes incorporated with CdS for photocatalytic hydrogen production from splitting water under visible light irradiation, International Journal of Hydrogen Energy, 2010, 35, 7073–7079.

[34] Pan C, Xu J, Wang Y, Li D, Zhu Y. Dramatic activity of C3N4/BiPO4 photocatalyst with core/shell structure formed by self-assembly, Advanced Functional Materials, 2012, 22, 1518–1524.

[35] Khan MR, Chuan TW, Yousuf A, Chowdhury MNK, Cheng CK. Schottky barrier and surface plasmonic resonance phenomena towards the photocatalytic reaction: study of their mechanisms to enhance photocatalytic activity, Catalysis Science & Technology, 2015, 5, 2522–2531.

[36] Dinh C-T, Yen H, Kleitz F, Do T-O. Three-dimensional ordered assembly of thin-shell Au/TiO2 hollow nanospheres for enhanced visible-light-driven photocatalysis, Angewandte Chemie International Edition, 2014, 53, 6618–6623.

[37] Wan J, Liu E, Fan J, et al. In-situ synthesis of plasmonic Ag/Ag3PO4 tetrahedron with exposed {111} facets for high visible-light photocatalytic activity and stability, Ceramics International, 2015, 41, 6933–6940.

[38] Qin L, Wang G, Tan Y. Plasmonic Pt nanoparticles – TiO2 hierarchical nano-architecture as a visible light photocatalyst for water splitting, Scientific Reports, 2018, 8, 16198.

[39] Reddy NL, Rao VN, Vijayakumar M, et al. A review on frontiers in plasmonic nano-photocatalysts for hydrogen production, International Journal of Hydrogen Energy, 2019, 44, 10453–10472.

[40] Chang X, Wang T, Zhang P, Zhang J, Li A, Gong J. Enhanced surface reaction kinetics and charge separation of p–n heterojunction Co3O4/BiVO4 photoanodes, Journal of the American Chemical Society, 2015, 137, 8356–8359.

[41] Ye L, Wang D, Chen S. Fabrication and enhanced photoelectrochemical performance of MoS2/S-doped g-C3N4 heterojunction film, ACS Applied Materials & Interfaces, 2016, 8, 5280–5289.

[42] Suyana P, Ganguly P, Nair BN, Mohamed AP, Warrier KGK, Hareesh US. Co3O4–C3N4 p–n nano-heterojunctions for the simultaneous degradation of a mixture of pollutants under solar irradiation, Environmental Science: Nano, 2017, 4, 212–221.

[43] Zhang J, Qiao SZ, Qi L, Yu J. Fabrication of NiS modified CdS nanorod p-n junction photocatalysts with enhanced visible-light photocatalytic H2-production activity, Physical Chemistry Chemical Physics: PCCP, 2013, 15, 12088–12094.

[44] Bard AJ. Photoelectrochemistry and heterogeneous photo-catalysis at semiconductors, Journal of Photochemistry, 1979, 10, 59–75.

[45] Sasaki Y, Iwase A, Kato H, Kudo A. The effect of co-catalyst for Z-scheme photocatalysis systems with an Fe3+/Fe2+ electron mediator on overall water splitting under visible light irradiation, Journal of Catalysis, 2008, 259, 133–137.

[46] Maeda K, Higashi M, Lu D, Abe R, Domen K. Efficient nonsacrificial water splitting through two-step photoexcitation by visible light using a modified oxynitride as a hydrogen evolution photocatalyst, Journal of the American Chemical Society, 2010, 132, 5858–5868.

[47] Tada H, Mitsui T, Kiyonaga T, Akita T, Tanaka K. All-solid-state Z-scheme in CdS-Au-TiO2 three-component nanojunction system, Nature Materials, 2006, 5, 782–786.

[48] Zhang N, Xie S, Weng B, Xu Y-J. Vertically aligned ZnO–Au@CdS core–shell nanorod arrays as an all-solid-state vectorial Z-scheme system for photocatalytic application, Journal of Materials Chemistry A, 2016, 4, 18804–18814.

[49] Durán-Álvarez JC, Méndez-Galván M, Lartundo-Rojas L, et al. Synthesis and characterization of the all solid Z-scheme Bi2WO6/Ag/AgBr for the photocatalytic degradation of ciprofloxacin in water, Topics in Catalysis, 2019, 62, 1011–1025.

[50] Yu J, Wang S, Low J, Xiao W. Enhanced photocatalytic performance of direct Z-scheme g-C3N4–TiO2 photocatalysts for the decomposition of formaldehyde in air, Physical Chemistry Chemical Physics, 2013, 15, 16883–16890.

[51] Zhang J, Hu Y, Jiang X, Chen S, Meng S, Fu X. Design of a direct Z-scheme photocatalyst: preparation and characterization of Bi_2O_3/g-C_3N_4 with high visible light activity, Journal of Hazardous Materials, 2014, 280, 713–722.

[52] Tian N, Huang H, He Y, Guo Y, Zhang T, Zhang Y. Mediator-free direct Z-scheme photocatalytic system: BiVO4/g-C3N4 organic–inorganic hybrid photocatalyst with highly efficient visible-light-induced photocatalytic activity, Dalton Transactions, 2015, 44, 4297–4307.

[53] Ma SS, Maeda K, Hisatomi T, Tabata M, Kudo A, Domen K. A redox-mediator-free solar-driven Z-scheme water-splitting system consisting of modified Ta3N5 as an oxygen-evolution photocatalyst, Chemistry (Weinheim an der Bergstrasse, Germany), 2013, 19, 7480–7486.

[54] Wang X, Liu G, Chen Z-G, et al. Enhanced photocatalytic hydrogen evolution by prolonging the lifetime of carriers in ZnO/CdS heterostructures, Chemical Communications, 2009, 23, 3452–3454.

[55] Miyauchi M, Nukui Y, Atarashi D, Sakai E. Selective growth of n-type nanoparticles on p-type semiconductors for Z-scheme photocatalysis, ACS Applied Materials & Interfaces, 2013, 5, 9770–9776.

[56] Wang S, Zhu B, Liu M, Zhang L, Yu J, Zhou M. Direct Z-scheme ZnO/CdS hierarchical photocatalyst for enhanced photocatalytic H2-production activity, Applied Catalysis B: Environmental, 2019, 243, 19–26.

[57] Jia Q, Iwase A, Kudo A. BiVO4-Ru/SrTiO3:Rhcomposite Z-scheme photocatalyst for solar water splitting, Chemical Science, 2014, 5, 1513–1519.

[58] Hu T, Li P, Zhang J, Liang C, Dai K. Highly efficient direct Z-scheme WO3/CdS-diethylenetriamine photocatalyst and its enhanced photocatalytic H2 evolution under visible light irradiation, Applied Surface Science, 2018, 442, 20–29.

[59] Xu F, Xiao W, Cheng B, Yu J. Direct Z-scheme anatase/rutile bi-phase nanocomposite TiO2 nanofiber photocatalyst with enhanced photocatalytic H2-production activity, International Journal of Hydrogen Energy, 2014, 39, 15394–15402.

[60] Etebu E, Arikekpar I. Antibiotics: classification and mechanisms of action with emphasis on molecular perspectives. 2016.

[61] Kümmerer K. Antibiotics in the aquatic environment--a review--part I, Chemosphere, 2009, 75, 417–434.

[62] Kanakaraju DGBD, Oelgemöller M. Heterogeneous Photocatalysis for Pharmaceutical Wastewater Treatment, In: Lichtfouse ESJ, Robert D, ed, Green Materials for Energy, Products and Depollution, Springer, Dordrecht, 2013, 69–133.

[63] Guo F, Shi W, Wang H, et al. Facile fabrication of a CoO/g-C3N4 p–n heterojunction with enhanced photocatalytic activity and stability for tetracycline degradation under visible light, Catalysis Science & Technology, 2017, 7, 3325–3331.

[64] Cui D, Wang L, Du Y, Hao W, Chen J. Photocatalytic reduction on Bismuth-based p-block semiconductors, ACS Sustainable Chemistry & Engineering, 2018, 6, 15936–15953.

[65] Wen X-J, Niu C-G, Zhang L, Zeng G-M. Fabrication of SnO2 nanoparticles/BiOI n–p heterostructure for wider spectrum visible-light photocatalytic degradation of antibiotic

oxytetracycline hydrochloride, ACS Sustainable Chemistry & Engineering, 2017, 5, 5134–5147.

[66] Xu J, Feng B, Wang Y, Qi Y, Niu J, Chen M. BiOCl decorated NaNbO3 nanocubes: a novel p-n heterojunction photocatalyst with improved activity for ofloxacin degradation, Frontiers in Chemistry, 2018, 6.

[67] Guo L, Zhang K, Han X, Zhao Q, Wang D, Fu F. 2D in-plane CuS/Bi(2)WO(6) p-n heterostructures with promoted visible-light-driven photo-Fenton degradation performance, Nanomaterials, (Basel, Switzerland) 2019, 9.

[68] Shi W, Li M, Huang X, et al. Construction of CuBi2O4/Bi2MoO6 p-n heterojunction with nanosheets-on-microrods structure for improved photocatalytic activity towards broad-spectrum antibiotics degradation, Chemical Engineering Journal, 2020, 394, 125009.

[69] Obregón S, Ruíz-Gómez MA, Rodríguez-González V, Vázquez A, Hernández-Uresti DB. A novel type-II Bi2W2O9/g-C3N4 heterojunction with enhanced photocatalytic performance under simulated solar irradiation, Materials Science in Semiconductor Processing, 2020, 113, 105056.

[70] Xue J, Ma S, Zhou Y, Zhang Z, He M. Facile photochemical synthesis of Au/Pt/g-C3N4 with plasmon-enhanced photocatalytic activity for antibiotic degradation, ACS Applied Materials & Interfaces, 2015, 7, 9630–9637.

[71] He F, Wang S, Zhao H, et al. Construction of Schottky-type Ag-loaded fiber-like carbon nitride photocatalysts for tetracycline elimination and hydrogen evolution, Applied Surface Science, 2019, 485, 70–80.

[72] Cai T, Wang L, Liu Y, et al. Ag3PO4/Ti3C2 MXene interface materials as a Schottky catalyst with enhanced photocatalytic activities and anti-photocorrosion performance, Applied Catalysis B: Environmental, 2018, 239, 545–554.

[73] Xu W, Lai S, Pillai SC, et al. Visible light photocatalytic degradation of tetracycline with porous Ag/graphite carbon nitride plasmonic composite: degradation pathways and mechanism, Journal of Colloid and Interface Science, 2020, 574, 110–121.

[74] Wang K, Li Y, Zhang G, Li J, Wu X. 0D Bi nanodots/2D Bi3NbO7 nanosheets heterojunctions for efficient visible light photocatalytic degradation of antibiotics: enhanced molecular oxygen activation and mechanism insight, Applied Catalysis B: Environmental, 2019, 240, 39–49.

[75] Zhang S, Khan I, Qin X, Qi K, Liu Y, Bai S. Construction of 1D Ag-AgBr/AlOOH plasmonic photocatalyst for degradation of tetracycline hydrochloride, Frontiers in Chemistry, 2020, 8.

[76] Wang H, Yang X, Zi J, et al. High photocatalytic degradation of tetracycline under visible light with Ag/AgCl/activated carbon composite plasmonic photocatalyst, Journal of Industrial and Engineering Chemistry, 2016, 35, 83–92.

[77] Panneri S, Ganguly P, Nair BN, Mohamed AAP, Warrier KG, Hareesh UNS. Copyrolysed C3N4-Ag/ZnO ternary heterostructure systems for enhanced adsorption and photocatalytic degradation of tetracycline, European Journal of Inorganic Chemistry, 2016, 2016, 5068–5076.

[78] Luo B, Xu D, Li D, et al. Fabrication of a Ag/Bi3TaO7 plasmonic photocatalyst with enhanced photocatalytic activity for degradation of tetracycline, ACS Applied Materials & Interfaces, 2015, 7, 17061–17069.

[79] Deng F, Zhao L, Luo X, Luo S, Dionysiou DD. Highly efficient visible-light photocatalytic performance of Ag/AgIn5S8 for degradation of tetracycline hydrochloride and treatment of real pharmaceutical industry wastewater, Chemical Engineering Journal, 2018, 333, 423–433.

[80] Zhang G, Guan W, Shen H, et al. Organic additives-free hydrothermal synthesis and visible-light-driven photodegradation of tetracycline of WO3 nanosheets, Industrial & Engineering Chemistry Research, 2014, 53, 5443–5450.

[81] Li Q, Guan Z, Wu D, et al. Z-Scheme BiOCl-Au-CdS heterostructure with enhanced sunlight-driven photocatalytic activity in degrading water dyes and antibiotics, ACS Sustainable Chemistry & Engineering, 2017, 5, 6958–6968.

[82] Jiang D, Xiao P, Shao L, Li D, Chen M. RGO-promoted all-solid-state g-C3N4/BiVO4 Z-scheme heterostructure with enhanced photocatalytic activity toward the degradation of antibiotics, Industrial & Engineering Chemistry Research, 2017, 56, 8823–8832.

[83] Wen X-J, Niu C-G, Zhang L, Liang C, Guo H, Zeng G-M. Photocatalytic degradation of ciprofloxacin by a novel Z-scheme CeO2–Ag/AgBr photocatalyst: influencing factors, possible degradation pathways, and mechanism insight, Journal of Catalysis, 2018, 358, 141–154.

[84] Yang Y, Zeng Z, Zhang C, et al. Construction of iodine vacancy-rich BiOI/Ag@AgI Z-scheme heterojunction photocatalysts for visible-light-driven tetracycline degradation: transformation pathways and mechanism insight, Chemical Engineering Journal, 2018, 349, 808–821.

[85] Che H, Che G, Jiang E, Liu C, Dong H, Li C. A novel Z-Scheme CdS/Bi3O4Cl heterostructure for photocatalytic degradation of antibiotics: mineralization activity, degradation pathways and mechanism insight, Journal of the Taiwan Institute of Chemical Engineers, 2018, 91, 224–234.

[86] Chen F, Yang Q, Li X, et al. Hierarchical assembly of graphene-bridged Ag3PO4/Ag/BiVO4 (040) Z-scheme photocatalyst: an efficient, sustainable and heterogeneous catalyst with enhanced visible-light photoactivity towards tetracycline degradation under visible light irradiation, Applied Catalysis B: Environmental, 2017, 200, 330–342.

[87] Yan M, Zhu F, Gu W, Sun L, Shi W, Hua Y. Construction of nitrogen-doped graphene quantum dots-BiVO4/g-C3N4 Z-scheme photocatalyst and enhanced photocatalytic degradation of antibiotics under visible light, RSC Advances, 2016, 6, 61162–61174.

[88] Wang K, Zhang G, Li J, Li Y, Wu X. 0D/2D Z-scheme heterojunctions of bismuth tantalate quantum dots/ultrathin g-C3N4 nanosheets for highly efficient visible light photocatalytic degradation of antibiotics, ACS Applied Materials & Interfaces, 2017, 9, 43704–43715.

[89] Shao B, Liu X, Liu Z, et al. A novel double Z-scheme photocatalyst Ag3PO4/Bi2S3/Bi2O3 with enhanced visible-light photocatalytic performance for antibiotic degradation, Chemical Engineering Journal, 2019, 368, 730–745.

[90] Li M-Y, Tang Y-B, Shi W-L, Chen F-Y, Shi Y, Gu H-C. Design of visible-light-response core–shell Fe2O3/CuBi2O4 heterojunctions with enhanced photocatalytic activity towards the degradation of tetracycline: Z-scheme photocatalytic mechanism insight, Inorganic Chemistry Frontiers, 2018, 5, 3148–3154.

[91] Guo F, Shi W, Wang H, et al. Study on highly enhanced photocatalytic tetracycline degradation of type II AgI/CuBi2O4 and Z-scheme AgBr/CuBi2O4 heterojunction photocatalysts, Journal of Hazardous Materials, 2018, 349, 111–118.

[92] Chen D, Xie Z, Zeng Y, et al. Accelerated photocatalytic degradation of quinolone antibiotics over Z-scheme MoO3/g-C3N4 heterostructure by peroxydisulfate under visible light irradiation: mechanism; kinetic; and products, Journal of the Taiwan Institute of Chemical Engineers, 2019, 104, 250–259.

[93] Zhang Z, Pan Z, Guo Y, Wong PK, Zhou X, Bai R. In-situ growth of all-solid Z-scheme heterojunction photocatalyst of Bi7O9I3/g-C3N4 and high efficient degradation of antibiotic under visible light, Applied Catalysis B: Environmental, 2020, 261, 118212.

[94] Huang S, Wang G, Liu J, Du C, Su Y. A novel CuBi2O4/BiOBr direct Z-scheme photocatalyst for efficient antibiotics removal: synergy of adsorption and photocatalysis on degradation kinetics and mechanism insight, Chemical Catalysis Catalytical Chemistry, n/a, 2020, 12, 4431.

[95] Xiao T, Tang Z, Yang Y, Tang L, Zhou Y, Zou Z. In situ construction of hierarchical WO3/g-C3N4 composite hollow microspheres as a Z-scheme photocatalyst for the degradation of antibiotics, Applied Catalysis B: Environmental, 2018, 220, 417–428.

[96] Fei W, Song Y, Li N, et al. Hollow In2O3@ZnFe2O4 heterojunctions for highly efficient photocatalytic degradation of tetracycline under visible light, Environmental Science: Nano, 2019, 6, 3123–3132.

[97] Wang T, Quan W, Jiang D, et al. Synthesis of redox-mediator-free direct Z-scheme AgI/WO3 nanocomposite photocatalysts for the degradation of tetracycline with enhanced photocatalytic activity, Chemical Engineering Journal, 2016, 300, 280–290.

[98] Song S, Wu K, Wu H, Guo J, Zhang L. Synthesis of Z-scheme multi-shelled ZnO/AgVO3 spheres as photocatalysts for the degradation of ciprofloxacin and reduction of chromium (VI), Journal of Materials Science, 2020, 55, 4987–5007.

[99] Ji R, Ma C, Ma W, Liu Y, Zhu Z, Yan Y. Z-scheme MoS2/Bi2O3 heterojunctions: enhanced photocatalytic degradation performance and mechanistic insight, New Journal of Chemistry, 2019, 43, 11876–11886.

[100] Ali I. New generation adsorbents for water treatment, Chemical Reviews, 2012, 112, 5073–5091.

[101] Kanakaraju D, Kockler J, Motti CA, Glass BD, Oelgemöller M. Titanium dioxide/zeolite integrated photocatalytic adsorbents for the degradation of amoxicillin, Applied Catalysis B: Environmental, 2015, 166–167, 45–55.

[102] Panneri S, Ganguly P, Mohan M, et al. Photoregenerable, bifunctional granules of carbon-doped g-C3N4 as adsorptive photocatalyst for the efficient removal of tetracycline antibiotic, ACS Sustainable Chemistry & Engineering, 2017, 5, 1610–1618.

[103] Hao R, Xiao X, Zuo X, Nan J, Zhang W. Efficient adsorption and visible-light photocatalytic degradation of tetracycline hydrochloride using mesoporous BiOI microspheres, Journal of Hazardous Materials, 2012, 209–210, 137–145.

[104] Niu J, Xie Y, Luo H, Wang Q, Zhang Y, Wang Y. Cobalt oxide loaded graphitic carbon nitride as adsorptive photocatalyst for tetracycline removal from aqueous solution, Chemosphere, 2019, 218, 169–178.

[105] Wang H, Yuan X, Wu Y, et al. In situ synthesis of In2S3@MIL-125(Ti) core–shell microparticle for the removal of tetracycline from wastewater by integrated adsorption and visible-light-driven photocatalysis, Applied Catalysis B: Environmental, 2016, 186, 19–29.

[106] Wang D, Jia F, Wang H, et al. Simultaneously efficient adsorption and photocatalytic degradation of tetracycline by Fe-based MOFs, Journal of Colloid and Interface Science, 2018, 519, 273–284.

[107] Panneri S, Thomas M, Ganguly P, et al. C3N4 anchored ZIF 8 composites: photo-regenerable, high capacity sorbents as adsorptive photocatalysts for the effective removal of tetracycline from water, Catalysis Science & Technology, 2017, 7, 2118–2128.

[108] Hou X, Shi J, Wang N, et al. Removal of antibiotic tetracycline by metal-organic framework MIL-101(Cr) loaded nano zero-valent iron, Journal of Molecular Liquids, 2020, 313, 113512.

[109] Zhang C, Liu J, Huang X, Chen D, Xu S. Multistage polymerization design for g-C3N4 nanosheets with enhanced photocatalytic activity by modifying the polymerization process of melamine, ACS Omega, 2019, 4, 17148–17159.

[110] Zhou B, Yang B, Waqas M, Xiao K, Zhu C, Wu L. Design of a p–n heterojunction in 0D/3D MoS2/g-C3N4 composite for boosting the efficient separation of photogenerated carriers with enhanced visible-light-driven H2 evolution, RSC Advances, 2020, 10, 19169–19177.

[111] Yang Y, Zeng G, Huang D, et al. In situ grown single-atom cobalt on polymeric carbon nitride with bidentate ligand for efficient photocatalytic degradation of refractory antibiotics, Small, 2020, 16, 2001634.

[112] Wang Q, Zhang D, Chen Y, Fu W-F, Lv X-J. Single-atom catalysts for photocatalytic reactions, ACS Sustainable Chemistry & Engineering, 2019, 7, 6430–6443.

[113] Chen F, Wu X-L, Yang L, Chen C, Lin H, Chen J. Efficient degradation and mineralization of antibiotics via heterogeneous activation of peroxymonosulfate by using graphene supported single-atom Cu catalyst, Chemical Engineering Journal, 2020, 394, 124904.

Saeed Punnoli Ammed, Anupama R. Prasad, Sanjay Gopal Ullattil

Chapter 5
Self-Cleaning Surfaces: Experimental Advances and Surface Model Controversies

5.1 Self-Cleaning Surface, Wettability, and Contact Angle

Self-cleaning surfaces are a class of materials with an inherent ability to retain its clean surface in the presence of water, bringing down the maintenance cost, effort, and use of detergents. The materials have gained wide attention in recent years in different fields of indoor applications as well as exterior construction materials. Self-cleaning materials are used for a wide range of applications such as antifouling, anti-smudge, low drag, low adhesion, antireflection, self-sterilizing, anisotropic wetting, photocatalysis, and directional adhesion. Particularly, there are two types of self-cleaning surfaces: hydrophilic surfaces in which the spreading process of water drops on the surface, the contaminants are washed away [1]. Secondly, due to the water-repellent property, some hydrophobic surfaces rapidly roll of the water droplets and thus remove the sticky contaminants at the surface [2].

Surface wettability and water contact angle (WCA) are the two terms that are considered as the backbone of a system in determining the nature of the surface whether it is hydrophobic or hydrophilic. The surface wettability is the affinity between liquid and solid surface that can be measured by the WCA [3]. WCA is the angle delimited by the water droplet and the solid surface or the angle between the tangent line drawn to the droplet and surface. Generally, measuring contact angle (Θ) reflects the wettability of a surface. In short, different surfaces such as hydrophobic ($\Theta > 90°$), superhydrophobic ($\Theta > 150°$), hydrophilic ($0° < \Theta < 90°$), and superhydrophilic ($\Theta < 150°$) can be determined by measuring the angle as displayed in Figure 5.1. In a superhydrophilic surface, the water droplets spread quickly under the pollutant and wash it away and finally water produced at the surface may evaporate rapidly. Still the surface becomes wet for a long period which limits its viability for some applications. Secondly, a superhydrophobic surface will repel the water molecules and

Saeed Punnoli Ammed, Department of Chemistry, Government Engineering College, Kozhikode, Kerala, India – 673 005
Anupama R. Prasad, Department of Chemistry, University of Calicut, Calicut 673 635, Kerala, India
Sanjay Gopal Ullattil, Laboratory of Environmental Sciences and Engineering, Department of Inorganic Chemistry and Technology, Hajdrihova 19, Ljubljana, Slovenia – 1000, e-mail: sanjay.gopal.ullattil@ki.si

https://doi.org/10.1515/9783110668483-005

roll off the surface, carrying the impurities with the droplet [2]. On the other hand, the oil-based contaminant cannot be removed by the water droplet in a superhydropho-bic surface because the pollutants and the surface will repel the water droplets [2]. Generally, the oil-based impurities can also be washed away from a superhydrophilic surface. Moreover, oil-based contaminants can be effectively expelled from the hydro-philic and superoleophobic surfaces. Due to the superoleophobicity in water, the water droplets efficiently go under an oil-based contaminant and raise it up and thereby wash the contaminants away [2].

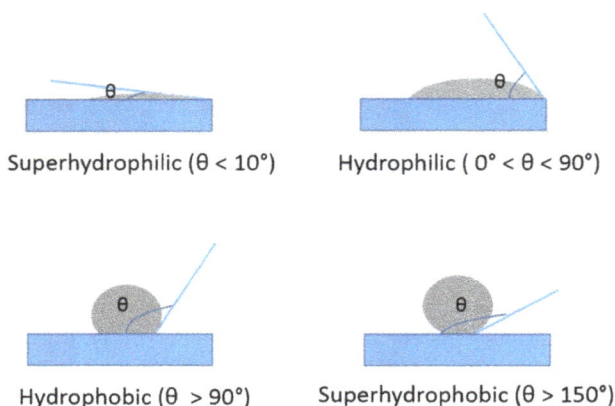

Superhydrophilic (θ < 10°)   Hydrophilic (0° < θ < 90°)

Hydrophobic (θ > 90°)   Superhydrophobic (θ > 150°)

Figure 5.1: Representation of superhydrophilic, hydrophilic, hydrophobic, and superhydrophobic surfaces.

5.2 Self-Cleaning Models and Controversies

Solid surface wettability has been well studied for a long time since 1805. In that year, Thomas Young has proposed a fundamental equation to quantify the static contact angle of a liquid droplet on a flat solid surface. The droplet forms a three-phase contact line (TPL) at the solid–liquid, liquid–vapor, and solid–vapor interfa-ces where they intersect. The shape of the droplet is defined by the forces created by the surface energy per unit area or surface tension at each interface. According to the Young's equation, the cosine of the WCA (θ_e) which is independent of the droplet size is directly proportional to the difference between the surface energy of solid–vapor (γ_{SV}) and solid–liquid (γ_{SL}) interfaces and inversely proportional to the surface energy of the liquid–vapor (γ_{LV}) interface as follows [4]:

$$\cos\theta_e = \frac{\gamma_{SV} - \gamma_{SL}}{\gamma_{LV}} \tag{5.1}$$

This is a concept based on the single equilibrium contact angle. In this, the droplet motion to its resting state through receding or advancing on the surface and hysteresis are not taken into account. In an inclined surface, on each side of the drop, the water droplet has a different contact angle. The angle in front of the droplet motion and the angle on the other side are termed as the advancing angle (θ_A) and the receding angle (θ_R), respectively. The difference between these angles is termed as the contact angle hysteresis (H) which has to be low enough to make rolling the droplets possible. Surface morphology and chemical composition play a pivotal role in determining the adhesion strength of a droplet. There is no ideal flat surface exists in real life. Hence, the surface topological structures on WCA must be taken into consideration and it is noted that the most influential factors that determine the overall wettability are surface energy and surface roughness [5].

Surface energy is the measurement of the interaction between the water droplet and the solid surface [6]. If the contact angle is invariant, that is, if the droplet is in equilibrium, the surface energies in and out of the droplet are the same. The droplet is liable to expand if the energy of the surface inside the droplet is less than the surface energy outside and tends to lower the contact angle leading to hydrophilicity. For inducing hydrophobicity, there should be the surface energy greater inside the water droplet than that of outside. Surface roughness is determined from the difference between the actual surface with deformation such as valleys and peaks, and the geometric or planar area. The geometric or planar area is the area of contact of the water droplet with the surface assuming a totally smooth surface. The roughness factor (r) can be calculated as [6]

$$r = \frac{\text{actual surface area}}{\text{planar (geometric) area}} \tag{5.2}$$

5.2.1 The Wenzel, Cassie–Baxter, and Miwa–Hashimoto Models

On a rough surface, the interaction of water droplet with the solid surface is defined either by Wenzel [7] or by Cassie–Baxter model [8]. In the Wenzel model, the roughness factor is incorporated directly, whereas in the Cassie–Baxter model, a comparable measurement is used. Although they deal with the same type of interactions, there are some fundamental differences and these models can be compared. In the Wenzel model, the droplet makes intimate contact with the surface by penetrating into the surface asperities leading to higher surface area as compared to the ideal flat surface (Figure 5.2A). Wenzel assumption is also called homogeneous wetting and he formulated an equation as follows [7]:

$$\cos \theta_w = r \times \cos \theta_e \tag{5.3}$$

where θ_w is the actual WCA on a rough surface, r is the roughness factor, and θ_e is the Young's contact angle or the equilibrium contact angle. If the roughness factor is unity, the surface is considered as an ideal surface and then $\theta_w = \theta_e$, that is, the Wenzel equation transforms to Young's equation. For an actual surface, the value is always greater than unity leading to increase in contact angle (>90°) for a hydrophobic surface and decrease in contact angle (<90°) for a hydrophilic surface [5].

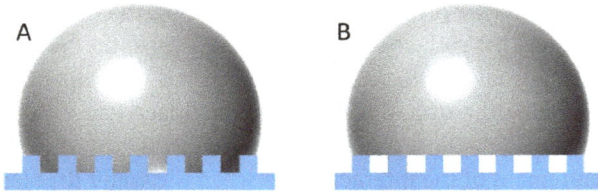

Figure 5.2: The equilibrium (static) contact angle of a liquid droplet on a rough surface: (A) Wenzel and (B) Cassie–Baxter models.

In the Cassie–Baxter model, the droplet sits at the top of the protrusions and air is entrapped by water in surface cavities (Figure 5.2B). In this case, the interaction between water and solid is minimized and that with air is maximized. It enables more spherical shape of the water droplet and thereby higher contact angle and in general the Cassie–Baxter equation can be described as

$$\cos \theta_c = f \times \cos \theta_e + f - 1 \tag{5.4}$$

where f is the ratio of solid–liquid interface area to the area of geometrical projection of water droplet. When the roughness increases, that is, more air is trapped in the surface cavities, f decreases and thus WCA increases.

Miwa et al. [9] investigated the effect of the surface roughness on WCA of superhydrophobic surfaces by comparing various roughness parameters. As explained in Cassie–Baxter model, the air trapped within the surface structure plays an important role for surfaces with low sliding angles (SA). So, achieving more space in a deformed surface is beneficial to trap enough air for a better (low) SA and the condition can be fulfilled by a needle-like surface as shown in Figure 5.3. The SAs decrease with increase in the contact angle depending on the surface roughness. This model combines the formulations of Wenzel and Cassie–Baxter models and can be described as Miwa–Hashimoto model [9]. The equation is as follows:

$$\cos \theta_{MH} = r_{MH} f_{MH} \cos \theta_S + f_{MH} - 1 \tag{5.5}$$

where θ_{MH} and θ_S represent the static (equilibrium) contact angles on a rough and a flat surface, respectively. Υ_{MH} is the ratio of the side area to the bottom area of the needle (represented by a/b) and f_{MH} is the fraction of surface area of the material in contact with the liquid represented in one dimension to be $\sum b/(\sum b + \sum c)$ as shown in Figure 5.3 [9].

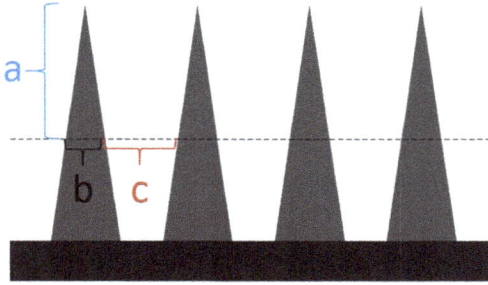

Figure 5.3: Schematic illustration of the Miwa–Hashimoto surface model with a series of uniform needles.

5.2.2 Drawbacks and Controversies on Wenzel and Cassie–Baxter Models

Based on the Wenzel model, water adhesion is more and thus high tilting angle is required to roll of the water droplet. While in the Cassie–Baxter model, the less water adhesion gives rise to a slippery superhydrophobic surface facilitating the rolling of water droplets even at no tilting. However, both these models do not consider the dynamic behavior of the water droplet. In other words, these models were accurate only along the solid–liquid–vapor TPL of the system instead of the whole contact region between the liquid droplet and the surface [10]. David Quere investigated the dynamic behavior of liquid droplets on hydrophobic surfaces and a series of papers have been published [11–14].

Öner and McCarthy [15] proved that this TPL defines the dynamic behavior of the water droplet on a solid surface. The water molecules on this contact line are the only solid–water interface molecule that moves during the movement of the water droplet on the solid surface.

In 1945, Pease clearly suggested the ideas to question Wenzel model [16], and a few years later, Bartell directly questioned the Wenzel model [17]. He explained that, whether it is a smooth or rough surface, the contact angles of droplets within the contact line were identical. C. W. Extrand and the team of Gao and McCarthy experimented on the chemically heterogeneous surfaces. According to Extrand, contact angles are determined by interactions at the contact line and not within the interfacial contact area. In addition, the heterogeneous surface buried under the liquid droplet cannot influence hysteresis [18]. Based on the investigations of Gao and McCarthy, the contact angle is determined by the interactions of the liquid and the solid at the TPL alone. Also they state that the contact area plays no role in determining the contact angle; thus they support the arguments put forwarded by Bartell and Extrand [19]. Their statement had made controversy and a few publications have commented to the arguments posed by Gao et al. [20–23]. McHale reported that the Wenzel and Cassie equations can be applied when local values of the roughness ratio are used instead of global values. He explained that with the sufficiently large relative drop size, the surface appears to be uniform to the drop and hence the averaging approach of Wenzel and Cassie can be justified is justified. Also, it does not have to be truly uniform. He also proposed that it is only applicable to apply the Wenzel and Cassie equations for the surfaces with similar and isotropic distribution everywhere [20].

According to Nosonovsky, the local values of the roughness ratio and the area fractions of chemical heterogeneity should be included in the Wenzel and Cassie equations [21]. Panchagnula et al. similarly suggested that it is necessary to calculate the area fractions in the Cassie equation only for the immediate vicinity of the contact line for getting meaningful results and they did not comment on the Wenzel equation [22]. Later, Marmur et al. fairly explained the context. According to Marmur et al., the experiments those refuted the Wenzel and Cassie–Baxter equations were carried out with drop sizes of the same order of magnitude as the wavelength of heterogeneity and thus no generalization is possible from those results [23]. Hejazi et al. explained the second-order effects of the surface by correlating that the effect of various interactions occurs during wetting such as the effect of the disjoining pressure and of crystal microstructure, defects, and grains. They termed these effects as the second-order effects and presented a model depicting how the disjoining pressure isotherm affects the wettability due to the formation of thin liquid films [24]. Very recently, Bell et al. introduced a volume-corrected Wenzel model where they applied a thermodynamic model for the wetting of surfaces (which are periodically patterned) to introduce the volume correction in the Wenzel equation. According to them, the corrected equilibrium contact angle was less than that predicted by the classic Wenzel equation. The reduction was occurred because of the non-negligible volume of liquid within the asperities compared to the total droplet volume [25].

5.3 Introducing TiO$_2$ Self-Cleaning Coatings

Nature has a plethora of hydrophilic, superhydrophilic, and superhydrophobic surfaces in its account. The contact angles of butterfly wings and lotus leaves are measured about 152° and 164°, respectively, are some of the unique among them

[2]. The drive of current research in self-cleaning surfaces is to mimic such inspired designs of nature; in other words, bioinspired surfaces are surfaces in which the fabrication of highly efficient self-cleaning surfaces is enthused from our living nature. So far, various approaches have been introduced to develop self-cleaning biomimetic surfaces showing antireflective and antifouling properties [26, 27]. Among these, photocatalytic TiO_2-based surfaces are considered the most efficacious, and they have energy and environmental applications including self-cleaning surfaces for air and water purification. Commercial TiO_2 thin films are frequently used for self-cleaning and antifogging surfaces. TiO_2-based self-cleaning surfaces work by two photoinduced phenomena; initially via photocatalysis, where the organic pollutants accumulated on the film surface are decayed on exposure to UV/visible light, and it helps for purification and self-cleaning properties [28]. Secondly, the photoinduced superhydrophilicity, where the pollutants and dust are expelled from the surface by the water film on it. However, the key challenge in the usage of TiO_2 thin films is its high refractive index which can reduce the glass transmittance [29] and quick reinstatement of hydrophobicity in dark environments which retards the self-cleaning efficacy [30]. Moreover, the wide bandgap of TiO_2 and electron–hole recombination causes severe limitations in abruption of light by the material and thereby converting undoped TiO_2 as superhydrophilic and this has been a critical issue in the practical application of self-cleaning materials [31].

Extensive researches are ongoing for the fabrication of metal- and nonmetal-doped TiO_2 to overcome the limitations such as the white color of TiO_2 that limits solar spectrum utilization beyond the UV region and rapid charge carrier recombination. In particular, metal doped systems (especially with 3d transition metal [32, 33], lanthanides [34, 35], and noble metals [36, 37]) achieve the preferred properties by narrowing the bandgap and thereby formation of intrinsic defects. For example, the addition of Y_2O_3 into TiO_2 during UV exposure results in a high number of oxygen vacancies, which accounts for the hydrophilicity of prepared nanocomposite film [38]. Additionally, Zn^{2+}-modified TiO_2 lattice shows high surface oxygen vacancies supported by different subenergy levels resulting in high visible light absorption [39].

While looking into nonmetal-modified TiO_2, carbon, nitrogen, and sulfur doping are the most promising and show superior upgrades in visible light absorption of TiO_2 by bandgap narrowing [40–42]. These can be achieved by the mixing of TiO_2 orbitals and impurity, resulting in oxygen vacancies, and the establishment of localized energy levels in between the valence band (VB) and the conduction band (CB) [41]. N-doped TiO_2 films promise that noticeable conversion to hydrophilic surface especially results from the narrowed bandgap reflecting the mixing of 2p orbitals of nitrogen and oxygen and also the formation of oxygen vacancies [42]. Irie et al. studied the effect of concentration of carbon for UV–visible light-irradiated hydrophilic conversion in C-doped TiO_2 thin films [43]. Quite a lot of methods have been explored to produce S–N [44], C–N [45], and N–F [46, 47] Co-doped TiO_2 to improve the photocatalytic activity under visible light and hinder the recombination of charge carriers. In N, S-co-doped TiO_2 films by mixing of N 2p, S 3p, and O 2p orbitals exhibit better photoinduced hydrophilic conversion [48]. Xu et al. introduced C–N–F-co-doped TiO_2 coatings possess the self-cleaning activity and more photocatalytic activity for the degradation of stearic acid accomplished by the synergistic effect of doped C, N, and F atoms as well as high surface area [49].

Furthermore, dispersed PdO nanoparticle-based N, F-co-doped TiO_2 nanotube arrays have been prepared with fast translation to superhydrophilic surface on exposure to visible light irradiation which provides combined synergistic effect of N, F co-doping, and higher crystal lattice distortion of the nanotube morphology [50]. Promoted visible light activity and inhibited electron–hole recombination result from photogenerated Pd^0, surface plasmon resonance of Pd^0 nanoparticles, and optoelectronic coupling of N, F-co-doped TiO_2 with PdO nanoparticles. Metal and nonmetal Co-doped TiO_2 systems such as N–Cu [51], N–Fe [52], N–W [53], V–N [54], C–Mo [55], C–Nd [56], and S–Fe [57] have promising synergistic effects. Modifications with metals and nonmetals can generate added energy levels within the energy bandgap, in which the metal element can replace at the Ti lattice site while the nonmetal can exist as a surface species thereby increasing the overall visible light absorption [52–56]. The hydrophilic surfaces are proved to be remarkable in water purification. They usually clean the surface by photocatalysis followed by sheeting of water [58]. The surface wettability is very high for such surfaces and thus have a contact angle nearly $0°$ which can be termed as a superhydrophilic surface as described earlier.

5.4 Photocatalytic Superhydrophilic Surfaces and Mechanism

In a superhydrophilic surface, water droplets spread rapidly, and the water is removed from the surface with substantial speed. TiO_2-coated surfaces are one of the most widely used photoinduced superhydrophilic self-cleaning surfaces based on the photooxidation, which effectively degrade the organic pollutants and dust. Photocatalytic TiO_2 has attracted tremendous attention due to its exceptional physical and chemical stability, nontoxicity, cheap, photostable, and high reactivity [59, 60]. Particularly, TiO_2-based self-cleaning surfaces have been widely used in outside materials such as glass, plastic films, and tiles. The mechanism of photocatalytic self-cleaning can be explained as follows.

5.4.1 Photon Absorption

It is a well-accepted mechanism that during the absorption of photon at suitable energy, that is, higher or equal to its bandgap leads to the primary reactions wherein electron transmits from VB to CB of TiO_2 and this leads to generation of hole in the VB. In short, electron–hole pairs are generated in the outer surface region of the material because of the short penetration depth of the UV light [61, 62]. The light absorption truly depends on the bandgap energy of absorbing materials. Simply, the generation of the electron–hole pair in photocatalytic TiO_2 can be represented as follows [63]:

$$TiO_2 + h\nu \rightarrow e^- (TiO_2) + h^+ (TiO_2) \qquad (5.6)$$

However, the major challenge is the rapid recombination of electron–hole pairs resulting in reduced efficiency of TiO_2 photocatalysis. Invariably, the reunion of charge carriers brings about heat evolution due to nonirrradiative pathways or through light emission and irradiative ways [64]. Generally, the defect sites control as recombination centers on the surface or in the bulk and thereby the energy of the charge carriers is converted to the vibrational energy of lattice atoms [63]. The charge carriers that are escaped from the charge recombination reaction migrate to the surface, inducing oxidation and reduction [41]. In general mechanism, atmospheric oxygen is reduced by the photoexcited electrons resulting in the formation of superoxide radicals ($\bullet O_2^-$) or hydroperoxyl radicals ($HO_2^\cdot$) [31]. On the other hand, the photogenerated hole can oxidize surface adsorbed water or hydroxide ions (OH^-) and produce hydroxide radicals ($\bullet OH$) [31]. Repeated attacks of these reactive oxygen species eventually lead to the degradation of organic contaminants on the surface into CO_2 and water, and the surface gets cleaned [31]. Besides, the positions of VB and CB edges truly determine the redox capability of TiO_2 photocatalysts:

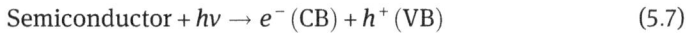

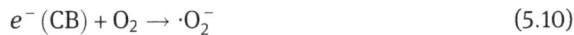

$$Semiconductor + h\nu \rightarrow e^- (CB) + h^+ (VB) \qquad (5.7)$$

$$h^+ (VB) + e^- (CB) \rightarrow Energy \qquad (5.8)$$

$$h^+ (VB) + H_2O \rightarrow \cdot OH + H^+ \qquad (5.9)$$

$$e^- (CB) + O_2 \rightarrow \cdot O_2^- \qquad (5.10)$$

The wide bandgap of TiO_2 is undesirable in photocatalysis since its absorption in the UV region of the solar spectrum which is only available in 5% [65] and thereby restricting the worth of pure TiO_2 in the production of self-cleaning materials for outdoor application such as glass, tiles, and plastic films. So far, various strategies are developed to upgrade the efficiency of TiO_2 to effectively utilize the visible region of the solar spectrum. Wang et al. reported the superhydrophilic and oleophilic character of TiO_2 under UV irradiation, whose contact angle was decreased to 0°

and the water droplets spread on the surface. Noticeably, the wettability of anatase and rutile TiO_2 was changed regardless of their photocatalytic activity. The anatase TiO_2 thin films show high transparency, photocatalytic properties, and superhydrophilicity under UV light, thereby improving its self-cleaning coatings [66]. Additionally, electrospinning technique was effective to develop photocatalytic, superhydrophilic, and optically transparent TiO_2 films suitably for photovoltaic cells and window coating applications [31, 67]. In short, the exhibited photoinduced hydrophilicity is due to the formation of light-induced vacancies on the surface and photoinduced reformation of surface hydroxyl groups [68], and finally the removal of carbonaceous layer on the TiO_2 surfaces exposed to air [31, 69].

5.4.2 Light-Induced Surface Vacancies

The photoinduced hydrophilicity comprises the creation of defects at the TiO_2 surface on UV light illumination and thereby employing the interfacial force along the solid–liquid boundary that results in the change of the contact angle [70]. Consequently, the sixth position of the photocatalytic surface is occupied by H_2O or OH^- since the surface of TiO_2 contains five coordinated Ti atoms. According to the literature, UV illumination generates oxygen vacancies at the two coordinated oxygen bridging sites, thus changing Ti^{4+} to Ti^{3+}. Finally, create hydrophilic domains due to the formation of adsorbed hydroxyl ions resulting from the dissociation of chemisorbed water molecules and these are more efficient in crystal planes (1 1 0) and (1 0 0) of rutile TiO_2 [71, 72]. However, the hydrophilicity reduces gradually in the dark atmosphere due to the slow replacement of the chemisorbed groups by oxygen molecules from the air and mechanistically can be reversed by further UV illumination. Moreover, prolonged UV irradiation generates hydrophilic–oleophobic surface owing to the rate difference in conversion of TiO_2 grains to hydrophilic [73].

In general, the rate of photoinduced hydrophilic conversion of TiO_2 film was promoted by high positive electrode potentials and retarded by the presence of hole scavenging agents. Further, the key step is the diffusion of photogenerated holes to the surface and subsequently the reconstruction of hydroxyl groups at the surface which closely depend on the contact angle [68]. After the diffusion, the holes get trapped at lattice oxygen, thereby weakening the binding energy between the Ti and lattice oxygen and finally the formation of new hydroxyl bonds from water molecules. Although the hydroxyl groups get desorbed from the surface during the absence of light, the surface gets regenerated to the original less hydrophilic state. Fortunately, the superhydrophobic state and superhydrophilicity can be reversibly attained by UV illumination and surface alteration [74].

5.4.3 Photooxidation of Adsorbed Hydrocarbon

Conversely, Zubkov et al. suggested the rapid initiation of surface wetting induced under UV light and highly controlled atmospheric conditions [69]. Also, the organic pollutant is photocatalytically degraded under UV light on a water droplet and is gradually increased by irradiation time, resulting in the formation of free surface without contaminant. When the surface coverage approaches zero, the rapid spreading of water droplets occurs. Usually, under UV irradiation, water gets desorbed via heating effect due to the evaporation which promotes the reduction in the size of the water droplet instantaneously with the photodegradation of organic pollutants resulting in the hydrophilic conversion [75]. Also, electronic photoexcitation is inevitable for the improvement of the hydrophilic conversion rate of TiO_2 under UV irradiation and high electrode potential.

Later, the scientific community has studied the combined result of photocatalysis and hydrophilicity in the resulting long-term self-cleaning properties [28]. Authors suggested that initially water molecules get chemisorbed on the TiO_2 surface and lead to physisorption of new water molecules via van der Waals forces or hydrogen bonds, thereby developing a barrier that avoids the contact between the surface and contaminants. Ultimately, the immediate contact of pollutants gets arrested easily by these loose water molecules (Figure 5.4). Besides, SiO_2-added TiO_2 coatings reduce the contact angle than pure TiO_2 due to the high surface acidity of Si cations that absorb more hydroxyl groups, extensive research has been focused which verifies the photodegradation of surface adsorbed contaminants, and not from the surface wetting phenomenon [76]. Moreover, the efficiency of photoinduced superhydrophilic conversion has strong influence on the intensity and wavelength of the light source, which implies that the photoexcitation and generation of charge carriers are the crucial phases [77]. Concisely, the temperature and surface acidity truly influence the hydrophilic conversion due to changes in the surface energy and entropy of the hydrate layers.

Some research groups have focused on photoabsorbing dye sensitization of TiO_2 for the semiconductor photocatalysis, photodegradation of organic pollutants, and dye-sensitized solar cells [31]. The key process is sensitization in which photoexcitation of a sensitizer molecule to the singlet or triplet electronically excited state. Further, the transfer of an excited electron from the sensitizer molecule to the CB of semiconducting material thereby generates the electron–hole pair with several reactive oxygen species and finally resulting in the degradation of organic contaminants. Generally, the sensitizer molecules are the complexes of metals such as ruthenium(II), osmium(II), platinum(II), and rhenium(II) or an organic dye especially metalloporphyrins due to its strong absorption in the visible region beneficially from its delocalized electron systems and high thermal and chemical stability [78–80]. Thiophene–catechol-based systems are also widely used as sensitizer molecules [81].

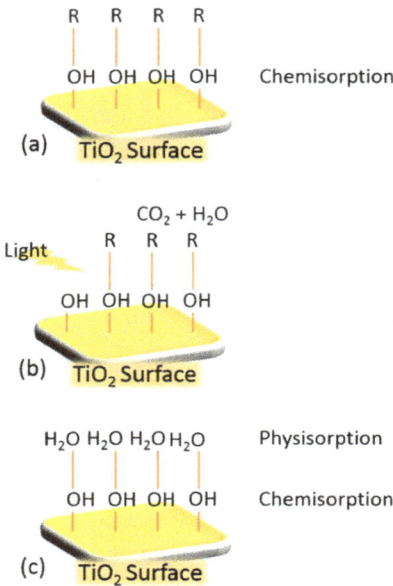

Figure 5.4: Self-cleaning mechanism of hydrophilic TiO2 thin films.

Some studies were conducted to develop the heterojunction structures of combined TiO_2 and new narrow bandgap semiconductor materials such as ZnO, WO_3, SiO_2, and CNT resulting from attractive photostability and photocatalysis due to the effective separation of charge carriers [82–85]. Concisely, the improved hydrophilicity of the composite film mainly reflects its enhanced surface area and effective interfacial charge separation ensuing from the transfer of photoexcited electron of TiO_2 to the CB of ZnO and finally photogenerated holes of ZnO get injected ZnO to the TiO_2 VB. The TiO_2/Carbon nano tube (CNT) prepared by chemical vapor deposition (CVD) technique shows photoinduced superhydrophilicity resulting from the slow electron–hole recombination at the interface [84].

5.5 Self-Cleaning Superhydrophobic Surfaces

As stated earlier, hydrophobic surface is a surface in which the interaction of the water droplet with the self-cleaning surface (contact angle) is >90°. In superhydrophobic surfaces, the rolling motion of water droplets fulfills the self-cleaning effect and it is the need to have high WCA > 150°, and the water droplets should weakly hold onto the surface such that it can be removed easily at low inclination (<10°); in other words, the water SA or contact angle hysteresis (*H*) should be less than 10°. Water droplets should pick up the dust or dirt while rolling and this is possible only when the adhesion between water and the dust particle is higher than that in between the dust particle and the solid surface [86, 87].

Superhydrophobic surfaces generally having low free energy surfaces with greater WCA (>150°) shows photoinduced self-cleaning properties. Conversely, very limited studies are done but promising benefits include reduced microbial adhesion, super-hydrophobicity – lotus effect, and anti-misting action [88, 89]. The most accepted mechanism is its photoinduced amphiphilic activity, that is, dual behavior of hydrophilic and lipophilic characteristics. In these aspects, various approaches have been formulated for the designing functionalized TiO_2 surfaces with polytetrafluoroethylene (PTFE), polydimethylsiloxane (PDMS), and fluorinated alkyl moieties. After these modifications, surface gets upgraded with promoted surface roughness and retarded surface free energy. Fluoroalkylsilane-coated calcined mixture of aluminum acetylacetonate and titanium acetyl acetonate film was developed with correlated roughness and porosity of surface [90]. Different approaches have been proposed to develop efficient superhydrophobic surfaces of TiO_2 photocatalyst with calcium hydroxyapatite (HAP) [91], fluoroalkyl silane (FAS) [92], PTFE [93], PDMS [94], high-density polyethylene (HDPE) [95], and dodecylamine [96]. The FAS-treated Ti(IV)–HAP–polymer films show good hydrophobicity which is truly due to the FAS coating that can inhibit through interaction with Ti(IV)–HAP and polymer [90, 91]. PTFE-based superhydrophobic film exhibits photoinduced photodegradation of oleic acid with retained superhydrophobicity for five cycles of oleic acid adhesion [93]. Surprisingly, PDMS-supported film of TiO_2 nanoparticles shows efficient superhydrophobic nature owing to the high surface roughness and low surface energy of the polymer with attractive superhydrophobic stability [94]. Multifunctional HDPE-based nanocomposite coatings with surface roughness ranging from microscale to nanoscale was fabricated excellent superhydrophobicity with a low slip-off angle and noticeable self-cleaning properties and explained the reversible hydrophobicity by heating [95]. The improved hydrophobicity of the hybrid surfaces was ascribed to the apparent orientation of hydrocarbon chains. In conclusion, the restorability and reversibility of the super-hydrophobic surface ultimately reduce the mechanical impairment and promising applications in self-cleaning materials.

5.6 Self-Cleaning TiO$_2$: Recent Advancements

The self-cleaning TiO_2 coatings are widely accepted owing to favorable physico-chemical properties of the material, including benign, chemically nonreactive when not exposed to light, inexpensive, and easy to work with. Moreover, TiO_2 has already been used in household chemicals like paints and pigments. The strong oxidizing power and superhydrophilic nature of TiO_2 make it a good choice for self-cleaning coatings, particularly for outdoor applications [97]. Figure 5.5 illustrates the momentum of

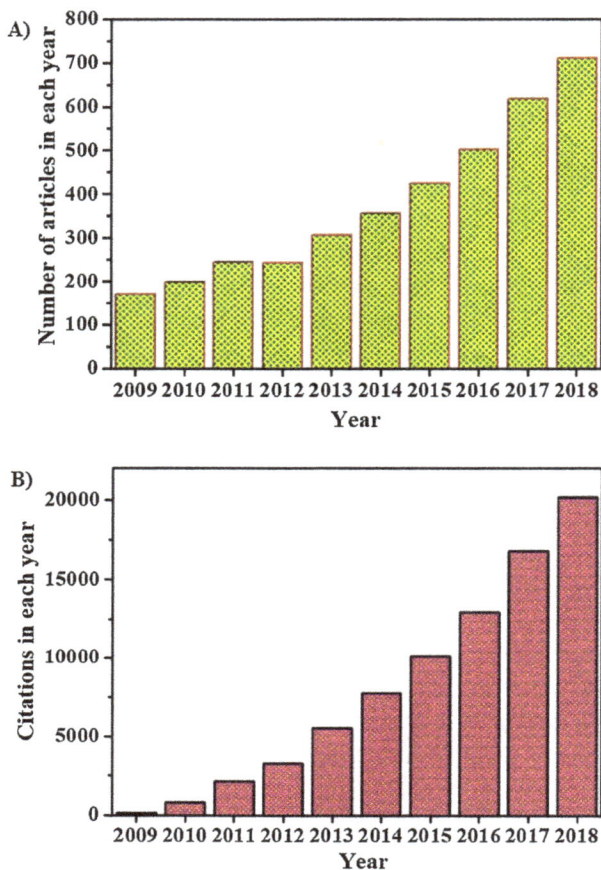

Figure 5.5: Numbers of papers published (A) and numbers of times cited (B) from 2009 to 2018 under the topics of self-cleaning (source: Web of Science).

research on self-cleaning surfaces during the last few years from the number of published papers and the number of times they were cited.

Nanostructured TiO_2-based coatings are applied to building materials and cultural heritages decisively for photocatalytic depollution and self-cleaning. Water and alcohol-derived simple TiO_2 coatings offered photocatalytic self-cleaning surface on limestones [98]. Nitrogen-doped TiO_2 coatings were fabricated for the protection of travertine stones [99]. The relative humidity (RH) of the occupied environment has an impact on the performance and life cycle of TiO_2 construction materials. Humidity influences the effective photocatalytic exclusion of organic stains. The effect was monitored for the photocatalytic performance of nonporous and porous TiO_2 self-cleaning coatings. The nonporous and porous coatings showed the finest performance at RH = 53.4% and RH ≥ 35.1%, respectively [100]. TiO_2 nanoparticles applied along with the acrylic (FosBuild) and fluorinated (Akeograd P) binders have shown high superhydrophilicity, extended durability, and effective self-cleaning action for the lithotypes Carrara marble and the Noto calcarenite stones that are widely employed in built heritages [101].

Along with the degradation of organic pollutants, the photocatalytic TiO_2 may deteriorate other coating components. To protect the sensitive layers, an interlayer is often applied to shield the undesirable effect of the photocatalyst. Perceptibly, this would complicate the practical application. Fei Xu et al. have successfully developed self-cleaning TiO_2 coatings consisting of a transparent TiO_2 upper layer deposited on an opaque latex film. These coatings were functional under outdoor exposure without producing any damage to the matrix [102]. Self-cleaning $PdO–TiO_2$ thin films were proposed for the survival of technical surface outdoors [103]. Hadis et al. were successfully established photocatalytic L-histidine–TiO_2–CdS self-cleaning composite membrane [104].

Superhydrophobic TiO_2-embedded fluoropolymer coating was developed for imparting self-cleaning property on stones. The field studies conducted revealed that the titania was not disturbing the hydrophobicity of fluoropolymer. Moreover, TiO_2/fluoropolymer coating conserved the stone surface cleaner than the bare polymer coating. Obviously, the photocatalytic property of the coating was dependent on the TiO_2 load [105]. Manganese-doped multifunctional photocatalytic and biocompatible TiO_2 has exhibited self-cleaning ability on cotton fabrics. They are intended for the self-cleaning of textile materials including for sports, medicinal uses, clothing, home furnishing, and leisure doings [106]. Similarly, Co-doped TiO_2 was useful for self-cleaning purpose with antibacterial properties [107].

In contrast to the wide applicability, pure TiO_2 has a drawback that quickly reinstates the WCA under dark conditions consecutively reducing the window transmittance. Intending to manage this downside, silica (SiO_2) supports have been introduced, and $TiO_2–SiO_2$ composites were extensively studied. For instance, self-cleaning photocatalytic $TiO_2–SiO_2$ coatings with different compositions of TiO_2 was made and supported on lightweight polycarbonates that could preserve the superhydrophilicity in dark environments [108]. Likewise, rod-like composites have produced with high transparency and mechanical stability on glass substrates for self-cleaning applications under visible irradiation [109]. Similarly, Au-TiO_2/SiO_2 coatings were developed with remarkable self-cleaning action. Here, SiO_2 could enhance the adherence of the coatings while the gold nanoparticles influence the cleaning action and photocatalytic decontamination [110].

Silica and KH570 (silane coupling agent)-modified rutile TiO_2 nanocomposites have exhibited highly superhydrophobic nature. The quantity of coupling agent had a noticeable impact on the superhydrophobicity of the nanocomposite. $TiO_2–SiO_2$ composite modified with 0.3 mL of KH570 showed the maximum hydrophobicity and WCA apposite for the application in pigment for exterior walls [111]. Manying et al. have developed optically transparent, superhydrophilic and self-cleaning $TiO_2–SiO_2$

composites on glass substrates via sol–gel method. The composite with a single layer of SiO_2 deposited onto the TiO_2 film holds the enhanced superhydrophilicity (WCA < 3°) than the bare TiO_2 thin films [112]. TiO_2–SiO_2–PDMS nanocomposite was coated on Modica stone for self-cleaning. The effect of variable amount of TiO_2 was investigated. About 23.7 g/m^2 of TiO_2 has imparted maximum photocatalytic performance [113].

Visible light–absorbing melon–TiO_2 composites incorporated into bare chitosan surface exhibited remarkable self-cleaning properties. Melon is a graphitic carbon nitride compound having visible light absorption. Malachite green was used as the test dirt and the composite window exhibited cleaning properties for 15 days of incessant indoor lightening. Even though the self-cleaning ability was consistent, changes were experienced in the physicochemical properties and surface morphology of the coatings. Also, the Malachite green molecule possibly endured an oxidative attack by the ROS generated resulting in N-demethylated by-products and 4-dimethylaminobenzophenone [114]. Hui Liu et al. fabricated self-cleaning coatings with higher efficiency than many of the commercially available nanofiltration membranes. They have implanted TiO_2 nanorods into the graphene-based membrane on a cellulose acetate mat. The nanoparticles could expand the interlayer spacing of graphene sheets which enhanced the separation and dirt removal efficiency [115].

Xiaoju et al. have developed graphene oxide (GO)–TiO_2 membrane having improved photocatalytic activity and self-cleaning properties. GO was assembled on a polyacrylonitrile surface interrelating the GO nanosheets with polyethylenimine via layer-by-layer method. Further, the GO sheets were deposited with nano-TiO_2 by ethanol/UV-assisted posttreatment [116]. Tunable hydrophilicity and photocatalytic activity were obtained with reduced GO (rGO)–TiO_2 hybrid films. At a particular composition, the WCA tends to decline below 5° under UV and visible light exposure. This superhydrophilicity disappears and the films restore the initial contact angle when the light source was removed [117]. S. Prabhu et al. have prepared highly transparent rGO–TiO_2 composite coatings with noteworthy self-cleaning capacity within 30 min under simulated solar irradiation. The coatings are proposed for the effective consumption of solar energy for indoor and outdoor applications [118]. Polyaniline/TiO_2 superhydrophobic-modified meshes showed high flux, stability, and durability when applied for oil–water separation. Also, the meshes were multifunctional with good self-cleaning ability, photocatalytic activity, and anticorrosion properties [119].

A ternary composite system Pt/MoS_2/TiO_2 was prepared by hydrothermal method and electrochemical deposition. Pt nanoparticles were allowed to grow vertically aligned MoS_2 nanosheets embedded on TiO_2 nanotube arrays (Pt/MoS_2/TiO_2). The composites displayed excellent self-cleaning performance under UV and visible light–assisted photocatalytic activity compared to pure TiO_2 nanotubes. The synergistic action of Pt-deposited MoS_2 moieties, the ultrafast electron transfer in the ternary system and favorable bandgap accounts for the enhanced performance

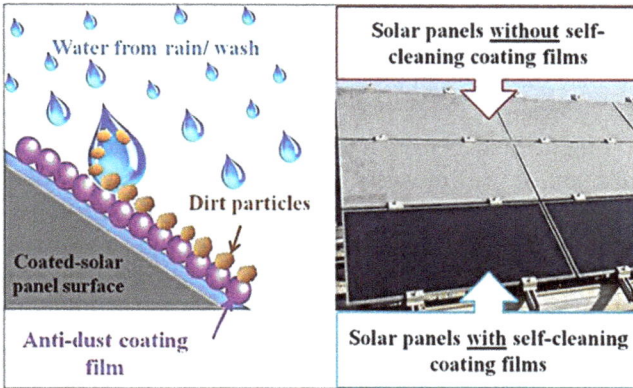

Figure 5.6: Schematic showing self-cleaning coating effect on solar panels in dusty regions [123].

[120]. TiO_2–polydopamine nanohybrid functionalized polysulfone membranes (TiO_2-PDA/PSF) showed self-cleaning and protective action on exposure to UV irradiation [121]. Yuan et al, have crafted TiO_2/CuO nanoneedle array dual-coated meshes potentially applicable for oil/water separation. The meshes are reusable, which offer superhydrophilic nature, underwater superhydrophilicity, and UV-assisted efficient self-cleaning properties [122].

Isaifan et al. have prepared photocatalytic TiO_2 nanocolloids via cost-effective polyol reduction method and was successfully applied as optically transparent and wettable thin-film coatings on borosilicate glass substrates for self-cleaning applications. The cleaning activity of the coating was monitored after several days of dust accumulation on the surface (antidust films). The schematic self-cleaning action has been depicted in Figure 5.6. The study proposes transparent coatings for self-cleaning applications in photovoltaic panels, lightweight windows, and door polycarbonates [123].

Here we have made an outlook of the recent research and practical applications of self-cleaning surfaces based on TiO_2 nanoparticles. The practice of self-cleaning technology for textile industry, toilet, solar cell, cotton fabrics, building construction, and flexible self-cleaning materials was successful. The development of some modified eco-friendly surfaces can keep better environmental conditions for long periods being time-efficient and reduced energy and laundry cost. The continuous practice of self-cleaning surfaces demands enhanced durability and stability. Several researchers have produced modified self-cleaning coatings based on TiO_2 that are multifunctional with appreciable durability and stability. However, the durability of the hydrophobic surfaces needs to be improved further in future works [124].

5.6.1 Commercialized TiO$_2$ self-cleaning materials

Self-cleaning surfaces are widely discussed in present days due to the cleaning action on materials without any external effort. TiO$_2$ photocatalytic products were commercialized in Japan during the mid-1990s. The photocatalytic property of TiO$_2$ was discovered by Akira Fujishima at the University of Tokyo in 1967. The phenomenon was entitled "Honda-Fujishima effect" and he made the first self-cleaning house, obviously his own. In cooperation with the University of Tokyo, Toto Ltd., a Japanese tile company, has constructed photocatalytic tiles, which has become a commercial product in 1994 and being successfully marketed today in Japan and under their license in various countries [125]. Later, Kazuhito Hashimoto and Toshiyo Watanabe have extended research on photocatalysis bringing the applicability in visible light. The business was fledged rapidly and more than 2,000 companies invested in the area in later decades [126]. The products were mostly intended for cleaning purposes for outdoor structural materials, interior materials, materials for road construction, purification equipment, and domestic properties (Table 5.1). The self-cleaning windows commercialized so far effectively employ transparent photocatalytic TiO$_2$ coatings such that the photodegradation of organic dirt in the presence of sunlight is viable. Furthermore, the superhydrophilicity (nearly zero contact angle) is achieved by the sheeting of water. Several companies have paid business in this area after Pilkington [127] has commercialized the first self-cleaning window in 2001. Titan Shield TM Solar Coat [128] operates with TiO$_2$-coated glass substrates in photovoltaic panels

Table 5.1: TiO$_2$-based photocatalytic products that have appeared on the market in Japan [126].

Category	Products	Properties
Exterior construction materials	Tiles, glass, tents, plastic films, aluminum panels, coatings	Self-cleaning
Interior furnishing materials	Tiles, wallpaper, window blinds	Self-cleaning, antibacterial
Road construction materials	Soundproof walls, tunnel walls, road blocks, coatings, traffic signs and reflectors, lamp covers	Self-cleaning, air cleaning
Purification facilities	Air cleaners, air conditioners, purification system for wastewater and sewage, purification system for pools	Air cleaning, water cleaning, antibacterial
Household good	Fibers, clothes, leathers, lightings, sprays	Self-cleaning, antibacterial
Others	Facilities for agricultural uses	Air cleaning, antibacterial

for enhanced transmittance and lower reflectance. Moreover, it is possible to enhance the efficiency of the solar panel with the superhydrophilic nature [129].

However, the hydrophilic nature of TiO_2 was noticed much later. TiO_2-based self-cleaning coatings are frequently used for facade sheets made up of glass or ceramic membranes. Since the self-cleaning effect is impossible with water molecules, it should be designed so that the rainwater should reach the façade. Also, it is needed in glazing to refrain from the silicon-based seals and sealants because the oils they contain were transferred to the glass. Moreover, they are unsuited with the surface coating, making it partly hydrophobic nature and subsequently unattractive streaky. Thus, the photocatalytic coatings require silicon-free sealants for practical use, and contact with the film-forming detergents should be avoided [125].
Owing to the mass production of glass materials in industrial scales, economically it is practical to apply coating via CVD, and such coatings cannot be retrofitted. It can be exclusively applied for large buildings like conservatries and winter gardens for noise barriers in rod building. In situ built tiles with durable coatings are available in market for indoor and outdoor applications. Similarly, the concrete baked with self-cleaning coatings are also being sold in our markets. One of the example is the self-cleaning roof of the famous opera house in Sydney. The photocatalytic materials are equipped with a variety of products that are available in market ranging from solar protection glasses, windows, vacuum cleaners to refrigerator or freezer deodorizing units [125].

The self-cleaning windows are not simply some pieces of glass but having a very thin outer layer coating of TiO_2, as this white powder gives amazing glow to paints, toothpaste, and all kinds of other bright white products. TiO_2 is an essential component of self-cleaning windows and it is the component that puts "white" in to the "white paint." In self-cleaning glasses, the TiO_2 outer layer is of 10–25 nm deep on the glass substrate which is about 4 mm thick such that it will not reduce the light transparency to more than 5% [130]. PURETi (pronounced "purity") Inc., a company in New York City, had designed a novel construction with the photocatalytic TiO_2. The company developed a liquid-based technique of growing TiO_2 nanocrystals that are highly photoactive. The nanocrystals are suspended in a durable and highly adhesive water-based solution. The method is useful to have several nontoxic, environmentally sound commercial photocatalytic designs for transforming any surfaces like buildings, textiles, glass products, and self-cleaning air purifier. It can keep the surfaces clean and can purify the surrounding air for at least 3 years [131].

5.6.2 Challenges, Future Perspectives, and Conclusions

The self-cleaning coatings have quite fair looks and they are designed to protect the surface for its lifetime. Since no special external effort for cleaning is needed, they are economic and save time. Moreover, the hydrophobic surfaces can avoid the risks due to slippery. The self-cleaning windows are produced with variable thickness typically in the range of 4–10 nm with blue shades to lessen the sunlight glare in places like conservatories, and with heat-reflecting inner films to enhance the energy efficiency [131]. However, there are challenges to be confronted in applying self-cleaning technology. First, the self-cleaning materials are nearly 15–20% more expensive than the traditional sealing. The self-cleaned surfaces do not look like clear and shiny as fresh planes, but definitely not dirty as an uncleaned one because the self-cleaning action occurs gradually and continually. The self-cleaning activity depends on sunlight and rain. The films need sometime to activate on the first-time installation depending on the climate. Another challenge is that the photocatalytic material only targets organic pollutants, and may not be effective for salts or sand dirt. The self-cleaning coatings may be damaged before the substrate materials' life span if we use some kind of abrasives or reactive chemicals [131]. The research on self-cleaning technology is now carrying a great momentum, and many new formulations have been put forward to overcome the limitations including stability and durability.

References

[1]   Fujishima A, Zhang X, Tryk DA. TiO$_2$ photocatalysis and related surface phenomena, Surface Science Reports, 2008, 63, 515–582.
[2]   Nishimoto S, Bhushan B. Bioinspired self-cleaning surfaces with superhydrophobicity, superoleophobicity, and superhydrophilicity, RSC Advances, 2013, 3, 671–690.
[3]   Duran IR, Laroche G. Current trends, challenges, and perspectives of anti-fogging technology: surface and material design, fabrication strategies, and beyond, Progress in Materials Science, 2019, 99, 106–186.
[4]   Young T. An Essay on the cohesion of fluids, Philosophical Transactions of the Royal Society, 1805, 95, 65–87.
[5]   Pakdel E, Wang J, Kashi S, Sun L, Wang X. Advances in photocatalytic self-cleaning, superhydrophobic and electromagnetic interference shielding textile treatments, Advances in Colloid and Interface Science, 2020, 277, 102116.
[6]   Crick CR, Parkin IP. Preparation and characterisation of super-hydrophobic surfaces, Chemistry – A European Journal, 2010, 16, 3568–3588.
[7]   Wenzel RN. Resistance of solid surfaces to wetting by water, Industrial and Engineering Chemistry, 1936, 28, 988–994.
[8]   Cassie AB, Baxter S. Wettability of porous surfaces, Transactions of the Faraday Society, 1944, 40, 546–551.
[9]   Miwa M, Nakajima A, Fujishima A, Hashimoto K, Watanabe T. Effects of the surface roughness on sliding angles of water droplets on superhydrophobic surfaces, Langmuir, 2000, 16, 5754–5760.
[10] Sun D, Böhringer KF. Self-cleaning: from bio-inspired surface modification to MEMS/ Microfluidics system integration, Micromachines, 2019, 10, 101.

[11] Richard D, Quéré D. Viscous drops rolling on a tilted non-wettable solid, Europhysics Letters, 1999, 48, 286.

[12] Richard D, Quéré D. Bouncing water drops, Europhysics Letters, 2000, 50, 769.

[13] Aussillous P, Quéré D. Liquid marbles, Nature, 2001, 411, 924–927.

[14] Richard D, Clanet C, Quéré D. Contact time of a bouncing drop, Nature, 2002, 417, 811.

[15] Öner D, McCarthy TJ. Ultrahydrophobic surfaces. Effects of topography length scales on wettability, Langmuir, 2000, 16, 7777–7782.

[16] Pease DC. The significance of the contact angle in relation to the solid surface, The Journal of Physical Chemistry, 1945, 49, 107–110.

[17] Bartell FE, Shepard JW. Surface roughness as related to hysteresis of contact angles. II. The systems paraffin–3 molar calcium chloride solution–air and paraffin–glycerol–air, The Journal of Physical Chemistry, 1953, 57, 455–458.

[18] Extrand CW. Contact angles and hysteresis on surfaces with chemically heterogeneous islands, Langmuir, 2003, 19, 3793–3796.

[19] Gao L, McCarthy TJ. How Wenzel and Cassie were wrong, Langmuir, 2007, 23, 3762–3765.

[20] McHale G. Cassie and Wenzel: Were they really so wrong?, Langmuir, 2007, 23, 8200–8205.

[21] Nosonovsky M. On the range of applicability of the Wenzel and Cassie equations, Langmuir, 2007, 23, 9919–9920.

[22] Panchagnula MV, Vedantam S. Comment on how Wenzel and Cassie were wrong by Gao and McCarthy, Langmuir, 2007, 23, 13242.

[23] Marmur A, Bittoun E. When Wenzel and Cassie are right: reconciling local and global considerations, Langmuir, 2009, 25, 1277–1281.

[24] Hejazi V, Moghadam AD, Rohatgi P, Nosonovsky M. Beyond Wenzel and Cassie–Baxter: second-order effects on the wetting of rough surfaces, Langmuir, 2014, 30, 9423–9429.

[25] Bell MS, Borhan A, Volume-Corrected Wenzel A. Model, ACS Omega, 2020, 5, 8875–8884.

[26] Zhao N, Wang Z, Cai C, Shen H, Liang F, Wang D, Wang C, Zhu T, Guo J, Wang Y, Liu X. Bioinspired materials: from low to high dimensional structure, Advanced Materials (Deerfield Beach, Fla.), 2014, 26, 6994–7017.

[27] Liu K, Tian Y, Jiang L. Bio-inspired superoleophobic and smart materials: design, fabrication, and application, Progress in Materials Science, 2013, 58, 503–564.

[28] Guan K. Relationship between photocatalytic activity, hydrophilicity and self-cleaning effect of TiO_2/SiO_2 films, Surface & Coatings Technology, 2005, 191, 155–160.

[29] Fujishima A, Liu Z, Zhang X, Murakami T. Sol–gel SiO_2/TiO_2 bilayer films with self-cleaning and antireflection properties, Solar Energy Materials and Solar Cells, 2008, 92, 1434–1438.

[30] Houmard M, Riassetto D, Roussel F, Bourgeois A, Berthomé G, Joud JC, Langlet M. Morphology and natural wettability properties of sol–gel derived $TiO_2–SiO_2$ composite thin films, Applied Surface Science, 2007, 254, 1405–1414.

[31] Banerjee S, Dionysiou DD, Pillai SC. Self-cleaning applications of TiO_2 by photoinduced hydrophilicity and photocatalysis, Applied Catalysis B: Environmental, 2015, 176, 396–428.

[32] Di Paola A, García-López E, Ikeda S, Marci G, Ohtani B, Palmisano L. Photocatalytic degradation of organic compounds in aqueous systems by transition metal doped polycrystalline TiO_2, Catalysis Today, 2002, 75, 87–93.

[33] Kitano S, Murakami N, Ohno T, Mitani Y, Nosaka Y, Asakura H, Teramura K, Tanaka T, Tada H, Hashimoto K, Kominami H. Bifunctionality of Rh^{3+} modifier on TiO_2 and working mechanism of Rh^{3+}/TiO_2 photocatalyst under irradiation of visible light, The Journal of Physical Chemistry C, 2013, 117, 11008–11016.

[34] Xu J, Ao Y, Fu D, Yuan C. A simple route for the preparation of Eu, N-codoped TiO_2 nanoparticles with enhanced visible light-induced photocatalytic activity, Journal of Colloid and Interface Science, 2008, 328, 447–451.

[35] Reszczyńska J, Grzyb T, Sobczak JW, Lisowski W, Gazda M, Ohtani B, Zaleska A. Visible light activity of rare earth metal doped (Er^{3+}, Yb^{3+} or Er^{3+}/Yb^{3+}) titania photocatalysts, Applied Catalysis B: Environmental, 2015, 163, 40–49.

[36] Nolan NT, Seery MK, Hinder SJ, Healy LF, Pillai SC. A systematic study of the effect of silver on the chelation of formic acid to a titanium precursor and the resulting effect on the anatase to rutile transformation of TiO_2, The Journal of Physical Chemistry C, 2010, 114, 13026–13034.

[37] Linic S, Christopher P, Ingram DB. Plasmonic-metal nanostructures for efficient conversion of solar to chemical energy, Nature Materials, 2011, 10, 911–921.

[38] Zhang X, Yang H, Tang A. Optical, Electrochemical and hydrophilic properties of Y_2O_3 doped TiO_2 nanocomposite films, The Journal of Physical Chemistry B, 2008, 112, 16271–16279.

[39] Jing L, Xin B, Yuan F, Xue L, Wang B, Fu H. Effects of surface oxygen vacancies on photophysical and photochemical processes of Zn-doped TiO_2 nanoparticles and their relationships, The Journal of Physical Chemistry B, 2006, 110, 17860–17865.

[40] Gole JL, Stout JD, Burda C, Lou Y, Chen X. Highly efficient formation of visible light tunable $TiO_{2-x}N_x$ photocatalysts and their transformation at the nanoscale, The Journal of Physical Chemistry B, 2004, 108, 1230–1240.

[41] Napoli F, Chiesa M, Livraghi S, Giamello E, Agnoli S, Granozzi G, Pacchioni G, Di Valentin C. The nitrogen photoactive centre in N-doped titanium dioxide formed via interaction of N atoms with the solid, Nature and energy level of the species, Chemical Physics Letters, 2009, 477, 135–138.

[42] Irie H, Washizuka S, Yoshino N, Hashimoto K. Visible-light induced hydrophilicity on nitrogen-substituted titanium dioxide films, Chemical Communications, 2003, 11, 1298–1299.

[43] Irie H, Washizuka S, Hashimoto K. Hydrophilicity on carbon-doped TiO_2 thin films under visible light, Thin Solid Films, 2006, 510, 21–25.

[44] Xu JH, Li J, Dai WL, Cao Y, Li H, Fan K. Simple fabrication of twist-like helix N, S-codoped titania photocatalyst with visible-light response, Applied Catalysis B: Environmental, 2008, 79, 72–80.

[45] Periyat P, McCormack DE, Hinder SJ, Pillai SC. One-pot synthesis of anionic (nitrogen) and cationic (sulfur) codoped high-temperature stable, visible light active, anatase photocatalysts, Journal of Chemistry C, 2009, 113, 3246–3253.

[46] Katsanaki AV, Kontos AG, Maggos T, Pelaez M, Likodimos V, Pavlatou EA, Dionysiou DD, Falaras P. Photocatalytic oxidation of nitrogen oxides on NF-doped titania thin films, Applied Catalysis B: Environmental, 2013, 140, 619–625.

[47] Hamilton JW, Byrne JA, Dunlop PS, Dionysiou DD, Pelaez M, O'Shea K, Synnott D, Pillai SC. Evaluating the mechanism of visible light activity for N, F- using photoelectrochemistry, The Journal of Physical Chemistry C, 2014, 118, 12206–12215.

[48] Sakai YW, Obata K, Hashimoto K, Irie H. Enhancement of visible light-induced hydrophilicity on nitrogen and sulfur-codoped TiO_2 thin films, Vacuum, 2008, 83, 683–687.

[49] Xu QC, Wellia DV, Sk MA, Lim KH, Loo JS, Amal R, Tan TT. Transparent visible light activated C–N–F-codoped TiO_2 films for self-cleaning applications, Journal of Photochemistry and Photobiology A, 2010, 210, 181–187.

[50] Li Q, Shang JK. Composite photocatalyst of nitrogen and fluorine codoped titanium oxide nanotube arrays with dispersed palladium oxide nanoparticles for enhanced visible light photocatalytic performance, Environmental Science & Technology, 2010, 44, 3493–3499.

[51] Ueda E, Levkin PA. Emerging applications of superhydrophilic-superhydrophobic micropatterns, Advanced Materials (Deerfield Beach, Fla.), 2013, 25, 1234–1247.

[52] Yang M, Hume C, Lee S, Son YH, Lee JK. Correlation between photocatalytic efficacy and electronic band structure in hydrothermally grown TiO_2 nanoparticles, The Journal of Physical Chemistry C, 2010, 114, 15292–15297.

[53] Thind SS, Wu G, Chen A. Synthesis of mesoporous nitrogen–tungsten co-doped TiO_2 photocatalysts with high visible light activity, Applied Catalysis B: Environmental, 2012, 111, 38–45.

[54] Liu J, Han R, Zhao Y, Wang H, Lu W, Yu T, Zhang Y. Enhanced photoactivity of V– N codoped TiO_2 derived from a two-step hydrothermal procedure for the degradation of PCP– Na under visible light irradiation, The Journal of Physical Chemistry C, 2011, 115, 4507–4515.

[55] Li YF, Xu D, Oh JI, Shen W, Li X, Yu Y. Mechanistic study of codoped titania with nonmetal and metal ions: a case of C+ Mo codoped TiO_2, ACS Catalysis, 2012, 2, 391–398.

[56] Wu X, Yin S, Dong Q, Guo C, Kimura T, Matsushita JI, Sato T. Photocatalytic properties of Nd and C codoped TiO_2 with the whole range of visible light absorption, The Journal of Physical Chemistry C, 2013, 117, 8345–8352.

[57] Niu Y, Xing M, Zhang J, Tian B. Visible light activated sulfur and iron co-doped TiO_2 photocatalyst for the photocatalytic degradation of phenol, Catal Today, 2013, 201, 159–166.

[58] Ragesh P, Ganesh VA, Nair SV, Nair AS. A review on 'self-cleaning and multifunctional materials', Journal of Materials Chemistry A, 2014, 2, 14773–14797.

[59] Keane DA, McGuigan KG, Ibáñez PF, Polo-López MI, Byrne JA, Dunlop PS, O'Shea K, Dionysiou DD, Pillai SC. Solar for water disinfection: materials and reactor design, Catalysis Science and Technology, 2014, 4, 1211–1226.

[60] Banerjee S, Pillai SC, Falaras P, O'shea KE, Byrne JA, Dionysiou DD. New insights into the mechanism of visible light, The Journal of Physical Chemistry Letters, 2014, 5, 2543–2554.

[61] Tachibana Y, Vayssieres L, Durrant JR. Artificial photosynthesis for solar water-splitting, Nature Photonics, 2012, 6, 511.

[62] Okumura H. Magnetic field effect (MFE) on heterogeneous photocatalysis and the role of oxygen, International Journal of Magnetics and Electromagnetism, 2016, 2, 1.

[63] Guo Q, Zhou C, Ma Z, Yang X. Fundamentals of TiO_2 photocatalysis: concepts, mechanisms, and challenges, Advanced Materials (Deerfield Beach, Fla.), 2019, 50, 1901997.

[64] Serpone NI, Lawless DA, Khairutdinov R, Pelizzetti E. Subnanosecond relaxation dynamics in TiO_2 colloidal sols (particle sizes Rp= 1.0–13.4 nm). Relevance to heterogeneous photocatalysis, The Journal of Physical Chemistry, 1995, 99, 16655–16661.

[65] Ullattil SG, Narendranath SB, Pillai SC, Periyat P. Black TiO_2 nanomaterials: a review of recent advances, Chemical Engineering Journal, 2018, 343, 708–736.

[66] Zhao Z, Tan H, Zhao H, Li D, Zheng M, Du P, Zhang G, Qu D, Sun Z, Fan H. Orientated anatase TiO_2 nanocrystal array thin films for self-cleaning coating, Chemical Communications, 2013, 49, 8958–8960.

[67] Ganesh VA, Nair AS, Raut HK, Walsh TM, Ramakrishna S. Photocatalytic superhydrophilic TiO_2 coating on glass by electrospinning, RSC Advances, 2012, 2, 2067–2072.

[68] Sakai N, Fujishima A, Watanabe T, Hashimoto K. Quantitative evaluation of the photoinduced hydrophilic conversion properties of TiO_2 thin film surfaces by the reciprocal of contact angle, The Journal of Physical Chemistry B, 2003, 107, 1028–1035.

[69] Zubkov T, Stahl D, Thompson TL, Panayotov D, Diwald O, Yates JT. Ultraviolet light-induced hydrophilicity effect on TiO_2 (110)(1× 1). Dominant role of the photooxidation of adsorbed hydrocarbons causing wetting by water droplets, The Journal of Physical Chemistry B, 2005, 109, 15454–15462.

[70] Wang R, Hashimoto K, Fujishima A, Chikuni M, Kojima E, Kitamura A, Shimohigoshi M, Watanabe T. Photogeneration of highly amphiphilic TiO_2 surfaces, Advanced Materials (Deerfield Beach, Fla.), 1998, 10, 135–138.

[71] Nakajima A, Koizumi SI, Watanabe T, Hashimoto K. Effect of repeated photo-illumination on the wettability conversion of titanium dioxide, Journal of Photochemistry and Photobiology A: Chemistry, 2001, 146, 129–132.

[72] Wang R, Sakai N, Fujishima A, Watanabe T, Hashimoto K. Studies of surface wettability conversion on TiO_2 single-crystal surfaces, The Journal of Physical Chemistry B, 1999, 103, 2188–2194.

[73] Nakajima A, Koizumi SI, Watanabe T, Hashimoto K. Photoinduced amphiphilic surface on polycrystalline anatase TiO_2 thin films, Langmuir, 2000, 16, 7048–7050.

[74] Zhang X, Jin M, Liu Z, Nishimoto S, Saito H, Murakami T, Fujishima A. Preparation and photocatalytic wettability conversion of TiO_2-based superhydrophobic surfaces, Langmuir, 2006, 22, 9477–9479.

[75] Takeuchi M, Sakamoto K, Martra G, Coluccia S, Anpo M. Mechanism of photoinduced superhydrophilicity on the TiO_2 photocatalyst surface, The Journal of Physical Chemistry B, 2005, 109, 15422–15428.

[76] Miyauchi M, Nakajima A, Watanabe T, Hashimoto K. Photocatalysis and photoinduced hydrophilicity of various metal oxide thin films, Chemistry of Materials, 2002, 14, 2812–2816.

[77] Emeline AV, Rudakova AV, Sakai M, Murakami T, Fujishima A. Factors affecting UV-induced superhydrophilic conversion of a TiO_2 surface, The Journal of Physical Chemistry C, 2013, 117, 12086–12092.

[78] Afzal S, Daoud WA, Langford SJ. Photostable self-cleaning cotton by a copper (II) porphyrin/ TiO_2 visible-light photocatalytic system, ACS Applied Materials & Interfaces, 2013, 5, 4753–4759.

[79] Kuang D, Ito S, Wenger B, Klein C, Moser JE, Humphry-Baker R, Zakeeruddin SM, Grätzel M. High molar extinction coefficient heteroleptic ruthenium complexes for thin film dye-sensitized solar cells, Journal of the American Chemical Society, 2006, 128, 4146–4154.

[80] Gholamkhass B, Mametsuka H, Koike K, Tanabe T, Furue M, Ishitani O. Architecture of supramolecular metal complexes for photocatalytic CO_2 reduction: ruthenium– rhenium Bi- and tetranuclear complexes, Inorganic Chemistry, 2005, 44, 2326–2336.

[81] An BK, Hu W, Burn PL, Meredith P. New type II catechol-thiophene sensitizers for dye-sensitized solar cells, The Journal of Physical Chemistry C, 2010, 114, 17964–17974.

[82] Xiao FX. Construction of highly ordered $ZnO–TiO_2$ nanotube arrays (ZnO/TNTs) heterostructure for photocatalytic application, ACS Applied Materials & Interfaces, 2012, 4, 7055–7063.

[83] Miyauchi M, Nakajima A, Watanabe T, Hashimoto K. Photoinduced hydrophilic conversion of TiO_2/WO_3 layered thin films, Chemistry of Materials, 2002, 14, 4714–4720.

[84] Abdi Y, Khalilian M, Arzi E. Enhancement in photoinduced hydrophilicity of TiO_2/CNT nanostructures by applying voltage, Journal of Physics D: Applied Physics, 2011, 44, 255405.

[85] Etacheri V, Seery MK, Hinder SJ, Pillai SC. Nanostructured $Ti_{1-x}S_xO_{2-y}Ny$ heterojunctions for efficient visible-light-induced photocatalysis, Inorganic Chemistry, 2012, 51, 7164–7173.

[86] Sas I, Gorga RE, Joines JA, Thoney KA. Literature review on superhydrophobic self-cleaning surfaces produced by electrospinning, Journal of Polymer Science Part B: Polymer Physics, 2012, 50, 824–845.

[87] Feng XJ, Jiang L. Design and creation of superwetting/antiwetting surfaces, Advanced Materials (Deerfield Beach, Fla.), 2006, 18, 3063–3078.

[88] Ragesh P, Nair SV, Nair AS. An attempt to fabricate a photocatalytic and hydrophobic self-cleaning coating via electrospinning, RSC Advances, 2014, 4, 38498–38504.

[89] Wooh S, Koh JH, Lee S, Yoon H, Char K. Trilevel-structured superhydrophobic pillar arrays with tunable optical functions, Advanced Functional Materials, 2014, 24, 5550–5556.

[90] Nakajima A, Hashimoto K, Watanabe T, Takai K, Yamauchi G, Fujishima A. Transparent superhydrophobic thin films with self-cleaning properties, Langmuir, 2000, 16, 7044–7047.

[91] Wakamura M, Hashimoto K, Watanabe T. Photocatalysis by calcium hydroxyapatite modified with Ti (IV): albumin decomposition and bactericidal effect, Langmuir, 2003, 19, 3428–3431.

[92] Watanabe T, Yoshida N. Wettability control of a solid surface by utilizing photocatalysis, The Chemical Record, 2008, 8, 279–290.

[93] Kamegawa T, Shimizu Y, Yamashita H. Superhydrophobic surfaces with photocatalytic self-cleaning properties by nanocomposite coating of TiO_2 and polytetrafluoroethylene, Advanced Materials (Deerfield Beach, Fla.), 2012, 24, 3697–3700.

[94] Crick CR, Bear JC, Kafizas A, Parkin IP. Superhydrophobic photocatalytic surfaces through direct incorporation of titania nanoparticles into a polymer matrix by aerosol assisted chemical vapor deposition, Advanced Materials (Deerfield Beach, Fla.), 2012, 24, 3505–3508.

[95] Xu QF, Liu Y, Lin FJ, Mondal B, Lyons AM. Superhydrophobic TiO_2–polymer nanocomposite surface with UV-induced reversible wettability and self-cleaning properties, ACS Applied Materials & Interfaces, 2013, 5, 8915–8924.

[96] Kartini I, Santosa SJ, Febriyanti E, Nugroho OR, Yu H, Wang L. Hybrid assembly of nanosol titania and dodecylamine for superhydrophobic self-cleaning glass, Journal of Nanoparticle Research, 2014, 16, 2514.

[97] Nakajima A, Koitzumi S, Watanabe T, Hashimoto K. Photoinduced amphiphilic surface on polycrystalline anatase TiO_2 thin films, Langmuir, 2000, 16, 7048–7050.

[98] Angela C, Mariateresa L, Maurizio M, Sudipto P, Antonio L, Valentina A. Limestones coated with photocatalytic TiO_2 to enhance building surface with self-cleaning and depolluting abilities, Journal of Cleaner Production, 2017, 165, 1036–1047.

[99] Bergamonti L, Predieri G, Paz Y, Fornasini, Lottici PP, Bondioli F. Enhanced self-cleaning properties of N-doped TiO_2 coating for Cultural Heritage, Microchemical Journal, 2017, 133, 1–12.

[100] Hernández-López JM, Ruiz-Valdés JJ, Virginia B, Cristina S. Self-cleaning coatings for building materials: the influence of morphology and humidity in the stain removal performance, Construction and Building Materials, 2020, 237, 117692.

[101] La-Russa MF, Natalia R, Alvarez De Buergo M, Cristina MB, Antonino P, Gino MC, Silvestro AR. Nano-TiO_2 coatings for cultural heritage protection: the role of the binder on hydrophobic and self-cleaning efficacy, Progress in Organic Coatings, 2016, 91, 1–8.

[102] Fei X, Tao W, Hong YC, James B, Alvin MM, Limin W, Shuxue Z. Preparation of photocatalytic TiO_2-based self-cleaning coatings for painted surface without interlayer, Progress in Organic Coatings, 2017, 113, 15–24.

[103] Veziroglu S, Hwang J, Drewes J, Barg I, Shondo J, Strunskus T, Polonskyi O, Faupel F, Aktas OC. PdO nanoparticles decorated TiO_2 film with enhanced photocatalytic and self-cleaning properties, Materials Today Chemistry, 2020, 16, 100251.

[104] Hadis Z, Ali Akbar Z, Sirus Z. Self-cleaning properties of L-Histidine doped TiO_2-CdS/PES nanocomposite membrane: fabrication, characterization and performance, Separation and Purification Technology, 2020, 240, 116591.

[105] Colangiuli D, Lettieri M, Masieri M, Calia A. Field study in an urban environment of simultaneous self-cleaning and hydrophobic nanosized TiO_2-based coatings on stone for the protection of building surface, Science of the Total Environment, 2019, 650(2), 2919–2930.

[106] Muhammad Z, Evie LP, Giulia S, Vassilios DB, George K, Iosifina G, Ourania M, Danae V, Ilker SB, Athanassia A. Fabrication of visible light-induced antibacterial and self-cleaning cotton fabrics using manganese doped nanoparticles, ACS Applied Biomaterials, 2018, 1(4), 1154–1164.

[107] Hosseini-Zori M. Co-doped TiO_2 nanostructures as a strong antibacterial agent and self-cleaning cover: synthesis, characterization and investigation of photocatalytic activity under UV irradiation, Journal of Photochemistry and Photobiology B: Biology, 2018, 178, 512–520.

[108] Takahiro A, Sanjay SL, Suresh WG, Nitish R, Norihiro S, Hiroshi I, Kazuki K, Ken-ichi K, Kazuya N, Manabu F, Tomohiro I, Takeshi K, Makoto Y, Akira F, Chiaki T. Photocatalytic, superhydrophilic, self coating on cheap, lightweight, flexible polycarbonate substrates, Applied Surface Science, 2018, 458, 917–923.

[109] Ren Y, Li W, Cao Z, Jiao Y, Xu J, Liu P, Li S, Li X. Robust TiO_2 nanorods-SiO_2 core-shell coating with high-performance self-cleaning properties under visible light, Applied Surface Science, 2020, 509, 145377.

[110] Manue L, María JM, Hilario V, José MG. Au-TiO_2/SiO_2 photocatalysts for building materials: self-cleaning and de-polluting performance, Building and Environment, 2019, 164, 106347.

[111] Pengqi C, Bangzheng W, Xi Z, Dalu G, Yufei G, Jigui C, Ye L. Fabrication and characterization of highly hydrophobic rutile TiO2-based coatings for self-cleaning, Ceramics International, 2019, 45, 6111–6118.

[112] Manying Z, Lei E, Ruirui Z, Zhifeng L. The effect of SiO_2 on TiO_2-SiO_2 composite film for self-cleaning application, Surface Interfaces, 2019, 16, 194–198.

[113] Vincenza C, Barbara F, Alessandro G, Zoltán K, La Russa MF, Domenico M, Claudio M, Michela R, Barbara R, Silvestro AR, Valentina V. TiO_2–SiO_2–PDMS nanocomposite coating with self-cleaning effect for stone material: finding the optimal amount of TiO_2, Construction and Building Materials, 2018, 166, 464–471.

[114] Vélez-Peña E, Pérez-Obando J, Pais-Ospina D, Marín-Silva DA, Adriana P, Antonela C, Jorge AD, Laura D, Luis RP, Osorio-Vargasa P, Rengifo-Herrera JA. Self-cleaning and antimicrobial photoinduced properties under indoor lighting irradiation of chitosan films containing Melon/TiO_2 composites, Applied Surface Science, 2020, 508, 144895.

[115] Hui L, Yawei F, Jiajia S, Yao C, Zhong LW, Li H, Xiangyu C, Zhenfeng B. Self-cleaning triboelectric nanogenerator based on TiO_2 photocatalysis, Nano Energy, 2020, 70, 104499.

[116] Xiaoju Y, Lu H, Cong M, Jinfeng L. Layer-by-layer assembly of graphene oxide-TiO_2 membranes for enhanced photocatalytic and self-cleaning performance, Process Safety and Environmental, 2019, 130, 257–264.

[117] Kavitha MK, Lia R, Liya J, Honey J, Jayaraj MK. Visible light responsive superhydrophilic TiO_2/reduced graphene oxide coating by vacuum-assisted filtration and transfer method for self-cleaning application, Materials Science in Semiconductor Processing, 2020, 113, 105011.

[118] Prabhu S, Cindrella L, OhJoong K, Mohanraju K. Superhydrophilic and self-cleaning rGO-TiO_2 composite coatings for indoor and outdoor photovoltaic applications, Solar Energy Materials and Solar, 2017, 169, 304–312.

[119] Jiarou W, Xiaofang W, Song Z, Bin S, Zhi W, Jixiao W. Robust superhydrophobic mesh coated by PANI/TiO_2 nanoclusters for oil/water separation with high flux, self-cleaning, photodegradation and anti-corrosion, Separation and Purification Technology, 2020, 235, 116166.

[120] Jianing D, Jianying H, Aurelia W, Biesold-mcgee GV, Xinnan Z, Shouwei G, Shanchi W, Lai Y, Lin Z. Vertically-aligned Pt-decorated MoS_2 nanosheets coated on TiO_2 nanotube arrays enable high-efficiency solar-light energy utilization for photocatalysis and self-cleaning SERS devices, Nano Energy, 2020, 71, 104579.

[121] Wu H, Liu Y, Mao L, Jiang C, Ang J, Lu X. Doping polysulfone ultrafiltration membrane with TiO_2-PDA nanohybrid for simultaneous self-cleaning and self-protection, Journal of Membrane Science, 2017, 532, 20–29.

[122] Shaojun Y, Chen C, Aikifa R, Ruixue S, Tie-Jun Z, Simo OP, Bin L. Nanostructured TiO_2/CuO dual-coated copper meshes with superhydrophilic, underwater superoleophobic and self-cleaning properties for highly efficient oil/water separation, Chemical Engineering Journal, 2017, 328, 497–510.

[123] Rima JI, Ayman S, Wafa S, Daniel J, Wubulikasimu Y, Amir AA, Brahim A. Improved self-cleaning properties of an efficient and easy to scale up TiO_2 thin films prepared by adsorptive self-assembly, Scientific Reports, 2017, 7, 9466.

[124] Sanjeev PD, Aly MAS, Sanjay SL, Ruimin X, Rajaram SS, Saravanan N, Chang-Sik H, Kishor KS, Shanhu L. Recent Advances in durability of superhydrophobic self-cleaning technology: a critical review, Progress in Organic Coatings, 2020, 138, 105381.

[125] https://link.springer.com/content/pdf/bfm%3A978-3-7643-8321-3%2F12%2F1.pdf, accessed on 27- 06-2020.

[126] Akira F, Xintong Z. Titanium dioxide photocatalysis: present situation and future approaches, Comptes Rendus Chimie, 2006, 9(5–6), 750–760.

[127] Parkin IP, Palgrave RG. Self-cleaning coatings, Journal of Materials Chemistry, 2005, 15, 1689–1695.

[128] http://www.titanshield.co.il/pdf/E/TSSolarCoat_V1_E.pdf, on 12th June 2020.

[129] Prathapan R, Ganesh VA, Nair SV, Nair AS. A review on 'self-cleaning and multifunctional materials, Journal of Materials Chemistry A, 2014, 2, 14773.

[130] https://www.explainthatstuff.com/how-self-cleaning-windows-work.html, accessed on 27- 06-2020.

[131] https://spinoff.nasa.gov/Spinoff2012/ee_5.html, accessed on 27- 06-2020.

A. Joseph Nathanael, Tae Hwan Oh

Chapter 6
Photodegradation of Air Pollutants

6.1 Introduction

Due to continually rising air pollution levels, it holds serious danger to our planet. Release of toxic particulate matters (PM) into the atmosphere by natural or man-made sources is defined as air pollution. In the contemporary society with extensive industry and plenty of automobiles, air is perilously polluted and it has drastic impact on both the environment and public health. Air pollution management systems aim at removal or reduction of pollutants (gaseous pollutants, suspended PM, and biological agents) to an acceptable level because these pollutants cause adverse effects on human health such as metabolic anomalies, malignant growths, central nervous system failure/disorder, and renal failures. They also bring harmful effects on animal or plant life and spoil the environment (e.g., climatic changes). According to the World Health Organization (WHO), air pollution itself kills around 7 million people every year [1]. In 2016, the WHO published a world map representing the deaths attributed by ambient air pollution [2] (Figure 6.1). This estimate is due to the direct and indirect effects of air pollution.

Unlike the primary air pollutants like carbon monoxide (CO), sulfur oxides (SO_x), nitrogen oxides (NO_x), volatile organic compounds (VOCs), and PM_{10} (particles <10 μm) and $PM_{2.5}$ (particles <2.5 μm) which are emitted directly into the atmosphere by the source, secondary air pollutants such as acid rain, nitric acid (HNO_3), sulfuric acid (H_2SO_4), and ozone (O_3) are produced by the reaction between the primary pollutants with other atmospheric components. The existence of both primary and secondary pollutants in the atmosphere has led to the global necessity for the advancement of novel, better quality, and innovative but effective technologies to successfully address the challenges of air quality. This chapter mainly concentrates on the photodegradation of air pollutants. Before that it would be better to know some basics of air pollutant detection and removal principles.

In this chapter, we provide brief discussion about the detection of air pollutants and photocatalytic removal of air pollutants and its general mechanism. Further down in the chapter, materials for photocatalytic degradations of air pollutants such as TiO_2, ZnO, WO_3, graphite and 2D materials, and MXenes were discussed. Finally, this chapter concludes with the future perspective.

A. Joseph Nathanael, Centre for Biomaterials, Cellular and Molecular, Theranostics, (CBCMT) Vellore Institute of Technology (VIT), Vellore, India
Tae Hwan Oh, Department of Chemical Engineering, Yeungnam University, Gyeongsan, South Korea

https://doi.org/10.1515/9783110668483-006

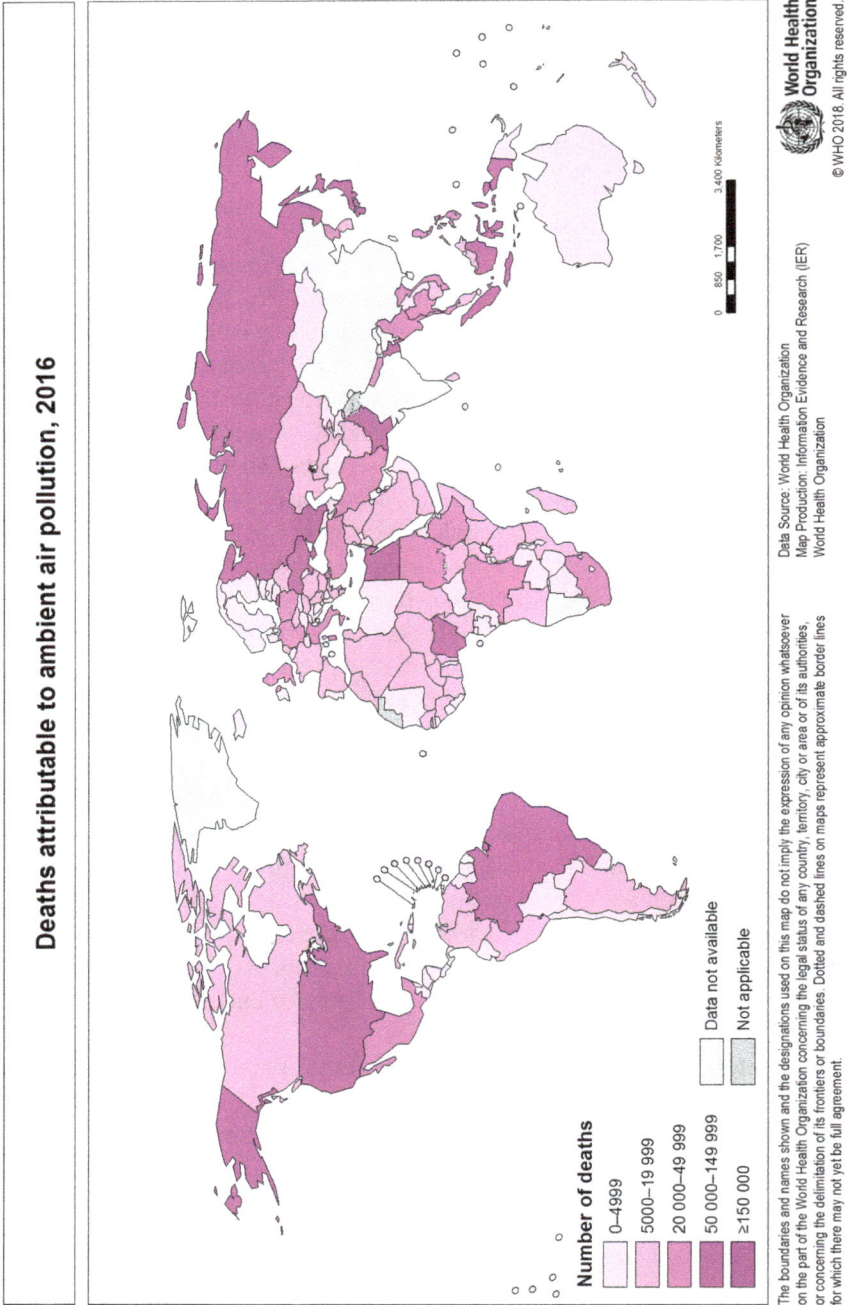

Figure 6.1: Deaths attributable to ambient air pollution in 2016. Reprinted with permission from World Health Organization© [2].

6.2 Detection of Air Pollutants

Pollution detection is of great implication in environment safety. Detection of air pollutants in both indoor and outdoor is a demanding task since the system should possess efficient detection system such as low response time, high resolution, automatic calibration, and stability in the long run even in a complex and harsh environment. In environmental protection, the contribution of pollution sensor networks is numerous due to its wide observation range, good distribution, and can work in any geographical locality. In air pollution control, the development of nanotechnology delivers conceivable benefit in the form of detection and sensing technologies [3]. Air pollution sensing arrays with higher sensitivity can be achieved through the high surface area of these nanostructured materials. In the last few years, sensing arrays have been fabricated using different kinds of nanomaterials and nanocomposites.

As a general principle, air or gas pollution detector systems usually have at least two roles: first to recognize a specific air or gas pollutant and then to convert the input signal into assessable sensing output signals. Receptor is the sensing element which is involved in some physical or chemical interactions when it is exposed to the air/gas. These signals will then be converted into a measurable output by the transducer. Based on the detection mechanism of the sensor, the output will be converted from either physical or chemicals signals. Likewise, based on the sensing material used in the sensor, the sensing mechanism will differ. For example, for sensors based on nanomaterials, atomic-level interaction takes place between the gas and the sensing element, that is, the receptor. Therefore, the air pollution sensors or gas sensors are mainly classified into semiconductor sensor, electrochemical sensors, optical sensors, and ionization sensors based on their sensing mechanism and the sensing materials used.

6.3 Removal of Air Pollutants by Photocatalytic Degradation

Removal of air pollution is the procedure employed to decrease or eradicate the harmful particulate and gaseous contaminants from the emission source, which can spoil the environment or human health as mentioned earlier. There are numerous air pollution removal methods available for controlling this global issue. Most common methods are photocatalytic degradation, adsorption, and biodegradation. Since its operational flexibility at ambient temperatures and by its unique redox properties, photocatalytic degradation is a more preferable and promising tool than other methods for the treatment of environmental pollutants such as SO_x, NO_x, CO), and VOCs [4, 5]. The principle of the photocatalysis process of the air pollution from the surface of the catalytic semiconductor materials can be classified into different phases: (1) dispersion of reactants to the surface, (ii) adsorption of reactants, (iii) reaction on the surface, (iv) desorption of products, and finally (v) diffusion of products [4]. The catalyst can be utilized to stimulate oxidation and reduction of substrates concurrently. Compared to conventional removal methods, heterogeneous photocatalysis are known to be more efficient for treating organic pollutants in the air. For example, photocatalytic degradation of air pollutants by TiO_2 can be explained as follows: electron–hole pairs are created on the conduction band (CB) and valence band (VB) when TiO_2 absorbed a photon with the energy that is higher than its band gap energy (>3.2 eV for anatase). These charge carriers form hydroxyl radicals (•OH) and superoxide anions (O_2^- on reaction with water molecules and adsorbed oxygen on the surface. These reactive oxygen species reacted with the pollutants adsorbed on the surface of TiO_2 and degrade them. The process of reactive oxygen species formation with the light irradiation on the surface of TiO_2 is schematically shown in Figure 6.2.

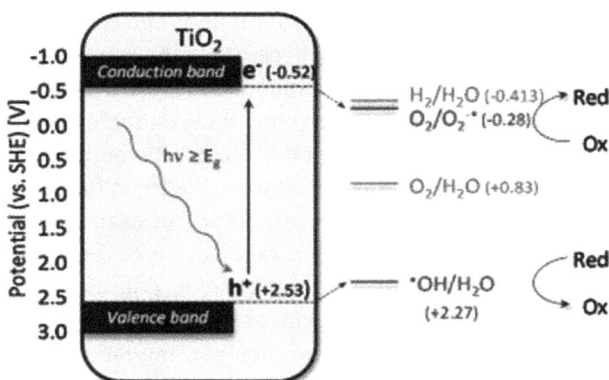

Figure 6.2: Schematic illustration of the photocatalytic degradation of air pollutants by TiO_2. Gray represents redox potentials for the water-splitting reaction; dark orange represents reactive oxygen radicals for photocatalytic degradation reactions. Reprinted with permission from Ref. [6]. Copyright (2015), Elsevier.

6.4 Materials for Photocatalytic Degradation

Several photocatalytic semiconductor materials such as TiO_2, ZnO, SnO_2, WO_3, Fe_2O_3, and CdS have been developed due to their unique photocatalytic properties for treating the air pollution. Photocatalytic degradation of a pollutant can be achieved by reactive oxidizing species produced by photogenerated electron–hole (e^-/h^+) pairs by the semiconductors (Figure 6.2). But the redox potential needed for pollution degradation is achieved only through the e^-/h^+ pairs produced by wide band gap semiconductors ($E_g > 3$ eV).

6.4.1 TiO$_2$

TiO$_2$ is one of the important photocatalysts due to its high chemical stability, availability, and brilliant catalytic properties. It has been employed in the photodegradation of compounds such as higher nitrogen containing organic compounds [7, 8], VOCs [9–12], hydrocarbons [13, 14], and other gases such as NO$_x$ [15, 16], SO$_2$ [17–20], etc. But due to its wide band gap energy (3.2 eV), TiO$_2$ is primarily triggered by ultraviolet (UV) light irradiation; hence, its application is limited. In order to improve its application to a wide range, continuous efforts are dedicated to modify its visible light absorption efficiency. Different methods such as doping and composite formation are adapted for this purpose. Nanocomposite materials and hybrid nanoparticles (NPs) are especially demonstrated to improve catalytic performances compared to its bulk counterpart. Properties of nanomaterials such as optical, catalytic, electrical, mechanical, and electrochemical are improved remarkably due to their higher surface to volume ratio. These properties are especially preferable for photocatalysis since they provide more active sites that are available for adsorption and catalysis.

Regarding the removal of VOCs by TiO$_2$, the low quantum efficiency and restricted adsorption efficiency allow it to remove only low concentrations. Hybrid or composite materials of TiO$_2$ and other materials mostly address these issues effectively. Hybrid NPs of TiO$_2$ shells on Au nanostructures by anisotropic growth were reported by Han et al. [21] and the catalytic activity of Janus-type hybrid particles showed higher catalytic activity than core–shell particles with whole TiO$_2$ shell. The geometry of this hybrid material with anisotropic structure was achieved by the step-by-step process first with seed-mediated growth of cetyltrimethylammonium bromide–coated Au nanorods (NRs) and then the Au NRs were coated with hydroxypropyl cellulose using ligand exchange process. Finally, the geometry-controlled Au–TiO$_2$ NPs were obtained by slowly hydrolyzed titanium precursor in the alkaline condition [21]. An excellent VOC decomposition material for indoor environments was reported by Qiu et al. [22] by grafting nanometer-sized Cu$_x$O clusters (Cu^I and Cu^{II} valence states) onto TiO$_2$. It was found that TiO$_2$ showed effective visible light photooxidation on VOCs by Cu^{II} species and antimicrobial properties under dark conditions with Cu^I species. Combination of VOC decomposition and anti microbial activities were achieved by balancing Cu^I and Cu^{II} species in Cu$_x$O cluster in the Cu$_x$O/TiO$_2$ hybrid nanocomposites (Figure 6.3) [22].

Photocatalytic degradation of toluene, which is a carcinogen, was reported by Zhang et al. [23] using heterostructured TiO$_2$/WO$_3$ nanocomposite under visible light condition. This nanocomposite was fabricated by electrospun of TiO$_2$ nanofibers and then WO$_3$ nanorods were grown on nanofibers by hydrothermal treatment. The mechanism of degradation of toluene was reported as follows: during the photogenerated carrier transfer between TiO$_2$ and WO$_3$, successive reduction of W^{6+} into W^{5+} also accompanied due to the capture of photogenerated electrons at the trapping sites in WO$_3$. But at the same time, due to the presence of oxygen (O$_2$) on the surface of WO$_3$,

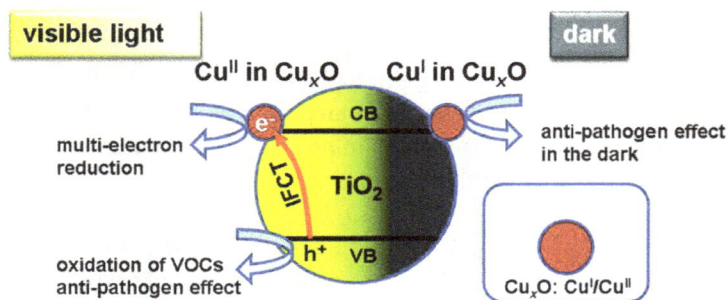

Figure 6.3: Schematic illustration of photocatalytic decomposition of VOCs and inactivation of microbes of Cu_xO/TiO_2 hybrid nanocomposites under visible light irradiation and dark conditions. Reprinted with permission from Ref. [22]. Copyright (2012), American Chemical Society.

W^{5+} were reoxidized into W^{6+}, and consequently O_2 is reduced into O_2^-. Then the hydroxyl groups (OH$^-$) captured these holes and generate hydroxyl radicals (OH$^.$). Degradation of toluene was achieved by either superoxide anions ($O_2^{.-}$) or by hydroxyl radicals (OH$^.$) [23].

Hybrid Pt–rGO (reduced graphene oxide)–TiO_2 nanomaterial was reported to have efficient VOC decomposition by highly active photothermal response catalyst under broad light wavelength absorption (800–2,500 nm) [24]. The photothermal conversion efficiency was reported as 14.1% with the conversion of toluene by 95% and 72% CO_2 yield in infrared (IR) irradiation. It also showed 50 h stability (Figure 6.4). The brilliant accomplishment of the Pt–rGO–TiO_2 nanomaterial for the VOC decomposition was expected due to the proficient translation of light into heat, improved adsorption ability, and efficient design of Pt–rGO–TiO_2 catalyst. This work demonstrated the success of the graphene-based composite to generate redox reactions through which it is possible to produce photothermal effect using solar light.

The oxidation of VOCs depends on activated forms of oxygen. Oxygen-temperature programmed desorption (O_2-TPD) measurement provided the adsorption, activation of oxygen, and lattice oxygen mobility. The study revealed that compared to 1% Pt–TiO_2 or TiO_2 NPs, 1% Pt–rGO–TiO_2 showed a powerful intensity for the O_2^- lattice desorption, which showed the enhanced surface lattice O_2 desorption of 1% Pt–rGO–TiO_2 (Figure 6.5a,b). Derived from O_2-TPD and other examination, it was concluded that via a Mars–van Krevelen mechanism (Figure 6.5c), 1% Pt–rGO–TiO_2 composite was able to react with adsorbed toluene with the help of surface lattice O_2 and produce carboxylate and CO_2 as an end product [24].

Toluene adsorption efficiencies of 98% and 77% for concentrations of <1,150 ppm and 6,900 ppm, respectively, and excellent photodegradation efficiency were achieved using nano-TiO_2/activated carbon fiber felt (TiO_2/ACFF) porous composite by Li et al. [25]. The outstanding adsorption and photodegradation efficiency were reportedly due to the synergetic effects involving the nanostructured TiO_2 and ACFF. Hierarchical

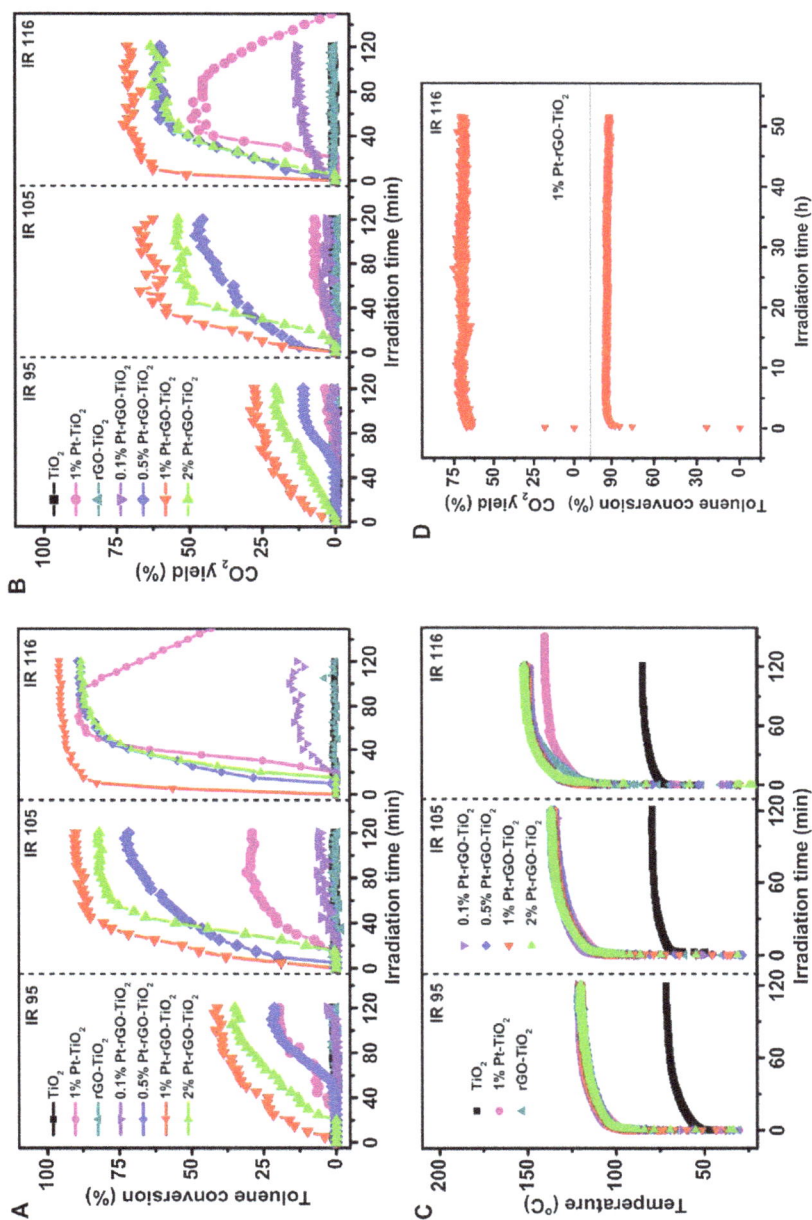

Figure 6.4: (A) Toluene conversion with respect to irradiation time; (B) CO_2 yield over various samples with different IR light intensities; (C) temperature versus irradiation time; and (D) the durability of a sample for toluene oxidation under steady-state conditions. Reprinted with permission from Ref. [24]. Copyright (2018), Elsevier.

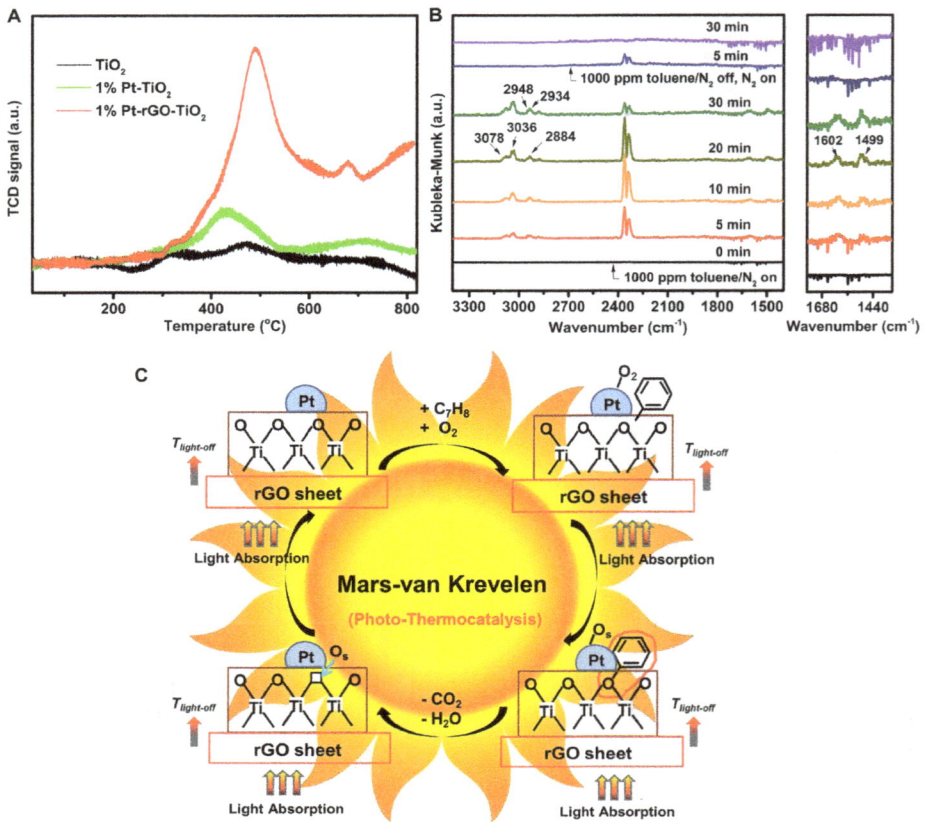

Figure 6.5: (A) O_2-TPD profiles of different samples; (B) in situ DRIFTS spectra of 1% Pt–rGO–TiO$_2$ composites heating at 150 °C; (C) schematic illustration of Mars–van Krevelen mechanism in the presence of Pt. Reprinted with permission from Ref. [24]. Copyright (2018), Elsevier.

TiO$_2$ arrangement offers room for toluene adsorption and TiO or OH groups in the TiO$_2$ supply-active sites to entice toluene. Due to large specific surface area, the ACFF acquires superior adsorption rates than TiO$_2$. Regarding photocatalytic activity, initially, the ACFF obstructs the electrons and hole recombination process and as a result increases the photocatalytic efficiency for toluene. Further, the ACFF changes the light adsorption property by shifting toward longer wavelength region (~334 nm) and decreases the band gap energy (E_g = 2.95 eV) (Figure 6.6). Due to these synergetic effects, TiO$_2$/ACFF composites provide tremendous adsorption and photodegradation efficiency for VOCs.

New kind of graphitic carbon nitride/titania (g-C$_3$N$_4$/TiO$_2$) composites together with calcium carbonate (CaCO$_3$) reported by Papailias et al. [15] showed notable enhancement in the photocatalytic degradation of NO under visible light and inhibit

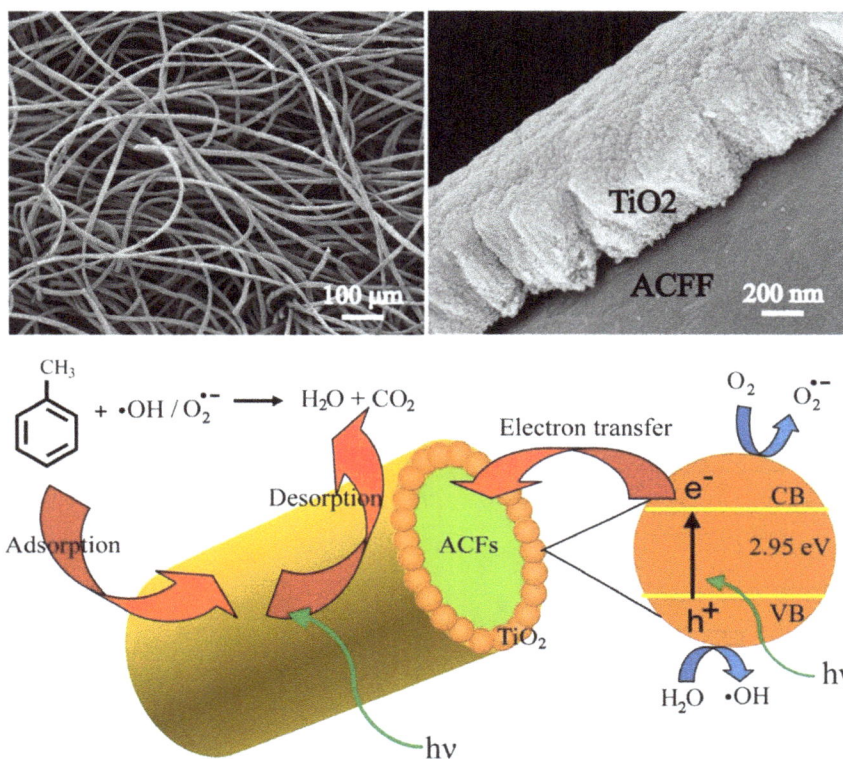

Figure 6.6: TiO_2/ACFF composites and schematic illustration of adsorption and photodegradation of TiO_2/ACFF composites. Reprinted with permission from Ref. [25]. Copyright (2017), Elsevier.

the release of NO_2 to the atmosphere. It was found that lowering of NO_2 emission was more prominent in the presence of $CaCO_3$ under both UV and visible light. The synergistic effect of the three components (TiO_2, g-C_3N_4, and $CaCO_3$) were considered to be attributing for the improved photocatalytic activity. TiO_2 and g-C_3N_4 were used to widen the light absorption from UV to visible light region. The role of $CaCO_3$ was to decrease the NO_2 product by assisting complete NO oxidation. Following figures represented the concentration curves and NO conversions of prepared composite photocatalyst in comparison with pure photocatalyst materials (Figure 6.7) (M_x – g-C_3N_4; Ti_x – TiO_2; MTi_x – g-C_3N_4/TiO_2; all samples were modified with $Ca(CH_3COO)_2$; x represents the percentage of $Ca(CH_3COO)_2$).

Similar to the above work with TiO_2 and g-C_3N_4, significance of rutile TiO_2 on the photocatalytic activity of g-C_3N_4/nitrogen-doped brookite TiO_2 composite was reported by Li et al. [16]. NO degradation under visible and UV light irradiation was performed for the photocatalytic analysis of the composite. Improved photocatalytic activity was observed on g-C_3N_4/rutile–brookite $TiO_{2-x}N_y$ over g-C_3N_4/single

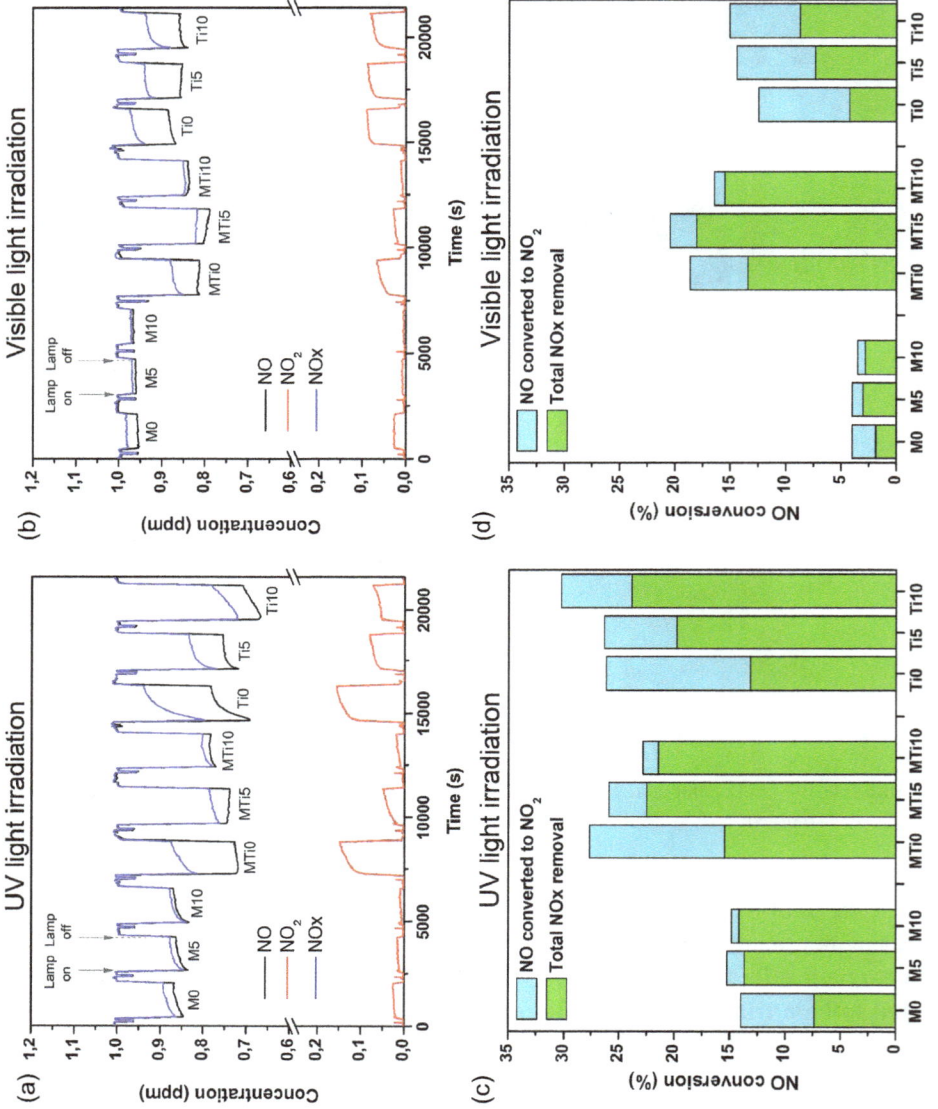

Figure 6.7: UV and visible light (a, b) concentration curves of NO, NO_2, and NO_x gases and (c, d) NO conversion of the sample groups. Reprinted with permission from Ref. [15]. Copyright (2017), Elsevier.

brookite $TiO_{2-x}N_y$ due to the formation of Z-scheme photocatalytic system. Efficient transfer of photogenerated electrons from rutile TiO_2 to g-C_3N_4 was achieved in this composite (Figure 6.8). Owing to the mixture of rutile and brookite, a heterojunction (semiembedded bicrystal structure) can be formed, which promote the separation of electron–hole pairs and hence the photocatalytic activity was enhanced.

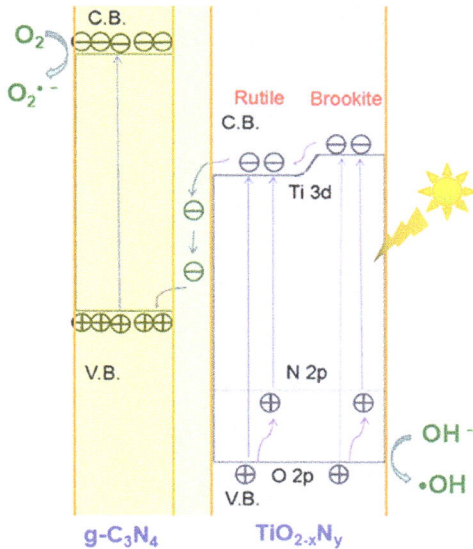

Figure 6.8: Schematic illustration of photogenerated charge carrier transfer in g-C_3N_4/rutile-brookite $TiO_{2-x}N_y$ composite. Reprinted with permission from Ref. [16]. Copyright (2017), Elsevier.

Enhanced photocatalytic performance of exfoliated graphite–TiO_2 nanocomposites was achieved through one-step process, which degraded 40% of pollutants and up to 70% of NO [26]. In this photocatalytic nanocomposite, reduction of charge recombination rate was observed within the first picoseconds of the relaxation dynamics due to an electron transfer from TiO_2 to graphite flakes. This transfer of electron increases the lifetime of trapped hole and inhibits the recombination process which contributed to the increased oxidative photocatalytic activity [26].

Composite photocatalytic material consists of TiO_2 onto the surface of recycled fine aggregates (RFA) of cementations material displayed 46.3% higher physiochemical absorption and 23.9% higher photocatalytic degradation than TiO_2/river sand composite mixture [19]. It was found that higher amount of TiO_2 can be loaded onto the surface of RFAs and ultimately it increases the photocatalytic activity. This work demonstrated the added value of green construction material and a way to produce low cost and highly effective method for SO_2 degradation. Similar kind of work was reported by Krishna et al. [20], where they used TiO_2 accommodating silicate as a coating material for buildings that degrade SO_2 effectively as the concentration of TiO_2 increases. It was reported that around 71–86% of SO_2 concentration was eliminated by 15% TiO_2 containing silicate coatings following a pseudo-first-order reaction kinetics.

TiO_2 was modified by loading Na_2CO_3 by wet coating, which showed enhanced photocatalytic oxidation of SO_2 [18]. About 0.05 and 0.2 M Na_2CO_3 loading onto TiO_2 showed photocatalytic oxidation efficiency of 1.6 and 10.6 times, respectively. Loading of Na_2CO_3 forms CO_3^- which restrains the recombination of electron–hole pairs but it simultaneously induces the formation of hydroxyl radical (•OH) through the photoreduction of CB O_2. Hence, the incorporation of Na_2CO_3 enhanced the adsorption of TiO_2 as well as supported the formation of •OH radicals, which rapidly reacted with SO_2 and effectively improved the photocatalytic oxidation.

Dry photocatalytic oxidation process to remove SO_2 using different TiO_2-based nanofibers with the loading of 5% Cu and 3% Ce and with different flue gas constituents was reported by Wang et al. [17]. SO_2 removal was 100% efficient in Ce–TiO_2 nanofibers.

6.4.2 ZnO

Apart from TiO_2, ZnO is the other mostly used semiconductor photocatalyst due to its photosensitivity, large band gap (3.37 eV), nontoxic nature, large excitation binding energy (60 meV), oxidizing power, and photochemical stability. Also compared to TiO_2, ZnO is more efficient in the absorption of deep violet/borderline UV solar spectrum at room temperature [27]. In this context, ZnO has been suggested to be used as a prominent heterogeneous photocatalyst. Similar to TiO_2, ZnO also followed the same photodegradation principle, as simply explained below: initially the pollutants diffuse to the surface adsorbed by ZnO. Then oxidation and reduction occurred in the adsorbed phase and then desorption, and finally the removal of pollutants (Figure 6.9) [27].

The photocatalytic property of ZnO can be enhanced by doping. Band gap energy and hydroxyl radicals are the main parameters for the photocatalytic activity. Using of ZnO for photocatalytic material has a shortcoming due to its low charge separation efficiency. This can be overcome by doping ZnO with metal or nonmetal impurities to adjust the band gap energy preferable for photocatalytic activity in the desired region [27, 28].

Normally, visible light photocatalytic reactions by ZnO need band gap reduction. This can be accomplished through doping by (i) raising the VB maximum, (ii) lowering the CB minimum, and (iii) establishing a local energy level inside the band gap (Figure 6.10) [29, 30].

Metal doping induces band gap narrowing in ZnO by creating intraband gap level and thus shifts the absorption edge to a visible range with <10 at.% cations [30]. A p-type and an n-type metal doping introduce a level below the original CB (Figure 6.10a) and a level above the original VB (Figure 6.10b), respectively, as an electron

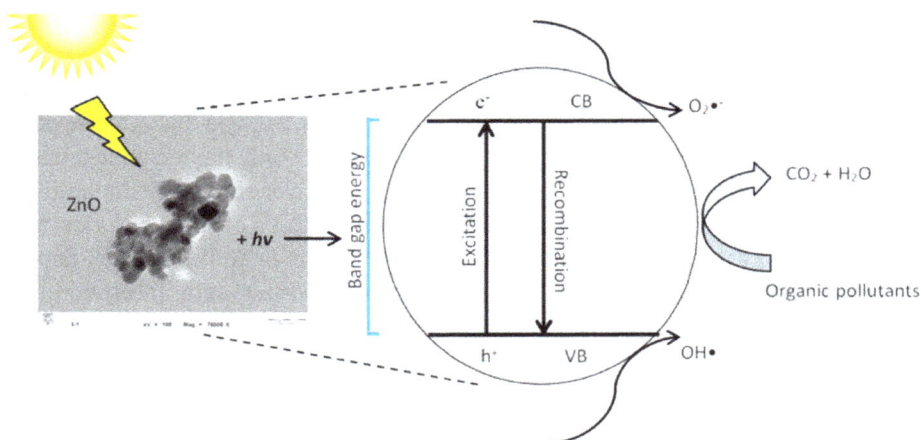

Figure 6.9: Schematic illustration of photodegradation of pollutants by ZnO in the presence of sunlight. Reprinted with permission from Ref. [27]. Copyright (2018), Elsevier.

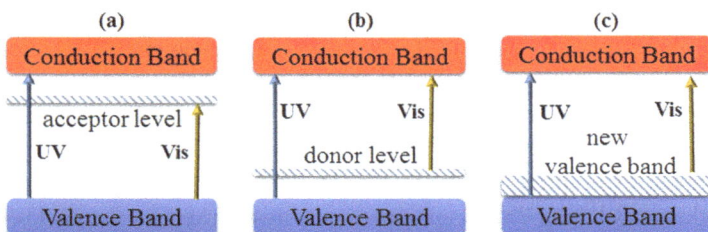

Figure 6.10: Schematic illustration of the energy levels of doped ZnO: (a) acceptor level and (b) donor level, and (c) introduction of new valence band by nonmetal doping. Reproduced with permission from Ref. [29]. Copyright (2016), Elsevier.

acceptor and an electron donor. In nonmetal doping, VB maximum is elevated by the formation of new VB state, which directs to band gap narrowing (Figure 6.10c). Hence, the chances of forming recombination centers by nonmetal doping are slim, which ultimately enhances the visible light photocatalytic activity of ZnO [29, 31–33].

Nonmetal doping such as nitrogen, sulfur, and carbon can diffuse through the lattice of ZnO due to its small sizes and shift the band gap of ZnO by replacing oxygen vacancies and provide higher number of defects on the surface of NPs [34]. For this reason, N-, C-, and S-doped ZnO displayed superior photocatalytic activity compared to pristine ZnO. The degradation efficiency of the organic pollutants also increased by the doping elements that can function as electron scavengers and can stop the electron–hole pair recombination and contribute in producing higher amount of OH$^\bullet$ radicals (free the positive hole (H$^+$)) [35–37].

Superior photocatalytic activity of N-, C-, and S-doped ZnO than pure ZnO was reported by Yu et al. [38]. Narrowing of band gap is achieved when the CB is lowered due to the doping of N and C which generated vacant state above the Fermi level. This allows visible and UV light absorption of the N- and C-doped ZnO. On the other hand, in S-doped ZnO, vacant states are absent and hence only limited enhancement of light adsorption was observed. Due to its considerable electron-deficient nature, C-doped ZnO displayed strong light absorption property [38]. Yu et al. reported the visible light photocatalytic activity of N-doped ZnO [28]. In their work, nitrogen doping was performed in semicrystalline ZnO NPs. Due to its visible light absorption, the photocatalytic degradation of organic pollutants was better in the visible light.

Kong et al. [39] developed a visible light–induced photocatalytic N-doped ZnO/g-C_3N_4 nanocomposite for the degradation of VOCs. Light absorption in the visible region was enhanced by the addition of g-C_3N_4 by creating additional charge carriers and supporting the electron–hole pair separation and movement. Addition of N to ZnO (N-ZnO) enhanced the absorption in the visible light region due to its narrow band gap energy, which enhances the photocatalytic degradation of phenol [39]. An N-doped carbon quantum dot (N-CQDs)/Ni-doped–ZnO nanocomposites for visible light photodegradation of organic pollutants was reported by Behnood et al. [40]. In this work, Ni-doped ZnO was surface modified by N-doped CQD. The existence of Ni resulted in the band gap reduction of ZnO and N-CQDs improved the electron–hole pair separation. This nanophotocatalyst revealed reusability and stability after four reaction cycles and suggested to be perfect materials for industrial applications.

Degradation of organic pollutant (4-nitrophenol) by ZnO under sunlight was reported by Rajamanickam et al. [41]. Various parameters were checked for the best results, and favorable pH, catalyst loading, and initial concentration of the pollutant (4-nitrophenol) were reported. Langmuir–Hinshelwood (L–H) kinetic model was used to discuss about the heterogeneous photocatalytic degradation kinetics.

Apart from ZnO, zinc with other materials also provides photocatalytic degradation of air pollutants. $ZnAl_2O_4$ NPs produced by solvothermal method reported to show around 90% of photocatalytic degradation of VOC (toluene) [42]. Photocatalytic degradation of VOC (gaseous toluene) by $ZnAl_2O_4$ was studied, and the efficiency was compared with the nanostructured TiO_2 samples under UV lamp irradiation in situ Fourier transform IR reactor. Result showed that after the degradation toluene was mineralized into CO_2 and H_2O [42].

6.4.3 WO$_3$

Zeolite/WO_3–Pt hybrid photocatalysts were developed for the photocatalytic degradation of air pollutants using visible light by Jansson et al. [43]. Pt-loaded WO_3 was dispersed onto zeolites (with different Si/Al ratios) and was treated by a lyophilization process. Pt was loaded by photodeposition method. WO_3–Pt-hybridized zeolites demonstrated photocatalytic degradation of acetic acid, acetaldehyde, and trichloroethylene under visible light, with enhanced performance than pure WO_3–Pt.

Enhanced detection of NO_2 gas using zeolite-modified WO_3 gas sensor was reported by Varsani et al. [44]. WO_3 was screen printed and modified with acidic and catalytic zeolite layers. Zeolite over layer was used to improve the selectivity of the sensor. This sensor can be tailored by changing different zeolites. This method showed higher stability and sensitivity, which can be used for electronic nose technology.

Zhu et al. constructed a visible light photocatalytic degradation of pharmaceuticals in the environment using WO_3-g-C_3N_4 composite [45]. Various dosages of g-C_3N_4 were used to optimize the composite, and sulfamethoxazole was used as a model antibiotic. Compared to WO_3, WO_3-g-C_3N_4 largely enhanced the photocatalytic activity efficiency of up to 91.7%. It was reported that this enhanced photocatalytic activity was attributed by the efficient separation of photogenerated electron–hole pair and redox capability due to the development of direct Z-scheme heterojunctions between g-C_3N_4 and WO_3.

Semiconductor materials combined with suitable cocatalysts (usually noble metals) enhanced the efficiency of photocatalysis. Kim et al. [46] reported about the visible light degradation of VOC using nanodiamond (ND) combined with WO_3. Nanocarbon material (ND) possesses unique sp^3 (core)/sp^2 (shell) structure and WO_3 acts as an alternative cocatalyst for Pt in their study. This combination (ND-loaded WO_3) demonstrated superior photocatalytic activity of VOC (acetaldehyde), which was 17 times superior to pure WO_3. This was reported mainly due to the structure of the carbon material, where graphitic carbon shell (sp^2) was on the diamond core (sp^3) that acts a vital role in charge separation and the successive interfacial charge transfer (Figure 6.11) [46].

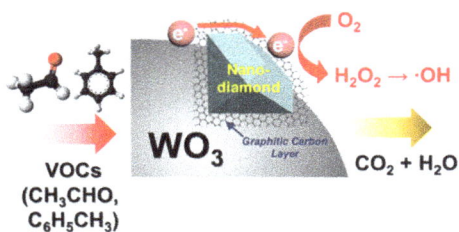

Figure 6.11: Schematic illustration of photocatalytic degradation of VOC by nanodiamond-loaded WO_3. Reproduced with permission from [46]. Copyright (2016), American Chemical Society.

Pt/Au/WO_3 photocatalyst was successfully prepared by Tanaka et al. [47] and establish that under visible light irradiation, prepared Pt/Au/WO_3 photocatalyst constantly generate H_2 and CO_2 from glycerin and O_2 by oxidation of H_2O. It was reported that due to the band gap excitation of WO_3 and the surface plasmon resonance of Au H_2 and O_2 were produced under visible light irradiation. H_2 and O_2 production by Pt/Au/WO_3 was illustrated in Figure 6.12.

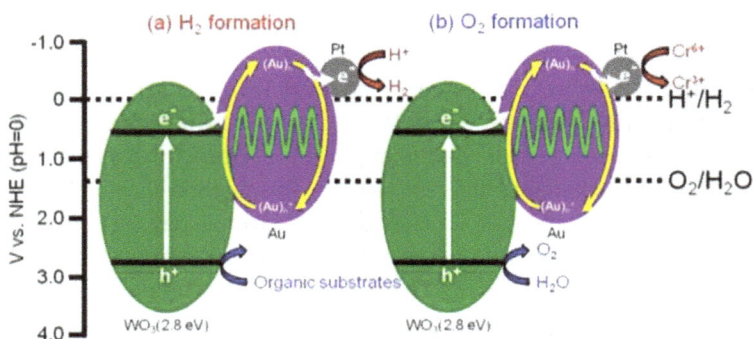

Figure 6.12: Schematic illustration of the generation of H_2 and O_2 by Pt/Au/WO$_3$. Reproduced with permission from Ref. [47]. Copyright (2014), American Chemical Society.

6.4.4 Graphite

In the past decade, graphene or reduced graphite oxide has gained much interest owing to its exceptional benefits such as huge specific surface area [48] and excellent chemical stability [49]. Two-dimensional (2D) honeycomb sp^2 carbon lattices are formed by a tightly packed single layer of carbon atoms [50–52]. Its large specific surface area makes it possible to add other inorganic photocatalyst nanomaterials, which provides outstanding adsorption property, conductivity, and controllability and thus acts as an excellent photocatalytic degradation material [53]. Together with the above-mentioned materials, graphite composites are shown to have enhanced photocatalytic degradation of pollutants.

Guidetti et al. [26] investigated the photocatalytic degradation of dye and NO$_x$ using graphite/TiO$_2$ nanocomposite (Figure 6.13). This composite was prepared by liquid-phase exfoliation of graphite in the presence of TiO$_2$ and showed enhanced photocatalytic degradation of dye up to 40% and NO$_x$ up to 70%. The enhanced photocatalytic degradation was observed due to the reduction in the charge recombination rate and an increase in the competence of the reactive species photoproduction.

Three-dimensional (3D) graphene-based gels (GBGs) such as aerogels and hydrogels attracted enormous research interest due to their intrinsic properties of 3D porous structure and single graphene. In photocatalytic degradation of pollutants, their application extends from water pollutants to air pollutants such as heavy metal ions [54–56], bacteria [57–59], organic pollutants [60–63], and gaseous pollutants [64–66]. A detailed review of the 3D graphene-based gel photocatalysts was given by Zhang et al. [67]. Three-dimensional GBGs acquire more active sites and its 3D framework offers perfect support for active compounds. Further, its porous structure supports photogenerated electron transfer (Figure 6.14).

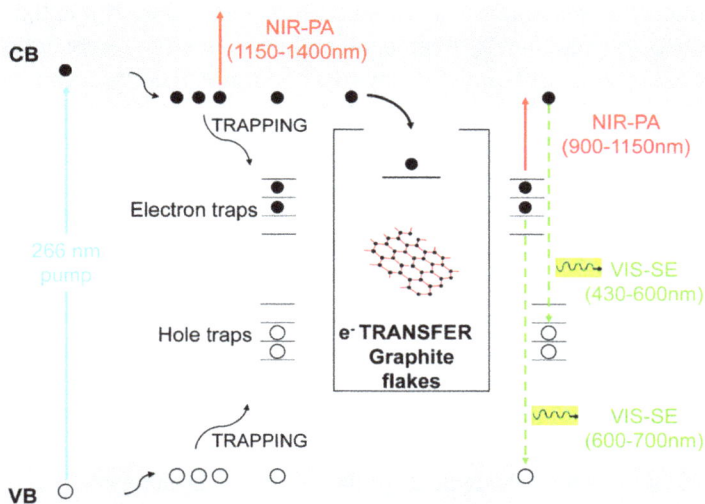

Figure 6.13: Schematic description of the optical transitions contributing to the transient absorption signals of TiO_2 and TiO_2–graphite. Reproduced with permission from Ref. [26] Copyright (2019), The Royal Society of Chemistry.

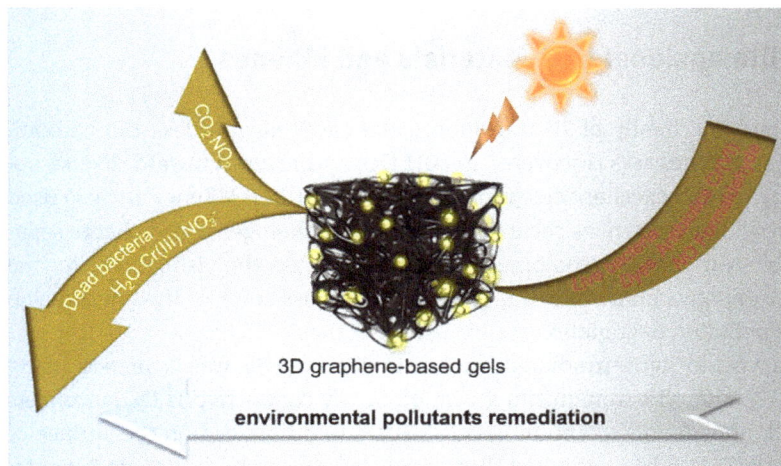

Figure 6.14: Three-dimensional graphene-based photocatalysis for environmental pollutants' remediation. Reproduced with permission from Ref. [67] Copyright (2019), Elsevier.

The 3D structure of silver phosphate/graphene hydrogel (Ag_3PO_4/rGH) exhibits enhanced visible light adsorption–photocatalytic degradation ability for the removal of bisphenol A [68]. It was found that BPA could remove 100% in 12 min under visible light irradiation. Similar BPA adsorption and photocatalysis removal in visible light using 3D GH–AgBr@rGO was reported by Chen et al. [63]. In their study, AgBr@rGO composite was prepared by encapsulating AgBr by rGO and it was incorporated into graphene and 3D rGH-AgBr@rGO hydrogel network was formed. This 3D network possessed enhanced adsorption and photocatalytic degradation of pollutants under visible light irradiation. It was reported that the BPA removal and degradation were maintained at 100% for the duration of the first 6 h continuous reaction condition.

Nitric oxide (NO) removal under visible light using g-C_3N_4 modified with a perylene imide (PI) and graphene oxide (GO) aerogels was reported by Hu et al. [66]. Results revealed that this photocatalyst material displayed enhanced activity for NO removal due to the strong visible light absorption of PI–g-C_3N_4, good charge transport properties of GO, and large specific surface area. The best photocatalytic removal of NO was reported to be 66%. These results were confirmed by density functional theory also.

rGO aerogel loaded with $BiVO_4$ quantum tube composite was reported as a proficient photocatalyst for gaseous formaldehyde degradation [69]. This enhanced photocatalysis effect was reported due to the combined effect of BVO, which can produce energetic photoelectrons necessary for oxygen reduction, and rGO–aerogel acts as a fast transfer path for photogenerated electrons. In 15 min, this catalyst degrades the formaldehyde from 1.0 to 0.4 ppm. It was found out that this photocatalyst is reusable, stable, and hence a promising photocatalyst for VOC removal.

6.4.5 Two Dimensional (2D) Materials and MXenes

A novel and budding family of 2D transition metal carbides, nitrides, and carbonitrides named as MXenes was discovered in 2011 [70] and focused toward diverse application owing to their excellent electronic properties. Further, MXenes are also used in some photocatalytic system as cocatalysts to increase photogenerated charge separation, restrain swift charge recombination, and also to use the visible light for the degradation of organic compounds [71]. Herein, we discuss some of the latest development in the pollution degradation system using MXenes.

Enhanced visible light–irradiated photocatalytic activity was achieved using Ti_3C_2 MXene [72]. TiO_2 NPs were in situ grown on highly conductive Ti_3C_2 nanosheets and then black phosphorus quantum dots (BPQDs) are introduced on the surface of TiO_2 NPs. BPQDs/Ti_3C_2@TiO_2 exhibited surprisingly improved photocatalytic degradation and hydrogen evolution reaction compared to BPQDs/Ti_3C_2 and Ti_3C_2@TiO_2 composites. Improved photocatalytic degradation was observed due to (i) the easier electron transfer of CB of TiO_2 from CB of BPQDs by the formation of intimate heterojunction by the deposition of BPQDs onto TiO_2 NPs; (ii) through electron transfer channel and with outstanding electron conductivity, migration of electrons to Ti_3C_2 nanosheets from TiO_2 NPs were swift, and hence they prevent the recombination of photogenerated carriers and extended the lifetime of electrons; (iii) added benefits for accelerated photocatalytic reaction were achieved from the superior visible light absorption and higher specific surface area of the BPQDs/Ti_3C_2@TiO_2 composite [72].

Another Ti_3C_2-based MXene modified with Bi_2WO_6 nanoplates was developed for photodegradation of VOCs by Huang et al. [73]. Bi_2WO_6 is known for its nontoxicity, stability, and visible light absorbance but it lacks electron/hole separation. In order to overcome that issue and for efficient electron–hole separation, Ti_3C_2 MXene NPs, which have impressive capability of trapping photoinduced electrons, were electrostatically adsorbed on Bi_2WO_6 nanoplates. Ti_3C_2 MXene has powerful chemical adsorption for VOCs due to charge transfer between these molecules. This photocatalyst revealed 2 and 6.6 times higher photocatalytic degradation for HCHO and CH_3COCH_3 compared to bare Bi_2WO_6 (Figure 6.15).

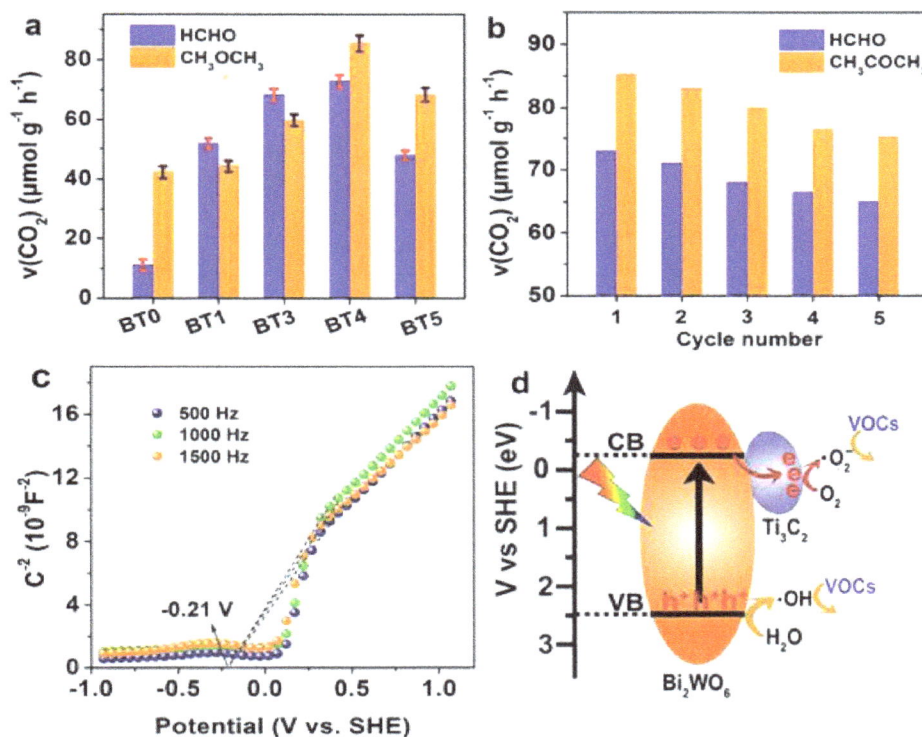

Figure 6.15: (a) Visible light photocatalytic oxidation of HCHO and CH_3COCH_3 and CO_2 production rate of Ti_3C_2 MXene modified with Bi_2WO_6 nanoplates, (b) photocatalytic activity with five cycles, (c) Mott–Schottky plots at various frequencies, and (d) schematic illustration of photocatalytic degradation mechanism of Ti_3C_2 MXene modified with Bi_2WO_6 nanoplates. Reproduced with permission from Ref. [73] Copyright (2020), Elsevier.

Photocatalytic reduction of CO_2 using TiO_2/MXene Ti_3C_2 composite for resolving both CO_2 pollution and energy crisis was prepared through in situ growth of TiO_2 on highly conductive Ti_3C_2 by the calcination method [74]. Even distribution of TiO_2 NP on Ti_3C_2 was attained. Compared to commercial TiO_2 (P25), this MXene composite (TiO_2/Ti_3C_2) showed higher photocatalytic reduction of CO_2 around 3.7 times. The main reason for the enhanced photocatalytic activity was reported owing to the large surface area, in situ growth which provides intimate contact that accelerated photogenerated electron transfer from TiO_2 to Ti_3C_2, and the inhibition of electron–hole pair recombination by T_3C_2, which acts as an electron reservoir and thus improves the accumulation of electrons.

Selective detection of VOCs was achieved by the combination of MXene and transition metal dichalcogenide nanohybrid material by Chen et al. [75]. In their work, $Ti_3C_2T_x$ and WSe_2 were used for hybridization and executed for the detection of various VOCs. This combination of hybrid sensor showed ultrafast response/recovery times, good flexibility, and lower noise level for a range of VOCs. The abundant heterojunction interfaces produced by the nanohybrids ($Ti_3C_2T_x$/WSe_2) were reported to be the reason for the improvement of sensing toward oxygen-containing VOCs. The sensitivity toward ethanol for this sensor was enhanced by over 12-fold compared to pure $Ti_3C_2T_x$ (Figure 6.16). Thus, the flexible sensors reported here have a high potential for use as practical gas-sensing devices.

MXenes can also be employed to construct sensors for VOC detection. It was reported that $Ti_3C_2T_x$ MXene showed superior gas-sensing properties with an ultra-high signal-to-noise ratio and a very low detection limit of 50–100 ppb for VOCs and ammonia at room temperature [76, 77]. Results revealed that this MXene has higher metallic conductivity and fully functionalized surface. These properties in the sensors are greater compared to conventional semiconductor sensors and other 2D materials. The adsorptive detection of the pollution was claimed to be mainly due to the terminal hydroxyl ($^-$OH) groups on the surface of $Ti_3C_2T_x$.

6.4.6 Conclusion and Future Perspective

Even after a long history, photocatalytic degradation of pollutants still has a bright future ahead. As a result, it appears relevant to discover the inclination of this technology in an effort to anticipate the progresses to come in the following decades. The photocatalytic degradation of air pollutants is vastly studied in the past few years using nanomaterials for visible light photodegradation. It is apparently not an insignificant mission considering the unpredictability built in any level of its development. But new approach and design of novel materials can be the two main aspects in moving forward with a reasonable degree of confidence. Development in nanotechnology provides an extraordinary control on the architecture of functional materials, especially in a bottom-up approach. Hierarchical nanomaterials

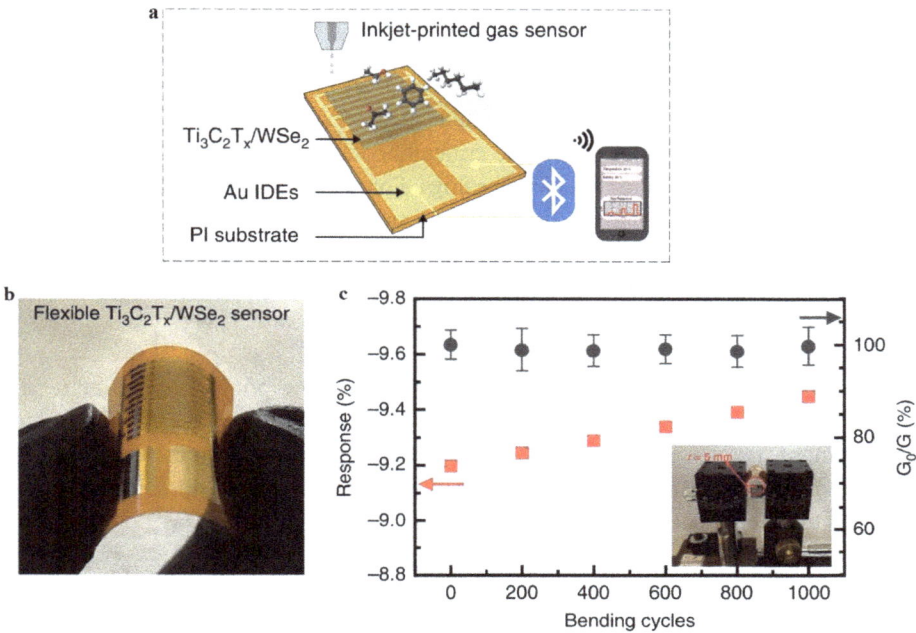

Figure 6.16: (a) Schematic diagram of printed gas sensor setup with wireless monitoring for VOC detection, (b) flexible $Ti_3C_2T_x/WSe_2$ hybrid sensor, and (c) ethanol sensing versus electrical conductivity for bending cycles. Reproduced with permission from Ref. [75] Copyright (2020), Springer Nature (under a Creative Commons Attribution).

can provide new prospects for improving the effectiveness of the photocatalyst. This can be achieved by combining the structures in very different scales. In order to overcome future challenges and improve the feasibility of the visible light photodegradation of pollutants, photocatalytic degradation experiments should be conducted using the pollutant of concern rather than a proxy. Model pollutants can be easily removable and they can be accomplished effortlessly compared to actual pollutant. For commercialization and bulk production, thorough understanding of the degradation mechanisms between the material and pollutants is essential. Enhancement in the nanostructure may still be needed for the development of the photocatalyst. New strategies should be developed for operating issues such as loss and recovery during posttreatment as well as photoactivity of recycled photocatalyst.

Graphene possesses better stability, higher surface area, and also environmentally friendly, which makes it a prospective material for environmental remediation. Graphene-based materials showed different toxicity levels by various studies based on their concentration, homogeneity, and interaction tissue type. Hence, graphene should be taken into consideration, and all its possibilities while using it for environmental applications have to be investigated. All these conditions should be considered for better and safety use of graphene materials' environmental remediation. For polar and ionic pollution indemnification, 2D materials such as MXenes have established encouraging outcome which is greater than conservative pollution absorbents and other low-dimensional materials like graphene and carbon nanotubes. But the stability of different MXenes in water should be critically considered and worth investigation their ability of absorption in that surroundings. For example, $Ti_3C_2T_x$ MXene, in water environment, can be oxidized into TiO_2 above a span of time [78]. Similarly, toxicity of MXenes also should be considered for all applications. A report found that normal as well as cancerous cells were affected by the delaminated Ti_3C_2 [76, 79]. However, these are all in the laboratory level, and still numerous open questions need to be addressed before these materials can be used commercially.

In conclusion, it is reasonable to consider that new photocatalytic materials will be effectively used in the future. In order to achieve this massive potential of development, research efforts should be investigated. It is still uncertain that how the photocatalysis research will direct in the coming years but will certainly direct to bright technological developments.

References

[1]   air-pollution @. www.who.int, n.d.
[2]   Global_aap_deaths_2016 @. http://gamapserver.who.int, Glob. Heal. Obs. Map Gall. World Heal. Organ. 2016.
[3]   Yunus IS, Kurniawan A, Adityawarman D, Sofian I, Kurniawan A, Adityawarman D. Nanotechnologies in water and air pollution treatment, Environmental Technology Reviews, 2012, 2515, 136–148. Doi: https://doi.org/10.1080%2F21622515.2012.733966.
[4]   Reddy PVL, Kim K, Kim Y. A review of photocatalytic treatment for various air pollutants, Asian Journal of Atmospheric Environment, 2011, 5, 181–188.
[5]   Hunger M, Hüsken G, Brouwers HJH. Photocatalytic degradation of air pollutants – From modeling to large scale application, Cement and Concrete Research, 2010, 40, 313–320. Doi: https://doi.org/https:%2F%2Fdoi.org%2F10.1016%2Fj.cemconres.2009.09.013.
[6]   Verbruggen SW. TiO2 photocatalysis for the degradation of pollutants in gas phase: from morphological design to plasmonic enhancement, Journal of Photochemistry and Photobiology C: Photochemistry Reviews, 2015, 24, 64–82. Doi: https://doi.org/10.1016%2Fj.jphotochemrev.2015.07.001.
[7]   Bamba D, Coulibaly M, Robert D. Science of the Total Environment Nitrogen-containing organic compounds: origins, toxicity and conditions of their photocatalytic mineralization over TiO2, Science of the Total Environment, 2017, 580, 1489–1504. Doi: https://doi.org/10.1016%2Fj.scitotenv.2016.12.130.
[8]   Jing J, Liu M, Colvin VL, Li W, Yu WW. Chemical Photocatalytic degradation of nitrogen-containing organic compounds over TiO2, Journal of Molecular Catalysis A: Chemical, 2011, 351, 17–28. Doi: https://doi.org/10.1016%2Fj.molcata.2011.10.002.
[9]   Shah KW. A review on catalytic nanomaterials for volatile organic compounds VOC removal and their applications for healthy buildings, Nanomaterials, 2019, 9, 910. (1–23).

[10] Dosa M, Piumetti M, Bensaid S, Andana T, Galletti C, Fino D. Photocatalytic abatement of volatile organic compounds by TiO2 nanoparticles doped with either phosphorous or zirconium, Materials (Basel), 2019, 12, 2121. (1–20).

[11] Sun. S, Zhang F. Insights into the Mechanism of Photocatalytic Degradation of Volatile Organic Compounds on TiO2 by Using In-situ DRIFTS. In: Cao W ed, Semicond. Photocatal. – Mater. Mech. Appl, Ed. by Wenbin Cao, Intech Open, London, 2016, 185–206. Doi: https:// doi.org/10.5772%2F62581.

[12] Jiang Z. Indoor and Built Kinetic studies on using photocatalytic coatings for removal of indoor volatile organic compounds, Indoor and Built Environment, 2019, 1, 1–12. Doi: https:// doi.org/10.1177%2F1420326X19861426.

[13] Li Y, Cai Y, Chen X, Pan X, Yang M, Yi Z. RSC Advances, 2016, 6, 2760–2767. Doi: https://doi. org/10.1039%2Fc5ra22459d.

[14] Bai H, Zhou J, Zhang H, Tang G. Biointerfaces enhanced adsorbability and photocatalytic activity of TiO2 -graphene composite for polycyclic aromatic hydrocarbons removal in aqueous phase, Colloids Surfaces B Biointerfaces, 2017, 150, 68–77. Doi: https://doi.org/10. 1016%2Fj.colsurfb.2016.11.017.

[15] Papailias I, Todorova N, Giannakopoulou T, Yu J, Dimotikali D, Trapalis C. Photocatalytic activity of modified g-C3N4/TiO2 nanocomposites for NOx removal, Catalysis Today, 2017, 280, 37–44. Doi: https://doi.org/10.1016%2Fj.cattod.2016.06.032.

[16] Li H, Wu X, Yin S, Katsumata K, Wang Y. Effect of rutile TiO2 on the photocatalytic performance of g-C3N4/brookite-TiO2-xNy photocatalyst for NO decomposition, Applied Surface Science, 2017, 392, 531–539. Doi: https://doi.org/10.1016%2Fj.apsusc.2016.09.075.

[17] Wang L, Zhao Y, Zhang J. Photochemical removal of SO2 over TiO2-based nanofibers by a dry photocatalytic oxidation process, Energy and Fuels, 2017, 31, 9905–9914. Doi: https://doi. org/10.1021%2Facs.energyfuels.7b01514.

[18] Wang H, You C, Tan Z. Enhanced photocatalytic oxidation of SO2 on TiO2 surface by Na2CO3 modification, Chemical Engineering Journal, 2018, 350, 89–99. Doi: https://doi.org/10. 1016%2Fj.cej.2018.05.128.

[19] Chen S-C, Xue-fei K. Sulfur dioxide degradation by composite photocatalysts prepared by recycled fine aggregates and nanoscale titanium dioxide, Nanomaterials, 2019, 9, 1553. (21).

[20] Krishnan P, Zhang M, Cheng Y, Riang DT, Yu LE. Photocatalytic degradation of SO2 using TiO2-containing silicate as a building coating material, Construction and Building Materials, 2013, 43, 197–202. Doi: https://doi.org/10.1016%2Fj.conbuildmat.2013.02.012.

[21] Seh ZW, Liu S, Zhang S, Bharathi MS, Ramanarayan H, Low M. Anisotropic growth of titania onto various gold nanostructures: synthesis, theoretical understanding, and optimization for catalysis, Angewandte Chemie International Edition, 2011, 50, 10140–10143. Doi: https://doi. org/10.1002%2Fanie.201104943.

[22] Qiu KHX, Miyauchi M, Sunada K, Minoshima M, Liu M, Lu Y, Li D, Shimodaira Y, Hosogi Y, Kuroda Y. Hybrid CuxO/TiO2 nanocomposites as risk-reduction materials in indoor environments, ACS Nano, 2012, 6, 1609–1618. Doi: https://doi.org/10.1021%2Fnn2045888.

[23] Zhang L, Qin M, Yu W, Zhang Q, Xie H, Sun Z, Shao Q, Guo X, Hao L, Zheng Y, Guo Z. Heterostructured TiO2/WO3 nanocomposites for photocatalytic degradation of toluene under visible light, Journal of the Electrochemical Society, 2017, 164, H1086–H1090. Doi: https:// doi.org/10.1149%2F2.0881714jes.

[24] Li J, Cai S, Yu E, Weng B, Chen X, Chen J. Efficient infrared light promoted degradation of volatile organic compounds over photo-thermal responsive Pt-rGO-TiO2 composites, Applied Catalysis B: Environmental, 2018, 233, 260–271. Doi: https://doi.org/10.1016%2Fj.apcatb. 2018.04.011.

[25] Li M, Lu B, Ke Q-F, Guo Y-J, Guo Y-P. Synergetic effect between and photodegradation on nanostructured TiO2/activated carbon fiber felt porous composites for toluene removal, Journal of Hazardous Materials, 2017, 333, 88–98. Doi: https://doi.org/https:%2F%2Fdoi.org %2F10.1016%2Fj.jhazmat.2017.03.019.

[26] Guidetti G, Pogna EAA, Lombardi L, Tomarchio F, Polishchuk I, Joosten RRM, Ianiro A, Soavi G, Sommerdijk NAJM, Friedrich H, Pokroy B, Ott AK, Goisis M, Zerbetto F, Falini G, Calvaresi M, Ferrari AC, Montalti M. Photocatalytic activity of exfoliated graphite–TiO2 nanoparticle composites, Nanoscale, 2019, 11, 19301–19314. Doi: https://doi.org/10.1039% 2Fc9nr06760d.

[27] Boon C, Yong L, Wahab A. A review of ZnO nanoparticles as solar photocatalysts: synthesis, mechanisms and applications, Renewable and Sustainable Energy Reviews, 2018, 81, 536–551. Doi: https://doi.org/10.1016%2Fj.rser.2017.08.020.

[28] Yu Z, Yin L, Xie Y, Liu G, Ma X, Cheng H. Crystallinity-dependent substitutional nitrogen doping in ZnO and its improved visible light photocatalytic activity, Journal of Colloid and Interface Science, 2013, 400, 18–23. Doi: https://doi.org/10.1016%2Fj.jcis.2013.02.046.

[29] Samadi M, Zirak M, Naseri A, Khorashadizade E, Moshfegh AZ. Recent progress on doped ZnO nanostructures for visible-light photocatalysis, Thin Solid Films, 2016, 605, 2–19. Doi: https://doi.org/10.1016%2Fj.tsf.2015.12.064.

[30] Coronado JM, Fresno F, Portela R. Design of Advanced Photocatalytic Materials for Energy and Environmental. 2013th ed, Springer, London, 2013.

[31] Taylor P, Lam S, Sin J, Abdullah AZ, Mohamed AR. Degradation of wastewaters containing organic dyes photocatalysed by zinc oxide: a review, Desalin, Water Treat, 2012, 41, 131–169. Doi: https://doi.org/10.1080%2F19443994.2012.664698.

[32] Chen X, Shen S, Guo L, Mao SS. Semiconductor-based photocatalytic hydrogen generation, Chemical Reviews, 2010, 110, 6503–6570. Doi: https://doi.org/10.1021%2Fcr1001645.

[33] Li J, Wu N. Semiconductor-based photocatalysts and photoelectrochemical cells for solar fuel generation: a review, Catalysis Science & Technology, 2015, 5, 1360–1384. Doi: https://doi. org/10.1039%2Fc4cy00974f.

[34] Di Valentin C, Pacchioni G. Trends in non-metal doping of anatase TiO2: B, C, N and F, Catalysis Today, 2013, 206, 12–18. Doi: https://doi.org/10.1016%2Fj.cattod.2011.11.030.

[35] Sirelkhatim A, Mahmud S, Seeni A. Review on zinc oxide nanoparticles: antibacterial activity and toxicity mechanism, Nano-Micro Letters, 2015, 7, 219–242. Doi: https://doi.org/10. 1007%2Fs40820-015-0040-x.

[36] Hosseini SM, Sarsari IA, Kameli P, Salamati H. Effect of Ag doping on structural, optical, and photocatalytic properties of ZnO nanoparticles, Journal of Alloys and Compounds, 2015, 640, 408–415. Doi: https://doi.org/10.1016%2Fj.jallcom.2015.03.136.

[37] Nenavathu BP, Rao AVRK, Goyal A, Kapoor A, Kumar R. Synthesis, characterization and enhanced photocatalytic degradation efficiency of Se doped ZnO nanoparticles using trypan blue as a model dye, Applied Catalysis A: General, 2013, 459, 106–113. Doi: https://doi.org/ 10.1016%2Fj.apcata.2013.04.001.

[38] Yu W, Zhang J, Peng T. New insight into the enhanced photocatalytic activity of N-, C- and S-doped ZnO photocatalysts, Applied Catalysis B: Environmental, 2016, 181, 220–227. Doi: https://doi.org/10.1016%2Fj.apcatb.2015.07.031.

[39] Kong J, Zhai H, Zhang W, Wang S, Zhao X, Li M, Li H, Li A, Wu D. Visible light-driven photocatalytic performance of N-doped ZnO/g-C3N4 nanocomposites, Nanoscale Research Letters, 2017, 12, 526. (1–10) Doi: https://doi.org/10.1186%2Fs11671-017-2297-0.

[40] Behnood R, Sodei G. Synthesis of N doped-CQDs/Ni doped-ZnO nanocomposites for visible light photodegradation of organic pollutants, Journal of Environmental Chemical Engineering, 2020, 8, 103821. (1–11) Doi: https://doi.org/10.1016%2Fj.jece.2020.103821.

[41] Rajamanickam D, Shanthi M. Photocatalytic degradation of an organic pollutant by zinc oxide – solar process, Arabian Journal of Chemistry, 2016, 9, S1858–S1868. Doi: https://doi.org/ 10.1016%2Fj.arabjc.2012.05.006.

[42] Li X, Zhu Z, Zhao Q, Wang L. Photocatalytic degradation of gaseous toluene over ZnAl2O4 prepared by different methods: a comparative study, Journal of Hazardous Materials, 2011, 186, 2089–2096. Doi: https://doi.org/10.1016%2Fj.jhazmat.2010.12.111.

[43] Jansson I, Yoshiiri K, Hori H, García-garcía FJ, Rojas S, Sánchez B, Ohtani B, Suárez S. Visible light responsive zeolite/WO3 – Pt hybrid photocatalysts for degradation of pollutants in air, Applied Catalysis A: General, 2016, 521, 208–219. Doi: https://doi.org/10.1016%2Fj.apcata. 2015.12.015.

[44] Varsani P, Afonja A, Williams DE, Parkin IP, Binions R. Zeolite-modified WO3 gas sensors – Enhanced detection of NO2, Sensors and Actuators B: Chemical, 2011, 160, 475–482. Doi: https://doi.org/10.1016%2Fj.snb.2011.08.014.

[45] Zhu W, Sun F, Goei R, Zhou Y. Construction of WO3–g-C3N4 composites as efficient photocatalysts for pharmaceutical degradation under visible light, Catalysis Science & Technology, 2017, 7, 2591–2600. Doi: https://doi.org/10.1039%2Fc7cy00529f.

[46] Kim H, Kim H, Weon S, Moon G, Kim J-H, Choi W. Robust co-catalytic performance of nanodiamonds loaded on WO3 for the decomposition of volatile organic compounds under visible light, ACS Catalysis, 2016, 6, 8350–8360. Doi: https://doi.org/10.1021%2Facscatal. 6b02726.

[47] Tanaka A, Hashimoto K, Kominami H. Visible-Light-Induced hydrogen and oxygen formation over Pt/Au/ WO3 photocatalyst utilizing two types of photoabsorption due to surface plasmon resonance and band-gap excitation, Journal of the American Chemical Society, 2014, 136, 586–589. Doi: https://doi.org/10.1021%2Fja410230u.

[48] Bruinsma RF, De Gennes PG, Freund JB, Levine D. A route to high surface area, porosity and inclusion of large molecules in crystals, Nature, 2004, 427, 523–527. Doi: https://doi.org/10. 1038%2Fnature02294.1.

[49] Loh KP, Bao Q, Ang PK, Yang J. The chemistry of graphene, Journal of Materials Chemistry, 2010, 20, 2277–2289. Doi: https://doi.org/10.1039%2Fb920539j.

[50] Ishigami M, Chen JH, Cullen WG, Fuhrer MS. Atomic structure of graphene on SiO2, Nano Letters, 2007, 7, 1643–1648. Doi: https://doi.org/10.1021%2Fnl070613a.

[51] Kim H, Bak S, Kim K. Li4Ti5O12/reduced graphite oxide nano-hybrid material for high rate lithium-ion batteries, Electrochemistry Communications, 2010, 12, 1768–1771. Doi: https:// doi.org/10.1016%2Fj.elecom.2010.10.018.

[52] Dong P, Hou G, Xi X, Shao R, Dong F. WO3-based photocatalysts: morphology control, activity enhancement and multifunctional applications, Environmental Science Nano, 2017, 4, 539–557. Doi: https://doi.org/10.1039%2Fc6en00478d.

[53] Zhang H, Lv X, Li Y, Wang Y, Li J. P25-Graphene Composite as a High Performance Photocatalyst, ACS Nano, 2010, 4, 380–386. Doi: https://doi.org/10.1021%2Fnn901221k.

[54] Chen Y, Xie X, Xin X, Tang Z, Xu Y. Ti3C2Tx-based three-dimensional hydrogel by a graphene oxide-assisted self-convergence process for enhanced photoredox catalysis, ACS Nano, 2019, 13, 295–304. Doi: https://doi.org/10.1021%2Facsnano.8b06136.

[55] Zou J, Liu H, Luo J, Xing Q, Du H, Jiang X, Luo X, Luo S, Suib SL. Three-Dimensional reduced graphene oxide coupled with Mn3O4 for highly efficient removal of Sb(III) and Sb(V) from water, ACS Applied Materials & Interfaces, 2016, 8, 18140–18149. Doi: https://doi.org/10. 1021%2Facsami.6b05895.

[56] Nasiri R, Arsalani N, Panahian Y. One-pot synthesis of novel magnetic three-dimensional graphene/chitosan/nickel ferrite nanocomposite for lead ions removal from aqueous

solution: RSM modelling design, Journal of Cleaner Production, 2018, 201, 507–515. Doi: https://doi.org/10.1016%2Fj.jclepro.2018.08.059.

[57] Chen Y, Wang P, Liang Y, Zhao M, Jiang Y, Wang G, Zou P, Zeng J, Zhang Y, Wang Y. Fabrication of a three-dimensional porous Z-scheme silver/silver bromide/graphitic carbon nitride@nitrogen-doped graphene aerogel with enhanced visible-light photocatalytic and antibacterial activities, Journal of Colloid and Interface Science, 2019, 536, 389–398. Doi: https://doi.org/10.1016%2Fj.jcis.2018.10.061.

[58] Zhang M, Chen Y, Chen B, Zhang Y, Lin L, Han X, Zou P, Wang G, Zeng J, Zhao M. Fabrication of a three-dimensional visible-light-driven Ag–AgBr/TiO2/graphene aerogel composite for enhanced photocatalytic destruction of organic dyes and bacteria, New Journal of Chemistry, 2019, 43, 5088–5098. Doi: https://doi.org/10.1039%2Fc8nj06057f.

[59] Xin X, Li S, Zhang N, Tang Z, Xu Y. 3D graphene/AgBr/Ag cascade aerogel for efficient photocatalytic disinfection, Applied Catalysis B: Environmental, 2019, 245, 343–350. Doi: https://doi.org/10.1016%2Fj.apcatb.2018.12.066.

[60] Zhang J, Fang S, Mei J, Zheng G, Zheng X. High-efficiency removal of rhodamine B dye in water using g-C3N4 and TiO2 co-hybridized 3D graphene aerogel composites, Separation and Purification Technology, 2018, 194, 96–103. Doi: https://doi.org/10.1016%2Fj.seppur.2017.11.035.

[61] Chen F, Li S, Chen Q, Zheng X, Liu P. 3D graphene aerogels-supported Ag and Ag@Ag3PO4 heterostructure for the efficient adsorption-photocatalysis capture of different dye pollutants in water, Material Research Bulletin, 2018, 105, 334–341. Doi: https://doi.org/10.1016%2Fj.materresbull.2018.05.013.

[62] Fan Y, Ma W, Han D, Gan S, Dong X, Niu L. Convenient recycling of 3D AgX/graphene aerogels (X = Br, Cl) for efficient photocatalytic degradation of water pollutants, Advanced Materials, 2015, 27, 3767–3773. Doi: https://doi.org/10.1002%2Fadma.201500391.

[63] Chen F, An W, Liu L, Liang Y, Cui W. Highly efficient removal of bisphenol A by a three-dimensional graphene hydrogel-AgBr@rGO exhibiting adsorption/photocatalysis synergy, Applied Catalysis B: Environmental, 2017, 217, 65–80. Doi: https://doi.org/10.1016%2Fj.apcatb.2017.05.078.

[64] Kemp KC, Seema H, Saleh M, Le NH, Mahesh K, Chandra V, Kim KS. Environmental applications using graphene composites: water remediation and gas adsorption, Nanoscale, 2013, 5, 3149–3171. Doi: https://doi.org/10.1039%2Fc3nr33708a.

[65] Wan W, Yu S, Dong F, Zhang Q, Zhou Y. Efficient C3N4/graphene oxide macroscopic aerogel visible-light photocatalyst, Journal of Materials Chemistry A, 2016, 4, 7823–7829. Doi: https://doi.org/10.1039%2Fc6ta01804a.

[66] Hu J, Chen D, Li N, Xu Q, Li H, He J. 3D aerogel of graphitic carbon nitride modified with perylene imide and graphene oxide for highly efficient nitric oxide removal under visible light, Small, 2018, 14, 1800416. (1–10) Doi: https://doi.org/10.1002%2Fsmll.201800416.

[67] Zhang F, Li Y, Li J, Tang Z, Xu Y. 3D graphene-based gel photocatalysts for environmental pollutants, Environmental Pollution, 2019, 253, 365–376. Doi: https://doi.org/10.1016%2Fj.envpol.2019.06.089.

[68] Mu C, Zhang Y, Cui W, Liang Y, Zhu Y. Removal of bisphenol A over a separation free 3D Ag3PO4 -graphene hydrogel via an adsorption-photocatalysis synergy, Applied Catalysis B: Environmental, 2017, 212, 41–49. Doi: https://doi.org/10.1016%2Fj.apcatb.2017.04.018.

[69] Yang J, Shi Q, Zhang R, Xie M, Jiang X. BiVO4 quantum tubes loaded on reduced graphene oxide aerogel as efficient photocatalyst for gaseous formaldehyde degradation, Carbon N. Y., 2018, 138, 118–124. Doi: https://doi.org/10.1016%2Fj.carbon.2018.06.003.

[70] Naguib M, Mashtalir O, Carle J, Presser V, Lu J, Hultman L, Gogotsi Y, Barsoum MW. Two-dimensional transition metal carbides, ACS Nano., 2012, 1322–1331. Doi: https://doi.org/10.1021%2Fnn204153h.

[71] Sun Y, Meng X, Agnese YD, Agnese CD, Duan S, Gao Y, Chen G, Wang XF. 2D MXenes as co-catalysts in photocatalysis: synthetic methods, Nano-Micro Letters, 2019, 11(79), 1–22. Doi: https://doi.org/10.1007%2Fs40820-019-0309-6.

[72] Yao Z, Sun H, Sui H, Liu X. Construction of BPQDs/Ti3C2@ TiO2 composites with favorable charge transfer channels for enhanced photocatalytic activity under visible light irradiation, Nanomaterials, 2020, 10, 452. (1–15).

[73] Huang G, Li S, Liu L, Zhu L, Wang Q. Ti3C2 MXene-modified Bi2WO6 nanoplates for efficient photodegradation of volatile organic compounds, Applied Surface Science, 2020, 503, 1–8. Doi: https://doi.org/10.1016%2Fj.apsusc.2019.144183.

[74] Low J, Zhang L, Tong T, Shen B, Yu J. TiO2/MXene Ti3C2 composite with excellent photocatalytic CO2 reduction activity, Journal of Catalysis, 2018, 361, 255–266. Doi: https://doi.org/10.1016%2Fj.jcat.2018.03.009.

[75] Chen WY, Jiang X, Lai S, Peroulis D, Stanciu L. Nanohybrids of a MXene and transition metal dichalcogenide for selective detection of volatile organic compounds, Nature Communications, 2020, 11, 1302. (1–10) Doi: https://doi.org/10.1038%2Fs41467-020-15092-4.

[76] Zhang Y, Wang L. Adsorptive environmental applications of MXene nanomaterials: a review, RSC Advances, 2018, 8, 19895–19905. Doi: https://doi.org/10.1039%2Fc8ra03077d.

[77] Kim SJ, Koh H-J, Ren CE, Kwon O, Maleski K, Cho S, Anasori B, Kim C, Choi Y, Kim J, Gogotsi Y, Jung H-T. Metallic Ti3C2Tx MXene gas sensors with ultrahigh signal-to-noise ratio, ACS Nano, 2018, 12, 986–993. Doi: https://doi.org/10.1021%2Facsnano.7b07460.

[78] Mashtalir O, Cook KM, Mochalin VN, Crowe M, Barsoum MW, Gogotsi Y. Dye adsorption and decomposition on two dimensional titanium carbide in aqueous media, Journal of Materials Chemistry A, 2014, 1, 14334–14338. Doi: https://doi.org/10.1039%2Fc4ta02638a.

[79] Gazzi A, Fusco L, Khan A, Bedognetti D, Zavan B, Vitale F, Yilmazer A, Delogu LG. Photodynamic therapy based on graphene and MXene in cancer theranostics, Frontiers in Bioengineering and Biotechnology, 2019, 7, 295. (1–15) Doi: https://doi.org/10.3389%2Ffbioe.2019.00295.

James A. Sullivan, Raphaël Abolivier

Chapter 7
Artificial Photosynthesis over Heterogeneous Photocatalysts

7.1 Introduction

7.1.1 The Greenhouse Gas Problem

Questions about the equilibrium temperature of the Earth can be traced back to J. Fourier in 1822 [1]. He understood that Earth's temperature was linked to a certain energy balance between incoming solar energy and outgoing blackbody radiation.

Fourier suggested a mechanism to explain the differences in between expected and measured temperatures that could be related to the mode of operation of a greenhouse.

Even though this analogy is incorrect, the name remained to describe the currently understood effect. The earliest genuine explanation of the greenhouse effect is attributed to the American scientist Eunice Foote, who described, in 1856, the light-harvesting properties of CO_2 and the subsequent heating effect on the environment of the molecules for the first time [2].

Those results stimulated further study of the impact of CO_2 concentration on the temperature of the surrounding atmosphere. In a series of papers starting in 1860 the Irish physicist John Tyndall further developed these ideas, and he is also recognized as a founding figure for the modern description of the greenhouse effect.

In 1861, he proposed, a model of the atmosphere as built of a mixture of different compounds, each having different radiative properties. In order to do so he published a series of experiments for different gas absorptions [3]. In this paper, he understood the predominant role of CO_2 as an IR radiation absorbent within the different gases in the atmosphere and linked it to the atmospheric temperature.

Arrhenius published in 1898 [4] a quantitative analysis of the effect of CO_2 on the Earth's energetic budget. He calculated the outcome of either a halved or doubled CO_2 atmospheric concentrations. In the first case he concluded a reduced temperature in the range of 5–6 °C, in the second case an increased temperature of 4–5 °C.

Acknowledgments: This publication has been supported in part by funding from Science Foundation Ireland (SFI) under grant number 16/RC/3889 and is cofunded under the European Regional Development Fund and by BiOrbic industry partners.

James A. Sullivan, UCD School of Chemistry, Belfield, Ireland, e-mail: james.sullivan@ucd.ie
Raphaël Abolivier, UCD School of Chemistry, Belfield, Ireland

https://doi.org/10.1515/9783110668483-007

Those calculations, while not exact, were a first step toward the quantitative comprehension of the influence of greenhouse gases (GHGs) on global warming. In later work, he suggested that burning of fossil fuels could become a significant source of CO_2.

However, at this time in history, the danger of such an increase in CO_2 production was not fully understood, in part due to the idea that whatever CO_2 quantities were humanly produced, the absorption of this compound by the oceans would completely cancel out the problem.

This possibility had been addressed by Plass, in 1955 [5]. His work aimed to understand the link between the fluctuations of glacial and inter-glacial areas and the CO_2 content in the ocean/atmosphere system.

Plass applied the CO_2 theory to deep ocean floor sediments analysis performed by Wiseman[6] that concluded the existence of, at least, 10 distinct temperature minima during the last 620,000 years and to the four known (at that time) glaciation periods, relative to his measurement of the radiation flux, at different temperature and heights in the atmosphere.

Plass concluded that the CO_2 equilibrium between the atmosphere and the ocean could be the reason for the fluctuations between glacial and interglacial periods.

Briefly, Plass proposed that formation of glaciers (reducing the ocean volume) would slowly release CO_2 in the atmosphere (from the release of absorbed CO_2 in the oceans). The CO_2 content in atmosphere would grow until a critical point is attained, slowly leading to the glaciers melting, therefore increasing the ocean volume, in turn leading to higher absorption of CO_2 in the oceans. This then results in a reduced level of in the atmospheric CO_2, which in turn results in a decrease in temperature, which again in turn leads to formation of glaciers, forming a freeze-thaw-freeze cycle.

Plass proposed that anthropogenically produced CO_2 could not be efficiently absorbed in the oceans within short (on geological scales) time periods and that, given it affected temperature, global warming would continue as long as fossil fuels are used. In his conclusion he proposed nuclear technologies as a possible solution.

In order to determine the atmospheric concentration of CO_2 at times that precede direct measurements (see above) a methodology has been developed as explained by Raynaud and Barnola [7]. This paper also clearly quantified the variation in CO_2 concentrations during the preindustrial area and the relative important increase seen in modern days, associated with anthropogenic factors (deforestation, use of fossil fuel, etc.).

Briefly, during the firnification process (transformation of snow into ice) air is trapped in inter-grain space; this air is therefore isolated from the external atmosphere. The composition of this air, including CO_2 content, can be extracted and analyzed therefore providing information on the associated area atmosphere.

The increase in atmospheric CO_2 was confirmed in 1960 by Keeling, who published [8] a study of CO_2 concentration fluctuations in atmosphere, taken in different places

(Mauna Loa (Hawaii), Antarctica and La Jolla (California)) and observed the systemic seasonal variations in concentration.

Concurrent measurements in Antarctica, taken over a 2-year span, showed that the average concentration increased in the second year. This increase was consistent with industrial levels of CO_2 production without subsequent removal, proving the inability of the oceans to rapidly absorb the added CO_2 in the atmosphere.

To this date, uninterrupted measurements of CO_2 have been done at the Mauna Loa station (chosen for small local variations due to lack of vegetation and facile correction for nearby volcanic activity), installed by Keeling, to quantify levels of CO_2 in the atmosphere (and monitor recent increases following industrial activities).

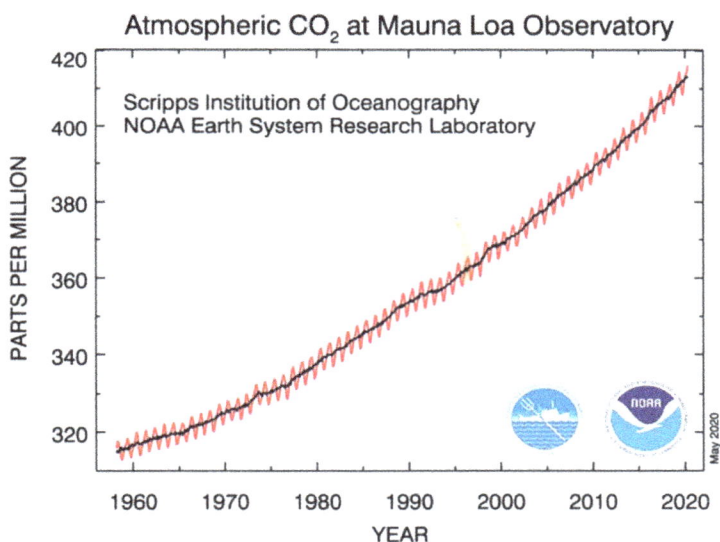

Figure 7.1: The famous Keeling curve showing measurements of atmospheric CO_2 at Mauna Loa between 1960 and 2009 – from https://commons.wikimedia.org/wiki/File:Mauna_Loa_Carbon_Di oxide-en.svg. Daily updated data and historical recording can be found at https://sioweb.ucsd. edu/programs/keelingcurve/.

A dramatic and continuous increase in the average CO_2 concentration in the atmosphere is recorded, from 313 ppm in 1958 to 410 ppm in 2019 (seasonal variations are of the order of 5 ppm/year).

In the meantime, while the exact Earth's average temperature modifications related to CO_2 are difficult to assess due to the different factors including natural variation (El Nino, volcanic activity, etc.) a clear unprecedented increase has been recorded, for which the consensus is that the increased CO_2 concentration has played a major role.

This plot (the Keeling curve) is considered as one of the most important contributions to the dissemination of knowledge about CO_2 levels and associated global warning to the public.

Although the effect was initially studied for water vapor and carbon dioxide numerous gaseous molecules in the atmosphere are now known to play a major role in atmospheric temperature regulation; for example, CH_4, nitrous oxide, and a range of fluorinated compounds [9].

Conceptually, the Earth absorbs light in the visible range from solar radiation (daytime), then as a "hot" body and this energy is reemitted from the surface (and through to the atmosphere) to space in the form of IR light. This process acts as a temperature regulation mechanism.

It is this IR energy that is absorbed by GHG. While the most abundant gases in the atmosphere, oxygen and nitrogen, are transparent at these wavelengths, GHGs absorb IR radiation (resulting in molecules bending and stretching as they are promoted to higher-energy vibrational modes).

GHG are classified by their relative impacts, and calculations of these depend on parameters that include the broadness of their light absorption potential, their lifetimes in the atmosphere, and their abundance.

The absorbed energy within vibrating GHG will eventually be radiatively reemitted either out to space or toward the surface of the Earth (recycling the energy of the photon into the planet's energy budget), the GHG therefore acting as a blanket through which some of the emitted radiation is not permitted to initially exit the planet [10].

Without this back-radiation the Earth's temperature would be ca. 33 °C lower than it currently is and (given water would be frozen) life as we currently understand it could not have evolved on the planet.

However, a constantly increasing average temperature is (and will be) a problem which will lead to climate change, which has (and will have) direct and indirect consequences. These include changes in precipitation patterns (leading to flooding, drought, crop failure, etc.), acceleration of the melting of glacial ice, leading to sea levels rising and endangering coastal areas and islands.

The relative stability of the environment is of foremost importance for the wellbeing of any ecosystem and rapid changes in global temperatures, driven by changes in CO_2 levels in the atmosphere should be, where possible, limited.

7.1.2 The Carbon Dioxide Cycle

Carbon atoms form the building blocks of living species as well as being intimately involved in the reactions we use to generate energy, through either combustion or respiration. Although crucial for all life (since photosynthesis is an essential part of the cycle) presence of CO_2 in the atmosphere at the elevated levels discussed above also presents major drawbacks, namely, its important role in climate change (see above).

In order to understand the recent increases in CO_2 concentration it is necessary to consider the carbon dioxide cycle.

7.1.2.1 Sources and Sinks of Carbon Dioxide

An analysis of the carbon dioxide cycle can be split into two main aspects: the sources that emit carbon derivatives (principally CO_2) and the sinks that trap or process carbon compounds. Both aspects can be either natural or related to human activity and are numerous and evolving.

The "carbon cycle" outlines the dynamics of exchange between different carbon-containing reservoirs that, added together, compose the entirety of available carbon-containing materials.

Briefly, CO_2 atmospheric content can change in different manners. First, through photosynthesis, consuming CO_2 to produce glucose for plant growth, the latter being further metabolized to other organic compounds (pyruvate, glycogen, etc.). Second, plant decomposition or combustion (biomass-based energy) generates CO_2 that is released back to the atmosphere to be absorbed and used again by other photosynthetic plants (this is known as the short CO_2 cycle).

Dissolution of CO_2 in the ocean can be a reversible or an irreversible process; in the latter CO_2 is fixed as insoluble carbonates and this is an important sink for atmospheric CO_2 on geological timescales.

These cycles are mostly self-equilibrated, enabling medium-term stability of the atmospheric CO_2 content (and therefore the planetary temperature and associated environmental well-being).

However, another important CO_2 long-term (millions of years) sink is the decomposition of organic components (living matter, plants/animals) to fossil fuels under high pressures and temperatures. Following formation, these have been stored within the Earth's successive sedimentary layers.

The energy stored within these fuels is currently released through combustion to CO_2 (and H_2O) and used for industrial and transportation purposes. CO_2 is released to the atmosphere through this combustion, and this has significantly perturbed the short cycle resulting in the currently observed (and future predicted) current climate change problems.

Similarly, rapid deforestation leads to release (as CO_2) of the carbon trapped within wood as well as reducing the number of sinks, that is, growing forests, again artificially impacting the CO_2 cycle and increasing the CO_2 content in the atmosphere. The constant need for further agricultural space is necessary to feed a rapidly growing population but needs to be accompanied by reforestation policies.

Other significant aspects of the carbon cycle include respiration processes in living organisms producing CO_2 through the reaction: $C_6H_{12}O_6 + O_2 \rightarrow CO_2 + H_2O + ATP$

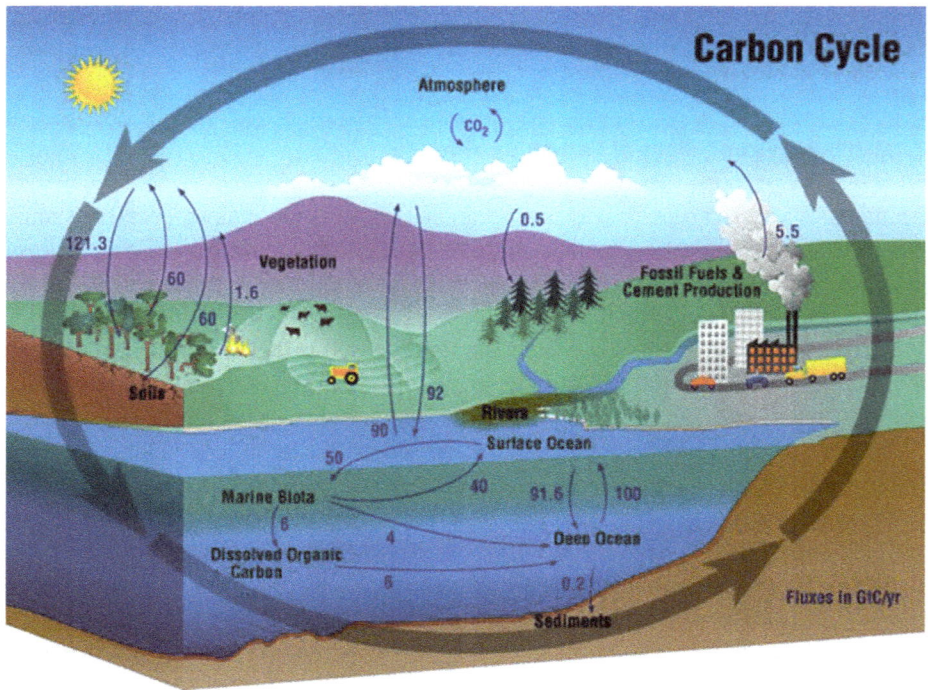

Figure 7.2: A depiction of the slow and fast carbon cycles (in GtC/year) – from https://www.flickr. com/photos/atmospheric-infrared-sounder.

and the decay of dead organisms. These specific emissions are, however, difficult to quantify for a complete analysis.

GHG emission are commonly estimated as CO_2 equivalents (using the parameters defined earlier to account for other the different behavior of other GHGs) and carbon-related emissions in gigatons of carbon units (GtC).

The Global Carbon budget 2019[11], published by the Global Carbon Project consortium, aims at the assessment of anthropogenic CO_2 emissions and their redistribution (e.g. in oceans and terrestrial biosphere). This report was published with Friedlingstein as lead author along with numerous other scientists with a broad range of scientific specialities relating to Earth science.

While it is difficult to exactly assess the amount of carbon involved in each part of the complex cycles, this document provides an estimation based on different data sets. The presented data corresponds to averaged yearly values over the period 2009–2018.

Overall, the natural (ocean (90 GtC) and land (120 GtC)) production of carbon is around 20 times higher than the anthropogenic production (9.5 GtC, plus 1.5 GtC arising from land use change).

No anthropogenic sink currently exists, while the natural sinks (oceans and lands) are estimated to fully counterbalance the naturally produced carbon and, arising from their calculations, to also account for absorption of part of the anthropogenically produced one (3.2 GtC of land uptake; 2.5 GtC of ocean uptake).

The latter is not fully included in a cycle, therefore leading to an overall increase in the atmospheric carbon concentration of 4.9 GtC. This addition of carbon to the atmosphere content is the core of the problem and is due to the rapid release of carbon that has been slowly "stocked" in the form of fossil fuels during previous millennia.

7.1.2.2 CO_2 as a Significant Indicator of Anthropic Effects

As described earlier, human activities have released a significant amount of CO_2 and this has impacted the composition of the atmosphere.

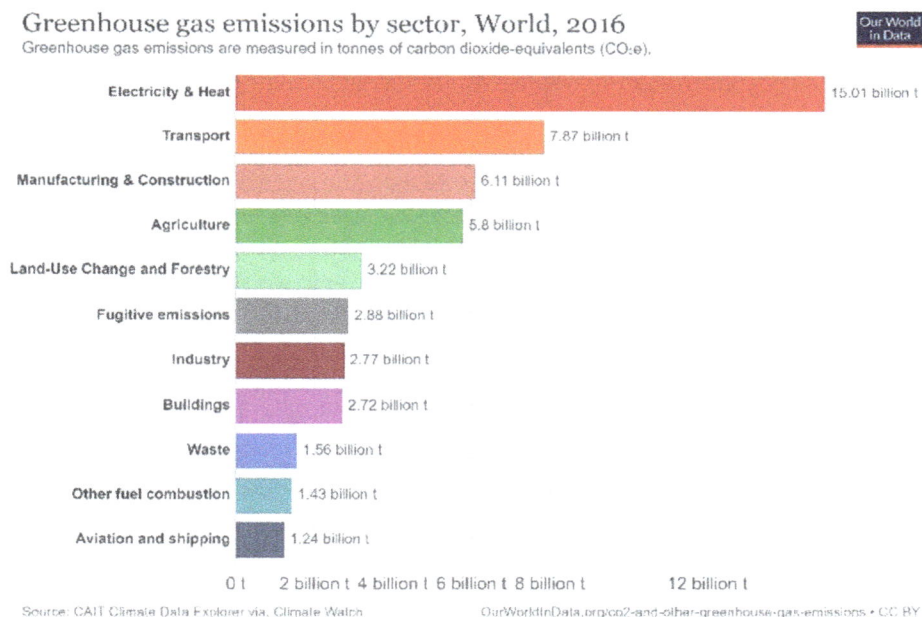

Figure 7.3: A bar chart showing anthropogenic CO_2 emissions by sector in 2016 – from https://ourworldindata.org/grapher/ghg-emissions-by-sector.

The data presented by NASA is unambiguous concerning the anthropic role involved in the constantly increasing atmospheric CO_2 concentrations in air since the 1950s and the third industrial revolution.

The level of carbon dioxide has almost doubled in the last century and is currently (even with seasonal fluctuations) over 400 ppm.

Historically recorded values and data extrapolated from CO_2 concentrations in polar ice caps and sediments display a clear breathing cycle between 180 ppm and 260 ppm CO_2 that pertained over millions of years. This cycle is no longer followed since the wide-scale combustion has been developed over the past 150 years (see above). Numerous international organizations have been created in order to understand, follow, and propose solutions to the increasingly environmental problems associated with these levels of CO_2. Notably, Agenda 21 [12], created during the Earth Summit in 1992 in Rio de Janeiro, signed by 178 countries, was the first important international agreement relating to the effects of CO_2 on the environment.

Other important international summits includes the Paris agreement[13] aiming at limiting the temperature increase associated with CO_2 release to under 2 °C until 2100, and various UN summits [14] that included not only governmental representatives but also industry representation and sub-national actors, among others.

The UN Sustainable Development Goals (SDGs) [15] while aiming at working on 17 different problems (not all related to the science-related factors) also include major sections concerning the wellbeing of the environment (and specifically carbon emissions).

Interestingly, the COVID-19 crisis led to significantly reduced human activity in the beginning of 2020 (and therefore to decreased anthropogenic CO_2 emissions).

Notwithstanding this decrease, the record for measured CO_2 levels in the atmosphere has recently been set (May 2020). This shows that there is a lag (possibly of significant duration) between decreasing anthropogenic emissions of CO_2 and decreases in atmospheric levels of CO_2 (and therefore a lag in any corrective action to remedy increased global temperatures).

This suggests that decarbonization of the world's economy will not immediately reset the CO_2 levels to those of preindustrialization. In turn, this suggests that we may be living with the effects of increased levels of atmospheric CO_2 for hundreds of years following decarbonization (unless some other method is found to remove CO_2 from the atmosphere).

7.1.2.3 Mitigation of CO_2 Production Through Novel Technologies

There will be two approaches used to tackle the problem of atmospheric CO_2.

First one is to reduce the industrial production of CO_2. This approach falls within the field of green chemistry described by Anastas and Warner [16].

One of the principles of green chemistry involves "prevention of pollution rather than remediation" and this suggests we will need to decarbonize our sources of energy.

The rising interest in renewable electricity generation coupled with the increased production of electrical (rather than internal combustion driven) vehicles and a societal change toward more efficient consumption habits could significantly reduce levels of CO_2 emissions (and in time affect the levels of atmospheric CO_2 – causing these to re-decrease).

However, these measures, while important and obviously part of the solution, are also faced with increased consumption related to population growth and demands for improved standards of living. This means that other innovative solutions will be required.

The second approach to further decrease CO_2 emissions (and subsequently CO_2 levels) will involve use of the molecule as feedstock for further chemical processing. This should have a double impact, first, as stated earlier, to reduce emissions of CO_2 to atmosphere, and second to develop new, clean, and efficient ways to produce carbon-based fuels and chemicals.

Artificial photosynthesis (AP) is expected to be an efficient reaction for the second approach. However, the use of CO_2 as a feedstock requires significant energy for the actual thermochemistry of the chemical reaction itself, and the capture of CO_2 from exhaust emissions or atmosphere.

7.1.3 Photosynthesis

Photosynthesis is a growth mechanism shared by most plants (photoautotrophs) in which glucose is formed (using light as an energy source) from H_2O and CO_2. Glucose is the primary substrate for the formation of all organic materials. The phenomenon is mostly understood but involves numerous complex enzymatic mechanisms.

The photosynthesis reaction is $6CO_2 + 12H_2O \rightarrow C_6H_{12}O_6 + 6O_2 + 6H_2O$ and is an endothermic reaction (where strong C=O and O–H bonds are broken, and weak C–O, C–H, and C–C bonds are formed). The energy is provided in the form of light.

Differing from CO_2 light absorption (see above), photosynthesis requires electromagnetic radiation in the visible range (400–700 nm). Within plants, chlorophyll acts as a light harvester and as a site for the reactions to initiate.

Because of the benefits, both in removing CO_2 that would be emitted to atmosphere and in generating fuels and chemicals that would normally be derived from fossil fuels, the pursuit of artificial systems to react CO_2 and H_2O under promotion of freely available solar energy has become an objective for the scientific community.

7.2 Novel Technology to Solve the Issues Arising from CO_2

7.2.1 CO_2 as a Feedstock

Carbon dioxide is a colorless, odorless gas, at room temperature. It is a linear compound formed of two strong C=O bonds (804.4 kJ/mol) (O=C=O) presenting a $D_{\infty h}$ symmetry [17].

The processing of CO_2 belongs in the domain of "C1 chemistry" [18], the upgrade of compounds that contain only one carbon atom to generate fuels or platform chemicals.

Commercial scale use of CO_2 as a feedstock requires large volumes of high-purity CO_2. Fundamentally this will require its capture (more than likely from combustion exhausts) and technologies studying this are known as carbon capture and storage (CCS) (to essentially generate an artificial carbon sink).

Reducing the concentration of anthropic CO_2 in the atmosphere is one of the key aspects of the European Union, Research, Innovation, and Competitiveness Priorities program [19] and CCS (and subsequently carbon capture and utilization (CCU)) is expected to contribute to reaching this goal CCS consists of capturing industrially produced CO_2 at the source, purifying it to remove $N_2/O_2/H_2O$ and other contaminants) and eventually storing it by pumping it within geological reserves [20].

Current CCS plants are already in use within different countries (Iceland (Carb-Fix), the United States, Canada, etc.) but are still scarce due to problems associated with the increased necessity of fossil fuel to power the process (it requires an energy input to purify waste streams) [21] and relies on major trade-offs as presented in the Europe Environment Agency (EEA) report from November 2011 [22].

CCU differs from CCS in that the CO_2 is directly used as a resource, preventing the need for transportation or storage (although still requiring the purification step) and its conversion to potentially useful products such as CH_3OH, C_{2+}, and C_nH_{2n} would lead to a reduced reliance on fossil-based chemicals or fuels [23], as well as decreasing emissions of CO_2 to atmosphere (with all the attendant problems that brings (see above)).

It appears interesting to mention that in this context unusually, certain industries focus on the production of carbon dioxide for direct use in different spheres; for example, carbonated drinks, refrigerants, and fire extinguishers.

This is normally produced in petrochemical plants through the combustion of CH_4 (with the exotherm generated along with CO_2 being used in other areas of the plant). Widespread and economic CCS/CCU should end this practice.

In this chapter we will solely focus on the upgrade of CO_2 through light-promoted reactions over solid-phase materials. Successful uses of these processes using (currently classified as) waste CO_2 to generate precursors for valuable fuels and chemicals would rapidly become a useful addition to the circular economy.

7.2.2 Light-Based Energy Production

The total solar energy reaching the Earth's surface annually is estimated to be around 4.0×10^{24} J [24], while the total energy consumed in the world in 2019 was almost four orders of magnitude lower, 5.8×10^{20} J [25].

However, this value accounts for consumed energy only, that differs significantly from the total produced energy (around 30% higher in 2014 [26]) as a significant amount of generated energy is wasted, through processing (e.g., petroleum refining), transmission, or other losses (e.g., heating).

One of the major problems with harnessing solar energy is its intermittency and a second relates to power density. Both problems can be solved by using the solar energy to promote the formation of energy dense fuels through the artificial photosynthesis reaction. This also applies to the storage of other renewable energies.

Overall, the desire is to harness the energy of sunlight to decrease the generation of energy through the combustion of fossil fuels (as this depletes a potentially valuable resource and also generates CO_2, which is released to the atmosphere).

A broad range of successful applications relying on solar light harvesting to reduce fossil fuel use (and concomitant CO_2 production) have already been developed.

These include photovoltaic processes to generate electricity [27] and solar thermal devices to store heat production [28] and are notable industrially scaled projects that are now both commercial and impacting on society.

Those systems prove the validity of the concepts related to the use of solar light, and the emergence of these technologies does lower the need for fossil energy and therefore reduces the pollution arising from both the extraction and use such energy sources.

Storage of solar energy using these approaches, however, is a problem and, while battery (Goodenough, Whittingham, Yoshino shared the 2019 Chemistry Nobel prize for development of the Li-ion battery [29, 30]) and heat storage technologies are evolving rapidly they cannot yet replace the energy densities of hydrocarbon fuels.

Furthermore, these approaches do not generate any chemical precursors for the pharmaceutical/fine or bulk chemical industries and therefore these industries would still rely on fossil-fuel feedstocks.

Ideally there would be mechanisms to trap solar energy by using it carrying out endothermic reactions, following which the carbon-containing liquid products could be easily stored.

The energy could then be released through carrying out the reverse reaction. One example of this kind of "solar energy pump" is the photosynthesis reaction to generate biomass and O_2, coupled with the reverse of this reaction (combustion of biomass).

The latter reaction releases heat energy (essentially "pumping" the solar energy from where the biomass grew to where it was combusted).

The reaction we are interested in is the solar light–promoted redox reaction between CO_2 and H_2O where the CO_2 is reduced and H_2O oxidized. The reaction releases O_2 and generates reduced carbon-containing species such as CO, CH_4, and CH_3OH; that is, an artificial photosynthesis reaction analogous to the photosynthesis reactions that take place in plants.

A beneficial extra to this is that the products (as well as being considered as fuels) could also be used as feedstocks for the chemicals industries discussed above. Strategies for photoconversion of CO_2 using biomimetic approaches where the natural photosynthetic machinery is simulated using molecular analogues have proven to be successful (e.g., mimicking photosystem II for water splitting [31] or the use of RuBisCo for CO_2 fixation [32]). However, this chapter focusses on the use of solid-phase catalysts to promote the $CO_2 + H_2O$ reaction.

Industrially relevant uses of light harvesting through photocatalysis to form fuels or chemicals could become significant in industrial catalysis, but still requires major technological improvements for scaling up and the achievement of significant product yields.

The promotion of organic reactions using light as an energy source is not as recent as one would think. The first example was developed by G. Ciamician in 1912 [33] and since then numerous other examples have been developed [34], and it is to be expected that in future these will include efficient and industrially relevant catalyzed CO_2 upgrade reactions.

7.3 CO_2 Conversion into Valuable Fuels and Chemicals Through Photocatalysis

7.3.1 CO_2 Conversion

The upgrade of CO_2 can be performed through numerous methodologies [35] and notably using traditional paths such as thermal catalysis [36] and electrochemistry [37]. Those approaches are not as cheap or as environmentally respectful (requiring significant amounts of thermal or electrical energy to proceed) as photocatalysis should be, but they are efficient and deserve mention.

Electrochemical processing of CO_2 typically goes as follows: oxidation of H_2O (generating H^+ through electrolysis) at the anode and subsequent electron and proton transfer to CO_2. The reaction produces different compounds with the selectivity depending on of the reductive potential of the system (e.g., $CO_2/CHOH$: -0.51 V; CO_2/CH_4: -0.24 V vs SHE (standard hydrogen electrode) at pH = 7) and the number of H^+ species involved.

This methodology gains from the spatial separation of the two half reactions, reducing the possibility of parasitic reactions (and increasing the selectivity toward the desired product); however, only low volumetric conversion can be obtained.

Thermally induced catalytic upgrade reactions of carbon dioxide such as dry reforming of methane ($CO_2 + CH_4 \rightarrow 2\,CO + 2H_2$) to form syngas[38] require high temperatures of >600 °C (which in turn requires a significant energy input). This also leads to low yields and catalysts that suffer from coke deposition and sintering. Also, direct methanation ($CO_2 + 4H_2 \rightarrow CH_4 + 2H_2O$) [39], where the reactant reacts with one of the products, can take place (at around 350 °C).

Other thermally promoted catalytic paths for CO_2 activation can also be considered; for example, acid/base catalysis [40] or plasma-induced activation [41]. These are not further discussed here but are of interest for an overall view of CO_2 reduction chemistry.

Other considerations to be borne in mind when looking at the overall CO_2 reduction process include the nature of the reductant. These are vital for assessing the "greenness" of the overall procedure.

For example, the use of fossil-fuel-derived H_2 to reduce CO_2 is not a green CO_2 reduction (given CO_2 is generated and significant energy used during H_2 formation).

The origin of the reductant, the energy required to form the reductant, and the nature of any side-products (including toxicity) generated during both generation and use of the reductant all need to be considered before an overall method of reducing CO_2 to a useful fuel or chemical can be considered as environmentally beneficial.

It appears obvious that green reactants should be used to develop this technology. As a naturally occurring and highly available reactant, water can be used as a reductant for carbon dioxide (as is the case for photosynthesis itself).

The first example of a successful artificial photosynthesis (light-driven reaction) was shown by Inoue et al. in 1979 [42] using nanoparticles (NP) of semiconductor (SC) (TiO_2, WO_3, ZnO, CdS, GaP, and SiC) powders suspended in water for production of a diverse range organic compounds (CH_4, HCOOH, CH_3OH, and CH_2O were reported).

Cost assessment (energetic, environmental, and economic) of such research is a complicated task that requires knowledge in a broad number of domains (economy, ecology, and societal impact). For the latter, the objectives and the funding rationale and benefits behind the desire to develop artificial photosynthesis setups are clear; the acceptable financial and ecological costs, however, remain to be properly defined.

It is important to keep in mind the objective of such research, the environmental gain through reduced atmospheric CO_2 concentrations, and petroleum reliance. It is therefore important to include these parameters in reaction design.

The economic cost of the proposed processes cannot be more substantial than the outcome arising from the obtained products, and the environmental gain will be entirely fulfilled only if the technology does not rely on the use of rare compounds, critical elements, or extensive energy input.

To exemplify these somewhat abstract concepts, consideration of the use of noble metals to promote these reactions is interesting. While efficient for promoting these reactions in several cases, and valuable for the determination of mechanistic and fundamental insights, these elements, given their lack of abundance, should not be considered for application in scaled-up industries.

7.3.2 Value-Added Products

The desire and need for carbon-containing molecules are increasing. This increase is due to a range of reasons that include their use as fuels and platform chemicals in the domains of transport, electricity generation and the pharmaceutical, adhesives, fine chemicals and bulk chemicals industries, etc.

Such products are currently obtained through petroleum processing. That is neither sustainable (due to their finite nature) nor environmentally respectful (due to "new" CO_2 evolution upon their degradation or combustion).

In short, replacements for these fossil resources are required. The use of biomass as a feedstock is an area of active consideration – but a necessary approach to the circular economy will also use CO_2 as a feedstock.

On a laboratory scale, products that have already been synthesized from carbon dioxide include hydrocarbons (fuels) [43], CH_3OH [44], syngas [45] (CO/H_2 mixtures), polymeric and cyclic carbonates [46]. However, to date, only one bulk chemical process (the production of urea) uses CO_2 as a reactant feedstock.

Implementation of plants generating these products in large quantities could ameliorate the problems associated with fossil fuel use (see above) and efforts are growing from political entities to promote their development (e.g., components of the Horizon 2020 program in the EU).

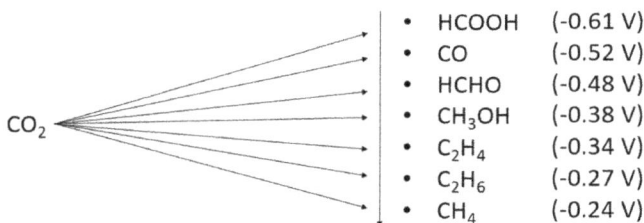

• HCOOH	(-0.61 V)
• CO	(-0.52 V)
• HCHO	(-0.48 V)
• CH_3OH	(-0.38 V)
• C_2H_4	(-0.34 V)
• C_2H_6	(-0.27 V)
• CH_4	(-0.24 V)

CO_2

Figure 7.4: Minimum required photogenerated electron reduction potential for CO_2 conversion (vs NHE; pH = 7) to different products.

It can be seen in Figure 7.1 that the artificial photosynthesis reaction can lead to different products (CH_4, CH_3OH, HCHO, etc.) as a function of the photogenerated electron's reduction potential.

It is, however, important to keep in mind that the selectivity of this reaction is highly difficult to control and is dependent on numerous other parameters [47] (e.g., number of electrons/protons involved (see below), the phase of the reaction [48], the nature and structure of the catalyst).

For the latter, it is proposed that the activation of CO_2 occurs following adsorption on the photocatalyst surface [49] through bending of the linear molecule [50] (reducing the CO_2 LUMO level energy as suggested by Álvarez et al. [51]). The adsorption energy of CO_2 is different on different catalysts, as shown by Liu et al. [52], and this parameter needs to be taken into account when selecting a photocatalyst.

In this paper, the adsorption, dissociation, and desorption catalytic steps of CO_2 reduction half reaction for transition metals are studied by DFT. It is shown that a clear trend can be obtained using different catalytic systems (in that case, earlier 3d metals were proven to lead to lower total energy barriers) and different surface structures for a given system (see examples in the TiO_2 photocatalysts section).

However, it is also straightforward that too high of a CO_2 adsorption energy will hinder the reaction and desorption steps; a proper equilibrium is therefore to be found for the design of an efficient system.

Different mechanistic paths have been proposed following the identification of intermediate species (through DRIFTS measurements) as reviewed by Lui and Li [53]. The nature and number of intermediate species in the reaction are a key point for the mechanistic understanding and for increasing selectivity toward desired product(s). Such studies will be highly impactful and help toward the scaling up of laboratory studies to industrial facilities.

The reaction conditions are also identified as significantly impacting the obtained products and selectivity toward their formation. Within these reaction conditions, the reactants' phase is known to considerably alter the selectivity, as detailed by Karamian and Shaifnia [48].

In this work, for the gas-phase reaction, $\cdot CO_2^-$ formation is proposed as the only intermediate arising directly from CO_2. For the aqueous-phase reaction, the formation of numerous first step intermediates is proposed (CO_3^{2-}, H_2CO_3, HCO_3^-, and $\cdot CO_2^-$), leading to a much more complex overall mechanism and a reduced selectivity toward any given product.

The complexity of the mechanism renders the clear and unambiguous comprehension of the path highly difficult. This will be an aspect of primary importance in future development of efficient system, and further studies will be required to fully uncover the complete mechanism of the aqueous-phase artificial photosynthesis reaction.

7.4 Underlying Principles of Artificial Photosynthesis

7.4.1 Photocatalytic Mechanisms

Before exploring the materials that are of use in the promotion of the photocatalytic conversion of CO_2 it is important to review the opto-electric effects that cause the expected efficiency of such systems.

These phenomena relate to the nature of the interaction between the light and the materials and the impact that this has on the reactions promoted by the materials.

This paragraph briefly describes the requisite for understanding the role of light in "photocatalysis"; the numerous concepts developed here are further explained along with examples in Section 7.4.4.

Even if etymologically described as such, light is not to be understood as a catalyst but as an energy source to power chemical reactions through activation of a catalytic material.

Within the large field of photocatalysis, electromagnetic radiation of interest (photons) is those arising from solar activity, typically either in the ultraviolet (1–400 nm; 1,240–3 eV, ca. 10% of solar rad.) or in the visible (400–800 nm; 3.09–1.55 eV, ca. 50% of solar rad.) light range.

Infrared radiation (800 nm to 25 μm; 1.55–0.05 eV, ca. 40% of solar rad.), while often considered too low in energy, can also feasibly be used to drive artificial photosynthesis, as presented by Liang et al. [54] using ultrathin WO_3 as a photocatalyst. This example, however, and to the best of the authors' knowledge, remains the only study of this sort.

Within heterogeneously catalyzed artificial photosynthesis the materials of interest are therefore the ones that can interact with light in these wavelength ranges (semiconductors). "Activation" of such materials involves the promotion of valence electrons to an excited state through absorption of the light's energy.

The electrons in the atomic orbitals in semiconducting materials are organized as "bands" generated from the interactions of "infinite" numbers of discrete orbitals' whose energy levels are close to one another in terms of energy.

The highest energy electrons lie within the valence band (VB) (e.g., O *2p* orbital in TiO_2), with a maximal energy corresponding to the valence band maximum (VBM). This is normally filled in a semiconductor. The next highest (unoccupied) energy level is referred to as the conduction band (CB) (e.g., Ti *3d* orbital in TiO_2). Its minimum energy is known as the conduction band minimum (CBM).

The difference in energy (between CBM and VBM) is the bandgap (BG) energy, and has a fixed value for a given material (e.g., 3.2 eV for Rutile TiO_2) and therefore defines the minimum energy (e.g., <387.45 nm for rutile TiO_2) that must be absorbed, from incident light for an electron to be excited from the VB to the CB (leaving a hole in the VB).

This is true only for direct BG materials; indirect BG materials have much more complex electron promotion mechanisms and are rarely used as photocatalysts.

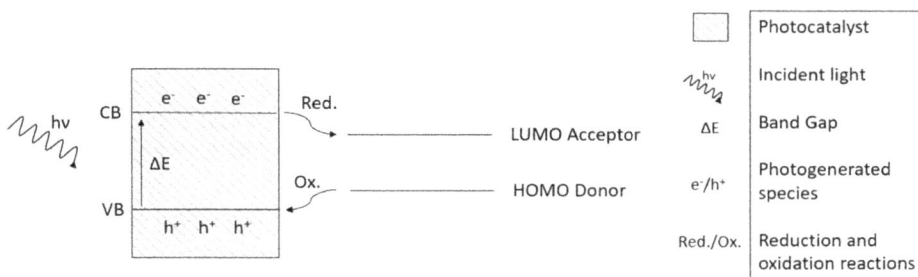

Figure 7.5: General photocatalytic mechanism. Following excitation by light and promotion to the CB, the photogenerated electrons will travel to the acceptor LUMO (CO_2) while the electrons in the donor HOMO (H_2O) will combine with the photogenerated holes.

The formed electron/hole pair is called an exciton. Those two charge carriers have reductive (corresponding to CBM energy) and oxidative (corresponding to VBM energy) potentials, respectively.

Once formed, there are several possible fates for the generated exciton. A principal reaction can be recombination of the species (i.e., the electron falling back from the CB to the VB and filling the generated positively charged hole). If this takes place the energy difference can be released as light or heat. The lifetime of an exciton is the length of time before this recombination (between several ps and μs), which results in the emission a photon or phonon of energy equivalent to the BG.

On the other hand, from a photocatalysis perspective, a desired reaction would be transfer of the electron from the CB into the LUMO (that must be of lower energy) of an adsorbed molecule (reducing this molecule) and transfer of an electron from the HOMO (that must be of higher energy) of an adsorbed molecule to the hole in the VB (oxidizing the adsorbed molecule and filling the hole in the VB). Thus, the excited semiconductor is essentially acting as a catalyst to transfer an electron from the oxidized molecule to the reduced one.

The longer the lifetime of the exciton, the better the chance that the material can find its use in catalysis. There are several techniques to encourage charge separation and transport and hence increase the exciton lifetime and thus the number of "active site" at a given time. These include BG engineering and the generation of heterojunctions between materials.

CO_2 reduction by H_2O is an uphill ($\Delta G > 0$) endothermic ($\Delta H > 0$) reaction [55]. The use of photocatalysts enables the overall reaction (as is the case in plants) to be divided into two "half-reactions": the reduction of the CO_2 from electrons in the CB and the oxidation of water from holes in the VB.

Each redox reaction has a given required potential to proceed, in general expressed as function of either the SHE, or the normal hydrogen electrode (NHE).

The reduction or oxidation reaction's potentials for the following reactions are expressed vs NHE [55].

H_2O/H^+: 0.82 V ($H_2O \rightarrow O_2 + 4H^+ + 4e^-$), the VB position of the photocatalyst must be more positive than this potential for the H_2O oxidation to happen. It should be noted that, while often omitted for the efficiency assessment of the artificial photosynthesis reaction, the production rate of O_2 evolution can also be an interesting parameter to follow.

Regarding the CO_2 reduction reaction, this can lead to formation of numerous products, associated with different reduction potentials; however, these different reactions require different numbers of both electrons and protons (see below). Except for the case of the formation of anionic CO_2^-, H_2O is always formed as by-product to balance these electrochemical equations.

CO_2/CH_4: −0.24 V (8 H^+ + 8 e^-); CO_2/C_2H_4: −0.34 V (12 H^+ + 12 e^-); CO_2/CH_3OH: −0.38 V (6 H^+ + 6 e^-); $CO_2/HCHO$: −0.48 V (4 H^+ + 4 e^-); CO_2/CO −0.52 V (2 H^+ + 2 e^-); $CO_2/HCOOH$: −0.61 V (2 H^+ + 2 e^-) and CO_2/CO_2^-: −1.9 V (1 e^-), The CB potential of the used photocatalyst must be higher in energy (i.e. have a more negative reduction potential) than at least one, of those potentials for the CO_2 reduction to take place.

The formation of products that require a relatively lower number of electrons and protons is easier than those requiring higher number of these charged species, even if the required potential conditions are fulfilled. This leads to complex reaction paths that are, to date, not clearly identified.

A more recent form of solid-phase photocatalysts involves the use of plasmonic systems [56]. In these (conductive materials), electrons are not promoted between bands following irradiation, but rather the incident light energy couples with a plasmonic electron wave in a metallic nanoparticle – causing the electrons to oscillate through the particle.

This moves their energies to levels above E_f. As these electrons relax back to E_f (through collisions) some can be increased in energy (becoming "hot" electrons with energies significantly higher than E_f and these can be pulled from the solid into a LUMO of appropriate energy on an adsorbed molecule (reducing this species).

This generates a hole in the metallic nanoparticle (below E_f) [57] that can extract electrons from an appropriately energetic HOMO (oxidizing this species).

The reporting of plasmonic related phenomena in photocatalysis is currently relatively scarce and examples of its operation in any reaction are in the minority compared to examples using various semiconductor systems.

7.4.2 Reaction Conditions

While a proper knowledge of the artificial photosynthesis reaction mechanism is crucial for their development, other parameters must also be accounted for.

Relying on efficient photocatalysts is important for the effective generation of products but (oftentimes ignored by chemists and more studied by chemical engineers) the reaction conditions and nonpurely chemical parameters are equally important both in driving conversion and in eventual industrial application.

In this section we briefly list some important aspects that need to be considered in any research aiming at the design of an efficient artificial photosynthesis reaction. This, while important, is not the core subject of this work and therefore does not fully cover the complexity of such aspects. For such complexity, readers are directed to the work of Martin et al. [58] and Ipek et al. [59].

The parameters of interest discussed here are the reaction environment (gas- or liquid-phase reactions and the mode of irradiation or light harvesting), the reaction time, and stability of the catalysts and the product characterization.

Two different reactor setups are commonly used in AP. One is used for the liquid-phase reaction, with catalysts being NP suspensions in water, the CO_2 being injected in form of a gas. Portions of this will be dissolved in H_2O.

This setup has been shown to provide efficient results but suffers from the low solubility of CO_2 in H_2O (0.0286 mol/kg at 303 K and 1 bar) [60]; this in turn limits the possibility of adsorption of CO_2 onto the catalyst within the suspension (and therefore the eventual reactivity of the catalysts).

The other approach (which can be considered as more recent) is a gas-phase reaction, within which the catalyst can be deposited on a nonreactive surface (e.g., silicon) or just inserted as a powder within a batch or continuous flow reactor. In that case, pure CO_2 (g) is passed through a water bubbler to produce a gaseous CO_2/H_2O mixture.

While these setups can both lead to the progress of the reaction, certain catalysts are limited by their stability (e.g., metal organic frameworks (MOFs)) within the aqueous environment.

The choice of one or the other will often be dependent on the available equipment and on the studied catalyst but can influence the obtained reaction rates and precautions are to be taken when seeking to compare results with those from literature.

A complete review of these aspects including reactor design, operating parameters, and catalysts support has been published by Ola and Maroto-Valer [61].

Optimizing the input and use of light energy requires an efficient reactor design as shown by Kočí et al. [62]. The most commonly used reactors are either fixed or fluidized bed reactors with transparent (in the UV–vis range) windows; however, certain novel ideas can improve the efficiency of these systems; for example, the use of optical fibers as reported by Wu [63].

In photocatalysis, the concept of quantum yield is used to assess the efficiency of light usage. This is defined as the number of reacted electrons (delivered from the

CB of the catalyst and leading to products) divided by the number of absorbed photons (as a percentage). The higher the quantum yield, the more efficient the systems are; that is, the more promoted electrons lead to reaction rather than recombine with holes to annihilate the exciton.

Proper assessment of the results is also a critical aspect of any reaction design. Within artificial photosynthesis, most products are low molecular weight compounds (CH_4, CH_3OH, HCOOC, etc.) and the most used product analysis methodology is gas chromatography (GC), or, in the case of liquid-phase analyses, ion chromatography (IC).

While most correctly designed chromatography systems can lead to an efficient separation of most products, the artificial photosynthesis reaction has a complex mechanism, as developed above, and a broad range of products can be obtained.

A broad range of products can be obtained from the artificial photosynthesis reaction and therefore precautions are to be taken when determining the conversion yields or selectivity of a reaction as numerous products could remain undetected due to low concentrations, inadequate analyses, or adsorption on the catalyst. Furthermore, catalyst contamination, such as might arise if organic compounds are used in catalyst synthesis, can lead to parallel reactions producing molecules (CH_4, CO, etc.) that are not directly attributable to the artificial photosynthesis reaction.

Most photocatalysts are also known to have catalytic properties over and above their photocatalytic properties (metal nanoparticles, TiO_2, etc.), and, therefore, for an efficient assessment and comparison of photogenerated results, a series of test reactions are highly desirable (reactions in the dark, without catalysts, without reactant input).

The reaction conditions, including parameters such as the phase of the reactants, the reactor design, and reaction time, are of primary importance for the design of an efficient process and are known to impact the reaction paths as well as the obtained yields. Therefore, comparing catalytic results is difficult since all these considerations need to be considered, in making comparisons.

Improvements in the domain of reaction engineering as well as the actual catalysts used will be required for major breakthroughs in the application and scale-up of the AP reaction.

7.4.3 Photocatalyst Characterization

The photocatalyst properties (BG, surface area, morphology, etc.) affect all catalyzed processes and it is therefore important to be able to fully and correctly characterize them. The principal characterization methodologies that are commonly used to assess photocatalysts (and reasons for their activity) are briefly discussed in this section. These are relatively common techniques in the field and their use is widespread.

Photoluminescence (PL) [64] measurements are used to comparatively analyze exciton lifetimes in different systems. After illumination (by visible light) excitons are formed in the system. Their radiative recombination is measured in a plot of the emitted photon energy vs intensity. Higher intensity is attributed to an increased number of radiative recombination phenomena.

Therefore, lower intensity is generally attributed to an enhanced photogenerated charge carrier separation and subsequent longer exciton lifetime, features that are important in promoting photocatalyzed reaction.

UV–vis diffuse reflectance [65] measurements are commonly used to determine the absorption properties of a sample. The BG of a system can be obtained by analyzing the absorption spectrum. The absorbance (in a.u.) is plotted vs wavelength (nm) of the incident light. A steep and linear increase in light absorption with increased incident energy is characteristic of an SC material.

Methodology based on the Tauc and Kubelka-Munk [66] equations proposes the following equation: $(F(R_\infty) \cdot h\nu)^{(1/\gamma)} = B \cdot (h\nu - E_g)$. R_∞ is the reflectance of an infinitely thick specimen; γ:1/2 for a direct BG; 2 for an indirect BG; and B is a constant. Reflectance is therefore a function of the incident light energy. The BG energy (E_g) corresponds to the intersection of the linear fit of the Tauc plot with the abscissa axis. In photocatalysis, the BG of the photocatalyst is a crucial parameter for the reaction and lower BG values lead to increased solar light-harvesting ranges.

The CBM position (E_{CB}) can subsequently be obtained using the following formula: $E_{CB} = \chi - E_C - \frac{1}{2}E_g \cdot x$ is the Mulliken electronegativity of the system; E_C is the energy of free electron on the hydrogen scale (4.5 eV). The VBM position (E_{VB}) can therefore be obtained by subtracting E_g from E_{CB}.

X-ray diffraction (XRD) [67] is one of the most commonly used characterization methods in material science. It is used to assess the presence of different phases in a solid material and can also be used to relatively quantify the percentage, if any, of amorphous material within a system (through definition of the "amorphous" background and calculation of the areas ratio (crystalline/amorphous)).

Diffraction pattern of the incident light (X-ray) following interactions with crystalline planes of as sample's structure generates a "fingerprint" of each type of structures (anatase, rutile, etc.). It can also be used to estimate crystallite size through use of the Scherrer equation ($\tau = (K\lambda)/(\beta \cos(\theta))$). τ refers to the crystallite size; K is the shape factor; β is FWHM; θ is the Bragg angle.

In photocatalysis, it can be used to verify the successful synthesis of a desired crystallographic system and to gain insight on the impact of loading or doping of a system on the structure (comparatively analyzing modified and unmodified patterns).

However, for the latter, peaks related to new crystallographic structures are not always observed if the dopant affected structure is present in too low a concentration and/or highly dispersed within the material matrix; other methods are therefore necessary to properly assess their presence; see above.

N_2 absorption–desorption (N_2 abs–des) [68] measurements are commonly used to quantify material surface areas and pore volumes distribution through Brunauer–Emmett–Teller (BET) and Barret–Joyner–Halenda (BJH) calculations. The adsorbed volume of gaseous N_2 (at 77 K) on the material is plotted as a function of the relative pressure oftentimes leading to hysteresis in the measurements depending on whether $P(N_2)$ is being increased or decreased. The shape of said hysteresis provides information on the porosity of the material (macro-, meso-, or micro-porosity). BET and BJH models are applied to the isotherm data to obtain the surface areas and pores volume distribution.

In photocatalysis (and catalysis in general) higher surface areas and/or pore volume systems generally have more active sites on their surfaces and therefore, theoretically, should be more efficient. Within loaded systems blocking effects that can hinder reactivity from deposited active-phase materials within pores can be observed.

X-ray photoelectron spectroscopy (XPS) is a surface analysis method based on the emission of core electrons from the sample following light stimulation (X-ray); commonly Kα radiation from Al (1,486.7 eV) or Mg (1,253.0 eV) anode is used. When expelled, the kinetic energy (KE) of the photogenerated electron is measured. It is directly proportional to the electron binding energy (BE) in the material and to the energy of the incident light ($h\nu$) ($KE = h\nu - BE + \Phi$) (where Φ is the work function of the material). The fastest electrons will arise from the VB – giving a measure of the energy of this feature.

A given element (and its oxidized species) has a specific and unique set of BE corresponding to each of its electrons. Therefore, the measured KE are used to gain information of the composition of the surface of material and on the presence of oxidized species that cannot always be observed by XRD.

Scanning and transmission electron microscopy (SEM/TEM) [69] are imaging techniques for morphological characterization of samples. They both rely on the use of electrons for the construction of those images. While in SEM the backscattered or secondary electrons are measured (providing a 3D image of the surface morphology), in TEM the transmitted (through the sample) electrons are measured (providing a 2D projection of the 3D material).

Therefore, for information regarding the surface, SEM is preferred while "bulk" information is obtained through TEM. The latter can also provide crystallographic information through selected area electron diffraction (SAED) [70]. The latter can be used as a phase determination characterization method (similar to XRD, except using electron diffraction in place of light diffraction) and instruments for this analysis are commonly installed within a TEM apparatus. The measurements are done on the area selected for TEM imaging and therefore the analyst is enabled to select a specific portion of the sample ("selected area") to gain insight on its crystallographic structure.

Energy-dispersive X-ray spectroscopy (EDS) [71] is, like SAED, an added characterization method that can be included within SEM or TEM apparatus. Briefly, each element will have a specific and unique reaction following excitation with electrons in the incident beam. An inner shell electron will be ejected and this leads to the relaxation of an outer shell electron to the orbital from where the core electron was removed. This will produce a photon (in the X-ray range), with an energy equal to the difference in energy between the outer and inner electron orbitals. This energy is measured by the EDS. As each element has a specific and unique arrangement of electron energies, so too it will have a unique spectrum; the identification of the chemical composition and mapping of the measured area can be obtained. However, it is important to keep in mind that this information is highly localized (and relatively surface sensitive) and therefore cannot be used to assess of the average loadings of these elements over the entire sample.

Electrochemical impedance spectroscopy (EIS) [72] is a complex characterization method based on the impedance (effective resistance (including both the resistance and reactance)) of a system. Briefly, the measurements are plotted in the so-called Nyquist plot, corresponding to the negative imaginary component of impedance versus the positive real component of the impedance.

In photocatalysis, this method is used to assess the difference in charge separation between systems. The plotted spectra are commonly shaped as arcs; the smaller the radius of said arc, the lower the resistance of the system, and therefore the higher the exciton lifetime (and this is assumed to relate to better charge separation).

Density functional theory (DFT) [73] calculation is a computational tool that is used to predict properties and applicability of given materials. Briefly, the calculations are based on the resolution of Schrödinger's equation $\left(\hat{H}\Psi = E\,\Psi \right)$, providing insight on the material properties. However, as this equation cannot be exactly solved for non-single electron systems, the level of approximation (and therefore the accuracy) of the obtained results relies on the choice of the most accurate type of "exchange and correlation" functional (LDA, GGA, etc.). This is used in photocatalysis for comparative purposes, between the obtained theoretical results and the observed experimental results and for design of predicted systems based on theoretical efficiency.

7.4.4 Photocatalysts

A broad range of different types of materials can be used for light harvesting and energy conversion via photocatalysis (oxides (TiO_2, ZnO, etc.) [74], sulfides (ZnS, CdS, WS_2) [75], (Pd, Pt, Ag, Au, Cu, etc.) [76], etc.) for a broad range of reactions (organic compound degradation [77], water splitting [78], artificial photosynthesis (see above), NO_X degradation [79], etc.).

In the case of the artificial photosynthesis process within heterogeneous catalytic materials, studies mostly focus on semiconductors and derivatives of these. Homogeneous catalytic reduction of CO_2 using metal complexes is also an important field but is not addressed here. For an example of a thorough review see the recent work of Zang et al. [80].

Semiconductors are the most commonly studied systems for heterogeneous photocatalysis due to their well-known light-harvesting properties, their relatively low cost, and the simple synthesis procedures required for their preparation.

The following sections focus on systems promoted by different classes of solid-phase materials; for example, oxides, MOF, plasmonic materials, and a variety of others.

Considering the number of publications in the domain this cannot be considered as an exhaustive list of systems currently studied but simply an overview of what we consider to be the most important and most interesting.

7.5 TiO_2-Related Compounds

7.5.1 Pure TiO_2

TiO_2 is nontoxic, inexpensive, chemically stable, insoluble in water, and abundant. It is used in a broad range of applications (pigments, sunscreen, cosmetics, etc.) [81], but we here focus on its use as a photocatalyst for the artificial photosynthesis reaction.

By far the most studied photocatalyst for all types of heterogeneously catalyzed processes, including for CO_2 photoreduction, TiO_2 can be used in three different phases, namely, anatase, rutile, and brookite. These phases have different direct BG energies, anatase (3.2 eV), rutile (3.0 eV), and brookite (3.45 eV). These correspond to photon wavelengths of ~413, 387, and 359 nm, respectively, meaning that they can be activated only through absorption of near-UV photons. This means, regarding solar activation, that they can only work with ~3% of the sun's incident radiation (the bulk of which is in the visible and IR ranges of the spectrum).

TiO_2 photocatalysts also fit the band energy level requirements to promote artificial photosynthesis described earlier in terms of its CB and VB energies (relative to the LUMO and HOMO of CO_2 and H_2O, respectively).

Within the following paragraphs TiO_2 has been classified as a function of the obtained powder's color. "White TiO_2" refers to the most commonly studied TiO_2 and it is now named as such following the formation of recently produced "Black TiO_2." The latter obviously enables a further light harvesting in the visible range (see below).

7.5.1.1 White TiO_2

Commonly used in the form of the commercially available "Degussa P25" (composed of both anatase and rutile phases) TiO_2-based systems have also been widely studied in both single-phase and biphasic systems.

Anatase and rutile-phase TiO_2 are the most commonly studied photocatalysts; however, rutile is known for its lower photocatalytic activity despite it having a lower energy BG (thereby allowing it to generate excitons from a larger component of incident solar radiation).

Some of the reasons for this have been related to the typically smaller surface area, longer "bulk" exciton lifetime (i.e., the time it takes for the exciton to arrive at the surface where it can react with adsorbed species), and less efficient charge carrier transport than anatase (as shown by Luttrel et al. [82]).

Tuning the properties of TiO_2 materials to enhance catalytic reactivity has been of interest for a long time within photocatalysis. Novel TiO_2 structures and synthesis methods have been systematically studied to generate materials that obtain higher yields of solar fuels (e.g., CO, CH_4) or fine chemical precursors (C_2H_4, C_2H_6, etc.) from the substantive reaction.

Pure anatase TiO_2 and commercial P25 (a mixture of anatase and rutile) are often used as reference photocatalysts when modified TiO_2 systems are studied. Brookite and brookite-based biphasic systems have been less studied due to difficulties in the synthesis process and relatively reduced stability.

Kočí et al. [83] studied the effect of TiO_2 particle's size on the reactivity in artificial photosynthesis for the production of gas-phase CH_4 and liquid-phase CH_3OH. The studied systems were in a particle size range between 4.5 and 29 nm (4.5, 6, 8, 14, and 29 nm).

XRD measurements confirmed that the synthesized particles were solely in the anatase phase, except for within one sample (6 nm) that contained both amorphous and anatase components. Interestingly, this sample had been prepared through a sol–gel methodology, while all the others had been synthesized by precipitation.

The impact of the particle size, within crystallite samples, on the material's BG was measured. As expected, smaller particles had lower BG energies ([3.08–3.14] eV) and interestingly the semicrystalline sample had further reduced BG (3.0 eV). BET measurements also showed, as expected, a reduced surface area for larger particles. While the same overall trends are observed for all catalysts in promoting the artificial photosynthesis reaction, the 14 nm sample displayed the best yield for both CH_4 and CH_3OH.

Explanations provided by the authors were that there was a trade-off between various parameters including available surface area (where the reaction happens), charge carrier dynamics, and light absorption efficiency as function of the system's size.

This publication provides interesting information regarding the role of the particle size of TiO_2 anatase for artificial photosynthesis. However, the obtained yields (maximum of 10 µmol/g_{cat} for CH_4 and an order of magnitude less for CH_3OH) after 24 h UV-light irradiation were low.

Zhao et al. [84] have studied the photocatalytic activity of anatase/brookite systems in artificial photosynthesis. Only CO and CH_4 formation were observed (and only in marginal quantities). Pure anatase (A_{100}) and brookite (B_{100}) photocatalysts were synthesized, along with a range of biphasic systems ($A_{96}B_4$, $A_{75}B_{25}$, $A_{50}B_{50}$, and $A_{37}B_{63}$).

While brookite-containing photocatalysis on their own are rarely studied (due to the aforementioned difficulties in brookite synthesis), the authors have here shown that its presence enhances photocatalytic activity for artificial photosynthesis within certain biphasic TiO_2-based systems. As measured by UV–vis diffuse reflectance spectroscopy and N_2 abs–des measurements, all biphasic systems have both higher BG and smaller surface areas with respect to those of pure anatase. This, theoretically, should be detrimental to photocatalysis.

However, all anatase-rich (up to 50% of A) biphasic systems have increased CO production rates with respect to A_{100}, B_{100}, or commercial P25. This yield from the $A_{75}B_{25}$ photocatalyst was 100% better compared to that from pure anatase sample and 300% higher than from B_{100}. These results highlight the importance of the biphasic nature of the photocatalysts in enhancing artificial photosynthesis reactivity. The authors attributed this increase in activity to interfacial charge transfer effects between the A and B phases.

Low catalyst lifetimes in artificial photosynthesis are currently a significantly hindering factor for development of efficient systems. In this particular study, after 4 h of illumination, highly reduced photocatalytic activity was measured. In situ DRIFTS (diffuse reflectance IR Fourier transform spectroscopy) was performed to gain insight on the reactivity and to understand possible paths of catalyst deactivation. Following this the authors suggested that both CO_2^- and HCO_3^- are important reaction intermediates while the formation of CO_3^{2-} could be a factor in TiO_2 poisoning.

Liu et al. [85] comparatively studied the reactivities of the three different TiO_2 polymorphs for artificial photosynthesis. Anatase, rutile, and brookite have been studied both as defect-free and as defective structures to gain insight on the impact of both the TiO_2 structure and the potential roles of defects on this reaction. The defective structures were obtained through high-temperature post-synthetic treatments in helium.

CO and CH_4 were the only carbon-containing products noted. For untreated catalysts only minor amounts of CH_4 were produced, while for their He-treated equivalent much higher amounts were formed. Eight and ten times higher production of CH_4 and CO were obtained in the case of brookite, and four and five times higher over the anatase samples. However, for heat-treated rutile a much smaller enhancement of photoactivity was observed.

The untreated catalyst activity for artificial photosynthesis was classified in the following order: rutile > anatase > brookite while He-treated ones was ranked as: brookite > anatase > rutile. The proposed mechanism to explain the enhanced photocatalytic activity of the treated catalysts involved the formation of oxygen vacancies (V_O) in the TiO_2 framework and subsequent formation of active Ti^{3+} species.

V_O is known to form color centers in TiO_2. This was confirmed in this work by UV–vis diffuse reflectance spectroscopy. In the case of both anatase and brookite, absorbance peaks with tails toward visible wavelengths were obtained. Spectra of the heat-treated rutile sample showed no significant differences when compared to the spectra of the untreated sample.

DRIFTS measurements confirmed the presence of Ti^{3+} in the treated systems as a peak corresponding to Ti^{3+}-OH bonds appeared only on the spectra of those catalysts. This site can act as an absorption site for the reactants, increasing the catalytic activity. Furthermore, both UV–vis and FTIR spectroscopy confirmed the presence of V_O/Ti^{3+} within anatase and brookite heat-treated catalysts (as predicted by the modified color of the powders in those cases). These materials were yellow, while the modified rutile remained white.

Pure anatase TiO_2 with a BG in the visible range (~2.0 eV) was synthesized by Dette et al. [86]. A combination of DFT with STM and spectroscopic methods was used to characterize the catalyst. The presence of undercoordinated surface Ti atoms (Ti_{4c} and Ti_{5c} instead of Ti_{6c} in traditional anatase) was understood to be a key factor in BG tuning.

The presence of occupied d states within the VB/CB BG resulting from the presence of these surface species is suggested by the authors as the reason for the reduced BG. While this publication provides an efficient and novel methodology for surface tuning of TiO_2, the system remains to be used in the promotion of photocatalytic reactions and especially for artificial photosynthesis.

A further detailed and extensive review of the progress and current limitations of the use of TiO_2 for artificial photosynthesis has been published by Lui and Li [53]. Reaction mechanisms including adsorption, activation, and dissociation of CO_2 and H_2O on TiO_2 are addressed along with discussions of intermediate species and proposed reaction pathways.

7.5.1.2 Black TiO_2

A recent trend in the use of titanium dioxide as a photocatalyst involves the use of the so-called "Black TiO_2"; this novel material (discovered in 2011 by Chen et al. [87].) has differing properties than traditionally used "White TiO_2." The colored nature of the powder is responsible for significantly increasing visible light absorption (thereby allowing solar light utilization to a far higher extent) and this explains the recent enthusiasm of the scientific community. The nature of the differences with respect to "White TiO_2" is still currently under debate, but certain aspects have been agreed.

Rajaraman et al. [88] recently reviewed the knowledge gained on this material. This work provides theoretical explanations for the observed enhanced photocatalytic activity and focuses on these novel material properties.

While numerous synthetic paths exist for the synthesis of this material, hydrogenation of pristine TiO_2 remains the most commonly used. Hydrogen is understood to react with lattice oxygen, forming H_2O, the consequent V_O leading to formation of reduced Ti (from Ti^{4+} to Ti^{3+}). Incomplete oxidation of low-valence Ti (Ti metal, TiO, Ti_2O_3, etc.) can also be used to obtain the TiO_{2-x} material [89].

The term "Black TiO_2" is generic and encompasses numerous systems with a range of different colors and varying properties obtained through different preparation methodologies (mostly involving reductions of some type). However, independently of the synthesis path used, the studied materials are always TiO_{2-x} systems.

This material differs from traditional TiO_2 due the presence of "engineered defects" (both surface and bulk defects) that include formation of Ti-H bonds, the presence of hydroxyl (-OH) groups on the surface, along with oxygen vacancies and Ti^{3+} species. The role of the defects' position (surface and/or bulk), however, remains unclear. While it appears clear that surface defects are preferred to bulk ones (that can act as recombination centers), the importance of the latter has also been assessed.

Further work will be required to fully understand and quantify the relative impact of defects' position. Therefore, the concentration of those defects, along with the ratio of surface/bulk defects, is an important parameter in relation to the material's photocatalytic properties. Defect species are suspected to introduce energy levels between the CB and the VB of the TiO_2 semiconductor. This explains the eventual coloration of the powder (through the generation of reduced BG energies in the visible light range).

Used as a photocatalyst for artificial photosynthesis by Wang et al. [90] under simulated solar light irradiation, a sample of porous Black TiO_2 film displayed improved reactivity with respect to the same reaction on films of commercially available P25. The catalyst was prepared by a simple hydrothermal method using H_2O_2 as an oxidant on Ti plates. Two "Black TiO_2" systems were synthesized (by performing the oxidation at different temperatures, 110 °C and 130 °C). Both showed similar reactivities.

The products formed in this reaction were CO and CH_4 with production rates for the modified systems being up to 115 $\mu mol/g_{cat}$ h for CO formation and 12 $\mu mol/g_{cat}$ h for CH_4 formation (while only 0.28 $\mu mol/g_{cat}$ h and 0.019 $\mu mol/g_{cat}$ h of these species were obtained using P25). Extensive material characterizations to understand the observed results were performed, a porous grid-like structure was observed by SEM, and the Black TiO_2 film thickness was calculated to be between 0.8 and 1.5 μm.

XRD measurements have shown that the materials were composed of only anatase and rutile phases, with no brookite phase being observed. EDS showed a near 1 ratio of Ti and O on the surface. The oxygen deficiency here (relative to TiO_2) indicates the presence of surface oxygen vacancies (V_O). Levels of Ti^{3+} were measured on the surface by XPS and this also confirmed the presence of V_O. PL measurements also suggested there should be an enhancement in photocatalytic activity, as significantly reduced emission intensities were recorded from the modified systems with respect to P25 (confirming longer exciton lifetimes).

The presence of V_O is suggested to distort the TiO_2 lattice on the surface, producing localized states that, in turn, lead to the improved light-harvesting capabilities in the modified systems. Improved activity was also attributed to improved charge transportation and separation steps.

As a recently emerging and promising solutions the material has been widely studied, but conflicting results have been obtained and further study of "Black TiO$_2$" will be necessary in the coming years and to rationalize the observed effects if it is to be used to develop efficient systems.

In general, TiO$_2$ presents interesting photocatalytic properties, although to date no sufficiently efficient systems have been developed yet for artificial photosynthesis. Major flaws of unmodified TiO$_2$ as a photocatalyst include its large BG (that hinders visible light harvesting) and rapid exciton recombination (that limits the possibility of charge transfer from the TiO$_2$ bulk – where they are generated – to the TiO$_2$ surface where they are used).

Modified TiO$_2$-based materials have received significant interest as photocatalysts over the past 40 years with efforts designed toward improvements in efficiency.

7.5.2 Metal- and Nonmetal-Modified TiO$_2$ Systems

7.5.2.1 Metals

Metals are commonly used as for their own catalytic properties and synergize with photocatalysts in combined systems. Metal doping of TiO$_2$ is understood as cationic substitution of Ti^{4+} or interstitial insertion within the SC framework. This alteration results in the development of new electron energy levels between the CB and the VB of the SC and therefore a reduced (redshifted) BG that enables the absorption of visible light. These "dopants" are beneficial to photocatalysis (see below).

On the other hand, loading of metallic nanoparticles (NPs) onto the TiO$_2$ surface is also a profitable approach, principally due to the ability of such NPs to trap photogenerated electrons thereby reducing the rate of excitons' recombination and increasing the efficiency of the photocatalytic process. Metals also present inherent catalytic capability (e.g., for H$_2$ dissociation) and provide catalytically active sites.

7.5.2.1.1 Noble Metals

Schottky barrier and plasmonic effects are different and possibly synergic effects that are of interest in the cases of noble metal (NM) NP-loaded semiconductors. Those properties can arise from the use of numerous metals but are especially observed in the cases of NM loading. In this section, we focus on both NM doping and the Schottky barrier properties (NM "loading"); plasmonic materials are briefly discussed later in the text in a more general manner.

For n-type SC (like TiO$_2$) a Schottky barrier is formed at an SC–metal interface. In this case, promoted electrons commonly flow from the SC to the metal as its work function (the energy difference between E_f and the vacuum level) is typically larger than the SC electron affinity (the energy difference between CBM and the vacuum level). That is to say that E_f will be lower in energy than CBM.

This flow will be continued until an equilibrium is reached, leading to equal E_f level in both components of the interface. This results in a distorted band structure at the interface (and the generation of an electric field). The SC's bands are raised upward in energy at the interface creating a potential barrier that prevents the electrons from flowing back to the SC, therefore creating an electron rich zone in the metal at the interface and an electron depletion zone in the SC (for charge compensation) [91]. This effect is common with NM loading on TiO_2 and explains the observed increased lifetime of excitons in these systems; that is, essentially through electron trapping in the metal where the metal particles act as electron sinks.

Koči et al. [92] synthesized silver-enriched (0.72, 2.37, 3.38, and 5.19 wt% (exact Ag content determined by XRF)) TiO_2 systems (Ag/TiO_2) through a sol–gel methodology and applied them in the artificial photosynthesis reaction under UV-light illumination (8 W Hg lamp (254 nm)). Reaction products included significant H_2 production, CO (in small quantities), and high yields of $CH_{4(gas)}$ and $CH_3OH_{(liq.)}$. These doped systems showed increased photocatalytic activity with respect to the reactivity of pure TiO_2 at all loadings of Ag.

Production of both CH_4 and CH_3OH increased with the following order: unmodified TiO_2 < 2.37 wt% Ag < 0.72 Ag wt% < 3.38 Ag wt% < 5.19 wt%, and no significant impact on selectivity was observed. The highest obtained production rate was 8 µmol/g_{cat} (for CH_4) and 1.8 µmol/g_{cat} (for CH_3OH) (as monitored by CG) following 25 h of illumination.

Two distinct phenomena were proposed in the two different ranges of Ag loadings to explain the obtained results. Up to 3.38 wt% Ag, doping, the increased reactivity was ascribed to an absorbance redshift (increasing the quantity of harvested light) as seen in the UV–vis absorption spectra and caused by Ag impurity bands within the TiO_2 BG. Above this loading, the formation of Ag metallic clusters inside the TiO_2 crystals was proposed by the authors. Those would reduce the exciton recombination rate and enable efficient charge separation through formation of Schottky barriers.

Singhal et al. [93] published a comparative study for noble-metal-loaded TiO_2 prepared by a photo-deposition method, for gas-phase artificial photosynthesis under UV illumination (320–390 nm) in a flow system. Both monometallic (Pd, Pt, Ag, and Au) and bimetallic (Au/Pd, Ag/Pd, and Pd/Pt) TiO_2 loaded noble metal (NM-TiO_2) systems were studied.

Reaction optimization was performed for the Pd-loaded systems. Interestingly, the flow rate was found to significantly impact the obtained production rate of CH_4, from 365.4 nmol/g h at 4 mL/min to 801.9 nmol/g h at 8 mL/min for a catalyst containing 2 wt% Pd loading. Further increase of the flow rate to 15 mL/min led to a decrease in the CH_4 production rate to 772.7 nmol/g h. At this optimized flow rate, the reaction was also studied over a 1 wt% Pd loaded material, leading to a reduced rate of production of 363.5 nmol/g h. These workers implemented these optimized parameters for all subsequent measurements.

While both CO and CH_4 were produced in all cases (Ag, Au, Pd and Pt) a clear selectivity reliance on the NM nature was observed. Both Ag and Au directed the selectivity toward CO production, up to 1,153.17 nmol/g h for Au (ca. 10 times that of CH_4) while both Pd and Pt directed the selectivity toward CH_4 production and only small amounts of CO were detected.

Used as reference, pure TiO_2 only produced 67.31 nmol/g h of CH_4 and negligible amounts of CO. It was therefore demonstrated that NM loading is highly profitable for CO or CH_4 production and that the selectivity could be tuned as function of the NM used. Pd yielded the highest production rate of CH_4, and it was used in synergy with the three other NM within the bimetallic loaded TiO_2 systems.

The three systems ($Pd_{1\%}Pt_{1\%}$, $Pd_{1\%}Au_{1\%}$, and $Pd_{1\%}Ag_{1\%}$) displayed similar activity and selectivity with product formation rates between 1,300 and 1,700 nmol/g h for CH_4 and small amounts (>200 nmol/g h) of C_2H_6 were also obtained. CO production was not noted over the bimetallic systems. The bimetallic structures were therefore useful to increase production of CH_4 with respect to both pure TiO_2 and monometallic NM-TiO_2.

Overall, those results were attributed to enhanced charge separation (due to the presence of Schottky barriers) and easier multi-electron transfer processes due to the greater number of available free electrons (caught in the NP particle electron traps).

NM are rare and expensive elements, thus hindering their use on large scales; however, they present highly interesting properties and are frequently studied in photocatalysis.

7.5.2.1.2 Transition Metals

Transition metal "TM" doping of TiO_2 for artificial photosynthesis has attracted significant attention. The rationale for altered activity of materials containing TM has been ascribed to their $3d$ electronic structures. These enable relatively facile transfer of electrons from metals to TiO_2 CBs [94].

Doping with different metals leads to different band structures. In most cases (V, Cr, Mn, and Fe) a mid-BG electron containing level is created; however, in the case of Co doping, the created level lies just above the TiO_2 VB and for Ni doping a hybridization of Ni and Ti orbitals (Ni and Ti $3d$) leads to a higher VBM, as calculated through DFT by Umebayashi et al. [95].

Regarding TM-metal doping, Jeong Yeon et al. [96] have studied numerous 3d TMs (Mn, Fe, Co, Ni, Cu, and Zn) as substituents within the TiO_2 framework. These have been prepared by sol–gel and studied for activity in the artificial photosynthesis reaction under UV illumination (365 nm). Doping levels have semi-quantitatively been measured by EDS. The obtained M/Ti atomic ratio (%) were as follows: 10.43 for Mn; 3.75 for Fe; 5.90 for Co; 4.79 for Ni; 10.55 for Cu, and 5.70 for Zn.

Smaller crystallites have been observed for all TM-TiO_2 materials than for pure TiO_2 as well as small lattice distortions indicated by downward shifts in XRD peaks. While smaller crystallites could explain enhanced photocatalytic activity (through increased surface areas), we the authors focused on the effect of the metal dopants themselves.

Interestingly, for all materials containing doped TM (excepted Zn) reduced PL and BG were measured, reflecting metal-dependent enhanced light handling properties that should both lead to improved photocatalytic capabilities. XPS measurements confirmed the presence of stable oxidized metal species in all cases.

Insertion of TM into the TiO_2 framework significantly impacted the absorption properties. Reduced TiO_2 BG were measured for all doped systems, except in the case of Zn (TiO_2:2.93 eV; Mn: 2.47 eV; Ni: 2.38 eV; Fe: 2.72 eV; Co: 1.96 eV, and Zn: 3.06 eV).

More interestingly, doping also impacted the positions of the CB and VB in the different systems. The VBM positions were obtained by XPS measurements of the peak that the authors associated to the VB of each system. The obtained values were 2.40 eV, 2.36 eV, 1.75 eV, 1.60 eV, 0.80 eV, and 2.15 eV for Mn, Fe, Co, Ni, Cu, and Zn, respectively. CBM positions were subsequently calculated from UV spectroscopy.

For Co- and Mn-doping, the CB was too low in energy to promote any CO_2 reduction reaction; that is, the promoted electrons were too low in energy to move to the CO_2 LUMO and no reaction products were observed. The measured products were CO for which only Zn- and Cu-doped systems has showed improved production rates with respect to pure TiO_2, CH_3OH for which only Cu-TiO_2 showed significant production rates (up to 3.5 mmol/g_{cat} after 10 h, < 0.5 mmol/g_{cat} for the others), and CH_4 in which case both the Fe- and Ni-doped materials showed similar (toward each other) and increased production rates with respect to pure TiO_2 (around 8 mmol/g_{cat} after 10 h compared to around 7 mmol/g_{cat} for TiO_2). This study therefore provides an interesting take on possibilities for selectivity-tuning in artificial photosynthesis.

A comparative study involving three different TM (V-, Cr-, and Co-) dopants within the TiO_2 framework has been published by Ola and Maroto-Valer [94]. Observed results include a red shifted absorption (indicating a reduced BG) for all doped materials as well as again smaller crystallite sizes. All materials containing dopants displayed an optimal increased light absorption at 2.0 wt% doping but, interestingly, differing optimal doping percentages in terms of photocatalytic reactivity for both methanol and hydrogen formation (V (1.5 wt%); Cr (0.5 wt%) and Co (1.0 wt%)).

Among the tested materials, all those containing dopants gave increased conversions of CO_2 to products with respect to the activity of unmodified TiO_2. The material doped with cobalt was the most active of these three doped materials, producing up to 62.1 μmol/g_{cat} h and 26.12 μmol/g_{cat} h of hydrogen and methanol, respectively. Reduced yields were measured after 4 h of visible light illumination suggesting instability in the catalyst.

Overall, numerous studies in TM-TiO$_2$ modified systems (both doped and loaded) have been performed and it is expected that further investigations will lead to more efficient systems. TM are, for the most part, abundant and inexpensive; therefore, such systems could be opportune for scaling up in industrial facilities.

7.5.2.1.3 Rare Earth Metals

Rare earth (RE) metal doping in TiO$_2$ is of interest due to their incompletely filled 4f orbitals that can act as trapping sites for photogenerated electrons, following either transfer from the CB of TiO$_2$ or directly from the VB of TiO$_2$. This is suggested to enhance charge carrier separation (albeit through a different mechanism to the action of metallic nanoparticles), thereby lengthening exciton lifetime and enhancing the possibilities for electron and hole reactivity.

Relatively few papers have been published studying artificial photosynthesis using these types of systems; however, promising results have been observed and these should lead to further studies. Chun-Ying et al. [97] prepared different Europium (Eu)-doped TiO$_2$ systems (0.15; 0.25, and 0.40%) using a sol–gel process and studied these in the artificial photosynthesis reaction under simulated sunlight illumination (Xe arc lamp).

The catalyst was present as suspended particles in an aqueous environment, and CO and CH$_4$ were the main products in all cases. The obtained results have been compared with those noted over pure, anatase, TiO$_2$. While XRD did not reveal the presence of Eu$_2$O$_3$, this was, however, confirmed by XPS, and again (as was seen for TM doping above) a clear trend in reduced crystallite size was observed with increasing doping level of Eu (broader and weaker peaks).

N$_2$ abs–des measurements were also performed; a clear increase in BET surface area was observed following increased doping (as would be expected from smaller crystals), from ~46 m^2/g for pure TiO$_2$ to ~86 m^2/g for 0.25% Eu doping. However, further doping led to reduced surface areas; that is, ~75 m^2/g for 0.40% doping.

A similar trend was observed for BG values, measured by absorption spectroscopy, from 3.22 eV for the pure system to 3.00 eV for 0.25% Eu, increasing again to 3.05 eV for 0.40% doped Eu. Again, PL measurements have shown the same trends in intensity. Therefore, optimal doping for photocatalytic purposes was considered to be 0.25% Eu.

This optimized system has shown clearly improved results in photocatalytic activity, 14-fold and 2- fold production of CH$_4$ and CO, at ~66 μmol/g$_{cat}$ and ~43 μmol/g$_{cat}$ (after 9 h), respectively, with respect to their production over unmodified TiO$_2$. The author proposed that the observed efficiency was linked to increased charge separation, as confirmed by electrochemical impedance spectroscopy (EIS) and photocurrent response measurements as well as to more efficient light harvesting and increased surface area. It was also concluded that over-doping would lead to creation of recombination centers, limiting the charge carrier's lifetime and thereby decreasing reactivity.

Hao et al. [98] prepared Samarium(Sm)-doped TiO_2 photocatalysts for liquid-phase artificial photosynthesis under Xe arc lamp illumination by the sol–gel method. This study is highly similar to the previously presented paper, and the exact same trends were observed for Sm-doped TiO_2 as for Eu-doped TiO_2. An optimal doping of 0.5 wt% Sm was found (0.25 and 1 wt% were also studied), with a reduced activity above and below the optimum.

Morphological characterization also yielded similar results; that is, increased surface area and smaller crystallite size (XRD) for doped systems, relative to those of pure TiO_2. The author proposed that introduction of Sm within the TiO_2 framework led to distortions, with this added strain energy in turn inhibiting anatase growth.

The presence of Sm_2O_3 (Sm^{3+}) and of Sm^{2+} species was also confirmed by XPS, and optical analysis showed the same BG trend as seen earlier, from 3.11 eV for pure TiO_2 to 2.95 eV for the optimized 0.5 wt% Sm-doped material. The 0.5 wt% Sm catalyst outperformed in photoreduction compared to TiO_2 by ~5 times for CO production (~55 μmol/g_{cat} after 6 h) and ~3 (~4 μmol/g_{cat} after 6 h) times for CH_4 production. The author proposed that aside of the increased surface area, lower recombination rates (electron trapping) in Sm (formation of Sm^{2+} species from Sm^{3+}) and reduced BG (due to the Sm $4f$ level positioned beneath the TiO_2 CB) were the reasons for the observed increased activity.

Overall RE-TiO_2 systems are proposed as materials to promote efficient ways to solve issues related to both short exciton lifetime (fast recombination) and low light-harvesting properties (by moving absorption onto the visible region) seen in pure TiO_2; however, the exact catalytic mechanisms of promotion remain unambiguously identified. In a wider discussion, RE-TiO_2 systems have been shown to be efficient for promotion of different photo-catalyzed reactions (e.g., organic compound degradation) as recently reviewed [99, 100] and these materials are also expected to become increasingly important for the promotion of artificial photosynthesis.

7.5.2.2 Nonmetals

Nonmetal doping differs from above mentioned studies in that the substituent or interstitial incorporated ions are anionic and therefore normally sit in O sites. N- and C-doped TiO_2 have been studied for promotion of model photo-catalyzed process (aqueous pollutant oxidation) such as for malathion [101] or trichlorophenol [102] degradation as reviewed by Varma et al. [103]. The rationale behind the efficiency of N-doped-TiO_2 on increased photoreactivity has been studied by Rumaiz et al. [104].

Nonmetal doping is known to impact the optical properties of the catalyst. In this article it was concluded (through DFT calculation) that N doping of TiO_2 was directly responsible decreasing the BG through VBM tailoring through Ti O $2P$ mixing with N $2P$ orbitals. It was also shown that N-doping facilitates the formation of oxygen vacancies (V_O), through "merging" of two oxygen position into one, partially filling Ti $3d$ orbitals, with respect to pure TiO_2. Their formation energy drops from 4.6 eV to 0.2 eV in the case of N doping [105]. V_O themselves are responsible for the impurity states above the VBM and are also of interest in terms of facilitating increased CO_2 absorption of the surface.

A further study of the impact of N doping on the formation of V_O in TiO_2 was presented by Peng et al. [106] through the study of substitutional and interstitial N insertion in the TiO_2 framework. Two different N-doped systems were synthesized from P25, one through microwave (N-M)-assisted synthesis (P25 heated with urea), and a second through annealing (N-NH$_3$) of P25 under an NH$_3$ flow. The first method is thought to lead to insertion of interstitial N atoms and the second to substitutional N insertion.

The nature of N doping was assessed by XPS. An N-related peak at 396.5 eV was attributed to substitutional nitrogen and another one, at 399.9 eV was attributed to electron emission from interstitial nitrogen. XPS measurements were also used to relatively assess the N contents within each sample, higher in N-M than in N-NH$_3$.

No significant structural modification of TiO_2 was observed by either XRD or N_2 abs–des measurements. However, in both cases increased light range absorption in the visible region (400–600 nm) was obtained for the N-doped systems.

Interestingly, instances of these systems being applied to artificial photosynthesis are relatively scarce. Selected examples of nitrogen, fluorine, and iodine doping of TiO_2 to generate catalysts for artificial photosynthesis have been reported and are discussed below.

Michalkiewicz et al. [107] synthesized N-doped TiO_2 systems through hydrothermal reaction for liquid-phase artificial photosynthesis under UV–vis illumination.

CH_3OH was the only observed product for the doped system, with production rate up to 23 μmol/h g. For comparative purposes the reaction was performed in similar conditions using P25, over which no products were obtained. Elemental analysis revealed a 0.24 wt% content of nitrogen in the synthesized sample. Morphological characterization showed an increased BET surface area (from 61 m^2/g in P25 to 285 m^2/g in the doped sample) along with increased micropore volume (from 0.019 to 0.121 cm^3/g).

XPS measurements confirmed interstitial insertion in the TiO_2 framework (through the deduction of O-Ti-N linkages). Unusually, given the discussion above, an increased BG was measured for the doped sample with respect to P25, from 3.22 to 3.31 eV, and this was attributed to the higher rutile content in P25. Interestingly, CO_2 adsorption studies showed increased CO_2 absorption on the N-doped sample, increasing from 70 mg/g on P25 to 120 mg/g. The Freundlich isotherm was applied to this measurement, and this clarified the nature of CO_2 chemisorption on the doped material.

The authors proposed that the basic N group on the catalyst surface interacted with the acidic CO_2. Along with the superior morphological properties of the doped material this was proposed as an important factor in rationalizing the obtained results.

Phongamwong et al. [108] synthesized a range of N-doped (sol–gel) and *Spirulina*-N-doped TiO$_2$ catalysts (using sol–gel processes for N-doping of the TiO$_2$, followed by *Spirulina* loading through Incipient Wetness Impregnation). They studied these in the artificial photosynthesis reaction under visible light illumination (400–800 nm). The use of *Spirulina* in this context was motivated by a "Biomimicking" logic due to the chlorophyll content of this organism, the latter being known for its impact on natural photosynthesis.

As seen numerous times above, an optimal level of loading (or doping) is often crucial. In this study the 10 wt% N-TiO$_2$ for N doping (5, 10, and 15 wt% were studied) was the best-performing material and was therefore used for further functionalization by chlorophyll-containing *Spirulina* (Sp) (known for visible light absorption). A 0.5 wt% loading of the latter was found to be the most efficient system (0.1, 0.5 and 1 wt% were studied).

In the following the most efficient systems are compared, *p*-TiO$_2$: pure TiO$_2$; 10N-TiO$_2$: 10 wt% N-TiO$_2$ and 0.5Sp10N-TiO$_2$: 0.5 wt% Sp to 10 wt% N-TiO$_2$. The presence of both N and Sp was confirmed by FTIR, through the identification of peaks related to vibrations of CN and C = O bonds.

Interstitial N insertion was confirmed by XPS (Ti-N-O linkages), as well as the generation of V$_O$ through the presence of Ti^{3+}-related peaks. This led to formation of sub-BG N-O mixed states above the VBM of TiO$_2$ (O 2P states); for 10N-TiO$_2$ a 2.18 eV (569 nm, visible light range) sub-BG was measured while *p*-TiO$_2$ the BG was measured at 3.06 eV.

Interestingly, the insertion of N in the TiO$_2$ framework improved the system's crystallinity, with bigger crystals prepared (as seen by XRD). The increased activity of the doped systems cannot therefore be associated with a higher surface area.

While the loading with *Sp* was aimed at enhanced production of photogenerated charge carriers, it was also noted that enhanced electron transfer (to TiO$_2$ CB) was obtained (as measured using PL), and this also enhanced the photocatalytic activity.

The reaction products used to assess the photocatalytic activity of the different systems were CH$_4$ as an undesired product and C$_{2+}$ (C$_2$H$_4$ and C$_2$H$_6$) as desired products (H$_2$ was also produced as a by-product). The catalyst's activity increased from *p*-TiO$_2$ < 10N-TiO$_2$ < 0.5Sp-10N-TiO$_2$ for all products.

Selectivity was defined in this instance as the ratio of the sum of the desired product production rates by the undesired product production rate. The highest selectivity was found to be for the 0.5Sp-10 N-TiO$_2$ system at 0.6. For the three-component system the following product rates were obtained: 0.12, 0.17, 0.48, and 145 µmol/g$_{cat}$ after 6 h of illumination for C$_2$H$_4$, C$_2$H$_6$, CH$_4$, and H$_2$, respectively. Production rates for C$_{2+}$ products obtained on p-TiO$_2$ were negligible, while on 10N-TiO$_2$ they were found to be in a range between 0.03 and 0.06 µmol/g$_{cat}$.

The authors proposed that the increased light-harvesting range and production of V$_O$ (linked to N-doping), acting as active sites for adsorption and dissociation, were important parameters in the observed activity.

Zhang et al. [109] synthesized I-doped TiO_2 with different dopant levels (2.5, 5, 10, and 15 wt%) through a hydrothermal method (titanium isopropoxide and iodic acid were used as starting materials). Furthermore, different calcination temperatures (T_C) (375, 450, and 550 °C) were applied to the 10 wt% I-TiO_2 samples. The activity of these samples has been measured in the artificial photosynthesis reaction under visible (>400 nm) and under UV light (Xe lamp). CO was the only detected carbon-containing product in all cases; however, a significant improvement of the production rate was obtained for the 10% I-TiO_2 (with $T_C = 375$ °C), with up to 2.4 μmol/g_{cat} h (670 ppm after 210 min) CO being produced under visible light illumination. Interestingly, the optimal doping was different under UV illumination, with the 5 wt% I-TiO_2 being the most efficient photocatalyst, however with a smaller CO production rate (yielding 600 ppm after 210 min).

XRD measurements showed that while all the systems were a mix of brookite and anatase (the latter increasing in content with increasing I doping levels and T_C), samples calcined above $T_C > 450$ °C had small rutile contents. Crystallite size (XRD) increased with higher T_C while almost no difference was observed following increases in I doping. This explained the observed reduced activity for higher T_C and the optimal performance of the material calcined at 375 °C. BET surface area (N_2 abs–des) increased after doping (from ~123 to 138 m²/g) but was not dependent on the I content. As expected, the surface area decreased with increasing T_C.

The BG of the doped systems was seen (through UV–vis absorption) to decrease with increasing I content up to 10 wt%, from 3.13 eV for the undoped catalyst to 3.00 eV. XPS measurements confirmed the presence of Ti^{3+} species in the doped material, and the authors proposed that I^{5+} was inserted in the TiO_2 framework, leading to the formation of Ti^{3+} to provide charge balance. The Ti^{3+}/Ti ratio increased with increasing I content, being 10 for the 5 wt% sample, 18.4 for the 10 wt% analogue, and 19.8 for the sample doped with 15 wt% I.

7.5.3 Different Morphologies

Another methodology currently studied to improve the efficiency of TiO_2-based systems as photocatalysts is the use of reduced dimension systems. Razzaq and In [110] have written a comprehensive review on these systems. We here focus on a selection of the materials to provide insight on the impact of certain parameters on the efficiency of the materials in the artificial photosynthesis reaction.

Particle shapes and sizes can be directly related to both charge migration and separation processes that are clearly of great influence within photocatalysts. These features were well reviewed by Chaitanya et al. [111]. Improvements both in the directional transport of photogenerated charges and in the active surface area increase the number of active sites available.

7.5.3.1 One-Dimensional Structures

Reñones et al. [112] used a combination of sol–gel and electrospinning processes to synthesize TiO_2 1D nanofibers ("A": treated at 500 °C under high Ar air flow; "B": treated at 500 °C under a static Ar atmosphere). Both were analyzed for reactivity in the gas-phase artificial photosynthesis under UV-light irradiation. A sol–gel method was been used to synthesize TiO_2 nanoparticles for reference.

Syngas (CO and H_2) was the main product of the reaction in all cases seen along with minor production of CH_4 and CH_3OH and other compounds (which were not detailed in the publication). However, a greatly improved production of carbon-containing compounds was observed when using sample A (converted carbon (sum of carbon-containing products after 20 h of irradiation): 253 $\mu mol/g_{cat}$, mostly CO, 204 $\mu mol/g_{cat}$ and CH_4, 27 $\mu mol/g_{cat}$) and B (converted carbon: 99 $\mu mol/g_{cat}$, mostly CO: 55 $\mu mol/g_{cat}$; CH_4: 18 $\mu mol/g_{cat}$ and CH_3OH: 15 $\mu mol/g_{cat}$). Regarding TiO_2 NPs the activity was lower (converted carbon: 59 $\mu mol/g_{cat}$, mostly CO: 37 $\mu mol/g_{cat}$).

XRD measurements were performed on all three samples, while the samples A and B were composed of both anatase and rutile phase, the NPs contained only anatase. However, similar crystal sizes were calculated (using Scherrer's equation). A small decrease changes in the measured BG (from 3.2 eV for the anatase nanoparticles to 3.1 eV for samples A and B) was attributed to the presence of rutile in the fibers.

TEM images showed large grain boundaries (between anatase and rutile phases) in the nanofiber samples (A and B), while the NPs were arranged as a randomly distributed agglomeration. These large grain boundaries between different phases were also suggested by a highly reduced surface area (96 m^2/g for NP, 35 m^2/g for A, and 41 m^2/g for B). This large inter-particle interface resulted in greater interconnections between crystals in the nanofiber morphologies than in the NPs, which in turn lead to the formation of rutile.

Electrochemical impedance spectroscopy (EIS) was used to gain insight on the charge transfer mechanisms within the different systems. A Nyquist plot showed that the electrospun samples had higher conductivity (reduced charge transfer resistance) than the nanoparticulate analogue at the SC–electrolyte interface. This means that the fiber configuration enables better charge transport and transfer properties than the nanoparticles (which should influence photocatalysis efficiencies) and better charge separation and enhanced charge transfer were proposed as explanations for the improved photocatalytic activity of the fibers sample compared to the NPs.

Ong et al. [113] synthesized Ni-loaded TiO_2 NPs through co-precipitation and then used CVD to grow CNTs onto their surfaces. Efficient CNT growth was confirmed by FE-SEM (with tube diameters in a 30–40 nm range) and this was quantified by TGA to be at a 14.2 wt% loading. This system has been used to study visible light promoted, gas phase artificial photosynthesis, converting CO_2 into CH_4. The results have been compared with those over both pure TiO_2 (anatase)- and Ni-doped TiO_2 (without CNTs). Unsurprisingly, the photocatalytic activity was observed in the following order: $TiO_2 <$ Ni-$TiO_2 <$ CNT/Ni-TiO_2 with the highest production of 0.16 µmol/g_{cat} h after 4.5 h for the latter (< 0.06 in both other cases). After this time, a decrease in the production rate was observed, attributed to catalyst surface poisoning by adsorption of reaction products on the surface and therefore a reduced number of adsorption sites.

In order to explain these results, optical characterizations were performed. A clear modification in UV–vis absorption properties was observed. While the Ni loading led to a reduced BG (from 3.32 to 3.15 eV) a much higher improvement toward visible light absorption was obtained following the growth of CNTs on the catalyst's surface (2.22 eV; i.e., a wavelength of ~558 nm).

As the Fermi levels of both Ni and CNTs are lower in energy than the TiO_2 CBM, these act as reservoirs of promoted electrons, reducing the recombination rate and facilitating multi-electron processes such as CH_4 formation. A strong chemical interaction (formation of Ti-O-C and Ti-C bonds) between CNT and Ni-TiO_2 was observed by Raman spectroscopy and is also suggested to significantly increase the photocatalytic activity through easing of electron transfer.

The 1D nature of CNTs along with their long-range π-electronic configurations was understood to enable facile electron transport along the lengths of the structure, and therefore enhance charge separation, in turn leading to reduced recombination rates.

7.5.3.2 Two-Dimensional Structures

Xu et al. [114] synthesized TiO_2 nanosheets with 95% of the surface exposing the {001} facet. SAED confirmed this orientation. TEM confirmed that the particles on average were 70 nm in length, 70 nm in width, and 2 nm thick. Based on the calculated surface energies within the three principal TiO_2 surface facets; that is, {101} (calculated surface energy (Wulff's construction) 0.53 J/m^2), {001} (0.90 J/m^2), and {100} (0.44 J/m^2), the {101} face has the greatest of all three and thus might be considered to be the most active for catalysis and photocatalysis.

The three faces are present in a typical cuboid TiO_2 structure and differ from one another in the nature of the surface termination species. While {101} exposed facets are composed of both Ti_{5C} and Ti_{6C} atoms, {100} surfaces are formed by Ti_{5C} only. The presence of Ti^{3+} species on the nanosheet surface was confirmed by EPR but not observed on the cuboid surface. This indicates a much higher level of V_O (and subsequently a higher CO_2 absorption) on these sites.

This material has been used to promote gas-phase artificial photosynthesis under UV-light irradiation. The obtained results for CH_4 production rates were compared with activity over a cuboid TiO_2 system (53% {100}). The nanosheet-based systems displayed up to 5 times higher CH_4 yield than the cuboid materials (from 1.2 to 5.8 ppm/g h). Morphological characterization showed that the surface area changed from $3.9\,m^2/g$ (for the cuboids) to $57.1\,m^2/g$ (for the nanosheets) and could well explain the observed enhancement in production rate.

To rule this out as the sole reason for the observed difference in reactivity, the authors compared the two systems at equivalent total surface areas. The yield of CH_4 over the nanosheet catalyst was still 2.8 times higher than that generated over the cuboids, confirming that parameters other than surface area also have a role in improving reactivity. UV–vis diffuse reflectance measurements showed a relatively minor redshift it the BG of each sample (from 3.17 eV (cuboids) to 3.12 eV (nanosheets)).

The VB position in both structures is thought to be equivalent and therefore the minor BG increase is due to a more negative CB in the case of the nanosheets (with respect to NHE). This suggests that the photogenerated electrons in the nanosheets (once formed) are more reductive; the authors proposed that this could also explain the observed efficiency of the nanosheet structure.

He et al. [115] synthesized a range of anatase TiO_2 (confirmed by XRD)-based compounds with different morphologies (Nanotube (NT); different nanosheets (fluorinated surfaces and unmodified NS) and nanoparticles (NP)), through hydrothermal methods. These have been used for liquid-phase artificial photosynthesis under UV-light illumination. The compounds formed included CH_4 and CH_3OH as major products along with HCHO, CO, and H_2 as minor products. Other carbon-containing molecules (CO_2H, C_2H_4, and C_2H_6) in amounts too small to accurately quantify were also observed.

In order to compare the efficiencies of each system, quantum yields (QY = moles of electrons reacted/moles of incident photons) and EROEI (energy return on energy invested) calculations were performed. Both QY and EROEI trends were the same; that is, in decreasing order of reactivity were F-NS, NS, NT, and NP. The most efficient system, F-NS, gave a QY of 0.755 and an EROEI of 0.336, while the least efficient (NP) gave values of 0.372 and 0.288 respectively. Different surface areas were measured for the different systems, 202, 253, and $126\,m^2/g$ for NP, NT, and 6HF-NS, respectively, and it should be noted that the previously presented calculations had been normalized to account for these differences.

Fluorination of the NS surface resulted in a photocatalytic activity enhancement, and the authors suggest this is through control over selective growth of surface facets. An optimal F/Ti = 1 ratio was noted, and this contained a {001}/{101} ratio of 72/28 (calculated using FE-SEM images).

Due to the strong electronegativity of the F atoms, these were also understood to play a role as electron acceptors, leading to reduced exciton recombination rates, along with increased generation of Ti^{3+} species. The latter concentrations were higher in the NS samples than within any other morphology ($Ti^{3+}/Ti^{3+}+Ti^{4+}$ ratio of 21.6% in NS; 22.6% in F-NS and around 10.4% in NP and NT, calculated from XPS measurements), and it is suggested that this is due to favored {001} surface facet growth.

Other possible parameters potentially responsible for the enhanced activity were the enhanced reactivity of photogenerated holes with water on the {001} TiO_2 facets.

This removal of holes (as well as oxidizing H_2O) increased the lifetime of the photo-generated electrons by removing exciton recombination partners (essentially hole traps analogous to the metallic NPs that can act as electron traps).

7.6 Other Oxides

Metal oxide semiconductors are of particular interest for photocatalytic CO_2 processing reactions due to their convenient BG values as well as their surface chemistries. As discussed, these systems also tend to be relatively cheap and easy to synthesize. Chen et al. [116] have synthesized an ultrathin (6 repeating units) WO_3 nanosheet and studied its activity for promotion of artificial photosynthesis under visible light. A solid–liquid-phase arc discharge was used to grow WO_3 seeds that then extended laterally into a 2D nanosheet structure, with thicknesses defined by the seed diameter (4.5 nm) as confirmed by HR-TEM and AFM images. These measurements also confirmed the single crystal nature of the monoclinic WO_3 nanosheets.

While the BG of commercial WO_3 (2.63 eV) is more adapted for visible light harvesting than that of TiO_2 (3.2 eV), the position of this material's CB (0.05 V vs NHE) does not enable any conversion of CO_2, the lowest reductive potential for the CO_2/CH_4 conversion being −0.24 V vs NHE.

The thickness of the film (being close to the exciton's Bohr's radius in WO_3 (3 nm)), resulted in size quantization effects along the vertical axis of the materials (and this increased the BG). Commercially available WO_3 does not promote the artificial photosynthesis reaction, due to the CBM (0.05 V vs NHE) being lower than all the required reduction potentials for electron transfer to CO_2.

Using UV–vis diffuse reflectance, the blue-shifted BG of the thin film, with respect to commercial WO_3, was calculated to be 2.79 eV. The position of the valence band maximum VBM, obtained from electron emission spectroscopy, was 2.37 V vs NHE, and therefore the CB edge was at -0.42 V vs NHE. The photogenerated holes and electrons are therefore sufficiently reductive and oxidative respectively for the production of CH_4 from CO_2 and H_2O (CO_2/CH_4 potential: −0.24 V vs NHE and H_2O/H^+ potential: 0.82 V vs NHE).

Production of CH_4 up to 16 μmol/g_{cat} following 14 h of illumination was obtained over the thin film while, as expected, commercial WO_3, used as reference, led to negligible CH_4 production. Along with the quantum confinement effect, the authors proposed that the ultrathin geometry of the material was also favorable for charge carrier transfer to the surface, and this could also have resulted in enhancement of the photocatalytic activity.

Morais et al. [117] synthesized Bi_2WO_6 through a hydrothermal method (using Bi_2O_3 and Na_2WO_4 as precursors), a modified version of this catalyst was also synthesized by condensation of 3-(aminopropyl)trimethoxysilane on the Bi_2WO_6 surface, leading to surface functionalization by n-propyl amine groups (named mod- Bi_2WO_6 in the following).

These were used for promoting the gas-phase artificial photosynthesis reaction under simulated solar light (Xe arc lamp). The obtained products were CH_4 and CO; however, FTIR measurements on the used catalysts also showed the presence of adsorbed alcohol and carbonylated species on the catalyst's surface. Interestingly, the surface functionalization of the material led to an increased selectivity of the reaction toward the production of CH_4, here calculated as CH_4/CO ratio (from 4.6 for the unmodified catalyst to 9.1). Production rates of 17 and 7 µmol/g and 80 and 65 µmol/g after 24 h illumination were obtained for CO and CH_4, on unmodified and NH_2-modified catalysts, respectively.

Morphological characterization of the samples showed that neither the average particle size (19 nm, obtained from SEM and TEM images) nor the BG (3.4 eV) was impacted by the modification of the system's surface while the surface area decreased from 176 to 114 m^2/g. XRD measurements showed that the obtained samples were poorly crystallized, and this is commonly understood to reduce photocatalytic activity.

FTIR measurements were used to rationalize the observed difference of activity between the two systems. It was shown that CO_2 interacted with the surface n-propyl amine groups to form carbamate species. The authors proposed that this specific CO_2 coordination on the modified catalyst resulted in a decrease in the rate of CO production while not affecting rates of CH_4 production and therefore significantly modifying the reaction selectivity.

Some more complex oxide-based complex structures have also been successfully applied to the artificial photosynthesis reaction, including perovskites (ABO_3) [118] and spinels (AB_2O_4) [119].

7.7 Heterogeneous Composite Materials

7.7.1 Heterojunctions

Large BG SCs (typically TiO_2) have been widely studied for promotion of artificial photosynthesis under UV-light irradiation (a limited component of incident solar light at around 5% of the total number of photons [120]; see above). However, for efficient solar-driven chemistry, systems that work under a wider light range, that is, visible light, are required.

Different types of systems that meet this requirement of promoting light-driven reactions under lower energy irradiation have been studied (see above for examples of doped TiO_2 systems). Typically, lower BG semiconductors have VBM and CBM energies; that is, promoted electrons or generated holes energies that are insufficient to result in the desired reactions.

A proposed solution to this is the synergistic use of two SCs (named SC I and SC II in the following discussion), one of these, at least, having a narrow BG. A combination of such materials will result in a heterojunction within which excitons are produced in both SCs (and can be transferred between SC) following illumination.

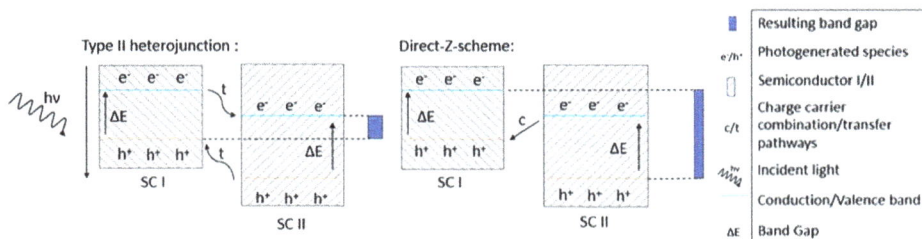

Figure 7.6: Type II heterojunction mechanism (left). Following excitation by light, photogenerated electrons promoted to SC I CB travel to SC II CB. Remaining electrons in the SC I VB, combine with photogenerated holes in SC II, the latter therefore travel to SC I VB. Direct-Z-scheme mechanism (right). Following excitation by light, photogenerated electrons in SC II CB combine with photogenerated holes in SC I VB.

Three types of such heterojunctions exist, differentiated by the relative positions of the VB and CB of each SC.

"Type I" heterojunctions are based on a straddling BG; the entirety of the SC I BG is contained within the BG of SC II ($CB_{SC\ I} < CB_{SC\ II}$ and $VB_{SC\ I} > VB_{SC\ II}$). This arrangement, as well as leading to one large BG SC, leads to increased PL and reduced charge carrier lifetime (so in essence is not valuable for photocatalysis). On the other hand, "type III" heterojunctions are based on a broken BG gap, the BGs of the two SCs do not overlap ($CB_{SC\ I} > VB_{SC\ I} > CB_{SC\ II} > VB_{SC\ II}$). This configuration leads to a high potential barrier for the movement of photogenerated charge carriers from one SC to the other and again this configuration is not useful for photocatalysis.

"Type II" heterojunction materials are based on staggered gaps, the SCs BGs partially overlap ($CB_{SC\ I} > CB_{SC\ II} > VB_{SC\ I} > VB_{SC\ II}$), and materials containing this type of heterojunction are especially interesting within photocatalysis [121, 122]. Excitons are produced within both SCs (generating electrons in $CB_{SC\ I}$ and $CB_{SC\ II}$ and holes in $VB_{SC\ I}$ and $VB_{SC\ II}$); however, the BG structure in those materials now enables novel transportation paths for the photogenerated carriers.

The electrons will tend to travel to lower energy levels; therefore, those present in the higher energy CB (SC I) will, if possible, travel to the lower-energy CB (SC II), while the holes in the lower-energy VB (SC II) will combine with electrons from the higher-energy VB (SC I), leaving extra holes in the latter. Two electrons (at the lower CBM) and two holes (at the higher VBM) should be available following absorption of two photons.

While this methodology enables the harvesting of lower energy light, ideally in the visible range, a major drawback is found in the reduced redox potentials of the charge carriers. The electrons will be less reductive in the final SC II CB than they were in SC I CB, and vice versa for holes that will be less oxidative in SC I VB than they were in the SC II VB. These features limit the possible reactions that can be driven by these composite materials.

P-N heterojunction composites contain both electron-deficient SC (p) and hole-deficient SC (n) components. These components neutralize at the composite's interface, creating a built-in electric field at the interface. This in turn leads to more efficient separation of the photogenerated charge carriers.

Spatial separation of the photogenerated charge carriers, and therefore spatial separation of the half reactions, coupled with longer exciton lifetimes and an enhanced light-harvesting range (with respect to each individual SC) can result in enhanced photocatalytic activity [123].

Such improvement was shown by Peng-Yao et al. [124] for MoS_2/TiO_2 systems in the substantive reaction under visible light irradiation. MoS_2 while having a small BG, suffers from rapid exciton recombination rates and therefore low photocatalytic activity. Conversely, as seen above, TiO_2 has a wide BG that restricts the wavelengths of useful light (but has relatively good charge separation characteristics.

When used in a composite material (MoS_2 (10 wt%)/TiO_2) containing a heterojunction, the observed yield for CO and CH_4 were 5.33 and 16.26 times higher than were seen over pure TiO_2 and 11.68 and 8.29 times higher than were seen over pure MoS_2 under the same reaction conditions.

Tan et al. [125] studied the impact of the calcination temperature as one step in a four-step synthesis of a Cr_2O_3/TiO_2 p-n heterojunction system on the CH_4 production rate in the artificial photosynthesis reaction. The developed synthesis protocol was as follows: pristine, raspberry-like Cr_2O_3 (p, SC I) microspheres were successively coated with SiO_2 (using TEOS as precursor) and titanium (IV) butoxide (TBT), then a hydrothermal alkali etching treatment was used to remove SiO_2 and excess TBT, and finally the obtained material was calcinated to promote further crystallization. In this composite TiO_2 is considered as SC II (n, SC II).

TEM-EDX analysis confirmed a homogeneous distribution of highly dispersed TiO_2 on the Cr_2O_3 microspheres while the materials conserved the raspberry-like macrostructure. Pristine Cr_2O_3 as well as heterogeneous Cr-Ti-O systems obtained following calcination at different temperatures (400, 550, 700, and 850 °C) were used as photocatalysts for the production of CH_4 (CO was also detected in trace amounts) through the artificial photosynthesis reaction under UV irradiation.

Powder XRD displayed typical TiO_2 monoclinic structure patterns following relatively low calcination temperatures (<550 °C); however, this structure was not stable above 550 °C and new peaks (indicative of new phases) arose in the XRD profiles following calcination at the higher temperatures. Meanwhile, PL measurements showed an increased charge carrier lifetime in the presence of the hetero-junction (confirming an expected effect of the p-n nature of the material).

An 82 µmol/h g_{cat} CH_4 production rate was obtained from the pristine Cr_2O_3 material, and coating TiO_2 onto the raspberry-like structure led to increased production rates (which further increased with calcination temperature). Materials that had been calcined at 400 and 700 °C produced 105 and 168 µmol/h g_{cat}, respectively, while further increases in the calcination temperature decreased CH_4 production (e.g., 63 µmol/h g_{cat} following calcination at 850 °C), and this decrease was explained as being related to the structural modifications following high temperature treatment (as seen by PXRD).

Overall, the observed enhanced production rate for the heterogeneous photocatalyst was attributed to reduced recombination rates arising from the p-n junction, as well as spatial separation of the different redox reactions.

Guo et al. [126] studied a type-II heterojunction photocatalyst, $Zn_{0.2}Cd_{0.8}S$ coated on g-C_3N_4 (as confirmed by HR-TEM). This was synthesized through sonication of cadmium acetate dihydrate; zinc acetate dihydrate and thiourea (as a precursor for ZnCdS) and previously prepared g-C_3N_4, followed by a hydrothermal treatment. Different levels of ZnCdS/CN mass ratios (10, 20, 30, 40, and 50) and both pure g-C_3N_4 and pure $Zn_{0.2}Cd_{0.8}S$ were analyzed for the production of CH_3OH through artificial photosynthesis under visible light illumination.

The BG of g-C_3N_4 (SC I) BG is 2.69 eV, with a CB positioned at −1.12 V vs NHE and a VB at 1.57 V vs NHE (H_2O/O_2: 0.81 V vs NHE). On the other hand, ZnCdS (SC II) has a BG of 2.45 eV, with its CB positioned at −0.52 V vs NHE (CO_2/CH_3OH: −0.38 V vs NHE) and a VB at 1.93 V vs NHE.

While at all levels of loading, the hetero-junction systems had increased activity (in comparison to the pure materials), the optimal mass ratio was found to be 30 (denoted as ZnCdS/CN30 in the manuscript), which produced 11.5 µmol/h g_{cat}. This was 2.6- and 2.7-times higher levels of CH_3OH production than was seen over $Zn_{0.2}Cd_{0.8}S$ and g-C_3N_4, respectively.

A relatively weak PL intensity from this material (compared to the component materials) suggested longer lifetimes of charge carriers, an aspect which the authors ascribed to the type II heterojunction structure aiding spatial separation. EIS measurements also showed a faster charge separation within the optimized material and generation of a higher photocurrent (as seen on an i–t plot).

A range of measurements of BG, CB, and VB energies have suggested that the photogenerated electrons travel to the ZnCdS CB (where they then take part in the reduction half-reaction) while the photogenerated holes travel to the g-C_3N_4 VB (where they extract electrons from H_2O as part of the oxidation half-reaction).

The potential for CO_2 conversion to CO is −0.54 V (vs NHE) and therefore here, g-C_3N_4 could produce CO through artificial photosynthesis but, on $Zn_{0.2}Cd_{0.8}S$, this reaction is not possible. In fact, this conversion is not observed on the composite ZnCdS/CN30 material, confirming that electrons in the ZnCdS material are performing the reduction and therefore the heterojunction nature is of type-II and not a direct-Z-scheme mechanism (detailed in the following paragraphs).

Structural and photochemical stabilities have also been measured for the ZnCdS/CN30 materials. After four AP reaction cycles, the XRD patterns do not display any significant changes, and the photocatalytic activity was only reduced to 93% of that observed during the first cycle. In contrast, this falls to 70.5% of the initial reactivity following recycling experiments over pure $Zn_{0.2}Cd_{0.8}S$.

7.7.2 Direct-Z-Scheme

The Z-scheme mode of operation of a heterojunction system for the AP reaction aims to mimic the natural photosynthesis process while also providing solutions for the principal issues with photocatalysts (i.e., rapid exciton recombination and low light-harvesting ranges). These operate in a manner similar to the heterogeneous junctions as described above.

While three types of Z-scheme exist, we here focus on the "Direct-Z-scheme" approach only. These systems do not require electron mediators between the two active SCs. As in the arrangement within a type-II heterojunction (see above), the systems are composed of two semiconductors (SC I and SC II) with staggered BG structures. However, while type-II heterojunction-based photocatalysts have a detrimental effect on the possible range of reactions that can be promoted (through reduced reductive and oxidative potentials for the photogenerated electrons and holes), direct Z-scheme approaches should not have these limitations.

Theoretically, direct-Z-scheme systems improve these potentials while conserving the advantages of such systems, namely, the increased light-harvesting range (with respect to both SCs individually) and the spatially separated oxidation and reduction reactions. This compatibility between light response and improvement in redox potentials is obtained through a different charge carrier transport mode with respect to the previously presented type-II heterojunction system.

Within the direct-Z-scheme we consider two SCs: the first SC (SC I) with a relatively high reduction potential (more energetic CB) and the second SC (SC II) with a relatively high oxidation potential (less energetic VB). Under visible light illumination (as above) the two SCs (1 oxidative and 1 reductive) produce excitons. From here the Z-scheme and the type II semiconductor differ. In the Z-scheme, electrons in the CB of SC II neutralize holes in the VB of SC I.

In other words, the less reductive electrons recombine with the less oxidative holes at the SCs interface, leaving behind highly reactive and spatially separated charge carriers (high-energy electrons available for reduction reactions in SC I and low-energy holes in SC II) [127]. This leaves a hole in the VB of SCII and an electron in the CB of SCI. So, one electron (at the higher CBM) and one hole (at the lower

VBM) should remain available for redox reaction following the absorption of two photons.

In practice the deliberate synthesis of a direct-Z-scheme or a type II heterojunction is complicated as shown by Jiang et al. [128]. They proposed that the synthesis method could be an important parameter controlling reactivity. In their work, CdS NPs were deposited on a g-C_3N_4 nanosheet through photodeposition and chemical deposition.

In the first case, the NPs were selectively deposited at the e^- transfer sites, while in the second approach they were randomly distributed, leading to type II heterojunction and direct-Z-scheme systems, respectively. Zhang et al. [129] selectively deposited CdS NPs onto TiO_2 nanosheet {001} and {101} surfaces, also leading to type II heterojunction and direct-Z-scheme systems, respectively.

Both those studies suggested that morphological and spatial parameters are more important than the relative position of the VB and CB of each SC for the specific construction of one or the other type of junction.

Lui et al. [130] used a simple hydrothermal methodology to synthesize a direct-Z-scheme Si/TiO_2 photocatalyst. As seen by SEM, TiO_2 nanosheets (10 nm in thickness) were coated on porous Si nanospheres (300 nm in diameter, obtained from magnesiothermic reduction of SiO_2 nanospheres). The reduction potential (−0.29 V vs NHE) of the photogenerated electrons within TiO_2 is not high enough to achieve the reduction reaction of CO_2 to CH_3OH (−0.38 V vs NHE), but its strong oxidation potential (2.91 V vs NHE) does enable the reduction of H_2O to H^+ and O_2 (0.82 vs NHE).

In contrast, within the Si nanosphere, the oxidation and reduction potentials are 0.72 V vs NHE and −0.38 V vs NHE respectively, and while the latter is sufficiently reducing to enable the reduction of CO_2 to CH_3OH, the former will not oxidize H_2O to H^+ and O_2. Therefore, a combination of these two materials (assuming successful operation go the Z-scheme) should provide be a good match for production of CH_3OH through the artificial photosynthesis reaction.

The composite heterogeneous catalyst was used for artificial photosynthesis reaction under highly monochromatic 355 nm (3.49 eV) laser illumination for 180 min and CH_3OH was the only product detected using the GC analysis system. Pure TiO_2 sheets and Si nanospheres used alone under similar reaction conditions and, as expected, no product was obtained, confirming the direct-Z-scheme mode of action of the composite Si/TiO_2 material.

The maximum CH_3OH concentration was observed after 150 min, after which reduced concentrations of CH_3OH were measured. It is proposed that above a certain concentration of CH_3OH, the molecule is converted on the catalyst surface to coke.

The illumination source, which is also known to impact the selectivity of the reaction, was also varied (a 300 W Xe lamp was used). In that case CH_4 was the main product observed from the direct-Z-scheme system. CH_4 could also be seen from this reaction over the TiO_2 catalyst (whose reduction potential is −0.24 V vs

NHE); however, in smaller concentrations than evolved from the Si/TiO_2 heterogeneous system.

A higher charge separation rate (and lower recombination rate) within the material was confirmed by PL measurements with respect to both pure materials and theorized by the authors as a confirmation of the Z-scheme operation and a contributing factor to explain the obtained results.

Jin et al. [131] synthesized direct-Z-scheme $CdS-WO_3$ photocatalysts, with a range of CdS contents [0, 0.47, 1.06, 2.71, 6.07, 12.03, and 100 mol%], by precipitation and used these for promotion of the artificial photosynthesis reaction under visible light illumination. The desired reaction is the formation of CH_4. The reductive potential of photogenerated electrons therefore must be higher than that of CO_2/CH_4 (−0.24 V vs NHE), and the oxidative potential of photogenerated holes must be lower than that of H_2O/O_2 (0.82 V vs NHE). The chosen material does fit those requirements as WO_3 (SC II; BG: 2.5 eV) CB and VB positions are 0.5 V vs NHE and 3.0 V vs NHE, respectively, while CdS (SC I; BG: 2.3 eV) CB and VB positions are −0.6 V vs NHE and 1.7 V vs NHE, respectively.

The formation of CH_4 was followed to quantitatively compare the reactivity of these systems with one another. The highest recorded CH_4 production (1.02 µmol/h g_{cat}) was over the 2.71 mol% (CdS) containing material. N_2 physisorption measurements on pure WO_3 and 2.71 mol% $CdS-WO_3$ confirmed a hierarchical (meso- and micro- porous) structure within the sample, and the authors suggested this promoted transfer of reactants and products to the active surface.

Surface area appeared to correlate with CdS content, and this in turn led to increases in CO_2 adsorption (but not in overall reactivity). Therefore, the observed enhanced results can also be attributed to structural changes along with the type II heterojunction nature of the system. UV-VIS diffuse reflectance measurements results were presented for pure WO_3 and CdS samples and for the optimally loaded system. These showed that depositing CdS did not affect the absorption properties of WO_3 confirming that no inter-material doping occurred (and that the two semiconductors were behaving essentially independently of one another for light absorption).

The reductive and oxidative potentials of WO_3 are 0.5 V and 3.0 V (vs NHE) respectively (leading to a BG of 2.5 eV). While the reductive and oxidative potentials of CdS are −0.6 V and 1.7 V (vs NHE) respectively (with a corresponding BG of 2.3 eV). The potentials of the reactions of interest are as follows: −0.24 V for CO_2/CH_4; 0.82 V H_2O/O_2 and 2.3 V for $OH^-/OH\cdot$. This arrangement, assuming operation of the Z scheme (where promoted electrons in WO_3 neutralized holes formed in CdS), generated reductive electrons in CdS and oxidative holes in WO_3, which in turn lead to spatially separated half reactions.

While both half reactions are theoretically possible on pure CdS, the rapid recombination rates measured in this material significantly hiders any respectable photocatalytic activity. The direct-Z-scheme enables charge carrier separation and therefore highly enhances the activity of the photocatalyst. The reaction rate was 10 times higher in the composite than over pure CdS. For pure WO_3, as expected, no products were detected.

Interestingly, the 1.06 and 6.07 mol% samples had similar reactivities to one another, as did the 0.47 and 12.03 mol% CdS/WO_3 composites. These are lower than the reactivity of the optimal 2.71 mol% CdS/WO_3 material, leading to a pyramidal form in a loading versus production rate plot.

Hydroxyl radical ($OH^-/OH^{\cdot}$: 2.3 V vs NHE) experiments were performed to successfully confirm the operation of the direct-Z-scheme within the photocatalyst and eliminate the possibility of a type-II hetero-junction (in which case this reaction would not have been possible, due to the insufficient oxidative power of the holes generated in this material).

7.8 Metal-Organic Framework Structures (MOFs)

MOFs are hybrid (organic–inorganic) materials, structured as a metal clusters linked by organic ligands that are of interest in catalysts and photocatalysis for their large surface areas, their well-ordered porous structure and the potentials for varying their composition and therefore their tuning [132]. On the other hand, numerous MOFs have been reported as having low stability under harsh conditions and in aqueous environments information about their capability as photocatalysts is limited.

Another drawback of MOFs is the common use of sacrificial agents to improve photocatalytic activity [133]; those are commonly electron donors (e.g., CH_3OH) added in order to scavenge holes, reducing the recombination rates of photogenerated charge carriers. On the other hand, electron acceptors (e.g., Ag^+) can also be used to reduce the recombination rate through electron scavenging.

However, it is expected that these will be compensated by the numerous advantages of such systems. In this section we focus on systems that are stable in aqueous environments and those systems that do not require the assistance of sacrificial agents.

The intrinsic structure of those compounds provides an impressively large potential for tuning, through the use of different metal clusters and/or organic ligands [134]. Their use as emerging photocatalysts for artificial photosynthesis is discussed here and several examples are presented. An exhaustive review of the different recent breakthroughs in MOF-based photocatalysis for artificial photosynthesis has recently been published by Kidanemariam et al. [135].

This work covers the use of different commonly used metal oxide clusters including Zn-, Zr-, Co-, Re-, Ni, and Fe-based compounds within different MOF architectures for the artificial photosynthesis reaction. However, to date, the instability of the MOFs in water has significantly hindered the "one-pot" artificial photosynthesis reaction (i.e., the operation of both half reactions (CO_2 reduction and H_2O oxidation)).

The light-harvesting mechanism of MOFs is generally understood as follows: the organic linkers act as an antenna, where light absorption and excitation of electrons occur; these are then transferred to the metal cluster through linker-to-metal Cluster Charge Transfer (LCCT) from whence they can be transferred to a reactant. Other less common paths have also been described; for example, use of surface functionalities (e.g., NH_2) as light harvesters or direct electron excitation within metal oxide clusters (e.g., Fe-O MOFs).

Ding et al. [136] studied the liquid-phase visible light artificial photosynthesis reactivity of MIL-125 (Ti) ($Ti_8O_8(OH)_4$ cluster and $[O_2CC_6H_3(NH_2)CO_2]_6$ ligands) and its derivative NH_2-MIL-125 (Ti) ($Ti_8O_8(OH)_4$ cluster and $[O_2CC_6H_3CO_2]_6$ ligands) MOFs for production of CH_4. The latter has shown an impressive 82% increased production rate (16 µmol/h g_{cat}) when compared to the same reaction over unmodified MIL-125 (Ti). This production rate has been compared with those of different systems in the literature, and this MOF-based system has excellent reactivities even with respect to some metal-TiO_2-based photocatalysts. This interesting comparative result justifies further interest in MOFs as promoters of artificial photosynthesis.

The optical properties of both systems described above have been studied using UV-diffuse reflectance PL and ESI spectroscopies. Both samples had BGs in the visible range, 2.58 eV (481 nm) for the unmodified and 2.44 eV (508 nm) for the modified materials. The latter also showed a reduced PL spectrum (indicating a reduced recombination rate) and a faster interfacial charge transfer (as seen by EIS measurements).

CO_2 and H_2O absorption capacities of the two catalysts were quantified through a combined TGA/MS approach. This confirmed the enhanced adsorption of both CO_2 and H_2O on the aminated photocatalyst. Amination of the MIL-125 (Ti) system has therefore been understood to promote CO_2 and H_2O adsorption and electron transfer, both contributing to the increased efficiency of the -NH_2 modified material. Recyclability tests have been performed on the amine-modified system, and after six cycles only a small reduction of the CH_4 production rate was observed (down to ca. 15 µmol/h g_{cat}) confirming the stability of the photocatalyst.

The reaction mechanism proposed by the authors involves a reduction of Ti^{4+} species to Ti^{3+} by the photogenerated electrons. These are highly reactive for CO_2 reduction through electron transfer to the adsorbed reactant, thereby reforming the Ti^{4+} original species.

Dong et al. [137] discussed different Fe_2M-based MOFs in the aqueous-phase artificial photosynthesis reaction under visible illumination. O_2 and HCOOH were the main products of the reaction. Fe_2-M(μ_3-O)(TGA)(H_2O)$_3$ (M = Co, Ni, and Zn) using TCA (4,4′,4″-tricarboxytriphenylamine) as photosensitive ligands were the three synthesized systems (referred to as NNU-31-M). The Fe/M 2:1 ratio was confirmed by EDS and ICP-AES and a $P_{CA}2_1$ crystal structure was determined by XRD.

The optical properties of the different systems were characterized using UV–vis diffuse reflectance and UPS spectroscopies. The obtained BGs all fall within the visible light range, 1.85 eV (Co, 670.18 nm); 1.82 eV (Ni, 681 nm), and 1.53 eV (Zn, 810 nm), while the ionization potential (equivalent to VBM) and the CBM were shown to be positioned correctly to allow both half reactions to proceed.

In general, difficulty with the identification of specific catalytic sites within most photocatalytic systems is problematic for the enhancement of catalyst design. In this study, DFT calculations were performed to gain insight on the reaction mechanism. This resulted in the identification of the two different sites for each half reaction (CO_2 reduction and H_2O oxidation). Both the metallic clusters and organic ligands act as sites for production of photogenerated electrons. Low-valent M^{2+} (e.g., Co, Ni, and Zn) metals were understood to act as electron reservoirs for CO_2 reduction, while high-valent Fe^{3+} acted as hole reservoirs for H_2O oxidation.

These studies act as proofs-of-concept for the possibility of designing and implementing working MOF systems for artificial photosynthesis. However, to date the photocatalytic reactivities of MOFs are not considered sufficient for application and further work will be required to develop efficient and stable systems that do not require the use of sacrificial agents.

7.9 Use of Physical Properties for Enhanced Photocatalysis Activity

Photocatalytic systems for artificial photosynthesis developed in recent years' reaction are so numerous that they cannot be exhaustively discussed here. However, in this section we give a brief description of certain atypical systems that have proved to be efficient and should be of interest regarding further investigation.

7.9.1 Plasmonic Materials

Plasmons are the result of the interaction between electrons in certain metallic structures (mostly within NM) that lead to the creation of localized surface plasmon resonance (LSPR). The latter is described as a charge density oscillation at the resonance frequency of those electrons resulting in enhanced local electromagnetic fields. They can promote the generation of "hot" electrons that can be used in photocatalysis, for example, as investigated by de Souza et al. for phenol degradation [138].

Morais et al. [139] synthesized three different photocatalytic systems through a solvothermal methodology, namely a pure Ru system ($RuCl_3 \cdot H_2O$ precursor), a 10 molar% Ru-TiO_2 ($TiCl_3$ precursor) ($10RuTiO_2$) and a 10 molar% TiO_2-Ru ($10TiO_2Ru$). The Ru nanoparticles can generate an LPSR. These have been applied to the artificial photosynthesis reaction under simulated solar illumination and the detected products were O_2, CH_4 and CO.

On pure Ru NPs only CH_4 (up to 15 µmol/g_{cat} after 24 h illumination) was produced, while TiO_2 modified catalysts lead to production of both CO (up to 5 µmol/g_{cat} for $10RuTiO_2$ and 23 µmol/g_{cat} for $10TiO_2Ru$ after 24 h) and CH_4 (up to 5 µmol/g_{cat} for $10RuTiO_2$ and 2.5 µmol/g_{cat} for $10TiO_2Ru$ after 24 h). P25 has also been used under similar conditions but no products were obtained from these reactions.

It was therefore understood that TiO_2 was not directly responsive to the applied light but was an active site promoting the production of CO through photogenerated species coming from Ru, itself promoting the production of CH_4. No conclusive differences were observed on UV–vis spectrum as the plasmonic response of Ru was present in all samples while TiO_2 was irresponsive to the visible light only irradiation.

XRD measurements confirmed the presence both Ru (P63/mnm, hexagonal) and TiO_2 (anatase, tetragonal I41/amd) materials in the modified systems, a small portion of oxidized Ru was also observed in all samples. TEM images were used to obtain morphological information, Ru particles were seen to be between 2 and 10 nm (on average 5 nm) in diameter, while the TiO_2 NPs were spherical and between 0.5 and 0.7 µm in diameter. However, the surface area decreased following modification, from 195 m^2/g for pure NPs to 177 m^2/g for 10% TiO_2 molar ratio and 68 m^2/g for 90% TiO_2 molar ratio. TPD measurements were performed to understand the interaction between the photocatalysts and CO_2. While two distinct desorption phenomena were observed from pure Ru (at 195 and 270 °C), the presence of TiO_2 suppresses the lower temperature phenomenon at 10 mol%, therefore increasing the interaction of the system with CO_2.

Low et al. [140] proposed direct evidences of the SPR impact of Ag NPs on the photocatalytic activity of Titanium Nanotubes Arrays (TNTAs) for visible light–driven artificial photosynthesis. Two different deposition methods were used to deposit Ag NPs on the TNTAs: a dip-coating method (Ag/TNTAs-C, 1.9 wt% Ag content) and an electrochemical method (Ag/TNTAs-E, 1.4 wt% Ag content).

SEM and TEM were used to characterize the morphologies of these catalysts and to differentiate the results obtained using the different preparation techniques. On the Ag/TNTAs-E system, a dispersion of the Ag NPs in the inner space of the TNTAs (80 nm in diameter, 15 nm in thickness and 2 µm in length) was observed, while on the Ag/TNTAs-C system the NPs existed mostly on the NTs surface. Increasing absorption intensities in the visible range (380–700 nm) and a new peak (450–550 nm), in Ag containing systems, were observed (in the intensity order, TNTAs < Ag/TNTAs-C < Ag/TNTAs-E).

CH_4 was the main reaction product, CH_3OH production was also observed but in negligible quantities. An increasing production rate up to 50 mmol/h m^2 after 4 h of illumination for Ag/TNTAs-E (30 mmol/h m^2 for Ag/TNTAs-C) was obtained, while no product was obtained over unmodified TNTAs or P25 as photocatalysts.

Two mechanisms were proposed to explain the observed enhancement of photocatalytic reactivity, that is, creation of a Schottky barrier, due to the metallic nature of Ag and plasmonic response producing hot electrons and a near field effect. In order to discriminate between those two possible paths synchronous-illumination XPS (SIXPS) measurements were performed.

A Schottky junction formation was observed in comparing TNTAs and Ag/TNTAs-E (both in the dark), through a shift toward higher binding energies, for the Ti $2p_{3/2}$ peak. This was understood as a reduction in electron density in TNTAs after loading of Ag NPs. However, when under illumination (520 nm, visible light), an opposite shift (toward smaller binding energies) was observed for the same peak in the Ag/TNTAs-E system.

This shift arose from the production of hot electrons that can overcome the Schottky barrier and migrate from the Ag NPs, where they are generated, to the TiO_2 surface. Ag was therefore proven to act as a light sensitizer and a source of hot electrons (via plasmon excitement) within this system.

The authors proposed that the strong near field effect due to the plasmonic response of Ag explained that modified intensity. This near field effect was also observed in EIS measurements, where an electron transport enhancement of the same order of magnitude was seen.

Collado et al. [141] studied the variation in obtained products from gas-phase artificial photosynthesis under UV light over bifunctional Au/TiO_2 photocatalysts. These were synthesized through a deposition-precipitation method to coat Au NPs, with different Au contents [0.29; 0.42; 0.56; 0.96 wt%] (measured by ICP-OES) on the TiO_2 (anatase) surface. To measure the photocatalytic activities of these systems the catalytic and characterization results were compared with those of pure TiO_2.

No significant structural modification arose from the deposition of Au. However, surface area and pore volumes slightly decreased due to partial pore blocking. A hierarchical (meso-, macro-) pore structure was also observed by N_2 abs–des measurements. Over pure TiO_2, CO, and H_2 were the main products (CH_4 and CH_3OH were also seen, in trace amounts), while over the Au-modified systems the selectivity was significantly impacted.

The same trends were observed regarding selectivity for all Au deposited materials; that is, up to ca. 12 times decreased CO production, 3 times higher H_2 production, while a selectivity from 30% (0.29 and 0.42 wt%) to 60% (0.56 and 0.96 wt%) was observed for CH_4.

The highest production rate was obtained with the 0.42 wt% Au loaded catalyst, up to 8.86 µmol/h g_{cat}, 12 times higher than that recorded over pure TiO_2. However, the production kinetics were also followed, and it was noted that the production rate significantly decreased after 6–7 h, following an initial rapid increase from t_0.

Transient absorption spectroscopy was used to gain insight on charge separation and transfer mechanisms. A faster electron transfer was suggested in the presence of Au as well as an Au acting as an electron reservoir, promoting such multi-electron processes as the formation of CH_4 or of C_{2+} products. The lifetime of the photogenerated carriers was also studied using this method and a 7-fold increase was obtained over the Au containing material with respect to the exciton lifetime in unmodified TiO_2 (from 50 μs to 350 μs for $t_{50\%}$).

The BG remained unchanged (3.3 eV); however, in the visible range an increased intensity band between 450 and 600 nm is clearly seen for Au-loaded materials that was attributed to the SPR of Au. PL measurements also suggest lower recombination rates; however, above 0.42 wt% Au a saturation is seen, and no significant further enhancements are observed with increased Au loading.

Visible light irradiation has also been used to excite the 0.42 wt% Au/TiO_2 catalyst. H_2 and CH_4 were the major products seen during this reaction, with CO also being detected as a minor product. This reactivity was attributed to the presence of Au NPs the surface, as TiO_2 itself will not be excited by visible light. However, the obtained production rates were extremely low (ca. 4 times lower than were observed using UV light).

It was therefore confirmed that the SPR in the visible range can have a critical role in enhancing the photocatalytic capabilities of a system for artificial photosynthesis and it should be expected that this will also be an area of more research in the future.

7.9.2 Ferroelectric Semiconductors

Ferroelectric materials are normally described as having a permanent spontaneous polarization linked to the spatial separation of positive and negative ion centers in the polar unit cell. For example, within the $BaTiO_3$ [142] perovskite-like structure, the central Ti atom is displaced along the a axis, breaking the polar symmetry and creating spatially separated centers for positive and negative electric charges.

This can be profitable in photocatalysis due to the expected enhanced photo-generated charge separation with respect to other systems; that is, it would be expected that as an exciton is generated, the oppositely charged carriers should flow in opposite directions. This effect is comparable to that seen following exciton generation close to p-n junctions in Si semiconductors. This would lead to both longer exciton lifetimes and to spatial separation of the two half reactions providing more efficient photocatalytic reactivity [143].

Tu et al. [144] studied layered $SrBi_4Ti_4O_{15}$ nanosheets for gas-phase artificial photosynthesis under simulated solar light irradiation and noted CH_4 as the main product and CO as a minor product. This material belongs to the non-centrosymmetric $A2_1am$ (orthorhombic) system, as confirmed by XRD. It is built as a typical Aurivillius layered crystal structure with $Bi_2O_2^{2+}$ slices and alternatively interleaved four layers of TiO_6 octahedra. Sr and Bi atoms co-occupy the interstitial caves of the four layers of octahedra.

The Ti-O bond distances within this structure (calculated by DFT) are in a range between 1.7 and 2.3 Å, while the sum of the ionic radii of Ti^{4+} (0.56 Å) and O^{2-} (1.40 Å) is 1.96 Å, therefore providing space for displacement of Ti^{4+}. Subsequent displacement of the positive and negative charge centers in TiO_6 octahedra produced the spontaneous polarization that enables efficient separation of the photogenerated charged species.

The BG of $SrBi_4Ti_4O_{15}$ is 3.0 eV (413 nm) with a highly negative CB (vs SHE) of -1.15 eV providing a strong driving force for CO_2 reduction with any promoted electrons The VBM position was also suitably positioned to enable the oxidation of H_2O. The measured production rate of CH_4 was up to 19 μmol/h g_{cat} and the selectivity to CH_4 was 93%. A QY of 1.33% following irradiation at 365 nm (3.4 eV) was obtained, reportedly outperforming most of the reviewed photocatalysts in the literature for this reaction.

Following the discussion above, it is suggested that photogenerated holes travel to the $Bi_2O_2^{2+}$ layers while electrons travel to the TiO_6 octahedral layers. This anisotropic charge separation, typical of ferroelectric materials, is of great help in photocatalysis, spatially separating the two half reactions (while also retarding exciton recombination). Recyclability experiments (x3) also showed the stability of the material.

The catalytic activity of this material was compared with that of with P25 and other Bi-based photocatalysts (SBTO and BiOBr) under conditions where all the systems had comparable specific surface areas and adsorption capacities. The layered ferroelectric $SrBi_4Ti_4O_{15}$ displayed by far the best production rate for both CH_4 and CO.

Stock and Dunn [145] studied the strong remanent polarization (70 μC/cm^2) ferroelectric semiconductor $LiNdO_3$ and TiO_2 for gas-phase artificial photosynthesis under both UV and visible light irradiation. While $LiNbO_3$ had a larger BG (3.78 eV) than TiO_2 (3.2 eV) it was a more active catalyst for the production of HCOOH and CH_2O (the principal products seen in this study).

The $LiNbO_3$ band structure is as follows—VB at 0.48 V and CB at −3.30 V (both vs NHE)—and these are clearly shifted to higher potential with respect to those of TiO_2 (VB at 2.7 V and CB at −0.5 V, both vs NHE). In theory, the position of the VB of $LiNdO_3$ does not allow the oxidation of H_2O to O_2 and H^+ (0.82 V vs NHE) while obviously this reaction is possible on TiO_2. This should significantly hinder the artificial photosynthesis reaction. The CB position, however, enables the reduction of CO_2 to CO_2^- (−1.9 V vs NHE) (a reaction that is not possible using TiO_2).

The formed CO^{2-} ions were understood to directly react with unoxidized H_2O leading to the formation of both HCOOH and C_2OH. However, the measured total products yield for the $LiNdO_3$ system was 7 times higher under UV light and 36 times under visible light illumination than observed over TiO_2.

The authors proposed two distinct mechanisms to explain the observed enhanced efficiency of the ferroelectric system. Firstly, the increased photogenerated charge carriers' separation directly linked to the intrinsic polarization of the material is responsible for a more efficient use of these (i.e., the recombination rate is slowed). Secondly, an altered adsorption energy of the reactants, linked to the charges experienced by the adsorbed reactants on the surface in the case of adsorption on $LiNdO_3$ (but not on TiO_2). This led to the formation of a physiosorbed reactant layer, modifying the bending angles of the adsorbed species leading to the observed nontraditional chemical pathway through easier electron transfer.

7.9.3 Quantum Dot (QDs) Semiconductors

Quantum dot (QDs) semiconductors differ from traditional SC due to their particle size. Those 0D materials are typically in a size range of 2–8 nm, which is typically smaller than the exciton's Bohr radius (distance between the electron and the hole) [146]. The bulk/surface ratio of QDs is therefore very low, increasing the probability for photogenerated carriers generated in the bulk to reach the surface of the catalyst and take part in redox reactions.

Their nanometric size also confers different optical and electronic properties to the materials, when compared to the properties in larger particles. Kumar et al. [147] have measured QD properties and synthesis methods along with certain applications of those materials. In brief, when a material is reduced to nanoscale size in a given number of dimensions its properties are significantly modified. "Nanowells" are materials within which the electron mobility is reduced to two dimensions; "nanowires" are the equivalent in 1 D and QDs are considered 0D materials.

For the latter, the dimensions of the material are lower than that of exciton's Bohr radius in all dimensions. This results in modified electronic structures, the bands forming the electronic structure within bulk materials, change to discrete energy levels within QDs, therefore significantly impacting BG values (increasing BG is always found when particle sizes decrease). This is referred to as "quantum confinement".

QDs also have the specific capacity of multiple exciton generation, meaning that while, in most materials, one photon can only generate one exciton (with any excess energy being lost as heat (phonon)), in QD materials, if the incident light energy is sufficient for the production of two exciton ($hv > 2\,E_g$) this can occur.

These properties are highly interesting for photocatalysis as they provide the possibility to tune the BG as a function of the particle size and to produce an increased density of photogenerated charged species. QDs are expected to become increasingly important as photocatalysts in the artificial photosynthesis reaction (and other photo-catalyzed reactions) due to the extensive recent knowledge gained about low-cost/large-scale, aqueous phase, synthesis, and characterizations. These materials are also noble metal–free and relatively environmentally friendly as suggested in a recent review by Xu-Bing et al. [148].

Hou et al. [149] have studied the effect of particle size of lead-based perovskite halide ($CePbBr_3$) QDs as colloidal suspensions for artificial photosynthesis under simulated solar light illumination. The studied QDs particle size's ranged between 3 and 12 nm and were obtained through the same solution-phase synthesis methodology with increased reaction temperatures leading to bigger particles; that is, 3.8 nm at 140 °C, 6.1 nm at 155 °C, 8.5 nm at 170 °C, and 11.6 nm at 185 °C.

The artificial photosynthesis reaction was performed with the catalyst as a colloidal suspension in a mix of ethyl acetate, as a solvent for CO_2, and H_2O. This solvent mix increased the stability of the catalyst (it is not stable in pure water). Reaction with the $^{13}CO_2$ isotope was carried out to ensure the validity of the process (and rule out the photo-conversion of the solvent).

CO (up to 34 µmol/g_{cat} after 8 h of illumination for the best system) was produced as the major product with CH_4 and H_2 detected as minor products (up to 11 and 1 µmol/g_{cat} after 8 h of illumination, respectively). PL spectra were measured for each system, and interestingly the longer lifetime was found in the 8.5 nm system (9.7 ns). The other systems had lifetimes of 6.6 ns (3.8 nm sample) 7.8 ns (for the 6.1 nm sample), and 8.9 ns for the 11.6 nm sample.

As previously mentioned, longer exciton lifetime is linked to reduced recombination and therefore enhanced photocatalytic activity. This is expected in QDs-based systems due to the discrete nature of the energy levels in the electronic structure; however, the larger particles (with smaller BG) are expected to have excitons with longer lifetimes [150]. This is only verified in QDs as within bulk SCs other parameters such as the density of charge carriers have significantly more influence on the carrier lifetimes.

The obtained BG from absorption data for this 8.5 nm system was 2.4 eV. UPS was used to obtain the VB position (−5.5 eV vs vacuum level) and the level of the CB was determined from this information; that is, −3.1 eV (vs vacuum level). This system was therefore suitable for promoting the artificial photosynthesis reaction and, as expected, was the most efficient for production of all reduced CO_2 products (CO, CH_4, and H_2).

DFT calculations were performed to obtain the DOS of both bulk and the QDs $CePbBr_3$ systems. An increased DOS at the CB of the QDs system was noted. This indicates that more photogenerated carriers could be efficiently transported to the CB (meaning that more excitons could be generated and therefore used) contributing to the observed increased activity of QDs.

Feng et al. [151] studied a metal-free and relatively environmentally friendly QD-based photocatalyst, CDQs-gC_3N_4 (carbon QDs-modified graphitic carbon nitride) for the production of CO and CH_4 through the gas-phase artificial photosynthesis reaction under visible light irradiation. Different levels of CQD loadings on C_3N_4 have been studied (1, 2, 3, and 4 wt%) and the optimal system was shown to be a 2 wt% loading. Lower loaded systems were less efficient, while it seemed that overloading led to creation of recombination centers. The results presented below are for the 2% loaded systems.

The obtained production rate with this system, ~21 µmol/h g_{cat} for CH_4 and ~23 µmol/h g_{cat} for CO were both significantly higher than seen over the pristine g-C_3N_4 material. The selectivity was also altered for CO production (at 6 times lower levels ~4 µmol/h g_{cat}), while no H_2 was produced on the modified sample. Interestingly, in the absence of the carbon QDs no CH_4 was produced.

An average particle size for CQDs of 2.5 nm was measured using TEM and decoration of the graphitic material with CQDs did not lead to significant structural changes or surface area modification. However, their presence on the g-C_3N_4 surface was also confirmed by Raman spectroscopy, with peaks in a range from 800–2,200 cm^{-1}. No peaks were observed in the spectrum of pristine g-C_3N_4, while, after CQDs loading, two distinct peaks arose (at 1,354 and 1,600 cm^{-1}). These were attributed to the D and G bands of carbon.

The CQDs are composed of sp^2-hybridyzed carbons and are understood to act as electron reservoirs, as well as modifiers of surface polarity and optical properties. They also accelerate the reaction kinetics (through the provision of more charge carriers) and increase CO_2 adsorption on the catalyst surface (through decreasing surface polarity). The optical properties were assessed by UV–vis diffuse reflectance spectroscopy, and an improved absorption in the visible and infra-red range was measured for the CQD-containing sample compared to the spectrum of the pristine g-C_3N_4 material.

The loaded sample had a 2.5 eV BG, while the pristine C_3N_4 had a BG of 2.8 eV and this feature was ascribed to the expected narrow BG of CQDs. The CQDs acted as an electron reservoir for the reaction, leading to a reduced electron hole recombination rate (which was observed by PL measurements).

The relative surface polarity of both the pristine and the 2 wt% CQDs loaded samples was measured through water contact angle measurements, and an increased contact angle was measured for the loaded sample indicating a decreased polarity of this system's surface. The level of CO_2 adsorption was also seen to be proportional to the CQDs content. Furthermore, reuse of the photocatalyst after five consecutives cycles showed no significant change in performance or structure.

7.10 Conclusion

This chapter discussed the origins and problems associated with the increased levels of atmospheric CO_2 that have arisen over the past 70 years.

It has also pointed out the conceivable benefits of the artificial photosynthesis reaction, the implementation of which would ameliorate increases in the levels of CO_2 both through decreasing direct emissions to atmosphere and through the generation of nonfossil-resource-derived fuels and chemicals (which invariably release CO_2 at end of life). Furthermore, the reaction could be considered as a method to condition and store solar energy, and in this way provides a buffer to the intermittency of solar power through the formation of energy dense fuels.

Different photocatalyst reactor configurations have been explained, the initial mechanism of photo-promotion clarified, and the optoelectronic features required to promote and govern the selectivity of the reaction have been expounded upon.

The application of a range of different catalyst families has been described, and it is clear that a significant amount of work on optimizing the reaction to condition solar power has been undertaken.

The application of families of catalysts based on TiO_2 (the most ubiquitous photocatalyst) in the reaction is described, as are the limitations on the reaction on these materials. Furthermore, the application of other (more novel) photocatalysts such as heterojunctions, MOFs, quantum dots, ferroelectric materials, perovskites, and plasmonic materials is also discussed.

However, it is also clear that currently the catalysts and processes described are not efficient enough to warrant industrial applications and therefore further work in both the area is to be expected. This work will concentrate on a number of aspects, including material efficiency (photon to chemical energy conversions), materials stability, enhancements of solar radiation absorption, and improvements in selectivity. The latter may well flow from theoretical understanding of the complex and convoluted mechanisms of the reaction – especially when products more complex than CO and CH_4 are sought.

The drivers for this future research will include national and international agreements relating to climate change (e.g., Kyoto, Paris, and future IPCC agreements), as well as the desire for governmental and nongovernmental funding agencies to direct research toward the fulfillment of the UN SDGs. Future prohibition on the use of fossil fuel, and costs associated with the emission of CO_2 that will flow from the aforementioned agreements will also drive this (and related) sustainable chemistry research in the future.

References

[1]   Fourier J. Théorie analytique de la chaleur. Landmark Writings in Western Mathematics 1640–1940, Elsevier Science, 2005.
[2]   Foote E Circumstance affecting the heat of sun's rays, American Journal of Art and Science, 1856, 382–383.
[3]   Tyndall J. On the Absorption and Radiation of Heat by Gases and Vapours, and on the Physical Connexion of Radiation, Absorption, and Conduction, Philosophical Transactions of the Royal Society of London, 1861.
[4]   Arrhenius S. On the influence of carbonic acid in the air upon the temperature of the ground, Philosophical Magazine and Journal of Science, 1896, 41, 237–276.
[5]   Plass GN. The carbon dioxide theory of climatic change, Tellus, 1956, 8, 140–154.
[6]   Wiseman JDH, Bullard EC. The determination and significance of past temperature changes in the upper layer of the equatorial Atlantic Ocean, Proceedings of the Royal Society of London. Series A, Mathematical and physical sciences, 1954, 222, 296–323.
[7]   Raynaud D, Barnola JM. An Antarctic ice core reveals atmospheric CO_2 variations over the past few centuries, Nature, 1985, 315, 309–311.
[8]   Keeling CD. The concentration and isotopic abundances of carbon dioxide in the atmosphere, Tellus, 1960, 12, 200–203.
[9]   Khalil MAK Non-CO_2 Greenhouse Gases in the atmosphere, Anual Review of Energy and the Environement, 1999, 645–661.
[10]   Strangeways I. The Greenhouse Effect: A Closer Look, Royal Meteorological Society, 2011.

[11]   Friedlingstein P. Global Carbon Budget 2019: Global Carbon Project, 2019.

[12]   United Nations. Agenda 21. United Nations Conference on Environment and Development. Rio de Janeiro, 1992.

[13]   United Nations. Paris Agreement. Paris, 2016.

[14]   United Nations. Report of the secretary-general on the 2019 climate action summit and the way forward in 2020: UN. 2019.

[15]   United Nations. The Sustainable Development Goals and Addressing Statelessness: UN, 2017.

[16]   Anastas PT, Warner JC. Green chemistry: Theory and practice, Oxford University Press., Oxford, 1998.

[17]   Database P. National Center for Biotechnology Information. Carbon dioxide, CID=280.

[18]   Mesters C A selection of recent advances in C_1 chemistry, Annual Review of Chemical and Biomolecular Engineering, 2016, 223–238.

[19]   Kapetaki Z. MBE. Carbon Capture Utilisation and Storage Technology Development Report 2018 (CCUS), European Commission, Luxembourg, 2019.

[20]   Arshad R, Raoof G, Reza R, Vamegh R, Minou R Significant aspects of carbon capture and storage – A review, Petroleum, 2019, 335–340.

[21]   Budinis S, Krevor S, Dowell NM, Brandon N, Adam. An assessment of CCS costs, barriers and potential, Energy Strategy Reviews, 2018.

[22]   European Environment Agency. Air pollution impacts from carbon capture and storage (CCS), 2011.

[23]   European Environment Agency. Transforming Industry through CCUS, Paris, 2019.

[24]   Kabir E, Kumar P, Kumar S, Adelodun AA, Kim K-H. Solar energy: Potential and future prospects, Renewable and Sustainable Energy Reviews, 2018, 82, 894–900.

[25]   bp. bp Statistical Review of World Energy 2020: Statistical Review of World Energy BP p.l.c., 2020.

[26]   Agency IE. Key World Statistics, 2017.

[27]   Chaar LE, Lamont LA, Zein NE Review of photovoltaic technologies, Renewable and Sustainable Energy Reviews, 2011, 2165–2175.

[28]   Modi A, Bühler F, Andreasen JG, Haglind F A review of solar energy based heat and power generation systems, Renewable and Sustainable Energy Reviews, 2017, 1047–1064.

[29]   Goodenough JB, Park K-S. The Li-Ion Rechargeable Battery: A Perspective, Journal of the American Chemical Society, 2013, 135, 1167–1176.

[30]   Whittingham M. Lithium batteries and cathode materials, Chemical Reviews, 2004.

[31]   Ye S, Ding C, Chen R, et al. Mimicking the key functions of photosystem II in artificial photosynthesis for photoelectrocatalytic water splitting, Journal of the American Chemical Society, 2018, 140, 3250–3256.

[32]   Satagopan S, Sun Y, Parquette JR, Tabita FR. Synthetic CO_2-fixation enzyme cascades immobilized on self-assembled nanostructures that enhance CO_2/O_2 selectivity of RubisCO, Biotechnology for Biofuels, 2017, 10.

[33]   Ciamician G. The photochemistry of the future, Sciences, 1912.

[34]   König B. Photocatalysis in organic synthesis – past, present, and future, European Journal of Organic Chemistry, 2017.

[35]   Boxun H, Curtis G, Steven LS Thermal, electrochemical, and photochemical conversion of CO_2 to fuels and value-added products, Journal of CO_2 Utilization, 2013, 18–27.

[36]   Galadima A, Muraza O Catalytic thermal conversion of CO_2 into fuels: Perspective and challenges, Renewable and Sustainable Energy Reviews, 2019, 115.

[37]   Jhong H-RM, Ma S, Kenis PJ. Electrochemical conversion of CO_2 to useful chemicals: Current status, remaining challenges, and future opportunities, Current Opinion in Chemical Engineering, 2013, 2, 191–199.
[38]   Jean-Michel L. Review on dry reforming of methane, a potentially more environmentally-friendly approach to the increasing natural gas exploitation, Frontiers in Chemistry, 2014, 2.
[39]   Lv C, Xu L, Chen M, et al. Recent progresses in constructing the highly efficient Ni based catalysts with advanced low-temperature activity toward CO_2 methanation, Frontiers in Chemistry, 2020, 8.
[40]   Hahn KR, Iannuzzi M, Seitsonen AP, Hutter J. Coverage effect of the CO_2 adsorption mechanisms on CeO_2(111) by first principles analysis, The Journal of Physical Chemistry C, 2013, 117, 1701–1711.
[41]   Bryony A, Xin T. Non-thermal plasma technology for the conversion of CO_2, Current Opinion in Green and Sustainable Chemistry, 2017, 3, 45–49.
[42]   Inoue T, Fujishima A, Konishi S Photoelectrocatalytic reduction of carbon dioxide in aqueous suspensions of semiconductor powders, Nature, 1979, 637–638.
[43]   Kunene T, Xiong L, Rosenthal J. Solar-powered synthesis of hydrocarbons from carbon dioxide and water, Proceedings of the National Academy of Sciences of the United States of America, 2019, 116, 9693–9695.
[44]   Marlin DS, Sarron E, Sigurbjörnsson Ó. Process advantages of direct CO_2 to methanol synthesis, Frontiers in Chemistry, 2018, 6.
[45]   Hernández S, Amin Farkhondehfal M, Sastre F, Makkee M, Saracco G, Russo N. Syngas production from electrochemical reduction of CO_2: current status and prospective implementation, Green Chemistry: An International Journal and Green Chemistry Resource: GC, 2017, 19, 2326–2346.
[46]   Kamphuis AJ, Picchioni F, Pescarmona PP. CO_2-fixation into cyclic and polymeric carbonates: Principles and applications, Green Chemistry: An International Journal and Green Chemistry Resource: GC, 2019, 21, 406–448.
[47]   Gao Y, Qian K, Xu B, et al. Recent advances in visible-light-driven conversion of CO_2 by photocatalysts into fuels or value-added chemicals, Carbon Resources Conversion, 2020, 3, 46–59.
[48]   Karamian E, Sharifnia S. On the general mechanism of photocatalytic reduction of CO_2, Journal of CO_2 Utilization, 2016, 16, 196–203.
[49]   Liu J-Y, Gong X-Q, Alexandrova AN. Mechanism of CO_2 photocatalytic reduction to methane and methanol on defected anatase TiO_2 (101): A density functional theory study, The Journal of Physical Chemistry C, 2019, 123, 3505–3511.
[50]   Rasko J, Solymosi F. Infrared spectroscopic study of the photoinduced activation of CO_2 on TiO_2 and Rh/TiO_2 catalysts, The Journal of Physical Chemistry, 1994, 98, 7147–7152.
[51]   Álvarez A, Borges M, Corral-Pérez JJ, et al. CO_2 activation over catalytic surfaces, ChemPhysChem, 2017, 18, 3135–3141.
[52]   Liu C, Cundari TR, Wilson AK. CO_2 reduction on transition metal (Fe, Co, Ni, and Cu) surfaces: in comparison with homogeneous catalysis, The Journal of Physical Chemistry C, 2012, 116, 5681–5688.
[53]   Liu L, Li Y. Understanding the reaction mechanism of photocatalytic reduction of CO_2 with H_2O on TiO_2-based photocatalysts: A review, Aerosol and Air Quality Research, 2014, 14, 453–469.
[54]   Liang L, Xiaodong L, Yongfu S, et al. Infrared light-driven CO_2 overall splitting at room temperature, Joule, 2018, 2, 1004–1016.
[55]   Wu J, Huang Y, Ye W, Li Y. CO_2 Reduction: From the electrochemical to photochemical approach, Advancement of Science, 2017, 4.

[56] Román Castellanos L, Hess O, Lischner J. Single plasmon hot carrier generation in metallic nanoparticles, Communications Physics, 2019, 2.

[57] DuChene J, Tagliabue G, Welch AJ, Cheng W-H, Atwater HA. Hot hole collection and photoelectrochemical CO_2 reduction with plasmonic Au/p-GaN photocathodes, Nano Letters, 2018, 18, 2545–2550.

[58] Angel M, Alexander N, Maria DB. Photocatalytic conversion of carbon dioxide into valuable chemicals in "Artificial Photosynthesis" systems: Recent developments and patents review, Recent Patents on Engineering, 2015, 9.

[59] Ipek B, Uner D. Artificial photosynthesis from a chemical engineering perspective, Artificial Photosynthesis Intechopen, 2012.

[60] Zhenhao D, Rui S. An improved model calculating CO_2 solubility in pure water and aqueous NaCl solutions from 273 to 533 K and from 0 to 2000 bar, Chemical Geology, 2003, 193, 257–271.

[61] Oluwafunmilola O, Maroto-Valer MM. Review of material design and reactor engineering on TiO_2 photocatalysis for CO_2 reduction, Journal of Photochemistry and Photobiology C: Photochemistry Reviews, 2015, 24, 16–42.

[62] Kamila K, Martin R, Ondřej K, et al. Influence of reactor geometry on the yield of CO_2 photocatalytic reduction, Catalysis Today, 2011, 176, 212–214.

[63] Wu JCS Photocatalytic reduction of greenhouse gas CO_2 to fuel, Catalysis Surveys from Asia, 2009, 30–40.

[64] Pelant I, Valenta J. Luminescence Spectroscopy of Semiconductors, Oxford University Press., Oxford, 2016.

[65] Bonnelle JP, Delmon B, Derouane E. UV-Visible Diffuse Reflectance Spectroscopy Applied to Bulk and Surface Properties of Oxides and Related Solids, Springer, Dordrecht, 1983.

[66] Makuła P, Pacia M, Macyk W. How to correctly determine the band gap energy of modified semiconductor photocatalysts based on UV–Vis spectra, The Journal of Physical Chemistry Letters, 2018, 9, 6814–6817.

[67] Shih K. X-ray Diffraction: Structure, Principles and Applications, Nova Publishers, New York, 2013.

[68] Thommes M, Cychosz KA. Physical adsorption characterization of nanoporous materials: progress and challenges, Adsorption, 2014, 20, 233–250.

[69] Egerton RF. Physical Principles of Electron Microscopy: An Introduction to TEM, SEM, and AEM, Springer, New York, 2005.

[70] Rong Y. Electron Diffraction. Characterization of Microstructures by Analytical Electron Microscopy (AEM), Springer, Berlin, 2012.

[71] Hodoroaba V-D, Unger WES, Shard AG Energy-dispersive X-ray spectroscopy (EDS), Characterization of Nanoparticles: Elsevier, 2020, 397–417.

[72] Mark EO, Bernard T. Electrochemical Impedance Spectroscopy, John Wiley & Sons, 2008.

[73] Gurdal Y. DFT-based theoretical simulations for photocatalytic applications using TiO_2, IntechOpen, 2017.

[74] Khan MM, Adil SF, Al-Mayouf A. Metal oxides as photocatalysts, Journal of Saudi Chemical Society, 2015, 19, 462–464.

[75] Hao H, Lang X. Metal sulfide photocatalysis: visible-light-induced organic transformations, ChemCatChem, 2019, 11, 1378–1393.

[76] Liu L, Zhang X, Yang L, Ren L, Wang D, Ye J. Metal nanoparticles induced photocatalysis, National Science Review, 2017, 4, 761–780.

[77] Muhammad U, Hamidi Abdul A. Photocatalytic degradation of organic pollutants in water, IntechOpen, 2013.

[78] Maeda K, Domen K. Photocatalytic water splitting: Recent progress and future challenges, The Journal of Physical Chemistry Letters, 2010, 1, 2655–2661.

[79] Angelo J, Andrade L, Madeira L, Mendes A. An overview of photocatalysis phenomena applied to NO_X abatement, Journal of Environmental Management, 2013, 129C, 522–539.

[80] Zhang B, Sun L. Artificial photosynthesis: Opportunities and challenges of molecular catalysts, Chemical Society Reviews, 2019, 48, 2216–2264.

[81] Ali I, Suhail M, Alothman ZA, Alwarthan A. Recent advances in syntheses, properties and applications of TiO_2 nanostructures, RSC Advances, 2018, 8, 30125–30147.

[82] Luttrell T, Halpegamage S, Tao J, Kramer A, Sutter E, Batzill M. Why is anatase a better photocatalyst than rutile? – Model studies on epitaxial TiO_2 films, Scientific Reports, 2014, 4.

[83] Kočí K, Obalová L, Matějová L, Plachá D, Lacný Z, Jirkovský J, Šolcová O. Effect of TiO_2 particle size on the photocatalytic reduction of CO_2, Applied Catalysis B: Environmental, 2009, 89, 494–502.

[84] Zhao H, Liu L, Andino JM, Li Y. Bicrystalline TiO_2 with controllable anatase–brookite phase content for enhanced CO_2 photoreduction to fuels, Journal of Materials Chemistry, 2013, 1, 8209–8216.

[85] Liu L, Zhao H, Andino JM, Li Y. Photocatalytic CO_2 reduction with H2O on TiO2 nanocrystals: Comparison of anatase, rutile, and brookite polymorphs and exploration of surface chemistry, ACS Catalysis, 2012, 2, 1817–1828.

[86] Dette C, Pérez-Osorio MA, Kley CS, et al. TiO_2 anatase with a bandgap in the visible region, Nano Letters, 2014, 14, 6533–6538.

[87] Chen X, Liu L, Yu PY, Mao SS. Increasing solar absorption for photocatalysis with black hydrogenated titanium dioxide nanocrystals, Science, 2011, 331, 746–750.

[88] Rajaraman TS, Parikh SP, Gandhi VG. Black TiO_2: A review of its properties and conflicting trends, Chemical Engineering Journal, 2020, 389.

[89] Wang B, Shen S, Mao SS. Black TiO_2 for solar hydrogen conversion, Journal of Materiomics, 2017, 3, 96–111.

[90] Wang Q, Zhang Z, Cheng X, et al. Photoreduction of CO_2 using black TiO_2 films under solar light, Journal of CO_2 Utilization, 2015, 12, 7–11.

[91] Khan M, Chowdhurya M, Chuan T, Cheng CK, Yousuf A. Schottky barrier and surface plasmonic resonance phenomena towards the photocatalytic reaction: Study of their mechanisms to enhance the photocatalytic activity, Catalysis Science & Technology, 2015.

[92] Kočí K, Matějů K, Obalová L, et al. Effect of silver doping on the TiO_2 for photocatalytic reduction of CO_2, Applied catalysis B: Environmental, 2010, 96, 239–244.

[93] Nikita S, Umesh K. Noble metal modified TiO_2: selective photoreduction of CO_2 to hydrocarbons, Molecular Catalysis, 2017, 439, 91–99.

[94] Oluwafunmilola O, Maroto-Valer MM. Transition metal oxide based TiO_2 nanoparticles for visible light induced CO_2 photoreduction, Applied catalysis A: General, 2015, 502, 114–121.

[95] Umebayashi T, Yamaki T, Itoh H, Asai K. Analysis of electronic structures of 3d transition metal-doped TiO_2 based on band calculations, Journal of Physics and Chemistry of Solids, 2002, 63, 1909–1920.

[96] Do JY, Kim J, Jang Y, Baek Y-K, Kang M. Change of band-gap position of $MTiO_2$ particle doped with *3d*-transition metal and control of product selectivity on carbon dioxide photoreduction, Korean Journal of Chemical Engineering, 2018, 35, 1009–1018.

[97] Chun-ying H, Rui-tang G, Wei-guo P, et al. Eu-doped TiO_2 nanoparticles with enhanced activity for CO_2 photocatalytic reduction, Journal of CO_2 Utilization, 2018, 26, 487–495.

[98] Hao P, Ruitang G, He L. Photocatalytic reduction of CO_2 over Sm-doped TiO_2 nanoparticles, Journal of Rare Earths, 2019.

[99]   Bingham S, Daoud WA. Recent advances in making nano-sized TiO_2 visible-light active through rare-earth metal doping, Journal of Materials Chemistry, 2011, 21, 2041–2050.

[100] Patrycja P, Paweł M, Wojciech L, Tomasz K, Joanna N, Adriana Z-M. A new simple approach to prepare rare-earth metals-modified TiO_2 nanotube arrays photoactive under visible light: Surface properties and mechanism investigation, Results in Physics, 2019, 12, 412–423.

[101] Kadam A, Dhabbe R, Kokate M, Gaikwad Y, Garadkar K. Preparation of N doped TiO_2 via microwave-assisted method and its photocatalytic activity for degradation of Malathion, Spectrochimica acta. Part A, Molecular and biomolecular spectroscopy, 2014, 133. 669–676.

[102] Malghe Y, Lavand A. Nano sized C-Doped TiO_2 as a visible-light photocatalyst for the degradation of 2,4,6- trichlorophenol, Advanced Materials Letters, 2015, 6, 695–700.

[103] Kiran SV, Rajesh JT, Kinjal JS, Pradyuman AJ, Atindra DS, Vimal GG. Photocatalytic degradation of pharmaceutical and pesticide compounds (PPCs) using doped TiO_2 nanomaterials: A review, Water-Energy Nexus, 2020, 3, 46–61.

[104] Rumaiz AK, Woicik JC, Cockayne E, Lin HY, Jaffari GH, Shah SI. Oxygen vacancies in N doped anatase TiO_2: Experiment and first-principles calculations, Applied physics letters, 2009, 95.

[105] Livraghi S, Paganini MC, Giamello E, Selloni A, Di Valentin C, Pacchioni G. Origin of photoactivity of nitrogen-doped titanium dioxide under visible light, Journal of the American Chemical Society, 2006, 128. 15666–15671.

[106] Peng F, Cai L, Yu H, Wang H, Yang J. Synthesis and characterization of substitutional and interstitial nitrogen-doped titanium dioxides with visible light photocatalytic activity, Journal of Solid State Chemistry, 2008, 181, 130–136.

[107] Beata M, Justyna M, Grzegorz K, Kamila B, Sylwia M, Antoni WM. Reduction of CO_2 by adsorption and reaction on surface of TiO_2-nitrogen modified photocatalyst, Journal of CO_2 Utilization, 2014, 5, 47–52.

[108] Thanaree P, Metta C, Jumras L. Role of chlorophyll in Spirulina on photocatalytic activity of CO_2 reduction under visible light over modified N-doped TiO_2 photocatalysts, Applied Catalysis. B, Environmental, 2015, 168–169, 114–124.

[109] Qianyi Z, Ying L, Erik AA, Marija G-J, Hailong L. Visible light responsive iodine-doped TiO_2 for photocatalytic reduction of CO_2 to fuels, Applied Catalysis. A, General, 2011, 400, 195–202.

[110] Razzaq A, In S-I. TiO_2 based nanostructures for photocatalytic CO_2 conversion to valuable chemicals, Micromachines, 2019, 10.

[111] Hiragond C, Ali S, Sorcar S, In S-I. Hierarchical nanostructured photocatalysts for CO_2 photoreduction, Catalysts, 2019, 9.

[112] Patricia R, Alicia M, Fernando F, Laura C, Juan JV, Víctor APOS. Hierarchical TiO_2 nanofibres as photocatalyst for CO2 reduction: Influence of morphology and phase composition on catalytic activity, Journal of CO_2 Utilization, 2016, 15, 24–31.

[113] Ong W-J, Gui MM, Chai S-P, Mohamed AR. Direct growth of carbon nanotubes on Ni/TiO_2 as next generation catalysts for photoreduction of CO_2 to methane by water under visible light irradiation, RSC Advances, 2013, 3, 4505–4509.

[114] Xu H, Ouyang S, Li P, Kako T, Ye J. High-Active anatase TiO_2 nanosheets exposed with 95% {100} facets toward efficient H_2 evolution and CO_2 photoreduction, ACS Applied Materials & Interfaces, 2013, 5, 1348–1354.

[115] He Z, Wen L, Wang D, et al. Photocatalytic reduction of CO_2 in aqueous solution on surface-fluorinated anatase TiO_2 nanosheets with exposed {001} facets, Energy & Fuels, 2014, 28, 3982–3993.

[116] Chen X, Zhou Y, Liu Q, Li Z, Liu J, Zou Z. Ultrathin, single-crystal WO_3 nanosheets by two-dimensional oriented attachment toward enhanced photocatalystic reduction of CO2 into hydrocarbon fuels under visible light, ACS Applied Materials & Interfaces, 2012, 4, 3372–3377.

[117] Morais E, Stanley K, Thampi KR, Sullivan JA. Scope for spherical Bi_2WO_6 Quazi-Perovskites in the artificial photosynthesis reaction – The effects of surface modification with Amine Groups, Catalysis Letters, 2020.

[118] Zeng S, Kar P, Thakur U, Shankar K. A review on photocatalytic CO_2 reduction using perovskite oxide nanomaterials, Nanotechnology, 2018.

[119] Chandrasekaran S, Bowen C, Zhang P, et al. Spinel photocatalysts for environmental remediation, hydrogen generation, CO_2 reduction and photoelectrochemical water splitting, J Mater Chem A, 2018, 6, 11078–11104.

[120] Cancer IAfRo. Solar and Ultraviolet Radiation. IARC Monographs on the Evaluation of Carcinogenic Risks to Humans, Lyon, 2012, 45–112.

[121] Wang Y, Wang Q, Zhan X, Wang F, Safdar M, He J. Visible light driven type II heterostructures and their enhanced photocatalysis properties: a review, Nanoscale, 2013, 5, 8326–8339.

[122] Guo L-J, Wang Y-J, He T. Photocatalytic reduction of CO_2 over heterostructure semiconductors into value-added chemicals, The Chemical Record, 2016, 16, 1918–1933.

[123] Wang H, Zhang L, Chen Z, et al. Semiconductor heterojunction photocatalysts: Design, construction, and photocatalytic performances, Chemical Society reviews, 2014.

[124] Jia P-Y, Guo R-T, Pan W-G, et al. The MoS_2/TiO_2 heterojunction composites with enhanced activity for CO_2 photocatalytic reduction under visible light irradiation, Colloids and Surfaces. A, Physicochemical and Engineering Aspects, 2019, 306–316.

[125] Tan JZY, Xia F, Maroto-Valer MM. Raspberry-like microspheres of core–shell Cr_2O_3@TiO_2 nanoparticles for CO_2 photoreduction, ChemSusChem, 2019, 12, 5246–5252.

[126] Haiwei G, Jie D, Shipeng W, Yanan W, Qin Z. Highly efficient CH_3OH production over $Zn_{0.2}Cd_{0.8}S$ decorated g-C_3N_4 heterostructures for the photoreduction of CO_2, Applied Surface Science, 2020, 528.

[127] Quanlong X, Liuyang Z, Jiaguo Y, Swelm W, Ahmed AA-G, Mietek J. Direct Z-scheme photocatalysts: Principles, synthesis, and applications, Materials Today, 2018, 21, 1042–1063.

[128] Jiang W, Zong X, An L, et al. Consciously constructing heterojunction or direct Z-scheme photocatalysts by regulating electron flow direction, ACS Catalysis, 2018, 8, 2209–2217.

[129] Zhang J, Zhou D, Dong S, Ren N. Respective construction of Type-II and direct Z-scheme heterostructure by selectively depositing CdS on {001} and {101} facets of TiO_2 nanosheet with CDots modification: A comprehensive comparison, Journal of Hazardous Materials, 2019, 366, 311–320.

[130] Ji G, Liu Y, Zhu L, et al. High-active direct Z-scheme Si/TiO_2 photocatalyst for boosted CO_2 reduction into value-added methanol, RSC Advances, 2014, 4.

[131] Jin J, Yu J, Guo D, Cui C, Ho W. A hierarchical Z-scheme $CdS-WO_3$ photocatalyst with enhanced CO_2 reduction activity, Material Views, 2015, 11, 5262–5271.

[132] Zhang T, Lin W. Metal–organic frameworks for artificial photosynthesis and photocatalysis, Chemical Society Reviews, 2014, 43, 5982–5993.

[133] Schneider J, Bahnemann DW. Undesired role of sacrificial reagents in photocatalysis, The Journal of Physical Chemistry Letters, 2013, 4, 3479–3483.

[134] Li Y, Xu H, Ouyang S, Ye J. Metal–organic frameworks for photocatalysis, Physical Chemistry Chemical Physics: PCCP, 2016, 18, 7563–7572.

[135] Kidanemariam A, Lee J, Park J. Recent innovation of metal-organic frameworks for carbon dioxide photocatalytic reduction, Polymers, 2019, 11.

[136] Ding J, Chen M, Du X, et al. Visible-light-driven photoreduction of CO_2 to CH_4 with H_2O over amine-functionalized MIL-125(Ti), Catalysis Letters, 2019, 149, 3287–3295.

[137] Dong L-Z, Zhang L, Liu J, et al. Stable heterometallic cluster-based organic framework catalysts for artificial photosynthesis, Angewandte Chemie International Edition, 2020, 59, 2659–2663.

[138] Lemos De Souza M, Pereira Dos Santos D, Corio P. Localized surface plasmon resonance enhanced photocatalysis: an experimental and theoretical mechanistic investigation, RSC Advances, 2018, 8, 28753–28762.

[139] Morais E, O'Modhrain C, Thampi K, Sullivan J Visible light-driven gas-phase artificial photosynthesis reactions over ruthenium metal nanoparticles modified with anatase TiO_2, International Journal of Photoenergy, 2019, 1–10.

[140] Low J, Qiu S, Xu D, Jiang C, Cheng B. Direct evidence and enhancement of surface plasmon resonance effect on Ag-loaded TiO_2 nanotube arrays for photocatalytic CO_2 reduction, Applied Surface Science, 2018, 434, 423–432.

[141] Collado L, Reynal A, Coronado JM, Serrano DP, Durrant JR, De La Peña O'shea VA Effect of Au surface plasmon nanoparticles on the selective CO_2 photoreduction to CH_4, Applied Catalysis. B, Environmental, 2015, 177–185.

[142] Tazaki R, Fu D, Itoh M, Daimon M, Koshihara S-Y. Lattice distortion under an electric field in $BaTiO_3$ piezoelectric single crystal, Journal of Physics: Condensed Matter, 2009, 21.

[143] Yogamalar N, Kalpana S, Senthil V, Chithambararaj A Ferroelectrics for photocatalysis, Multifunctional Photocatalytic Materials for Energy: Woodhead Publishing, 2018, 307–324.

[144] Shuchen T, Yihe Z, Ali HR, et al. Ferroelectric polarization promoted bulk charge separation for highly efficient CO_2 photoreduction of $SrBi_4Ti_4O_{15}$, Nano Energy, 2019, 56, 840–850.

[145] Stock M, Dunn S. $LiNbO_3$--a new material for artificial photosynthesis, IEEE Transactions on Ultrasonics, Ferroelectrics, and Frequency Control, 2011, 58, 1988–1993.

[146] Fu H, Wang L-W, Zunger A. Excitonic exchange splitting in bulk semiconductors, Physical Review B, 1999, 59, 5568–5574.

[147] Sumanth Kumar D, Jai Kumar B, Mahesh HM, et al. Chapter 3 – quantum nanostructures (QDs): an overview, Synthesis of Inorganic Nanomaterials: Woodhead Publishing, 2018, 59–88.

[148] Xu-Bing L, Tung C-H, Wu L-Z. Semiconducting quantum dots for artificial photosynthesis, Nature Reviews Chemistry, 2018, 2.

[149] Hou J, Cao S, Wu Y, et al. Inorganic colloidal perovskite quantum dots for robust solar CO_2 reduction, Chemistry – A European Journal, 2017, 9481–9485.

[150] Van Driel AF, Allan G, Delerue C, Lodahl P, Vos WL, Vanmaekelbergh D. Frequency-Dependent spontaneous emission rate from CdSe and CdTe Nanocrystals: Influence of Dark States, Physical Review Letters, 2005, 95.

[151] Feng H, Guo Q, Xu Y, et al. Surface Nonpolarization of g-C_3N_4 by decoration with sensitized quantum dots for improved CO_2 photoreduction, ChemSusChem, 2018, 11, 4256–4261.

Beatriz Villajos, Sara Mesa-Medina, Marisol Faraldos, Ciara Byrne,
Ana Bahamonde, Antonio Gascó, Daphne Hermosilla

Chapter 8
Degradation of Endocrine Disruptors, Pesticides, and Pharmaceuticals Using Photocatalysis

8.1 Introduction

Photocatalytic processes are currently gathering great attention and relevance within the physicochemical processes applied to the treatment of pharmaceuticals, pesticides, and EDCs mainly because of the economical possibility of applying solar radiation, or low-energy LED lamps, and the minimum to null use of chemical toxic products, therefore producing no process-intrinsic residues.

Trace organic chemicals, such as pharmaceuticals, pesticides, consumer care products, and industrial chemicals, have recently been detected present in water bodies. Wastewater treatment plant (WWTP) effluents, together with urban and agricultural run-offs, are considered the most significant emitters of trace organic chemical compounds. Many of the contaminants persist and are poured into surface water because WWTPs are only able to partially remove them. Farmers may furthermore valorize the fraction retained and adsorbed onto sewage sludge as agricultural soil fertilizer; hence, this fraction can finally end polluting soil and water as well. The main representative classes of pharmaceutical and personal care products (PPCPs), endocrine disrupting chemicals (EDCs), and pesticides that can most frequently be detected in surface water, wastewater, and WWTPs are shown in Table 8.1.

Beatriz Villajos, Sara Mesa-Medina, Marisol Faraldos, Ana Bahamonde, Instituto de Catálisis y Petroleoquímica, IPC-CSIC, C/Marie Curie 2, 28049 Madrid, Spain
Ciara Byrne, Department of Inorganic Chemistry and Technology, National Institute of Chemistry, Hajdrihova 19, SI-1001 Ljubljana, Slovenia
Antonio Gascó, Department of Forest and Environmental Engineering and Management, Universidad Politécnica de Madrid, Escuela Técnica Superior de Ingeniería de Montes, Forestal y del Medio Natural, C/José Antonio Novais 10, 28040 Madrid, Spain
Daphne Hermosilla, Department of Forest and Environmental Engineering and Management, Universidad Politécnica de Madrid, Escuela Técnica Superior de Ingeniería de Montes, Forestal y del Medio Natural, C/José Antonio Novais 10, 28040 Madrid, Spain; Department of Agricultural and Forest Engineering, University of Valladolid, Escuela de. Ingeniería de la Industria Forestal, Agronómica y de la Bioenergía (EIFAB), Campus Duques de Soria, 42005 Soria, Spain

https://doi.org/10.1515/9783110668483-008

Table 8.1: Classification of pharmaceuticals, endocrine disruptors, and pesticides [1–5].

Classes	Uses	Examples
PPCPs		
Analgesics	Pain reliever	Acetaminophen, acetylsalicylic acid
Antiepileptic drugs Antidepressants/ obsessive-compulsive regulators	Anticonvulsant Serotonin reuptake inhibition	Carbamazepine, primidone Fluoxetines, fluvoxamine
Antihyperlipidemics	Lipid regulators	Gemfibrozil, clofibric acid, fenofibric acid
Nonsteroidal anti-inflammatory drugs	Anti-inflammatory	Diclofenac, ibuprofen, ketoprofen, naproxen
Beta-blockers/ β_2-sympathomimetics	Cardioprotection	Atenolol, bisoprolol
Antimicrobials	Antibiotic	Erythromycin, sulfamethoxazole, tetracycline
	Antiseptic	Triclosan, biphenylol, chlorophene
Polycyclic musks	Fragrances	Hexahydrohexamethyl-cyclopentabenzopyran
Preservatives UV blockers	Cosmetics, toiletries Sunscreen agents	Parabens Methylbenzylidene camphor, avobenzene
Other	Insect repellent	DEET
	Fragrances	Acetophenone
	Stimulant	Caffeine
EDCs		
Synthetic hormones	Natural human estrogen	Estrone, 17β-estradiol, 17α-ethinylestradiol, atenolol
	Metabolite	Estrone
Alkylphenols	Manufacture of household and industrial products	ylphenol
Polyaromatic compounds	'	Polychlorinated biphenyls, brominated flame retardants
Organic oxygen compounds	Plasticizers	Phthalates

Table 8.1 (continued)

Classes	Uses	Examples
	Industrial production of polycarbonates and epoxy resins	BPA, phthalates, phenol
Pesticides	Insecticides, herbicides, fungicides	Atrazine chlordane, DDT, chlorpyrifos, 2,4-D, glyphosate, trifluralin
Others	By-products from diverse industrial and combustion processes	Dioxins, furans
Pesticides		
Carbamates	Herbicides, insecticides, and fungicides	Carbendazim, benomyl, carbaryl
Chloroacetaniles	Preemergent herbicides	Metolachlor, acetochlor, alachlor
Chlorophenoxy acids	Herbicides	Bentazone, 2,4-D, triclopyr
Organochlorines	Insecticides	DDT, aldrin, lindane, dieldrin, endrin, endosulfan
Organophosphates	Insecticides	Diazinon, malathion, ehion, chlorpyrifos
Pyrethroids	Insecticides	Biphenthrin, cypermetrhrin, esfenvalerate
Triazines	Herbicides	Atrazine, cyanazine, simazine
Other pesticides		Diuron, phenylurea herbicide isoproturon, prometon, mecoprop

8.1.1 Pharmaceuticals and Personal Care Products (PPCPs)

Pharmaceuticals are chemicals designed for the prevention, diagnosis, and treatment of diseases; as well as for enhancing the quality of human and animal lives [6, 7]. On the other hand, personal care products are substances of direct use on the human body aiming to alter odor, taste, touch, or appearance, which include active ingredients or preservatives as part of the formulae of fragrances, toiletries, and cosmetics. These chemicals are not intended for ingestion (except for food supplements), and they do not treat diseases, although they may prevent them, as it is the case of sunscreen agents [4].

Many PPCPs have been enlisted as emerging contaminants because of their recently reported detection in the environment in an increasing amount in levels of ng/L to mg/L [4, 8]. In fact, PPCPs have been identified within the variety of different water compartments (namely, WWTP effluents, surface water, seawater, groundwater, and even drinking water) [8]. The annual production of PPCPs can be higher than 2×10^7 tons, and the World Drug Report 2016 outlined a global annual drug consumption of 100,000–200,000 tons, pointing at Brazil, Russia, India, China, and South Africa as the countries holding a greater proportion of it [9]. This situation is the consequence of the widespread application of these products to protect and improve the health of the population and to sustain the development of the economy of activities such as aquaculture and livestock farming, which makes PPCPs indispensable to the current lifestyle [10, 11].

It is also well known that PPCPs are generally not covered by current regulations as environmental pollutants; thus, they are not listed in the World Health Organization (WHO) guidelines for drinking water quality [8]. Nevertheless, numerous publications have highlighted a great concern on water and soil pollution by pharmaceuticals, and, consequently, the European Commission intends to revise its current legislative framework. Currently, seven pharmaceuticals (17-α ethinylestradiol, EE2; 17-β estradiol, E2; macrolide antibiotics: erythromycin, clarithromycin, and azithromycin; amoxicillin, and ciprofloxacin), and one metabolite (estrone, E1) became part of the Water Framework Directive's watch list, established by Directive 2013/39/EU and amended by the European Commission Implementing Decisions EU 2015/495 and EU 2018/840 [12–14]. This log constitutes the first step to achieve a baseline data of pharmaceutical compounds present in the aquatic environment, and, thus, be able to assess public exposure and other subsequent effects.

Even though the concern on these emerging contaminants is growing within the water sector, the resulting harmful effects of some pharmaceutics are not well known yet because of this existing lack of regulation. However, it is already known that they are persistent and bioactive, as well as they bioaccumulate in aquatic ecosystems [15]. The main source of these contaminants of emerging concern (CECs) is not efficiently treated in water effluents outflowing WWTPs, where they enter because of human and animal consumption as part of their health care treatment (Figure 8.1). Particularly, non-steroidal anti-inflammatory drugs, anticonvulsants, antibiotics, and lipid regulators are the predominant pharmaceuticals that have mainly been detected present in water [4, 16].

In the case of PPCPs, most of them end in water as a result of household and industrial discharge. The effects of PPCPs are even more unknown to date, but some are suspected to potentially hold estrogen-like activity (parabens and UV blockers, for example), toxicity (UV blockers), and acute bioaccumulation (musks) [4, 17, 18].

Conventional WWTPs have been built aiming to remove easily biodegradable carbon, nitrogen, and phosphorous compounds, as well as microorganisms. As a result, they are very efficient in the treatment of homogeneous biodegradable organic waste. In contrast, biological processes are ineffective when it comes to toxic and recalcitrant compounds like PPCPs, especially in the case of pharmaceutical products that were designed to exert their biological activity in small concentrations; primarily

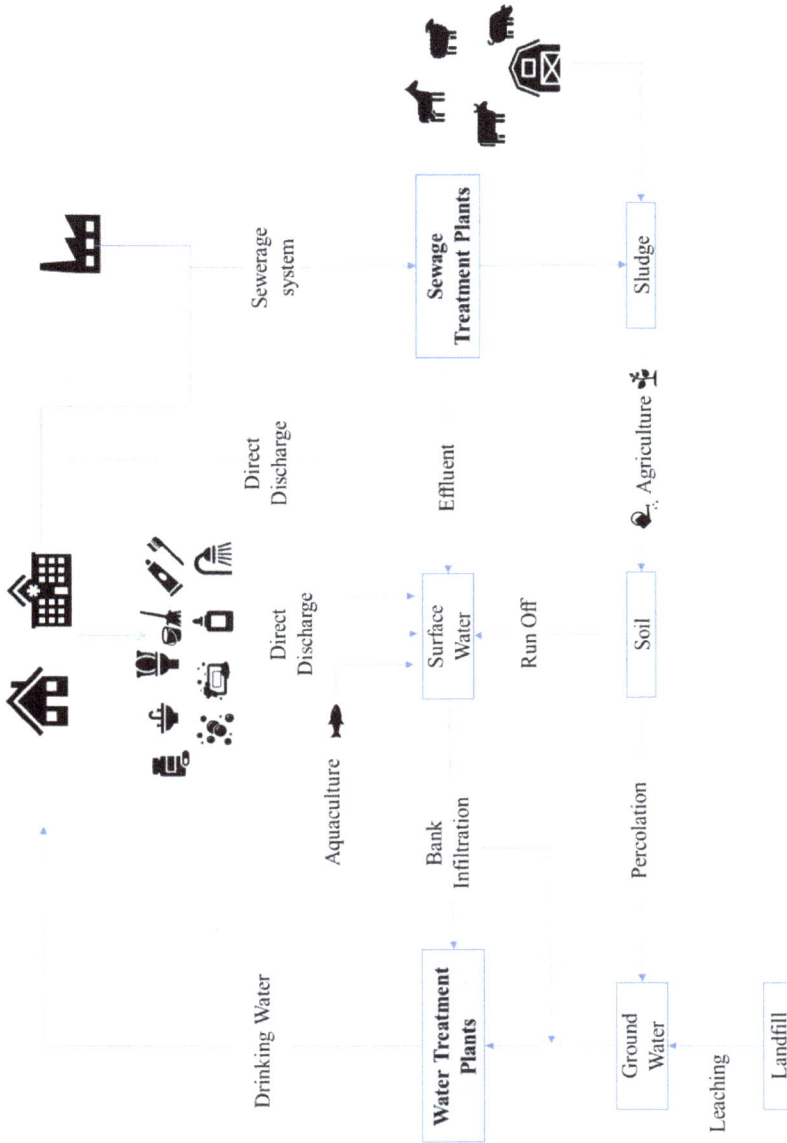

Figure 8.1: Potential sources of PPCPs to the environment. Adapted and modified from Yang et al. [10].

because most of these chemicals are polar, lipophilic, predominantly soluble in water, non-volatile, and bio-recalcitrant. Additionally, they are configured to be biologically active and persistent to keep their properties active until producing the expected effects on humans or animals [4, 19]. Thus, the removal efficiency of these compounds differs in a wide range depending on the class of compound, its different chemical and physical characteristics, and to operational conditions (mainly whether the reactor is aerobic, anaerobic, or anoxic; the pH, and water temperature). Moreover, the physicochemical processes that are developed in WWTPs, such as coagulation, flocculation, sedimentation, and filtration, are not appropriate for the removal of these compounds. In fact, these processes mainly concentrate the contaminants in the sewage sludge that is produced along treatment without actually eliminating them [6, 10, 20, 21]. Therefore, PPCPs end up in water bodies ultimately affecting aquatic and terrestrial ecosystems, causing immobilization, growth inhibition, reproduction problems, and mortality [19].

8.1.2 Endocrine-Disrupting Chemicals (EDCs)

The U.S. Environmental Protection Agency (EPA) has defined an endocrine-disrupting compound (EDC) as "an agent that interferes with the synthesis, secretion, transport, binding, or elimination of natural hormones in the body that are responsible for the maintenance of homeostasis, reproduction, development, and/or behaviour." From a physiological perspective, EDCs are natural or synthetic chemicals that, against an inappropriate environmental exposure to them, alter the hormonal and homeostatic systems that enable organisms to effectively communicate with and respond to their environment [22]. In short, EDCs are chemical compounds that interfere with a normal hormonal response [23]. Figure 8.2 shows a summary of the route of human exposure to EDCs.

Industrial areas are generally affected by contamination from a great variety of industrial chemical compounds that may percolate through soil to groundwater. As a result, complex mixtures of these substances enter the food chain and bio-accumulate in plants and animals higher up to predatory animals, including the human beings [2]. Animals and people contact EDCs through a diversity of processes and pathways. These include water drinking, food eating, inhalation, transfer from mothers to *foeti* or infants (via placenta or through lactation, respectively, when a woman has EDCs in her body), and through the skin by contact with polluted soil that has been exposed to chemicals of typical use in different agricultural, industrial, and household applications (such as pesticides, pharmaceuticals, plasticizers, alkylphenols, and flame retardants) [24]. Although EDCs may naturally appear in some cases, such as phytoestrogens, their presence in the environment is more frequent because of the human action, which has increased its synthesis, as it has already been mentioned [2].

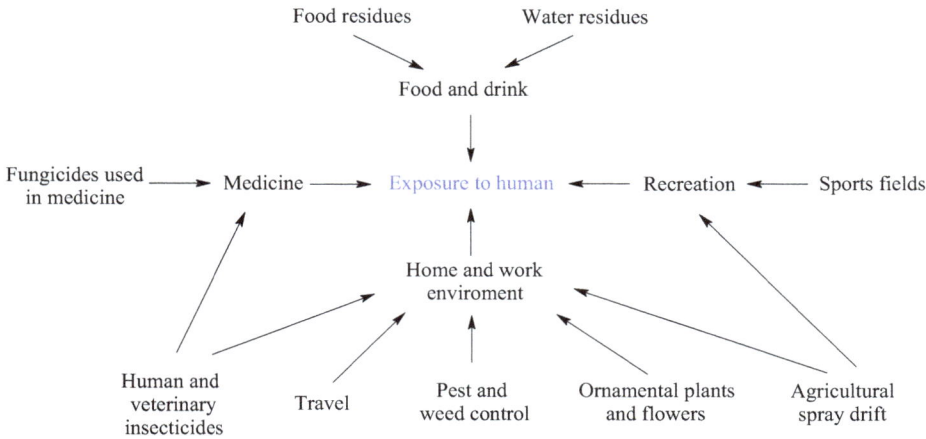

Figure 8.2: The pathways of human exposure to EDCs. Adapted and modified from Kabir et al. [23].

Many EDCs might therefore end up polluting the aquatic environment, turning water itself into a potential source of EDCs, which may further have toxic effects on other elements of the ecosystems (soil, plants, and animals), even though their presence is in a low concentration. Wastewater from the pharmaceutical and other industries may particularly act as a principal exposure source because of their potential negative effects on the environment, as they can contain natural human hormones, hormones from pharmaceutical products (such as birth control pills), and potential EDCs contained in different soaps, detergents, plastics, food, and personal care products (fragrances, cosmetics, etc.) [23].

Aside from these sources, livestock can also represent a very significant source of EDCs. Sex steroids (such as progesterone, testosterone, and estradiol; all EDCs) have shown promoting effects in humans and animals, and have regularly been applied in agriculture along many years, where they have externally been supplied to improve weight gain and feed efficiency of production animals. These compounds hold a hazardous big impact on the environment because numerous plants and wild animals may live in the area where they have been released into from these production activities, and are present at high concentration levels. In addition, these EDCs further contaminate the water, soil, and air of the environments of these agricultural areas and other zones where they might indirectly be released into [25]. Moreover, animals and humans that live in polluted areas carry personal body loads (amount of chemical compounds held in the tissues of each individual) from accumulated direct exposure along their live spans [24].

EDCs affect life beings in a number of different ways: mimicking natural hormones and binding to their cellular receptors; modifying the quantity of hormone receptors in the cells; binding to hormones themselves antagonizing their function; while others cause a similar antagonistic effect by binding to and disabling cellular receptors or hormone transport proteins. Deoxyribonucleic acid (DNA) may also be directly affected by some EDCs, causing changes that can cause cancer, deformity, disease, and infertility in affected organisms and their offspring [26].

Over 24% of human disorders and diseases have globally been estimated to be caused by environmental factors [27]. Furthermore, the environment is also estimated to play a significant role in 80% of the deadliest diseases, including cancer and cardiovascular ones [28]. EDCs primarily contribute to the development of these diseases because the perturbation of the endocrine system is key to the most prevalent ones. The occurrence of endocrine-associated pediatric disorders, including male reproductive ones (e.g. cryptorchidism, hypospadias, and testicular cancer), early female puberty, leukemia, brain cancer, and neuro-behavioral disorders, have all quickly increased during the last 20 years [24]. Finally, different essays performed on animal bodies, clinical observations, and epidemiological research have addressed the potential role of EDCs affecting reproductive systems, prostate, breast, lung, liver, thyroid, and metabolism (e.g., causing obesity) [22].

8.1.3 Pesticides

Pesticides are substances that are applied to kill pests; thus, diverse types of pesticides targeting specific pests include herbicides, insecticides, and fungicides, among other classes. Many pesticides can cause endocrine disruption, such us: (1) insecticides (e.g., lindane, malathion, parathion, chlorpyrifos, and chlordane); (2) herbicides (e.g., diuron, prodiamine, thiazopyr, and trifluralin); and (3) fungicides (e.g., vinclozolin, phenylphenol, and carbendazim) [26]. Pesticides are typically classified by their chemical design into: carbamates, chloroacetanildes, chlorophenoxy acids, organochlorines, organophosphates, pyrethroids, and triazines [3].

The application of agricultural pesticides is a clear source of pesticide exposure. The diversity of crop and livestock types generate, in part, the existing variety of pesticide types and their quantities to be applied in agriculture. For example, fruit and vegetables receive the major doses of insecticides and fungicides, and conventional cereal crops are greatly sprayed with herbicides [26].

The use of these chemicals has become a major environmental concern because they may be responsible for a number of diverse health disorders. The discharge of effluents from urban WWTPs is also a main contributor to the release of pesticides into water bodies because of their inefficiency to fully remove pesticides (for example, and particularly: atrazine, malathion, diazinon, diuron, and simazine) [1]. Moreover, several essays have reported that organochlorines and organophosphates hold a good potential to affect plants and human beings (e.g., Murray et al. [3] and Leong [29]). In particular, organochlorine pesticides represent a relevant group of persistent organic pollutants (POPs) because they are believed to be possible mutagens or carcinogens, besides their endocrine disrupting behavior as well. The United Nations Environmental

Program (UNEP) has enlisted 12 POPs that are chlorine-containing organic chemicals. Nine of them are pesticides. In order to make an effort to substitute persistent organochlorine chemicals, the agricultural sector has changed toward the application of organophosphate pesticides, which hold the advantage of being quickly biodegraded in the environment, but the fact is that organophosphate pesticides are, in general, much more toxic to vertebrates in comparison to other alternative insecticide types [29].

In addition, it has been well addressed that some pesticides are endocrine disruptors based on their elemental chemical configuration, including quantitative structure activity relationships (QSAR), experimental trials with lab animals, wildlife behavior evaluations, and human epidemiological essays [28]. The effects of diverse types of pesticides on hormone functioning, and their action mechanisms, are included in Table 8.2 [26].

Table 8.2: Typical endocrine disrupting pesticides: uses, effects, and action mechanisms [1, 26].

Pesticides	Uses	Hormones affected	Mechanism
Carbamates	Herbicides, insecticides, fungicides	Androgens, estrogens, steroids	Considered to affect androgen- and androgen-receptor-dependent mechanisms. Reported to interfere with cellular microtubule formation in estrogen-sensitive cells. Still widely unknown.
Organochlorines	Insecticides	Androgens, estrogens, prolactin	Competitive inhibitor for androgen receptors. Inhibits estrogen-sensitive reporter binding to androgen receptors. Some induce aromatase production. This enzyme transforms androgen into estrogen.
Organophosphates	Insecticides	Estrogens, thyroid hormones	Prevention of thyroid hormone binding to receptors. The expression of estrogen responsive genes is increased.
Pyrethroids	Insecticides	Estrogens, progesterone	Various chemicals potentiate or antagonize estrogen action by acting on estrogen receptors or possibly providing an alternative signaling pathway. Some of them affect progesterone itself inhibiting its action.
Triazines	Herbicides	Androgens	Inhibition of natural ligands that bind to androgen receptors and androgen-binding proteins. Some of them induce or prevent aromatase production. This enzyme transforms androgen into estrogen.

8.2 Endocrine Disruptors, Pesticides, and Pharmaceuticals That Have Efficiently Been Treated by Photocatalysis

Among the various AOPs, semiconductor-mediated photocatalysis, whether in suspension or immobilized, has recently gained a high significance because of its potential to degrade a great variety of bio-recalcitrant organic contaminants into harmless compounds at ambient temperature and pressure, including their complete oxidation to CO_2 and H_2O. Endocrine disruptors, pesticides, and pharmaceuticals are complex molecules that need a multi-step degradation pathway with many intermediates that may be as difficult to break down as the original chemicals. The presence of these contaminants, even at the very low concentration usually found in wastewater, has dramatic consequences for the environment, and can reach to human consumption; but this low concentration level may be an advantage for photocatalysis processes because they can precisely achieve a high treatment efficiency when very diluted contaminants are present. Representative examples of the most successful recent experiments addressing the photocatalytic treatment of endocrine disruptors, pesticides, and pharmaceuticals are shown in Table 8.3.

Although WWTPs have shown high removal efficiencies in the treatment of analgesics and anti-inflammatory substances, it has also been reported that, in the case of therapeutic types such as antiepileptics, antibiotics, and trimethoprim, its biological treatment efficiency is very low, even nil [30]. Hence, photocatalysis represents a promising alternative to achieve better results in compounds like paracetamol or ibuprofen. Borges et al. [31] achieved degradation efficiencies of up to 100% of the paracetamol (50, 500, and 1,000 mg/L) contained in a WWTP effluent using TiO_2 Degussa P25 whether in suspension or supported on glass spheres as the catalysts of the treatments. Regarding the removal of antibiotics and antiepileptics, heterogeneous photocatalysis has shown a great potential as well. Cai and Hu [32] achieved a degradation efficiency of up to 90% of sulfamethoxazole and trimethoprim with a LED lamp assisting the process and TiO_2 Degussa P25 as the catalyst in suspension. In addition, Carbajo et al. [33] treated a mixture of antibiotic and antiepileptic drugs (0.5 mg/L), using TiO_2 on a solar plant in Almería (PSA). The 100% photodegradation of these compounds was reported in less than 35 min. Thus, heterogeneous photocatalysis is likely a promising technology in the treatment of PPCPs.

Phthalates, phenolic compounds (e.g., bisphenol A, nitrophenol, alkylphenol, nonylphenol, and chlorophenols), triclosan, ethinylestradiol, 17β-estradiol, and diethylstilbestrol are included among the EDCs that have more recently attracted the attention of the scientific community because of their production and consumption dynamics [9]. Table 8.3 enlists some of these compounds, for which TiO_2 photocatalysis (UV lamps, LED included) has achieved degradation efficiencies of about the 90%. Nevertheless, optimal conditions for the degradation of EDCs may not be environmentally friendly if mineralization is considered. Assessing the estrogenicity and toxicity of the generated by-products in the photodegradation process of each compound should also be considered as relevant. Generally, mineralization is incomplete; thus, by-products could even be more harmful than parent EDCs themselves. For example, nonylphenol is a degradation product of nonylphenol ethoxylates [34].

Pesticides have become a main concern because they may cause a variety of health trouble [1], and the discharge from WWTPs is a principal contributor to their release into water bodies because of their inefficient treatment [1]. It is therefore expected that the development of effective photocatalytic treatments, including new more efficient photocatalysts, may result in a good alternative to achieve the total degradation of pesticides, if possible, from secondary streams or end-effluents of WWTPs. In general, the total mineralization of pesticides takes place slower than the degradation of the initial compounds when important intermediates are involved or produced at high concentration levels. However, pesticides are always totally mineralized by photocatalysis, except for the case of s-triazine herbicides because of the great stability of the triazine ring [35]. Some examples of the efficient photocatalytic treatment of pesticides are shown in Table 8.3. For example, Mahalakshmi et al. [36] achieved the total mineralization of carbofuran (200 mg/L) using TiO_2 (100 mg/L) as the catalyst during 300 min of irradiation with low-pressure mercury lamps. In addition, Carbajo et al. [37] studied the photodegradation of a mixture of pesticides (namely, atrazine, isoproturon, diuron, and alachlor) using different commercial TiO_2 catalysts. They reported the reduction of the 91% of the TOC from ultrapure water using the P25 catalyst, whereas a 68% TOC removal was just addressed from natural water using Sigma Aldrich anatase TiO_2 as the catalyst.

8.2.1 Efficiency of Different Semiconductor and Heterogeneous Photocatalysts Applied to the Treatment of EDCs, Pesticides, and Pharmaceuticals

Titanium dioxide (TiO_2) and its derivatives have been the most frequently used photocatalysts in the treatment of pesticides, pharmaceuticals, and EDCs, as collated in recent reviews [103–107] and summarized in the Table 8.3.

Table 8.3: Treatment of pharmaceutical and personal care compounds, EDCs, and pesticides by photocatalysis.

Compounds	[Compound]	[Catalyst]	Operating conditions	Results and comments	References
PPCPs					
Mixture of CECs (carbamazepine, ibuprofen, sulfamethoxazole, ofloxacin, flumequine)	0.1 mg/L each CEC; [CECs] = 0.5 mg/L	TiO$_2$ Degussa P25 (200 mg/L), sol–gel TiO$_2$ (TiEt-450) (500 mg/L)	CPC reactors of the PSA; matrix: deionized water (DW) and natural water (NW); V_T = 32 L, $V_{irradiated}$ = 6.5 L; solar irradiation: 30 W$_{UV}$/m^2	Degradation efficiency = 100% (t < 35 min with both catalysts). Photodegradation rate (DW): ofloxacin ≈ flumequine > ibuprofen >> sulfamethoxazole ≈ carbamazepine. Photodegradation rate (NW): ofloxacin >> flumequine ≈ sulfamethoxazole > carbamazepine ≈ ibuprofen.	[33]
Sulfamethoxazole, trimethoprim	0.4 mg/L	TiO$_2$ Degussa P25 50 mg/L	LED lamp; irradiation time = 20 min; system flow rate = 8 mL/min	Degradation efficiency > 90% of sulfamethoxazole and trimethoprim. An antibacterial activity essay with a reference *E. coli* strain indicated that with every removed portion of trimethoprim, the residual antibacterial activity also decreased by one portion.	[32]

Tetracycline	42 μM	TiO$_2$ Degussa P25 2,090 mg/L	UV lamp (Philips, PL-L 18 W); irradiation time = 20.95 min; pH 5.5	Degradation efficiency = 93.1%; k_{app} = 0.0604 min^{-1}; $t_{1/2}$ = 11 min	[38]
Chloramphenicol	10–80 mg/L	Suspended Degussa, anatase TiO$_2$, ZnO	UV-(320–400 nm); irradiation time = 90 min; pH = 5	Degussa TiO$_2$ and ZnO were equally active. Degradation efficiency = 100%; Mineralization efficiency = 70%.	[39]
Clofibric acid, carbamazepine, Iomeprol, iopromide	0.5–10 mg/L	Suspended Degussa, Hombikat TiO$_2$ 0.1–1,000 mg/L	Artificial sunlight at pH = 3.4–6.5	First-order kinetic rates increased with higher catalyst loading and lower initial concentration. Degussa TiO$_2$ was generally more active than Hombikat.	[40]
Flumequine	10 mg/L	Titanium tetrabutoxide (TBT, Sigma Aldrich, 97%) 4,000 mg/L	UV lamp (Spectroline XX-15 N); light intensity = 2 mW/cm^2; irradiation time = 240 min; pH 5	Degradation efficiency = 23%; k_{app} = 0.0009 min^{-1}	[41]
Cinnamic acid, ibuprofen, diatrizoic acid	10 mg/L	Spherical-shaped brookite TiO$_2$, 10 nm nanoparticles	*Photocatalysis system:* [TiO$_2$] = 40 mg/L; j = 3.2 mW/cm^2	Recalcitrance order: CA (100%) < IBP (85%) < DA. Scavenger experiments: h$^+$ and O$_2^·$ are the main active species.	[42]

(continued)

Table 8.3 (continued)

Compounds	[Compound]	[Catalyst]	Operating conditions	Results and comments	References
Progesterone, ibuprofen, naproxen	20, 40, 80 mg/L	ZnO 1–2 g/L	*Photocatalysis system:* batch stirred 100 mL photoreactor. UV lamp horizontally placed above the reactor	92.3%, 94.5%, and 98.7% removal of progesterone, ibuprofen, and naproxen after 120 min of treatment. Rate constant increased with higher initial contaminant concentration and decreased with higher catalyst loading above the optimal dose.	[43]
Cloxacillin	0.203 mM	TiO_2 Degussa P25 2,000 mg/L, Fe(II) (90 μM), TiIrO$_2$ anode	*Photocatalysis system:* [TiO_2] = 2.0 g/L^1; light power = 150 W; *photo-Fenton system:* [Fe(II)] = 90 μmol/L; [H_2O_2] = 10 mmol/L; light power = 30 W; *electrochemical process:* Ti/IrO$_2$ anode; [NaCl] = 0.225 mol/L; j = 30 mA/cm^2; pH = 6.0	Degree of mineralization: TiO_2 photocatalysis > photo-Fenton system > electrochemical oxidation; degradation of the pollutant requires longer treatment time in the synthetic pharmaceutical wastewater than in distilled water.	[44]

Cyclophosphamide	20 mg/L	Fe(II) (1.55 mM), TiO$_2$ Degussa P25 (500 mg)	UV-H_2O_2 system: $[H_2O_2]$ = 9.8 mM; $UVFe^{2+}$-H_2O_2 system: $[H_2O_2]$ = 9.8 mM; $[Fe^{2+}]$ = 1.55 mM; UV-TiO_2 system: $[TiO_2]$ = 500 mg; time = 256 min	DOC removal% = UV-H$_2$O$_2$ (72.5%) < UV-Fe^{2+}-H$_2$O$_2$ (87.2%) < UV/TiO$_2$ (89.6%)	[45]
Carbamazepine (CBZ), clofibric acid (CFA), sulfamethoxazole (SMX), diclofenac (DCF), ibuprofen (IBU)	2.50 mg/L, (500 µg/L of each target compound CBZ, CFA, SMX, DCF, IBU)	Powdered activated carbons (PACs)/TiO$_2$ composites, 510 mg/L equivalent to 500 mg/L of TiO$_2$, and 10 mg/L of PAC.	*Photocatalysis system:* 100 mL glass-bowl batch reactor (Ø$_i$ = 80 mm, H = 45 mm); continuously stirred (600 rpm); UVA (F18T8 BLB, 18 W, λ_{max} = 368 nm), UVB (G20T10E, 19 W, λ_{max} = 306 nm), and UVC (G20T10, 19 W, λ_{max} = 254 nm); j = 1.0, 2.5, and 3.65 mW/cm^2, respectively.	The fastest photoelimination kinetics was reported for DCF (from 0.1 to 0.4 min^{-1}, depending on the catalytic material form), whereas the lowest for SMX (0.01–0.03 min^{-1}). UVC light was the most effective in both deionized water and river water. Photocatalyst deactivation was observed for synthetic and environmental water, in comparison to deionized water. Seawater produced enhanced photodegradation yield, possibly because of its high ionic strength.	[46]

(continued)

Table 8.3 (continued)

Compounds	[Compound]	[Catalyst]	Operating conditions	Results and comments	References
Clofibric acid (CA)	20 mg/L	TiO$_2$ Degussa P25-rGO 0.5% 25–250 mg/L	Photoreactor Philips TL 4 W/08 Black light UVA (1 L); T = 298 K; flow rate = 1.5 L/m in; irradiation time = 360 min	CA < 0.01 mmol/L after 6 h; intermediate products 4-chlorophenol < 0.02 mM and p-benzoquinone < 0.003 mM (reached a peak after 2 h). Optimum photocatalyst loading = 100 mg/L	[47]
Ciprofloxacin (CIP), tetracycline hydrochloride (TCH), oxytetracycline (OTC)	10 mg/L	N,Fe-CDs/G-WO$_3$-0.6	*Photocatalysis system:* visible light source: 500 W gold halide lamp (420 nm cut-off filter), [catalyst] = 50 mg, [pollutant] = 10 mg/L, pH = 3, T = 25 ± 1 °C	Removal of CIP, TCH, and OTC achieved the 70.5%, 54.5%, and 47.8%, respectively, in a 3 h treatment. h$^+$ and ·OH as main ROS.	[48]
Theophylline, ibuprofen, bisphenol A (BPA), tetracycline, amoxicillin, sulfamethoxazole	5–66.7 mg/L	MOFs: Pd@MIL-100(Fe), In$_2$S$_3$@MIL-125, In$_2$S$_3$/UiO-66, NH$_2$-MIL-125(Ti)/BiOCl, MIL-101(Fe), MIL-100(Fe) MIL-53(Fe), MIL-101(Fe)/TiO2, MIL-68(In) eNH$_2$/GrO, Ag/AgCl@MIL-88A(Fe), AgI/UiO-66	*Photocatalysis system:* visible or sunlight, oxidants = air, H$_2$O$_2$	Photodegradation efficiency of 60–100% in 10–240 min of treatment	[49]

Tetracycline (TC), ciprofloxacin (CIP), bisphenol A (BPA)	15–45 mg/L for TC, 3–12 mg/L for CIP, 5–20 mg/L for BPA	Z–scheme of graphene layers anchored TiO2/g-C3N4 (GTOCN) 60 mg/100 mL	*Photocatalysis system:* visible light (>400 nm) 300 W Xenon lamp (CEL–HXF300, Beijing CEL Tech. Co., Ltd.) at 14 V and 21 A, $j = 300$ m W/cm^2. $V = 100$ mL	GTOCN3 shows the best performance. Removal efficiency of TC (80 min), CIP (60 min), and BPA (70 min) were 83.5%, 61.7%, 79.5%, respectively. The corresponding TOC removals were 66.3%, 41.8%, and 63.6%, respectively. Degradation rates: TC (0.02442 min^{-1}), CIP (0.01675 min^{-1}), BPA (0.01935 min^{-1}). Main species of $O_2^{\cdot-}$ and $OH\cdot$. Stability and reusability were certificated by recycling trials.	[50]
Aspirin and caffeine (methyl theobromine)	10 mg/L	WO$_3$, WO$_3$/TiO$_2$, and WTCN composite 1 g/L	Glass reactor; magnetically stirred; metal halide lamp of 500 W as visible light source	Visible light could be enhanced by the assimilation of g-C$_3$N$_4$ in WO$_3$/TiO$_2$ (WTCN) composite. About 98% and 97% removal efficiencies of aspirin and caffeine, respectively, could be achieved using WTCN composite material.	[51]

(continued)

Table 8.3 (continued)

Compounds	[Compound]	[Catalyst]	Operating conditions	Results and comments	References
15 CECs: pharmaceuticals, pesticides, and personal care products	100 µg/L each	TiO_2 immobilized on borosilicate glass spheres ($\varnothing = 6$ mm)	*Photocatalysis system at pilot-plant scale*: CPC photoreactor; $V_T = 10$ L; illuminated area: 0.30 m^2 (0.96 L); pump flow $= 3.65$ L/min	Removal of 15 CECs in simulated (SW) and spiked real effluents (RE). Five cycles = slightly longer time for complete removal. Most of the compounds were removed in SW After 120 min of treatment.	[52]
Metronidazole, atenolol, chlorpromazine	10 mg/L	Millennium TiO_2 PC-500, immobilized TiO_2 (supported on ceramic plates)	30-W UV-C lamp; light intensity = 38.45 W/m^2; irradiation time = 150 min	Degradation efficiency $= 95.32\%$, 87.02%, and 90.00% for metronidazole, atenolol, and chlorpromazine, respectively. [TOC]$_f$ removal = 90% after 16 h of treatment. Photocatalysis could efficiently remove the ecotoxicity of these compounds in water.	[53]

Paracetamol	50, 500, 1,000 mg/L	TiO$_2$ Degussa P25, immobilized TiO$_2$ (supported on glass spheres)	UV lamp; irradiation time = 240 min Matrix: WWTP effluents	Degradation efficiency = 99–100% for all the amounts of P25 tested. In TiO$_2$/UV system: k_{app} = 0.036, 0.045, and 0.042 min^{-1} for 50, 500 and 1,000 mg/L of paracetamol concentration, respectively.	[31]
Tylosin	0.0044–0.0327 M	TiO$_2$ Degussa P25, immobilized TiO$_2$	UV lamp (18 W/08); irradiation time = 420 min	Degradation efficiency = 90–96%; kinetic constant (k) = 0.000444 mol/L min; k_{app} = 0.00653 min^{-1}; and equilibrium constant of adsorption = 14.69 L/mol.	[54]
Norfloxacin	150 mg/L	TiO$_2$ immobilized glass beads (10 g), TiO$_2$ (0.3 g)	UV/H$_2$O$_2$ system: [H$_2$O$_2$] = 200 mg/L; UV/TiO$_2$-IGBT system: [IGBT] = 10 g; UV/ TiO$_2$ system: [TiO$_2$] = 0.3 g	Degradation efficiency = UV (45%) < UV/H$_2$O$_2$ (72%) < UV/TiO$_2$-immobilized glass beads (IGBT) (78%) < UV/ TiO$_2$ (90%)	[55]

(continued)

Table 8.3 (continued)

Compounds	[Compound]	[Catalyst]	Operating conditions	Results and comments	References
Antipyrine (AP)	50 mg/L	TiO$_2$-immobilized glass disk	UV/H$_2$O$_2$/TiO$_2$-immobilized process; spinning disk recirculating batch reactor (SDR); low-pressure-Hg UV lamp (20 W, λ = 254 nm)	At optimal conditions of pH = 4, [H$_2$O$_2$] = 1,500 mg/L, disk speed = 500 rpm; Q_R = 25 mL/s AP was totally degraded in 120 min of treatment. Disk regeneration allowed 10 cycles without efficiency loss.	[56]
Clofibric acid	9.3 × 10^{-8} mol/cm^3	TiO$_2$ slurry, TiO$_2$-coated window and TiO$_2$ immobilized on glass fibers	Batch recirculating reactor; TiO$_2$ slurry reactor (SR); fixed-film reactor with TiO$_2$ immobilized onto the reactor window (FFR); fixed-bed reactor with TiO$_2$-coated glass rings (FBR); halogenated-Hg lamp (λ = 350 nm, 400 nm); V_R = 100 mL	Slurry photoreactor was the most efficient configuration; a satisfactory quantum efficiency was reported in FBR.	[57]
Tetracycline hydrochloride, paracetamol, caffeine, and atenolol	35 mg/L	TiO$_2$ slurry; TiO$_2$-immobilized on titanium meshes	UV lamp (30 mW/cm^2) and simulated solar irradiation; halogen lamp (1 mW/cm^2; λ = 280–400 nm); (14 mW/cm^2; λ = 400–800 nm); V_R = 600 mL	High removal and mineralization degrees with TiO$_2$ slurry; 50% of pollutants mineralization level after 6 h of treatment with TiO$_2$ immobilized.	[58]

Tetracycline	35 mg/L	TiO_2 powder; TiO_2/ immobilized on laminas and pellets	Closed glass reactor; UV lamp (30 mW/cm^2); natural pH; V_R = 600 mL	Both immobilized substrates addressed notable photoactivity. TiO_2- coated alumina in pellets was the best performing immobilized photocatalyst possibly because of their higher available surface area.	[59]
CECs mixture: acetamiprid (ACP), imazalil (IMZ), bisphenol A (BPA), mercury-resistant bacteria (*Pseudomonas aeruginosa, Bacillus subtilis*)	ACP, IMZ, and BPA at 1 mg/L	TiO_2 Degussa P25 slurry and TiO_2 immobilized on glass spheres	V_R = 3 L; UVA-Hg lamp (125 W, λ = 365.4 nm, I = 29.8 W/m^2); UVA/TiO_2 P25 slurry, UVA/TiO_2 immobilized, UVA/TiO_2 immobilized/H_2O_2 (1.4 mg/L); deionized water and urban wastewater (UWWTP)	Immobilized TiO_2/H_2O_2 increased inactivation and removal of all these contaminants; the mixture of BPA, IMZ, and ACP decreased it a 62%, 21%, and <5%, respectively.	[60]
Ibuprofen	20 mg/L	Micro-TiO_2 Kronos (K1077) on glass Raschig rings	Batch slurry reactor; UV-C germicide immersion lamp (9 W, λ = 215 nm); irradiance = 55 W/m^2; V_R = 600 mL; [K1077] = 0.1 g/L; continuous reactor with K1077-coated Raschig rings; Q_R = 100 mL/min	Micro-TiO_2 mineralized 100% of IBP in 24 h. TiO_2- coated glass Raschig rings degraded 87% of IBP in 6 h of UV-C irradiation in a continuous reactor, with a mineralization of 25%	[61]

(continued)

Table 8.3 (continued)

Compounds	[Compound]	[Catalyst]	Operating conditions	Results and comments	References
Meropenem	100 mg/L	TiO_2 immobilized on fiberglass substrates	Batch lab reactor; two UV lamps (8 W); pH = 4.0, 5.7, and 7.9	25.80% and 29.60% COD and TOC removals at pH = 5.7. *Regeneration of the immobilized catalyst*: wash with 1% H_2O_2 solution in an ultrasonic bath. Four reuse cycles of the immobilized catalyst after regeneration.	[62]
CECs mix	CECs mix 20 ppb; *Escherichia coli* K12: 10^6 CFU/mL; real effluent UWWTP	TiO_2 Degussa P25; TiO_2 AQ-1; TiO_2 minclear WTT-P; TiO_2- immobilized on reticulated ZrO_2 3D foams	Annular photoreactor operating in a closed recirculating circuit; V_R = 00 mL 40 LED system; V_R = 00 mL	TiO_2 immobilized showed lower efficiencies than suspended. No significant differences were reported for the same treatment of real wastewater.	[63]
Carbamazepine, ibuprofen, and sulfamethoxazole.	5 mg/L	TiO_2-Fe and TiO_2-rGO immobilized on optical fibers (SOF)	*Irradiation:* (1) horizontally SOF with 25 mL water solution; (2) vertically SOF through stainless steel light channel with 4 mL water solution. Low-pressure-Hg UV lamp (160 W, λ = 254 nm) or visible light source (halogen lamp; 150 W). pH = 6.	TiO_2–rGO showed higher removal efficiency under UV irradiation. TiO_2–Fe was more suitable under visible light, addressing a 57% ibuprofen, 46% carbamazepine, and 35% sulfamethoxazole removal. The horizontal configuration of fibers was more effective than the vertical one.	[64]

Ciprofloxacin (CIP)	20 mg/L	N_2 TiO_2- immobilized glass spheres	Transparent glass tube reactor and xenon lamp (500 W, λ < 420 nm); V_R = 25 mL	N-TiO_2 immobilized got a 90% CIP removal in a 90 min treatment under visible light. Immobilized N-TiO_2 was well after five cycles of use. CIP degradation rate followed a first-order kinetic model.	[65]
CECs mix: carbamazepine, isoproturon, clopidogrel, diclofenac, atenolol, bezafibrate, tramadol, venlafaxine, fluoxetine	Real effluent from secondary UWWTP	Metal-free exfoliated graphitic carbon nitride (gCNT) and gCNT-immobilized on glass rings	Batch mode; V_R = 60 mL; 4 high-power visible LEDs (λ_{max} = 417 nm); Irradiance = 400–500 W/m^2; [catalyst] = 1.0 g/L^1; continuous flow reactor with 115 g CNT-coated glass rings. V_R = 31.5 mL; Q_R = 0.57, 1.26, 2.6, 6.3, and 15.7 mL/min	Complete removal in 10 min of treatment: carbamazepine > isoproturon > clopidogrel > diclofenac > atenolol > bezafibrate > tramadol > venlafaxine > fluoxetine. Lower efficiencies under continuous mode applying gCNT immobilized on glass rings.	[66]

(continued)

Table 8.3 (continued)

Compounds	[Compound]	[Catalyst]	Operating conditions	Results and comments	References
EDCs					
Estrone (E1), 17β-estradiol (E2), 17α-ethynylestradiol (EE2), and estriol (E3)	3.5×10^{-6} M	TiO$_2$ Degussa P25 0.4 g/L	UVA radiation (8 W, λ = 300–420 nm range with a peak at 355 nm, 115.6 W/m^2). UVC radiation (8 W, λ = 253.7 n m, 113.5 W/m^2).	UVA-TiO$_2$ $k_{T,i}$ (m$^{1.5}$ s$^{0.5}$/l$^{1.5}$) = 1.378×10^{-6} (E1), 1.313×10^{-6} (E2), 1.506×10^{-6} (EE2), 1.240×10^{-6} (E3). UVC–TiO$_2$ $k_{T,i}$ (m$^{1.5}$ s$^{0.5}$/l$^{1.5}$) = 1.932×10^{-6} (E1) 2.003×10^{-6} (E2) 1.772×10^{-6} (EE2) 2.250×10^{-6} (E3).	[67]
Bisphenol A (BPA)	2.5–10 mg/L	TiO$_2$ Degussa P25 100–500 mg/L	*Slurry photoreactor at lab-scale:* InGaN UVA emitter (UV-LED; 365 nm) and UV low-pressure BL lamp (UV-BL; 365 nm), V = 150 mL, 500 rpm. *CPC reactor:* light intensity of solar irradiation = 25–30 W/m^2, Flow rate = 30 L/min, V_T = 4.5 L; matrix: WWTP effluents	TiO$_2$/UV-LED and TiO$_2$/solar systems degraded up to 8.0 and 7.2 mg/L of BPA within 45 min of treatment. The corresponding figures for TiO$_2$/UV-BL was limited to 2.9 mg/L. Optimal photocatalyst TiO$_2$ concentration = 250 mg/L. Highest reaction rates for three systems: LED-driven photocatalysis > TiO$_2$/solar ≈ TiO$_2$/UV-BL.	[68]

Compound	Concentration	Catalyst	Conditions	Results	Ref.
Benzophenone-3 (BP3)	1 mg/L	TiO_2 Degussa P25 1,184 mg/L	Xe lamp (λ = 300–800 nm, I = 350 W/m²); [H_2O_2] = 128 mg/L; pH 9.0; T = 35 ± 2 °C	Degradation rate = 268.8×10^{-3} min⁻¹; Efficiency degradation >90%	[69]
Levonorgestrel (LNG)	3 mg/L	Potasium persulfate as catalyst	Matrix: WWTP effluents, UV-C lamps at λ = 265 nm	Degradation efficiency >90%	[70]
Bisphenol A (BPA)	10 mg/L	TiO_2 polymorphs (0–125 mg/L)	UVA high pressure Hg lamp (150 W, λ = 365 nm); pH 5.2, 20 °C	80% TOC removal with anatase/brookite TiO_2 in 60 min of treatment.	[71]
Atrazine (ATZ)	2 mg/L	N, F-co-doped TiO_2 NPs 500 mg/L	UV lamp (6–20 W, λ = 350 nm); visible light (Xe short arc lamp: 150 W, 1,000 nm with a UV cut-off filter at 420 nm); pH = 7.3	Degradation rates = 1.90×10^{-3} min⁻¹ (UV) and 1.58×10^{-3} min⁻¹ (Visible)	[72]
Methylparaben, ethylparaben, propylparaben, butylparaben and benzylparaben	10 mg/L each	Pd-TiO_2, 0.5% Pd and 0.07 g/L	3–6 W lamps, λ = 365 nm, Eo = 5.75×10^{-7} Einstein/Ls (8.9 Js/m²). 2 L, 180 min, 25 °C.	Methylparaben k = 0.00275 min⁻¹, ethylparaben k = 0.00295 min⁻¹, propylparaben k = 0.00336 min⁻¹, butylparaben k = 0.00377 min⁻¹, benzylparaben k = 0.00751 min⁻¹. 25% TOC removal.	[73]

(continued)

Table 8.3 (continued)

Compounds	[Compound]	[Catalyst]	Operating conditions	Results and comments	References
Diethyl phthalate (DEP)	5–25 mg/L	Nanorod (1:3) ZnO/SiC nanocomposite 0.5–1 g/L	*Annular batch type UV photoreactor:* 8 × 8 W lamps ($\lambda = 254$ nm); 4 h UV irradiation. *Annular-type visible photoreactor:* 500 W filament lamp; 10 h visible irradiation. $V_R = 10$ mL; pH = 2–12. Effect of added electrolytes: [NaCl, Na$_2$CO$_3$, NaHCO$_3$, KCl, MgSO$_4$] = 1%, 1.5%, 3%, 5%, 7%.	Degradation above 90% under UV and visible radiation. Maximum degradation at neutral pH under UV and visible light. Degradation decreased as [DEP] increased. Presence of Na$_2$CO$_3$ and NaHCO$_3$ proportionally increased degradation. Stability and reusability tested along three treatment cycles.	[74]
Ethyl paraben (EP)	0.21 mg/L	CuO$_x$/BiVO$_4$ with 0.75 wt% copper, 1,000 mg/L	100 W Xe lamp and filter simulating solar radiation ($\lambda > 280$ nm); UV region intensity ca. 7.3×10^{-7} E/Ls; irradiation time = 60 min	Degradation efficiency = 98%	[75]

Compound	Concentration	Catalyst	System	Results	Ref.
Ibuprofen (IBP), estrogens: 17 α-ethinyl estradiol (EE2) & estriol	10 mg/L	BiOCl, 25 mg/L BiOCl, BiOI	*Photocatalysis system:* 254 nm or 350 nm bulbs.	High photocatalytic degradation activity and IBP removal from water within 20 min of treatment. BiOI was the most effective under 350 nm radiation, and BiOCl for 254 nm.	[76]
Estrogens: estrone (E1), 17β-estradiol (E2), and 17α-ethinylestradiol (EE2)	$0.2 \ \mu L^{-1}$	TiO_2 P25-coated glass Rasching rings	Batch reactor; $V_R = 250$ mL; 125 W high-pressure-Hg lamp; UV-C radiation $= 11 \ mW/m^2$; UVA radiation $= 22 \ mW/m^2$	TiO_2 suspensions got degradation rates >90% for all estrogens in 30 min of treatment. Immobilized TiO_2 showed worse degradation kinetics (ca. 60 min to achieve the almost complete photodegradation of estrogens).	[77]
Synthetic hormones: progesterone and all types of estradiols	$2.9–190 \ \mu L^{-1}$	NnF Ceram TiO_2 immobilized on water glass	Recirculating batch reactor; $V_R = 5.23 \ dm^3$; 36 W UVA lamp ($\lambda = 368$ nm)	TiO_2 immobilized on water glass were shown as promising catalysts for the degradation of hormones in water by photocatalysis.	[78]

(continued)

Table 8.3 (continued)

Compounds	[Compound]	[Catalyst]	Operating conditions	Results and comments	References
Ceftriaxone	5.0 mg/L	N_2-doped TiO_2 coupled with ZnS blue phosphors immobilized on macroscopic polystyrene pellets	Laboratory reactor; $V_R = 50$ mL; UV–vis lamps ($\lambda = 368$ nm · $\lambda = 400$–800 nm); [photocatalyst] = 0.75–25 g/L; Solar compound triangular collector (CTC); solar UVA irradiance = 1.7–2.7 W/m^2; $V_R = 1$ L; $Q_R = 625$ mL/min	No catalyst deactivation after several reuse cycles for both distilled and real water. Under natural solar light, complete ceftriaxone degradation, and >95% mineralization (as TOC removal) after 150 min of treatment.	[79]
17α-ethynylestradiol (EE2), estriol (E3)	0.1–0.2 mg/L	TiO_2 Degussa P25 50 mg/L, Au-TiO_2 immobilized glass beads (4% wt Au-TiO_2)	Irradiation time = 120 min; *EE2 photodegradation:* optical fiber reactor, $V_R = 200$ mL; *E3 photodegradation:* LED strip reactor, $V_R = 45$ mL	EE2 degradation using 4% wt Au-TiO_2 ($t_{1/2} = 0.78$ h) > P25 TiO_2 ($t_{1/2} = 1.26$ h). About 100% degradation of E3 with 4% wtAu-TiO_2 photocatalyst ($k = 0.13$ h^{-1}; $t_{1/2} = 4.62$ h).	[80]
Pesticides					
Dimethoate	45 mg/L	TiO_2 Degussa P25 600 mg/L	High-pressure mercury lamp (300 W or 500 W, $\lambda = 365$ nm); irradiation time = 160 min; airflow = 400 mL/min	Degradation efficiency = 99%. Presence of low concentration H_2O_2 or $K_2S_2O_8$ (2.0 mmol/L). Enhanced photocatalytic degradation.	[81]

Carbofuran	200 mg/L	TiO_2 Degussa P25 (25–125 mg), ZnO (100 mg)	Low-pressure mercury lamps (8 × 8 W, λ = 254 nm or 365 nm); irradiation time = 300 min; V = 100 mL	Degradation efficiency = 100% (total mineralization). TiO_2 > activity than ZnO. Optimal conditions: 100 mg TiO_2, λ = 365 nm.	[36]
Isoproturon	10–20 mg/L	Sol–gel TiO_2 GICA-1 (500 mg/L), Degussa TiO_2 P25 (500 mg/L)	*Laboratory scale*: semi-continuous slurry photorreactor (1 L); 6 UV BBL lamps and 4 DL lamps; irradiation time = 300 min; I = 39.8 W/m², pH: 3, 5.4, 9. *Pilot plant scale*: CPC reactors (4.5 L); solar irradiation = 30 W_{UV}/m²	Degradation efficiency = 100% (GICA-1 and P25 in less than 60 min for pilot plant scale, 30 min for P25, and 180 min for GICA-1 at lab scale). Titania recovery (sedimentation) = 90% (10 h for GICA-1, and 17 h for P25). Optimal pH: 5.4	[82]

(continued)

Table 8.3 (continued)

Compounds	[Compound]	[Catalyst]	Operating conditions	Results and comments	References
Mixture of pesticides (diuron (DI), atrazine (ATR), isoproturon (IS) and alachlor (ALA))	[IS] = [ALA] = 12.5 mg/L; [DI] = 7.5 mg/L; [ATR] = 6.25 mg/L	TiO_2 Degussa P25, Degussa P25/20, Degussa P90, Hombikat UV100 (HBK), and anatase Sigma-Aldrich (SA)	Ultrapure water and Madrid's tap water. Irradiation time: 300 min; Semicontinuous slurry photorreactor (1 L); 6 UV black blue light lamps and 4 daylight lamps; $I = 38.4$ W/m², oxygen flow: 75 N cm³/min	*Ultrapure water:* TOC removal(%) = 91.02 (P25) > 88.69 (P90) > 82.25 (P25/20) > 68.71 (SA) ≫ 51.03 (HBK); degradation rate (mg/L min) = 0.103 (P25) > 0.090 (P90) > 0,075 (P25/20) > 0,072 (SA) ≫ 0.033 (HBK). *Natural water:* TOC removal (%) = 68.05 (SA) > 67.64 (P25/20) > 62.61 (HBK) > 62.49 (P25) > 60.98 (P90); degradation rate (mg/L min) = 0.062 (SA) > 0.053 (P25/20) > 0.051 (HBK) > 0.049 (P25) > 0.044 (P90).	[37]

| Diuron and monuron | 1.7×10^{-4} M | TiO$_2$ Degussa P25 1.0 g/L | *UV-mediated processes:* UV flurescent lamp (λ_{max} = 365 nm); photon flux = 1.2×10^{-5} Eins/s; centered in a water-cooled tubular glass reactor (L = 340 mm, $\varnothing_i$ = 46 mm); O$_2$ or air = 855 mL/min; T = 25 °C; V_T = 500 mL; recirculation rate = 375 mL/min | Initial rate of monuron transformation (air = 6.76×10^{-8} M/s; oxygen = 7.05×10^{-8} M/s) higher than for diuron (air = 4.33×10^{-8} M/s; oxygen = 4.15×10^{-8} M/s). Mineralization improved using the combined method and heterogeneous photocatalysis. Humic acids had negative effects on UV-mediated processes, but positive in ozonation. Negligible effects of natural water matrices and inorganic salts addition, except for heterogeneous photocatalysis. | [83] |

(continued)

Table 8.3 (continued)

Compounds	[Compound]	[Catalyst]	Operating conditions	Results and comments	References
Alachlor (ALC)	50 mg/L	TiO$_2$ Degussa P25 50 mg/L	*Photocatalysis, photo-Fenton, and photo-Fenton-like systems:* tubular photoreactor with a concentric low pressure mercury lamp (Atlantic UV, MP36B, 38 W), which emits at 254 nm. $T = 25$ °C; recirculation rate $= 300$ mL/min; $V_T = 10$ L. *Fenton:* pH $= 2.8$; [FeSO$_4 \cdot$ 7H$_2$O] $= 2$–10–56 mg/L; [H$_2$O$_2$] $= 250$–500–750 mg/L; [Na$_2$S$_2$O$_8$] $= 595$–$1,190$–$2,381$ mg/L	TiO$_2$ heterogeneous photocatalysis and photo-Fenton processes with 254 nm UV radiation showed promising alachlor removal efficiencies. Faster using Fenton reaction, even if H$_2$O$_2$ and Na$_2$S$_2$O$_8$ were added.	[84]
Lindane	100 mg/L	N-TiO$_2$ (40 mg/L)	Immobilized batch reactor: high-pressure visible lamp (500 W, $\lambda = 365$ nm (UV), $\lambda > 400$ nm (Vis)); irradiation time $= 330$ min, $V = 400$ mL; Oxygen flow rate $= 300$ mL/min	Degradation efficiency $= 37.5\%$ and 100% under UV and visible light, respectively.	[85]
Imidacloprid	10 mg/L	TiO$_2$/H-ZSM-5	UV lamp (0.5 mW/cm^2, $\lambda = 365$ nm); irradiation time $= 40$ min; $V = 500$ mL	Degradation efficiency $= 94.4\%$	[86]

Compound	Concentration	Photocatalyst	Conditions	Results	Ref.
Imidacloprid	100 mg/L	TiO$_2$, Degussa P25 (100 mg/L), ZnO (100 mg/L)	Low-pressure mercury vapor lamp (9 W, λ = 254 nm); irradiation time = 20 min; V = 100 mL	Degradation efficiency = 88.15% and 82.17% using TiO$_2$ and ZnO photocatalysts, respectively.	[87]
Thiamethoxam (TH), imidacloprid (IM), acetamiprid (AC),	0.1 mg/L of each one	ZnO or TiO$_2$ Degussa P25 50–300 mg/L; Na$_2$S$_2$O$_8$ 50–350 mg/L	*Photocatalysis system:* cylindrical glass photoreactor (L = 250 mm, $\emptyset_i$ = 100 mm); 8 W low-pressure mercury lamp; V_T = 2 L; $V_{irradiated}$ = 1,660 mL; j = 24 W/m^2; recirculation flow = 600 mL/min. *Natural sunlight photocatalysis:* Pyrex glass vessels (L = 110 mm, $\emptyset_i$ = 80 mm); V_T = 500 mL; Vis-NIR, UVA, and UVB lamps of 1,060.3 ± 56.1, 29.8 ± 3.5, and 2.1 ± 0.2 W/m^2, respectively, for 13 h.	Optimal catalysts loading = 200 mg/L; 250 mg/L for Na$_2$S$_2$O$_8$. Rate constants of pesticides degradation using ZnO/Na$_2$S$_2$O$_8$ and TiO$_2$/Na$_2$S$_2$O$_8$ systems under artificial light ranked as IM > TH > AC.	[88]
Diazinon	10 mg/L	ZnO 100 mg/L	Low-pressure mercury lamp (9 W, λ = 254 nm); irradiation time = 30 min; V = 100 mL	Degradation efficiency = 93.3%	[89]

(continued)

Table 8.3 (continued)

Compounds	[Compound]	[Catalyst]	Operating conditions	Results and comments	References
Diuron (DI), alachlor (ALC), isoproturon (IS) and atrazine (ATR).	[IS] = [ALC] = 12.5 mg/L; [DI] = 7.5 mg/L; [ATR] = 6.25 mg/L	GO-TiO$_2$, bare TiO$_2$ 500 mg/L	*Photocatalysis system:* semicontinuous slurry photoreactor with 10 fluorescent lamps enclosed: UV–visible experiments, 6 UV BBL lamps (Narva LT15 W/073; λ_{max} = 366 nm) and 4 daylight lamps (Narva LT15 W/865, λ = 400–700 nm); j = 40 W/m^2; Vis experiments: 10 daylight lamps.	Important inhibition of the degradation of pesticides (and TOC removal) always with bare TiO$_2$ in natural water ($r_{0,natural}/r_{0,ultrapure}$ < 1). Contrarily, GO-Ti O$_2$ $r_{0,natural}/r_{0,ultrapure}$ > 1. Under visible light, GO-TiO$_2$ required shorter reaction time to photodegrade 50% of the pesticides load compared to that of TiO$_2$ P25.	[90]

Compound	Concentration	Catalyst	System	Results	Ref.
Malathion (MT), fenitrothion (FT), quinalphos (QP), vinclozoline (VZ), dimethoate (DT), and fenarimol (FR)	Mixture of 0.3 mg/L of each pesticide	ZnO optimal concentration = 200 mg/L	*Photocalysis system:* cylindrical double wall photoreactor; 2 × 8 W medium pressure mercury lamps (366 nm); [catalyst] = 100–300 mg/L and [$Na_2S_2O_8$] = 100–400 mg/L; pH = 5.5–8.5 *Pilot plant (CPC technology):* sunlight, 100 L of sewage water effluent (pH = 7.2) polluted with 0.3 mg/L of each pesticide.	The residual levels were lower than 7% of the initial pesticide content after 4 h of radiation, except for QP (15%) and FR (56%). The initial content of DOC was strongly reduced. 90% dissipation time DT90 (t_{30w}, min) = 14 (VZ), 75 (FR), 11 (QP), 2 (MT), 15 (FT), 3 (DT) in the solar pilot plant.	[91]
Chloridazon	$5 \times 10-5$ M	Mesoporous TiO_2/ZrO_2 3 g/L	Slurry double-wall glass reactor with water cooling system; V_T = 100 mL; 150 W Xe-lamp with a 320 nm cut-off filter.	Sample with 0.14 mol% of Zr, and calcined at 450 °C, showed optimum activity.	[92]
Atrazine, methyl malathion, chlorpyrifos	–	*MOFs:* AB/Cu(tpa).GR, BiOBr/UiO-66, $AgIO_3$/MIL-53(Fe), and WO_3/MIL-53(Fe)	*Photocatalysis system:* visible or sunlight; oxidants: air or $Na_2S_2O_8$	Photodegradation efficiency of 85–100% in 50–240 min of treatment.	[49]

(continued)

Table 8.3 (continued)

Compounds	[Compound]	[Catalyst]	Operating conditions	Results and comments	References
Atrazine (ATZ), ibuprofen (IBP), estrogens: 17 α-ethinyl estradiol (EE2) and estriol	10–10,000 mg/L	BiOCl, Cu-BiOCl, 25 mg/L	*Photocatalysis system:* 254 nm	Cu-doped BiOCl chemicals interact with atrazine to aid photodegradation. About 65–71% of ATZ remained after 30 min of BiOCl and Cu-BiOCl photocatalysis. BiOCl showed photocatalytic degradation activity and IBP removal from water in 20 min of treatment. Estriol was totally removed by BiOI and BiOCl photocatalysis in 30 min.	[76]
Carbamazepine (CBZ), tetracycline hydrochloride (TC)	10 mg/L	2D/3D $Bi_5O_7Br/BiOBr$ 0.5 g/L	$V_T = 100$ mL; visible light irradiation by 500 W Xe-lamp with a 420 nm cut-off filter; $T = 25\,°C$	CBZ and TC conversions were 94% and 83% under 90 min of visible light radiation. TOC removal efficiency was ≈57%. Superior reuse cycling stability.	[93]

Atrazine (ATZ)	20 mg/L	CdO NPs or Zn-CdO NPs 50 mg	*Photocatalysis system:* 150 mL quartz reactor with water circulation; UV illumination by 125 W (311 nm) medium pressure Hg arc lamp; pH = 6.2	ATZ degradation rates were 1.41×10^{-3}, 1.79×10^{-3}, 2.48×10^{-3}, 3.33×10^{-3} min^{-1} for pure CdO NPs, 2.5%, 5.0%, and 7.5% Zn-CdO NPs, respectively.	[94]
Atrazine (ATZ)	20 mg/L	$In,S-TiO_2@rGO$ nanocomposite (0.5, 1.0, 2.0, and 3.0 g/L)	Batch-operated slurry glass circular photoreactor ($\emptyset_i = 10$ cm; $V_T = 100$ mL); 300 W Xe lamp; pH = 5.4; 500 rpm	Full degradation and 95.5% mineralization in 20 min of treatment. The photocatalytic activity of doped catalysts decreased in the following order: $In_3S_1\text{-}TiO_2@rGO > In_1S_1\text{-}TiO_2@rGO > In_3\text{-}doped\ TiO_2 @rGO > In_5S_1\text{-}TiO_2@rGO > S_1\text{-}doped\ TiO_2@rGO > un\text{-}doped\ TiO_2@rGO > TiO_2$.	[95]
Thiobencarb (TBC)	5 mg/L	MoS_2 microsphere 1.0 g/L	2× visible lamps (400–700 nm, F4T5/CW, Philips Lighting Co.); $j = 1{,}420$ lx; pH = 6–9; effect of Cl^- and NO_3^-; real natural river water; $V_T = 100$ mL	95% efficiency at 12 h of treatment. Minor effect of Cl^- and NO_3^-. Slightly lower activity in real water. No loss of catalytic activity in three successive runs.	[96]

(continued)

Table 8.3 (continued)

Compounds	[Compound]	[Catalyst]	Operating conditions	Results and comments	References
Atenolol, bezafibrate, carbamazepine, clopidogrel, diclofenac, fluoxetine, isoproturon, tramadol, venlafaxine, and azithromycin	Atenolol, 12.5 ± 1.2; bezafibrate 38.7 ± 6.0; carbamazepine 763 ± 18; clopidogrel 93.2 ± 10.7; diclofenac 1,102 ± 31; fluoxetine 21.7 ± 4.5; isoproturon 84.6 ± 7.4; tramadol 3,930 ± 244; venlafaxine 349 ± 47 (ng/L)	Metal-free exfoliated g-C_3N_4 1.0 g/L	Borosilicate reactor in batch mode, V_T = 60 mL; 4× high-power Vis LEDs (λ_{max} = 417 nm); j = 400–500 W/m^2, urban WWTP effluent.	Almost complete removal (carbamazepine > isoproturon > clopidogrel > diclofenac > atenolol > bezafibrate > tramadol > venlafaxine > fluoxetine) after 10 min of treatment. Just 5 min to remove atenolol, carbamazepine, clopidogrel, and diclofenac.	[66]
Terbuthylazine	2.5–8.0 mg/L	TiO_2 Degussa P25, chitosan film supported on glass fiber.	UV-C lamp (8 W, λ = 254 nm); pH = 5, 7, and 9; batch mode and annular photoreactor.	After pretreatment at pH = 5, TiO_2/chitosan immobilized catalyst exhibited good mechanical properties and accepted stability at different reaction conditions	[97]

Terbuthylazine	5 mg/L	TiO_2 Degussa P25, chitosan film supported on glass fiber woven.	UV-C lamp (8 W, λ = 254 nm); batch recirculating photoreactor operating at unsteady conditions; pH = 5; Q_R = 5; Q_R = 50–300 cm³/min; T = 35–65 °C	Developed a kinetic model of terbuthylazine degradation to cyanuric acid based on a simplified route of intermediate by-products. The proposed model describes a complex recirculating reactor system operating at unsteady conditions	[98]
Pyrimethanil	10^{-4}, 0.75×10^{-4}; 0.5×10^{-4}; 0.25×10^{-4} M	TiO_2 Degussa 3D Aeroxide P25, SiO_2 immobilized on aluminum foam. Two-step preparation.	Recirculating batch photoreactor; V_R = 1 L; UVA lamp (λ = 365 nm); UV radiation = 9–35 W_{uv}/m^2; pH = 8.5	Optimum TiO_2 layers at fifth successive deposition with good coating rate (40 + 3%). Complete degradation of pyrimethanil after 5 h of treatment. Pyrimethanil photodegradation followed the Langmuir-Hinshelwood apparent first-order kinetics.	[99]

(continued)

Table 8.3 (continued)

Compounds	[Compound]	[Catalyst]	Operating conditions	Results and comments	References
Pyrimethanil, Scala® BASF fungicide	10–30 mg/L	TiO$_2$ Degussa P25 supported on β-SiC foams	Recirculating mode photoreactor; $V_R = 375$ mL; two 18 W UVA lamps ($\lambda = 385$ nm); UV radiation $= 60$ W m^2; pH = 4–10; $Q_R = 26–78$ mL^3/min; $t_R = 4–12$ min	Pyrimethanil and Scala® BASF fungicide were efficiently photodegraded with TiO2 suspensions. 98% and 91% removals, respectively, were achieved after 30 min of treatment. Lower efficiencies with TiO2/SiC foam (88% and 74%, respectively, and mineralized 58% of Pyrimethanil and 47% of Scala® after 4 h	[100]
Cypermethrin	50 mg/L	Fe-TiO$_2$ Degussa P25 supported on coconut palm spathe	Lab-scale solar flat photoreactor, ($\lambda = 310$ nm); photoreactor module (100 cm^2); $V_R = 250$ mL; natural pH; solar radiation=12,000 J/m^2	Lower photoefficiency for Fe-TiO$_2$ immobilized catalyst. Degradation efficiency of about the 85% for TiO$_2$ Degussa P25 reference catalyst. Elimination of operating cost associated with catalyst powder separation	[101]

| Quinclorad (QNC) | 10–60 mg/L | Immobilized carbon-coated N_2-doped $CNTiO_2$/EP on glass spheres | UV–visible 45 W fluorescent lamp + UV filter glass; UV radiation = 2.9–300–400 W m^2; V_R = 30.5 cm^3; 40 mL/min air; H_2O_2 (0–40 μL); pH 3–8; $CNTiO_2$/EP-coated glass plate = 2.29 mg/cm^2 | Best photodegradation at pH = 3 with H_2O_2 30 μL, 94% and 87% QNC removal after 90 min of treatment with UV–visible and visible light, respectively. Sustainable reuse trial for 10 consecutive cycles resulted in 92% and 72% QNC removals under UV–visible and visible radiation treatments, respectively. | [102] |

8.2.1.1 Slurry Photocatalysts

TiO$_2$ semiconductor has attained great attention from the research community thanks to the possibility of its heterogeneous use, which allows its support and easy separation from aqueous solutions, enabling its further reuse. In addition, TiO$_2$ holds other advantageous properties, such as its high catalytic activity for a wide spectrum of contaminants, low toxicity, wide pH range of operation, good chemical and thermal stability, biocompatibility, good environmental behavior, abundancy, and low cost [108, 109].

Polymorphs of TiO$_2$ have shown differential photo-activity. Although brookite has been successfully used for the photodegradation of a number of pharmaceuticals (e.g. cinnamic acid, ibuprofen, and naproxen [42]), there have still been relatively few assays of its photocatalytic activity in comparison to anatase and rutile polymorphs because of the difficulty to synthesize brookite titania (high temperature and high pressure are required) [110–112]. Rutile is the least photocatalytically active, whereas anatase has widely been accepted as the most photocatalytically active phase; thus, it has been the most studied up to date [111, 113, 114].

A method that has been used with the aim of increasing the photocatalytic activity of anatase titania is using it in a mixture with other polymorphs (i.e., anatase-brookite or anatase-rutile heterojunctions) [115–117]. The TiO$_2$ anatase-rutile mixture has been the most studied one because commercial TiO$_2$ Degussa P25 (75% anatase and 25% rutile) is a very robust and versatile photocatalyst for the treatment of water containing CECs mixtures [33, 118], different drugs [32, 38–40, 44, 45, 119], EDCs [67–69, 71], or pesticides. Specifically: dimethoate [81] has been removed up to a 99% with the aid of low concentrations of H$_2$O$_2$ or K$_2$S$_2$O$_8$; carbofuran, imidacloprid, and a mix of some of these pesticides, were comparatively abated using TiO$_2$ P25, and ZnO, which resulted less active [36, 87, 88]; and analogously, isoproturon, diuron, alachlor, and mixtures of them, were quantitatively reduced [37, 82–84]. In addition, the coexistence of anatase and brookite TiO$_2$ polymorphs has been evaluated for the removal of bisphenol A (BPA) by photocatalysis [71], achieving an 80% TOC removal in the best case.

Many other efforts have been made to synthetize active TiO$_2$ with customized properties. Applications of these lab-made TiO$_2$ catalysts have been tested for the abatement of: a mixture of CECs (carbamazepines, ibuprofen, sulfamethoxazole, ofloxacin, and flumequine) [33, 46, 120]; different drugs [121–123]; EDCs standalone (bisphenol A, dimethyl phthalate esters, 17α-ethinylestradiol – EE2), in a mix in deionized water dissolution (10 ppm of BPA, BPS, TDP, DPSO, 2,4-DCP, 2,4,6-TCP, 4-CP, and 4-OP); or in acetonitrile:water (DNOP and TCS) [71, 124–127]; and pesticides (such as isoproturon and methomyl) [82, 128]. In general, lower efficiencies than with commercially available TiO$_2$ Degussa P25 were mostly obtained.

The development of doped photocatalysts widens the range of possibilities and it is a useful strategy for enhanced photocatalytic performance [129–131]. Particularly,

N,Cu-doped TiO_2, and N,Cu-doped TiO_2 associated to SWCNTs (single-walled carbon nanotubes) have been used for the treatment of sulfamethoxazole, and compared to the application of bare TiO_2. N,Cu-TiO_2@SWCNT nanocomposite addressed the complete degradation of sulfamethosazole [132].

Combinations of TiO_2 with graphene and its derivatives have been described as efficient photocatalysts for the degradation of multiple CECs [95, 133–136]. In particular, a significant inhibition of the degradation of diuron, alachlor, isoproturon, and atrazine pesticides in natural water, as well as for the removal of the TOC, has been reported consistent when using bare TiO_2 as the catalyst. In fact, the initial photodegradation rate in ultrapure water was higher than in natural water ($r_{0,natural}/r_{0,ultrapure} < 1$), contrarily to the application of GO-TiO_2 ($r_{0,natural}/r_{0,ultrapure} > 1$). Under visible light, semi-reaction times were shorter for GO-TiO_2 than for TiO_2 Degussa P25 [90]. Other photocatalysts that have been used for the photodegradation of CECs were developed to lessen TiO_2 drawbacks. The combination of TiO_2 with metals, oxides, or other compounds, aiming to develop alternative advanced photocatalysts or conform multifunctional composites, were among the ongoing strategies.

ZnO is a common photocatalyst that has been applied to the photocatalytic treatment of wastewater containing CECs [74, 137–140]. Conversions of the 92.3, 94.5, and 98.7% for progesterone, ibuprofen, and naproxen were respectively achieved after 120 min of treatment using UVA radiation and a high loading of ZnO (1–1.5 g/L) [43]. Furthermore, the use of the $ZnO/Na_2S_2O_8$ tandem greatly improved the photocatalytic response in the treatment at a solar pilot plant scale of a mix with endocrine-disrupting activity (two fungicides: vinclozoline and fenarimol; and four insecticides: malathion, fenotrothion, quinalphos, and dimethoate) in a wastewater effluent [91]. Less than 7% of the initial contaminants were observed after 4 h of solar photocatalytic treatment, except for quinalphos (15%) and fenarimol (56%), and a strong decrease of the initial DOC was also addressed. Searching for an enhancement under visible irradiation, sensitized-ZnO with different compounds (namely, heteropoly phosphotungstic acid (HPA), fluorescein, perylene-3,4,9,10-tetracarboxylic dianhydride (PTCDA), and porphyrin) were prepared. PTCDA-ZnO reported the best performance in the treatment of resorcinol, an EDC by solar photocatalysis. [141].

Novel TiO_2-ZnO/cloisite nanoarchitectures (0.5% ZnO) showed the best performance for the degradation of acetaminophen and antipyrine by solar photocatalysis; that is, almost the complete conversion of both target compounds was achieved in 10 h of treatment; although only less than 50% of the TOC was reduced [142]. In addition, the lower the chemical concentrations and higher the irradiation intensity, the better degradation efficiency was achieved for both compounds, and a higher degradation rate was addressed for antipyrine. A similar approach was reported by Akkari et al. [143].

Tungsten trioxide (WO_3) is a colored solid that has been used as photocatalyst in some applications [48, 109, 144–146]. Particularly, 98% and 97% of the content of aspirin and caffeine (methyl theobromine) in water have been already reported by photocatalysis using visible light and the WTCN (g-C_3N_4 in WO_3/TiO_2) composite [51].

Although the application of zirconium oxide (ZrO_2) has been sparsely reported, mesoporous TiO_2/ZrO_2 (optimal loading of 0.14 mol% Zr) showed better photocatalytic behavior under visible irradiation than bare TiO_2 and TiO_2 Degussa P25 when applied to the photocatalytic removal of chloridazon [92].

More recently, graphitic carbon nitride (g-C_3N_4) has been reported to be an effective photocatalyst standalone or in combination with other photoactive compounds [50, 147–151]. A slurry photocatalytic treatment using metal-free g-C_3N_4 as the catalyst completely removed the mix of organic micro-pollutants present in the biologically treated effluent of an urban WWTP in about 10 min. The photodegradation order of resulted as follows: carbamazepine > isoproturon > clopidogrel > diclofenac > atenolol > bezafibrate > tramadol > venlafaxine > fluoxetine [66]. The results were least favorable in the case of using this g-C_3N_4 catalyst to degrade a combination of pharmaceutical compounds. After 4 h of treatment, the following sequence of degradation was addressed: tetracycline (86%) > ciprofloxacin (60%) > salicylic acid (30%) > ibuprofen (20%); although mineralization efficiency resulted as follows: tetracycline > salicylic acid > ciprofloxacin > ibuprofen [152]. An analogous study comparing the performance of g-C_3N_4, TiO_2, and TiO_2 Degussa P25 in the photocatalytic degradation of paracetamol, ibuprofen, and diclofenac showed that treatment efficiency increased in the sequence P25 < TiO_2 < g-C_3N_4 when visible light was irradiated [153].

Fe_2O_3 applied in heterogeneous photocatalysis plays a double role. It is a photoactive element and a magnetic support to facilitate the recovery and reuse of the photocatalyst [109, 154–158]. In particular, a magnetically separable TiO_2/FeO_x microstructure decorated with poly-oxo-tungstate (POM) led to the 76% elimination of 2,4-dichlorophenol, which is an EDC, under solar radiation of low intensity. An optimal composition for a faster photodegradation was $TiO_2/FeO_x(25\%)/POM(1\%)$. The addition of FeO_x extended the visible light absorption band of TiO_2, although it slightly slowed down the photodegradation rate of the treatment [159].

Thiobencarb was eliminated a 95% from an aqueous effluent after 12 h of treatment using MoS_2 microspheres as catalyst and applying visible irradiation [96]. The stability and reusability of this photocatalyst were tested and preserved for three successive runs. Real water addressed a detrimental influence on treatment efficiency, although the effect of Cl^- and NO_3^- was reported as minor. Composites based on other chalcogenides have also been developed for an improved photodegradation of CECs [160–162].

Bismuth-based photocatalysts have attracted attention for promoting efficient photocatalytic degradation [163–168]. A new 2D/3D $Bi_5O_7Br/BiOBr$ heterojunction, where 2D Bi_5O_7Br nano-sheets are tightly attached to the surface of the BiOBr 3D structure holding nano-flakes self-assembled microspheres, showed a superior visible light photocatalytic performance applied to the degradation of carbamazepine and tetracycline hydrochloride. They were degraded in 94% and 82%, respectively, after 90 min of treatment, addressing a high mineralization efficiency (>57%) as well. Additionally, the $Bi_5O_7Br/BiOBr$ heterostructure displayed a good cycling stability [93]. In another case, the integration of organophosphorus hydrolase (OPH) immobilized on BiOBr (OPH@BiOBr) was synthesized with success to arrange a photo-enzymatic integrated nanocatalyst, which was successfully applied to the degradation in cascade of methyl parathion by visible light photocatalysis [169]. Firstly, the P-S bond was hydrolysed by OPH, and the product was photocatalytically degraded on BiOBr. The OPH@BiOBr nanocatalyst still retained 83% of initial activity after five cycles of reuse.

The photocatalytic performance of BiOX and Cu-doped BiOCl compounds applied to the treatment of some CECs (estrogens 17 α-ethinyl estradiol and estriol, ibuprofen, and atrazine, for example) has also been tested [76]. Estriol was completely removed in 30 min of BiOI and BiOCl photocatalysis; whereas a shorter time (20 min) was enough to remove ibuprofen with BiOCl. On the other hand, 65–71% of atrazine remained in aqueous solution after 30 min of BiOCl and Cu-BiOCl photocatalysis.

Recently, the most demanded photocatalysts are those conformed by assembling different components to provide multifunctional capabilities. Composites formed by TiO_2-/SiO_2-functionalized CNTs (carbon nanotubes) have been analyzed for the removal of pharmaceuticals and pesticides [170]. CNTs contribute to enhance pollutants removal because of their high specific area and high interaction with molecules, which increase residence time. SWCNTs were useful for EDCs withdrawal, showing high removal efficiencies of up to 95–98% [132]. The functionalities induced in the nanotubes walls (hydroxyl, carboxyl, and graphitized) promote this enhanced efficiency. In addition, MOFs could be considered in this group as multifunctional compounds, although their main drawback is their stability in water. MOFs based on Pd, Fe, Ag, or chalcogenides, among others, have been applied to the treatment of PPCPs and pesticides, but this is still a new field of research that requires further investigation. Finally, further knowledge on photodegradation pathways will help to develop this type of promising potential photocatalysts [49].

8.2.1.2 Immobilized Photocatalysts

The idea of an immobilized photocatalyst on an inert support began to be widespread from 1993 because it can eliminate the costly and impractical post-treatment recovery operation of already used photocatalysts in large-scale plants [171]. The elimination of micropollutants, pesticides and herbicides, CECs, PPCPs, EDCs, and trace organic chemicals is one of the most suitable applications for heterogeneous photocatalytic technologies using immobilized catalysts [171].

The knowledge that nano-sized TiO_2 could be harmful to the liver and heart in mice and potentially affect humans when remained in water after treatment led to intensification of research efforts on developing immobilized catalysis. Mejia et al. [172] reported the application of immobilized TiO_2 thick film on compound parabolic collectors (CPC) as an alternative way of catalyzing the process to TiO_2 powder. Seventy-five percent of resorcinol was efficiently removed in this trial at neutral pH. At solar scale, Jiménez-Tototzintle et al. [173] have also assessed the treatment of a real wastewater from the agro-food sector by a combined process based on an immobilized biomass reactor (IBR) followed by a photocatalytic tertiary treatment with supported TiO_2. Imazalil and thiabendazole were completely photodegraded, and >90% of acetamiprid was removed, by this treatment, but it required the complementary presence of hydrogen peroxide as a co-oxidant agent.

Sivagami et al. [174] developed a novel lab-scale photoreactor with TiO_2-coated polymeric beads as floating catalysts. This enabled an efficient penetration of UV radiation to remove an organophosphate pesticide (monocrotophos). Fifty to eighty percent degradation efficiencies were achieved depending on the following conditions: initial pesticide concentration, pH, and catalyst dosage.

Different advanced TiO_2-based immobilized photocatalysts such as TiO_2 supported on β-SiC foams have been applied to photodegrade the fungicide pyrimethanil and its commercial version (Scala®, from BASF) [100]; and 88% and 74% of pyrimethanil and Scala® were, respectively, removed after 4 h of treatment, which corresponded to 58% and 47% TOC removal efficiencies. In addition, Verma et al. [175] assessed the degradation of 4-chlorophenoxyacetic acid (4-CPA), which is a recalcitrant herbicide, by photocatalysis comparing the application of suspended and supported TiO_2. An approximately 97% removal of 4-CPA was reported after 5 h of treatment using novel TiO_2-coated clay beads as an immobilized photocatalyst. These beads were successfully reused more than 30 times without showing significant reduction of its photoefficiency. Similarly, good performance until the fourth cycle of reuse was obtained in the photocatalytic degradation of meropenem (MER) using TiO_2 immobilized on fiberglass substrates [62].

A novel 3D photocatalyst with a reticulated metallic foam functionalized by TiO_2 was applied in the photocatalytic treatment of pyrimethanil, where complete degradation was observed after 5 h of treatment [99]. In addition, and aiming to assess the effect of different reaction conditions, TiO_2 immobilized on glass fiber was used for the photocatalytic treatment of terbuthylazine (TBA) [97, 98]. Degradation was reported as more successful when the circulation rate of the mixture being treated was higher, and more cyanuric acid was consequently produced [98].

Perisic et al. [58] and Rimoldi et al. [176] compared slurry-TiO_2 with immobilized TiO_2 catalysts applied to the removal of some drugs. The first study proved to photodegrade diclofenac (DCF), whereas the latter targeted tetracycline hydrochloride, paracetamol, caffeine and atenolol, whether each one standalone or in mixtures. Although a lower mineralization degree was always achieved with the immobilized TiO_2 system, these pieces of research addressed enough efficiency for real scale applications. De Liz et al. [77] found less favorable photodegradation kinetics with TiO_2-coated glass Raschig rings applied to the photocatalytic removal of estrogens (estrone, 17β-estradiol, and 17α-ethinylestradiol) present in wastewater. In addition, Švagelj et al. [177] observed that the treatment efficiency and photodegradation rate of memantine using suspended TiO_2 nanoparticles were higher than when TiO_2-

coated Al_2O_3 foam was used as the photocatalyst. Nevertheless, TiO_2-coated Al_2O_3 foam was addressed a possibly more suitable as a valuable photocatalytic material from a practical point of view. Furthermore, the mineralization of ibuprofen (IBP) by photocatalysis using micro-TiO_2-coated glass Raschig rings, which was tested aiming to grant an easy catalyst recovery and reuse, was assessed by Cerrato et al. [61], in both batch and continuous mode reactors. Whereas the use of micro-TiO_2 particles was able to completely mineralize IBP in 24 h of treatment, TiO_2-coated glass Raschig rings addressed an 87% degradation in 6 h of UVC light application in a continuous reactor, which corresponded to a mineralization efficiency of the 25%.

Rimoldi et al. [59] compared two different substrate geometries (plates and pellets) to get highly active immobilized TiO_2 systems aiming to totally photodegrade tetracycline. Both immobilized systems proved good efficiency, but TiO_2-coated pellets particularly promoted the degradation and mineralization of this pollutant on a reasonable time scale. In addition, a novel nanofiber powder photocatalyst (NnF Ceram TiO_2) was immobilized on water glass by Zatloukalová et al. [78] addressing a very good performance for the simultaneous photodegradation of synthetic hormones (progesterone and all types of estradiol).

More technical analyses, including the configuration of the photoreactor, were carried out by Manassero et al. [57], aiming to estimate the efficiency of three diverse photocatalysis reactors in the treatment of the micro-pollutant clofibric acid, namely: (a) a slurry reactor with suspended TiO_2 particles; (b) a fixed-film reactor with immobilized TiO_2 on the window of the reactor; and (c) a fixed-bed reactor filled with TiO_2-coated glass rings. Although the slurry reactor was the most efficient configuration, the fixed-bed reactor achieved a quantum efficiency just one third lower than the suspended system, indication that this configuration is very convenient to perform photocatalysis processes.

Studies under more realistic conditions have been carried out by Moreira et al. [66], where heterogeneous visible light photocatalysis was applied to the treatment of organic pharmaceutical micropollutants (i.e., carbamazepine, diclofenac, atenolol, isoproturon, clopidogrel, bezafibrate, tramadol, venlafaxine, and fluoxetine) that are typically found in the biologically treated effluents of urban WWTPs. Although most of these chemicals were degraded in less than 10 min of photocatalytic treatment using TiO_2 powder as the catalyst, 25 min of minimum residence time were required to address significant treatment efficiencies with the tested metal-free exfoliated graphitic carbon nitride (gCNT) photocatalyst immobilized on glass rings, which was used in continuous mode. However, Martín-Sómer et al. [63] found that when a real wastewater effluent was applied to the simultaneous removal of different pharmaceutical micropollutant CECs, no differences were observed comparing TiO_2 suspensions with an immobilized TiO_2 photocatalyst on macroporous reticulated ZrO_2 3D. In this case, the presence of hydroxyl radical in the reaction solution enables immobilized-TiO_2 3D-foams to be as effective as control slurry catalysts.

The use of H_2O_2 as electron acceptor, and to promote higher oxidation rates, aiming to improve the photocatalytic treatment efficiency when immobilized TiO_2 catalysts are applied have thoroughly been addressed for many photocatalytic processes. For example, Santana et al. [178] reported that the best configuration to photodegrade metronidazole (MTZ) antibiotic was a catalytic system immobilized on the outer of an internal UVC lamp assisted with the addition of H_2O_2. Moreover, a heterogeneous process with TiO_2 immobilized on glass discs was optimized for the photodegradation of the CEC antipyrine (AP) with the addition of H_2O_2 as well [56]. Jiménez-Tototzintle et al. [60] performed a complete trial to simultaneously remove a mixture of contaminants (bisphenol A, acetamiprid, and imazalil), reporting that the immobilized TiO_2/H_2O_2 combination improved the inactivation and removal of all these compounds.

Other types of photocatalysts that are beginning to be applied for wastewater treatment include the novel immobilized cerium-doped zinc oxide (Ce-ZnO), which was developed by Zammit et al. [179] aiming to photodegrade two antibiotics (trimethoprim and sulfamethoxazole) as a potential tertiary treatment for urban WWTPs. Ce-ZnO immobilized on a metallic support led to somewhat slower kinetics, and >50% of its initial activity remained even after 5 cycles of reuse.

Considering the different types of doped TiO_2 catalysts with transition metals, the use of iron-titanium dioxide (Fe-TiO_2) nanoparticles on a coconut biomaterial has been assessed for the photocatalytic treatment of cypermethrin in a flat plate solar reactor at natural pH values [101]. Although better results were reported using Degussa TiO_2 P25, a 0.05 Fe:Ti ratio of active phases supported on the biomaterial addressed a > 80% photodegradation efficiency holding the advantage of minimizing the operational cost because filtering or centrifuging is not necessary to separate the catalyst. In addition, Lin et al. [64] carried out the comparison study of the photocatalytic activity of TiO_2-Fe and TiO_2-rGO nanocomposites immobilized on optical fibers that were synthesized by the polymer-assisted hydrothermal deposition method. TiO_2-rGO showed a higher photocatalytic activity in the treatment of pharmaceuticals under UV radiation. On the other hand, TiO_2-Fe was reported as more suitable for processes assisted by visible radiation. These results suggested that the improved photocatalytic performance of TiO_2-rGO may be attributed to a reduced recombination rate of photo-excited electrons–hole pairs, although a narrower band gap would have contributed to increase the photocatalytic activity of TiO_2-Fe nanocomposite.

New modifications of the structure of photocatalysts by doping with nitrogen have also been performed aiming to improve their photocatalytic performance under visible light. In this line, an immobilized carbon-coated nitrogen-doped TiO_2 (CNTiO_2/EP) immobilized on glass spheres was developed to improve its recyclability in the photocatalytic treatment of quinclorac (QNC, a herbicide) under UV-visible and visible light irradiations [102]. Better QNC removal results were reported at pH = 3 with the complementary addition of H_2O_2 (30 μL). Reusability tests showed a sustained CNTiO_2/EP performance in 10 serial cycles, for which the average QNC removal was of the 92% and 72%, respectively, under UV–visible and visible light applications.

Furthermore, Xing et al. [65] developed a novel N-TiO$_2$ catalyst immobilized on glass spheres to evaluate the photodegradation of the common antibiotic ciprofloxacin (CIP). Immobilized TiO$_2$ photocatalysts with a 0.34% N/Ti weight ratio showed the highest visible photocatalytic activity. CIP removal efficiency reached the 90% in 90 min of treatment under visible radiation. Finally, Sacco et al. [79] reported the assessment of a phosphor-based photocatalyst (N-doped TiO$_2$ coupled with ZnS blue phosphor) immobilized on macroscopic polystyrene pellets applied to the degradation of ceftriaxone (third-generation β-lactam antibiotic). The application of this so-structured photocatalyst was reported effective in the treatment of ceftriaxone in water, and showed no deactivation effects even after several cycles of reuse whether in distilled or real water.

8.2.2 The Influence of Operating Conditions on the Success of the Photocatalytic Treatment of EDCs, Pesticides, and Pharmaceuticals

Chemical, physical, and environmental parameters have a great impact on the design of photoreactors for photocatalysis performance [180]. It is imperative that, in each stage, there is a uniform distribution of light. This way, there will be an efficient absorption of the photons emitted to the system, therefore avoiding insufficient illumination that may limit the speed of the process [180, 181]. In addition, the design of a photocatalytic reactor can contemplate different alternatives, depending on the application and scale, which are: (a) artificial or solar light as irradiation source; (b) discontinuous (batch) or continuous mode of operation; (c) laboratory, pilot plant, or industrial scale; and (d) supported/immobilized or suspended photocatalyst. In the treatment of EDCs, PPCPs, and pesticides, the most studied configurations have been those applying artificial light, suspended photocatalysts in batch operation mode, and with recirculation sometimes at laboratory scale, as it can be checked in Table 8.3.

The radiation source (and its intensity) is a key factor affecting photocatalysis efficiency. The production of the electron–hole pair in the band gap of the photocatalyst is initiated because of the excitation of electrons by the emitted radiation [109, 112, 118, 182–185]. The most used wavelengths in the study of this type of contaminants are within the range of 254–400 nm, which corresponds to the UV light spectrum span, where TiO$_2$ photocatalysts are active. Table 8.3 shows some studies where it has been possible to degrade pesticides using visible light (400–700 nm) as well. Particularly, Huang et al. [96] were able to photodegrade thiobencarb by MoS$_2$ microspheres with efficiencies of the 95%, and Moreira et al. [66] almost achieved the complete removal of all the considered pesticides in the study by photocatalysis using metal-free exfoliated g-C$_3$N$_4$ as the catalyst.

Light intensity determines the quantity of light that would be absorbed by the photocatalyst, and the rate of the electron–hole formation [185–187]. Thus, this parameter is considered in most of the studies included in Table 8.3. The ability of light to spread within the reactor will also play a significant role in the photodegradation rate. Related to this affirmation, the catalyst loading is another parameter to take into account. Generally, it is considered that degradation rate increases as the photocatalyst concentration does until an optimum value is reached. After this optimum value, there is a significant decrease in the photocatalytic activity because of light scattering and screening effects [186]. Additionally, the agglomeration of photocatalyst particles may occur as the catalyst loading increases, which inconveniently results in the reduction of the overall surface area of the catalyst, which inevitably decreases its photo-activity. The increased amount of catalyst present in the reaction mixture will also cause the solution to become more turbid, which ultimately decreases the ability of light to penetrate into the solution. The review conducted by Ahmed et al. [185] stated that "The trade-off between these two opposing phenomena results in an optimum catalyst loading for the photocatalytic reaction" [155, 186–189].

Generally, the most used common catalyst concentration values are between 10 and 500 mg/L for the treatment of pesticides and pharmaceuticals by photocatalysis, as it is shown in Table 8.3; even reaching degradation efficiencies of the 99% for values of 1,000 mg/L of catalyst in the treatment of paracetamol [31]. In the case of EDCs at low concentration values, ranging between 0.2 µg/L and 10 mg/L, the catalyst loading in most studies is also much lower.

In addition, one parameter that also has a major impact on the photocatalytic activity of a catalyst is the pH of the solution [182, 186, 190], which can significantly alter the interactions among the surface of the semiconductor catalyst, solvent molecules, and free radicals along the photocatalytic process [186, 191], and, thus, the ability to photodegrade these organic compounds. Most EDCs, pharmaceuticals, and pesticides are uncharged in water (pH = 6–8) [182, 186, 190]; thus, most of the studies have been performed at pH = 5–8, as it can be checked in Table 8.3.

Finally, temperature is considered a minor influencing parameter in heterogeneous photocatalysis, because it is believed that it affects the rate of diffusion, but not the rate of reaction [192]. Chatzitakis et al. [39] studied the degradation of chloramphenicol, and concluded that the rate of degradation increases until dissolved O_2 is no longer present in the media ($\approx$45 °C) [39, 193]. Therefore, this parameter has hardly ever been assessed for its influence on the photocatalysis treatment of EDCs, pharmaceuticals, and pesticides, as it is shown in Table 8.3.

8.3 Pathways and Oxidation Routes, By-Products, and Intermediates

The degradation of EDCs, pesticides, and pharmaceuticals by photocatalysis treatment produces complex reaction routes. In addition, the generated reaction intermediates may be, in certain cases, even more toxic than the parent targeted compounds to remove [194]; hence, it is crucial to understand the degradation routes of standard compounds and chemical structures in order to provide further information to design treatment and understand the potential toxicity hazard of these compounds when they are going to be treated by photocatalysis. The total mineralization results of the photocatalytic treatment of these compounds may progress slower than the degradation of the parent compounds itself when certain intermediates are involved or are produced at high concentration levels. In some cases, total mineralization could be possible, although there are some structures that are very difficult to remove and mineralize, such as the triazine ring [35].

In general, the first oxidation step occurs by the attack of hydroxyl radical [195, 196]. Simultaneously, electron reduction and oxidation by reactive oxygen species are produced [197]. In addition, target compounds react with photogenerated holes and dissolved oxygen in water [195].

The structure of the compound to be treated is basic for the determination of the oxidation mechanism. The hydroxylation in polycyclic aromatic hydrocarbons is directly related to the localization energy of the different available positions in the structure of the compound [198]. In the degradation of alachor, for example, the oxidation of the arylethyl group has been reported as noticed, as well as the bond cleavage of the N-methoxymethyl group and *N*-chloroacetyl moiety [163]. In atrazine, dealkylation reaction and alkyl chain oxidation were described [196]. When BPA was degraded in the presence of cyclodextrins (CDs) applying UV radiation, the photocleavage effect was strengthened by the inclusion effect of BPA and b-CD [199]. From chlorinated compounds, Cl^- ions are easily released to the solution, which are the first ions that appear along the photocatalytic degradation process of these type of chemicals [200]. N-containing compounds are mineralized generating NH_4^+ and, mostly, NO_3^- ions [200–204]. Sulphur atoms produce SO_4^{2-} [204, 205], and organophosphorous chemicals are mineralized producing PO_4^{3-} ions [201, 206]. In general, it has been reported that nitrate presence has little effect on reaction kinetics; whereas sulphate, chloride, and phosphate ions, principally at concentrations greater than 10^{-3} mol/dm^3, can reduce degradation rate by 20–70% because of the competitive adsorption that unfolds at photo-activated reaction sites [35].

The main intermediates that are produced by the photocatalytic treatment of organic chemicals are: (a) products and derivatives of hydroxylation, which are typically generated after dehalogenation of the target compound if halogen substituents are present; (b) oxidation by-products of the alkali chain, if they are present; (c) ring opening intermediates for aromatic compounds; (d) products of decarboxylation; and (e) isomerization and cyclization products [35, 207].

The oxidation routes of some relevant EDCs, pesticides, and pharmaceuticals are described next in more detail.

8.3.1 Atrazine

Atrazine ($C_8H_{14}ClN_5$, 2-chloro-4-ethylamino-6-isopropylamino-1,3,5-triazine, ATZ) is the most representative herbicide of the organochlorine s-triazine group. It is applied as a selective pre- and post-emergence herbicide for weed control in agricultural fields and forest plantations [193, 208, 209]. Because of its high solubility, low adsorbance on soil particles, and low biodegradability in water, atrazine could easily percolate to underground water, or get to surface water by runoff. Moreover, this contaminant is considered a possible carcinogen and EDC by a number of assessment reports, as well as its chlorinated metabolites [210]. Hence, several countries, such as the United States, Australia, and some European countries, have restricted its use [193].

The degradation of atrazine may progress via two different initial pathways: (path I) an alkylic–oxidation process; and (path II) a dechlorination–hydroxylation process; as it is shown in Figure 8.3 [211]. HO· attacks aminoalkyl groups at side chains to produce free organic radicals, followed by alkylic oxidation to form intermediates 2-chloro-4-acetamino-6-isopropylamino-s-triazine and 2-chloro-4-diethylamino-6-(2-hydroxyethyl-amino)-s-triazine. These compounds suffer dealkylation until forming 2-chloro-4,6-diamino-s-triazine. Additionally, the C-Cl bond in atrazine is the most susceptible of being attacked by hydroxyl radical, so there is also a dechlorination–hydroxylation pathway (path II), in which 2-hydroxy-4-ethylamino-6-isopropylamino-s-triazine is formed. Subsequently, this photoproduct undergoes a two-step dealkylation process to form ammeline. Afterwards, 2-chloro-4,6-diamino-s-triazine and ammeline suffer a series of oxidations of the amino groups in their structure to turn into nitro groups, dechlorination and deamination–hydroxylation, or denitration–hydroxylation, to produce cyanuric acid [211–213].

Atrazine's total mineralization has not been reported to occur under conventional wastewater treatment methods, or under some AOPs like photocatalysis, especially involving TiO_2, graphene–titania composites, or silica. The high stability of the s-triazine ring may be responsible for this resistance to be fully mineralized. Thus, cyanuric acid has been detected as the final oxidation product of atrazine, and no further degradation occurs because of its stability to the attack of hydroxyl radical [212, 214, 215]. However, its total mineralization of atrazine is not actually necessary to report the detoxification of atrazine-containing water because cyanuric acid is a very stable, but non-toxic, compound by itself [212, 216].

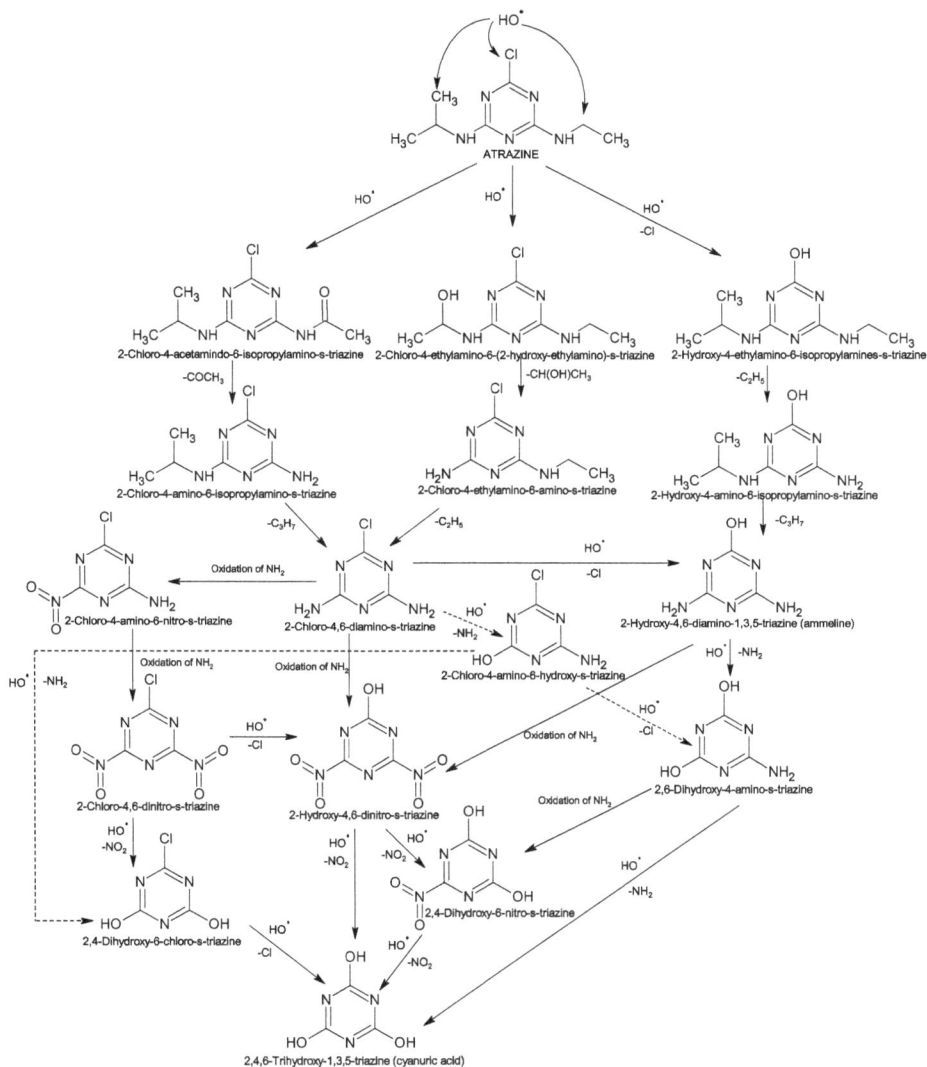

Figure 8.3: Possible photocatalytic degradation pathways of atrazine: an alkylic–oxidation process (path I), or a dechlorination–hydroxylation process (path II) [211–213].

8.3.2 Methomyl

Methomyl ($C_5H_{10}O_2N_2S$, S-methyl N-(methylcarbamoyloxy) thioacetimidate) is a carbamate insecticide of wide spectrum. It is a very toxic and hazardous substance, and a CEC because of its high solubility in water (57.9 g/L at 25 °C). In addition, it can easily contaminate surface and ground waters because its sorption affinity to soil matrices is rather low [217].

A tentative photocatalytic degradation pathway for methomyl is proposed in Figure 8.4, where principal intermediates are indicated inside blue frames, as reported in the studies of several researchers [204, 205, 218–222]. The oxidative environment that is generated in an irradiated mixture induces several molecular transformations; commonly, three possible transformations for methomyl: (a) the elimination of methylisocyanate (rupture of O=C–O bond) yields methomyl oxime ((E)-methyl-N-hydroxyethanimidothioate, photoproduct I) and methylamine;(b)the demethylation of methomyl yields (E)-methyl-N-carbamoyloxyethanimidothioate (photoproduct II); and (c) the initial hydroxylation of the methyl group attached to the N atom yields methomyl methylol ((E)-methyl-N-hydroxymethylcarbamoyl oxyethanimindothioate, photoproduct III). Indeed, the alcohol moiety formed from methomyl will be oxidized, in turn, into aldehyde and acid, which decarboxylates into CO_2 through the photo-Kolbe reaction, generating photoproduct II. Methomyl oxime may also be obtained from photoproducts II and III, in the same way as from methomyl, that is, by the rupture of the O=C–O bond by the attack of hydroxyl radical [204].

On the other hand, methomyl oxime can produce 2-imino-2-(methylthio)ethan-1-ylium ion by dehydration, which can be also obtained from photoproduct II by the elimination of carbamic acid [205]; and it can also be converted into acetonitrile and mehanethiol by Beckman rearrangement. Thereafter, a well-known series of reactions involving either the photo-Kolbe mechanism of decarboxylation or hydroxylation are going to facilitate the mineralization of the whole acetonitrile molecule after the formation of acetamide [204].

8.3.3 Antipyrine

Antipyrine is a common analgesic and anti-inflammatory drug; thus, it is present in water bodies (surface and ground water), and it is considered an emerging pollutant of EDC effect [194, 223].

The degradation of antipyrine, according to Tan et al. [224] and Kar et al. [194], begins with the attack of HO˙ to different possible positions of the benzene group and the pentacyclic ring [225]. As described by Kar et al. [194], the C=C bond in the pentacyclic ring will consequently be broken and the C-C bond will be attacked next, producing N-phenylpropanamide. Posteriorly, degradation progresses by the attack to C-N bonds, generating benzenamine. Exposito el al. [56] also reported the formation of N-phenyl propanamide, anthranilic acid, 1,4 benzenedicarboxylic acid, and benzenamine. These authors also reported the formation of 4-oxo-pentanoic and butanedioic acids after ring opening reactions (Figure 8.5). In addition, Tobajas et al. [142] concluded that in the photocatalytic oxidation of antipyrine, the formation of reactive species, such as H^+ and HO·, is more important than $O_2\cdot^-$.

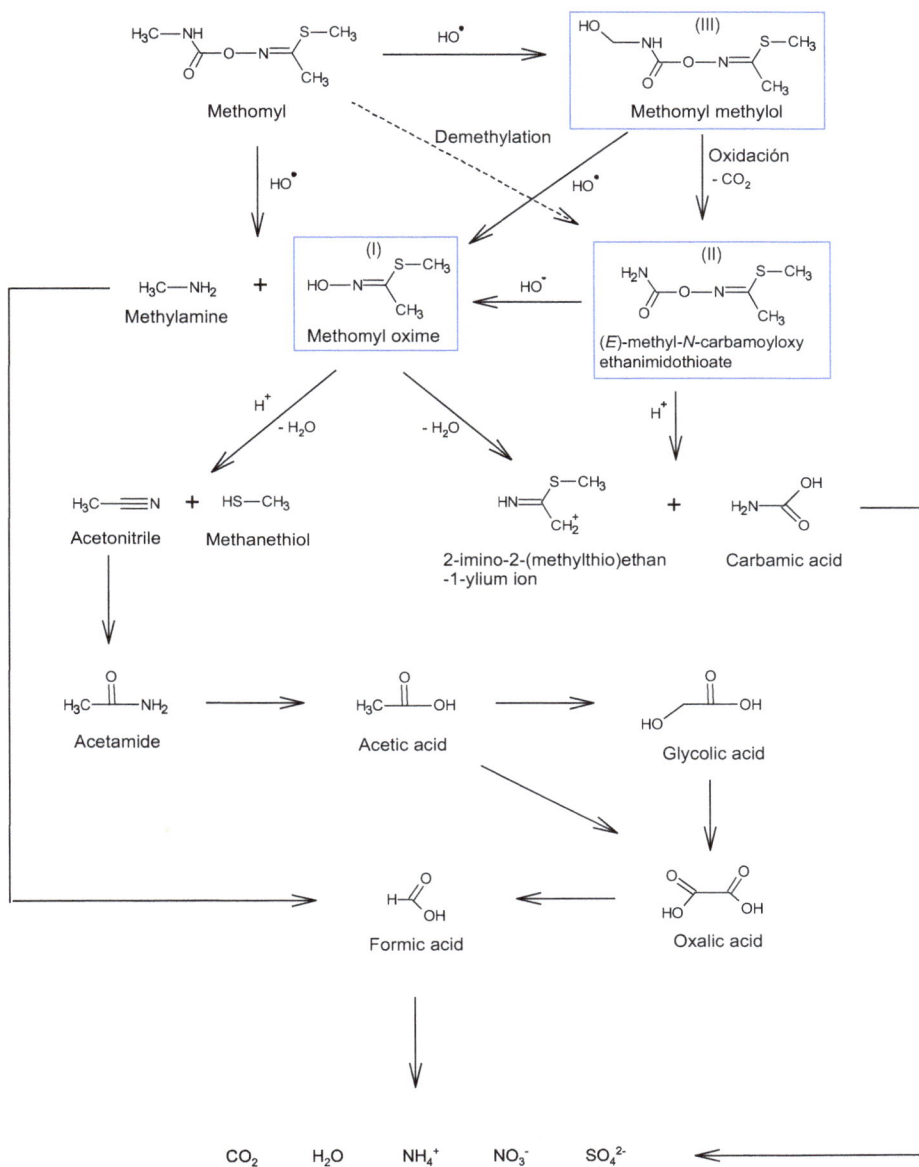

Figure 8.4: Tentative photocatalytic degradation pathway of methomyl [204, 205, 218–222]. Blue frames indicate major intermediates.

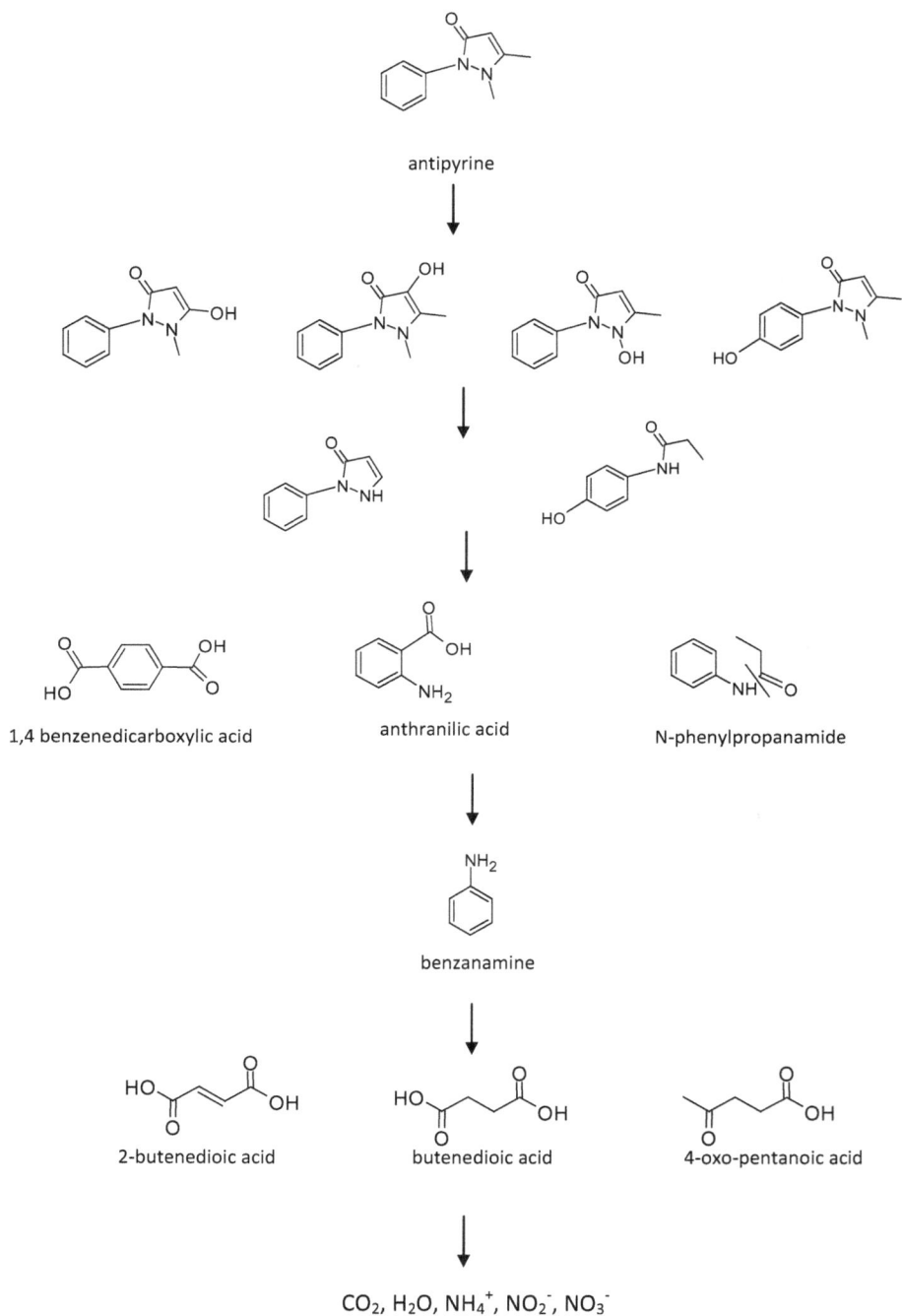

Figure 8.5: Antipyrine degradation route by photocatalysis [56, 194, 225].

8.3.4 Ciprofloxacin

Ciprofloxacin is one of the most used antibiotics of the fluoroquinolone family. Consequently, its presence is frequent in wastewater and, as a result, in nature. Therefore, the development of efficient and safe processes for its removal is of great interest [226].

The most probable photocatalytic oxidation route of ciprofloxacin involves the oxidation through defluorination, the attack to the piperazine ring via N-dealkylation, and hydroxylations by HO· additions, as it is shown in Figure 8.6 [227, 228]. The hydroxyl substitution of fluorine leads to a by-product that could be posteriorly oxidized by the addition of HO· to the piperazine ring, causing its opening. Next, radical attacks to the

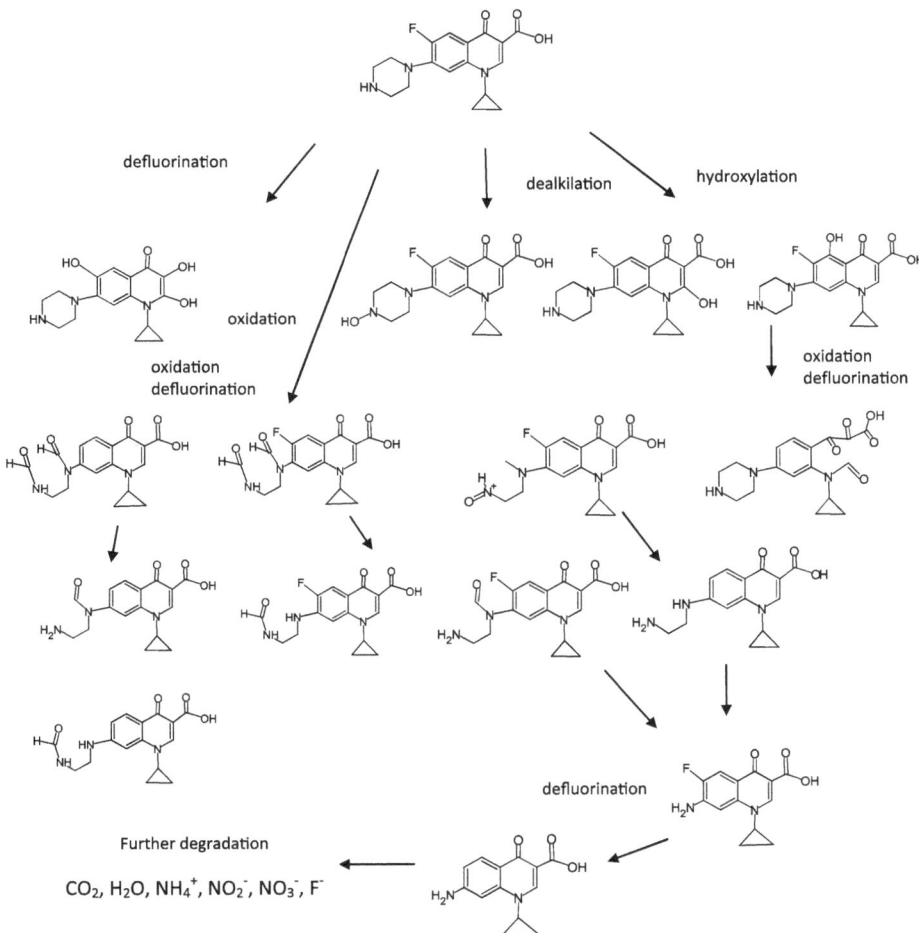

Figure 8.6: Degradation route of ciprofloxacin [227, 228].

piperazine ring will break it via N-dealkylation, and several hydroxilations will attack the molecule to break the produced quinolone.

As with many other compounds, it is important to consider that pH speciation, and its changes in absorption spectra, will influence its oxidation route [228]. In short, the generated by-products mainly keep the fluorine atom under acidic pH values, whereas they have no fluorine at basic pH ones; hence, the substitution of fluorine is just ensured in this situation [228]. As a result, the photolysis of ciprofloxacin is more efficient under basic pH conditions.

Similar degradation routes, including the three main pathways of fluoroquinolones degradation (namely, defluorination, ring conversion, and decarboxylation) have also recently been reported for other antibiotics of this family [194].

8.3.5 17β-Estradiol

Female hormones, such as estrone (E1) and 17β-estradiol (E2), are among the most potent EDCs, as far as the potential effects on fish physiology are concerned [229]. These compounds are, therefore, naturally present in the environment. Particularly, E2, which is one of the most detected EDCs in water bodies, is a compound of special concern because of its important role in the regulation and function of the female reproductive system [230]; besides it is an important CEC that is usually released from urban WWTPs [231, 232].

The estrogenic activity of E2 is attributable to its phenolic group; thus, a high residual estrogenic activity has also been reported when by-products that still hold the phenolic ring in their chemical structure remain in water [231]. Breaking this group is therefore key to reduce its estrogenic effects in water [233].

In the photocatalytic degradation of E2, and estrogens in general, the first oxidation reaction is hydroxylation (see Figure 8.7), and five possible pathways have been identified. For pathways I, II, and III, the reaction begins by the addition of HO· radical, and it is then followed by the ring opening mechanism and the generation of dicarboxylic acids [234]. For pathways IV and V, the reaction is initiated by the direct oxidation of E2 [235].

In pathways I and II, 2-hydroxyestradiol, or its resonance structure 10ε-17β-dihydroxy-1,4-estradien-3-one, are postulated as the first intermediates of E2 degradation [231] after an HO· radical is added to the C2 atom of the aromatic ring, or the C6 atom of cyclohexane ring, respectively. However, HO· may probably attack position 2 because of its high value of $FED^2_{HOMO} + FED^2_{LUMO}$. This addition of hydroxyl to the aromatic ring has previously been reported by several authors [231, 236]. The addition of HO· to position 2 will probably produce 2- hydroxyestradiol, which will be further attacked at the aromatic ring increasing its hydroxylated structure. This attack will ultimately produce the cleavage of the phenolic ring, thus reducing

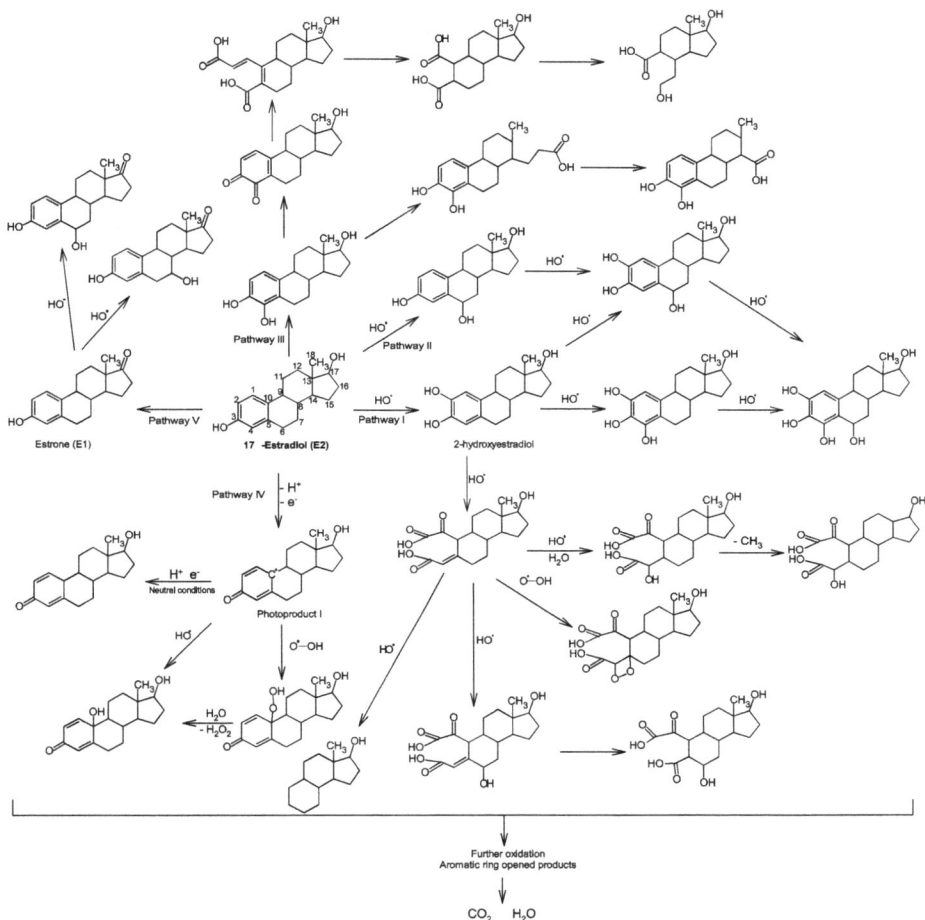

Figure 8.7: Proposed degradation pathway for the photocatalysis of 17β-estradiol [230, 231, 234, 235, 239].

estrogenicity [231, 232]. The resulting intermediates after ring opening are oxidized products holding dicarboxylic acids [234].

In pathway III, it may be possible that HO· attacks at the C4 atom position of the aromatic ring. The rupture of the aromatic ring, or the cyclopentane ring, is produced next [237].

In the case of pathway IV and according to the theoretically calculated frontier electron densities reported by Mai et al. [234], the highest $2FED^2_{HOMO}$ value is located in the phenol moiety at position C10 (Figure 8.7); thus, it will probably be attacked by the extraction of an electron by the hole in C10, modifying the phenol moiety into a phenoxide radical, thus yielding photoproduct I [235] and being favorable to the attack of HO· and HOO· radicals to yield other photoproducts. In addition, photoproduct I can be reduced to 17β-hydroxy-1,4-estradien-3-one in the presence of H^+ and an electron, which are formed in the presence of methanol and under neutral conditions [234].

In pathway V, E2 is directly oxidized by the -OH group of the cyclopentane ring and, as a result, estrone (E1) is produced [235]. This route is not complete, or an environmentally friendly process to remove E2 [230]. After suffering from photolysis and photooxidation, E1 is degraded into phenylacetic acid as the major intermediate of this pathway, and, then, total mineralization can be achieved [238].

In all pathways, further oxidation finally generates by-products of opened aromatic rings, and short-chain acids would ultimately be produced before its complete mineralization to CO_2 and H_2O is fully addressed.

8.3.6 Bisphenol A

Bisphenol A (2,2-bis(4-hydroxyphenyl) propane; BPA) is an industrial compound that is widely used as an additive or monomer in the production of polycarbonates (PCs), epoxy resins, and other polymeric materials. In fact, it is estimated that a global volume consumption of about 10.6 million metric tons will be reached by 2022 [240]. BPA is a synthetic chemical that is synthesized from the condensation of two phenol groups and a molecule of acetone under acid or basic conditions [240]; and it is a well-known EDC that has been reported to be able to bind to estrogen receptors and have shown estrogenic behavior in lab trials [241].

In the case of BPA degradation by TiO_2 photocatalysis, a negligible contribution of direct UV photolysis has been reported, and HO· radical and photogenerated holes have been addressed as the main oxidizing agents [242]. HO· attacks BPA by both addition and H-abstraction reactions [243]. The proposed mechanism for the photodegradation of BPA is shown in Figure 8.8.

Some studies have postulated that the attack of HO· radical and photogenerated holes to the phenyl groups of BPA succeed in adding one oxygen atom to produce bisphenol A catechol [242–245]. Then, bisphenol A dicatechol is produced by the addition of another hydroxyl group to the phenyl groups of bisphenol A catechol, and by the reaction of this intermediate with photogenerated holes. With the consequent loss of two hydrogen atoms, bisphenol 3,4–quinone is formed next. Both bisphenol A dicatechol and bisphenol 3,4–quinone, thereafter suffer a series of oxidations by HO· attack that achieve the cleavage of the C-C bond of the isopropyl group.

On the other hand, HO· radical might directly attack the C-C bond of the bisphenol A isopropyl group [243, 244, 246–249], which can cause the demethylation of bisphenol A to yield bis (4-hydroxyphenyl)methane [244]. The generation of p-hydroxybenzoic acid is enhanced by the attack of HO· radical to the bis(4-hydroxyphenyl)methane phenyl group. Besides, the attack of HO· to bisphenol A might

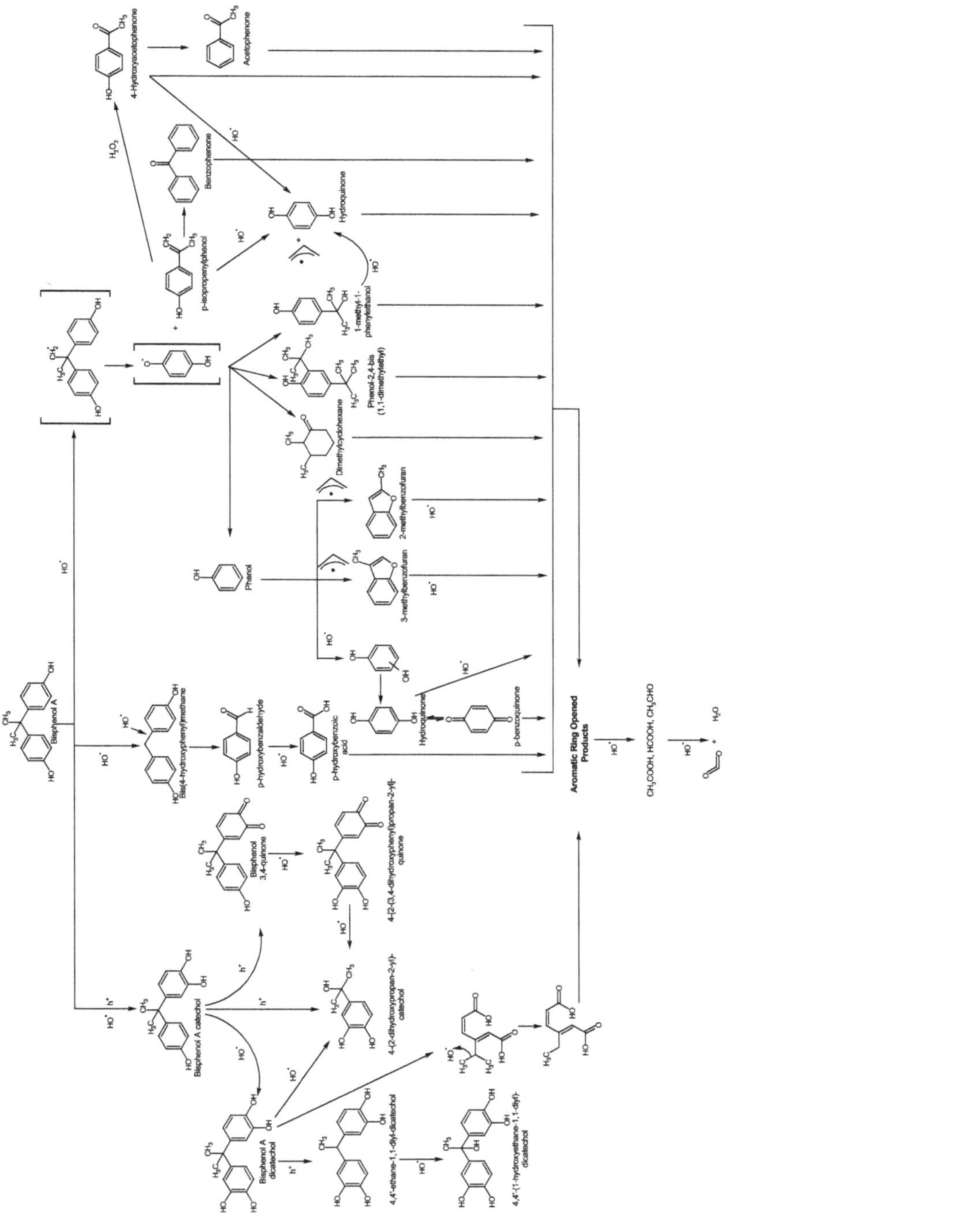

Figure 8.8: Proposed photocatalytic degradation route of bisphenol A [242–244, 246, 250–252].

lead to p-isopropenylphenol and a phenoxy radical [243]. Further oxidation of isopropenylphenol results in the formation of benzophenone, 4-hydroxyacetophenone and hydroquinone [243, 246, 250]. Moreover, the produced phenoxy radical contributes to produce 2,3-dimethyl-cyclohexanone, phenol, hydroquinone, 1-methyl-1-phenylethanol, and phenol-2,4-bis(1,1-dimethylethyl) [243, 246, 251]; whereas the attack of propenyl radicals to phenol would yield methylbenzofuran [246, 250], and the attack of HO· radical, once again, generates hydroquinone. Furthermore, these single-aromatic intermediate by-products are thereafter oxidized in ring-opening reactions that degrade them into aliphatic acids, including carboxylic acid, formic acid, and acetic acid, among others. Finally, the full mineralization reaction to carbon dioxide and water may eventually occur [247–249].

8.3.7 Nonylphenol

Alkylphenols (APs) and their polyethoxylated derivatives (alkylphenol ethoxylates, APEOs) are synthetic compounds that are used as antioxidants and plasticisizers in a number of agricultural, industrial, and household applications despite their high cost [253, 254]. In general, there are two groups of APs: nonylphenols (NPs) and octylphenols (OPs). NP is by far the most commercially relevant AP. NP is recognized for its stability, aquatic toxicity, and estrogenic activity; and it is primarily used to produce nonylphenol ethoxylate (NPEO) surfactants for a great variety of applications and market products, such as cosmetics, detergents, paint, and pesticides [254].

Among NPs, the nonyl group of nine carbons may be linear or ramified, as well as bound at various locations of the phenol ring (ortho, meta, and para). In fact, these variations induce confusion regarding the precise identity of the compound [254]. NP is not a single chemical structure, but a mixture of different highly branched NPs that are greatly mono-substituted in the para-position with small quantities of ortho- and di-substitutions. Theoretically, NP can consist of a mix of up to 211 possible constitutional isomers with a highly branched alkyl chain and 50–80 isomers that may exist in the proper environmental matrix [254, 255]. The molecular structures of some APs are shown in Figure 8.9.

In particular, the degradation route of 4-nonylphenol has already been investigated even knowing that NP is found as a mixture of many isomers, as it has just mentioned above. For this reason, the possible photodegradation pathway for the photocatalytic degradation of 4-nonylphenol is reported in Figure 8.10, where it is first shown that the structure of NP decomposes by dealkylation and hydroxylation to different intermediates of lower molecular mass, such as heptylphenol, pentylphenol, and propylphenol, which can generate phenol or benzoquinone [256–259]:

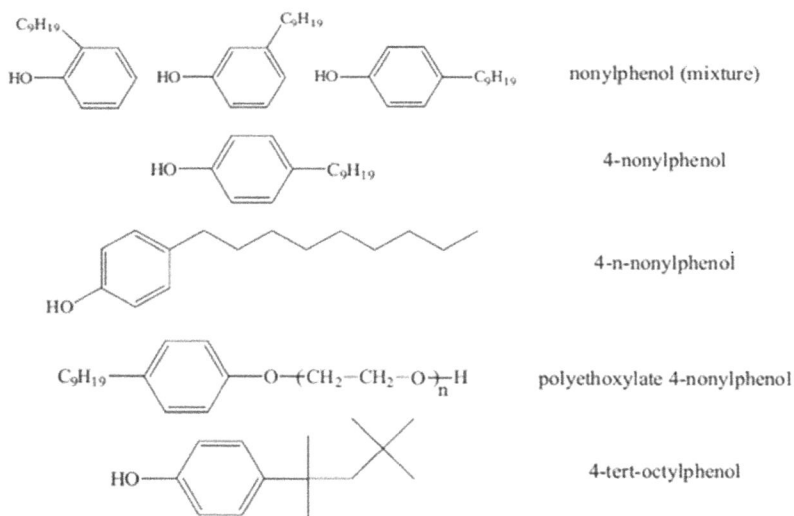

Figure 8.9: Molecular structures of some alkylphenols [254].

Phenol has also been reported to be an intermediate by-product of the oxidation pathway of high molecular weight aromatic hydrocarbons [260], and unraveling the oxidation pathway of phenol has been the subject of several pieces of work, which, in short, consider the hydroxylation of phenol to hydroquinone and catechol as the first step of this route. Then, further oxidation of the generated dihydroxylbenzenes produces benzoquinones and, finally, C1–C3 acids [260, 261].

Benzoquinone, which is a strongly electrophilic substance, rapidly reacts with hydroxyl radical to form (Z)-buta-1,3-diene-1,3-diol, which advanced oxidation and decarboxylation results in the formation of (Z)-3-hydroxyacrylaldehyde [262].

On the other hand, the generated holes and oxidative species may attack the aromatic ring of NP molecules by electrophilic addition at C=C double bonds to produce hydroquinone and other alcohol-containing alkyl chains of NP (R=C9H19) [263].

Finally, all these intermediates may eventually be oxidized to shorter chain acids, as well as may totally be mineralized to H_2O, CO_2, and inorganic salts along their further photocatalytic degradation.

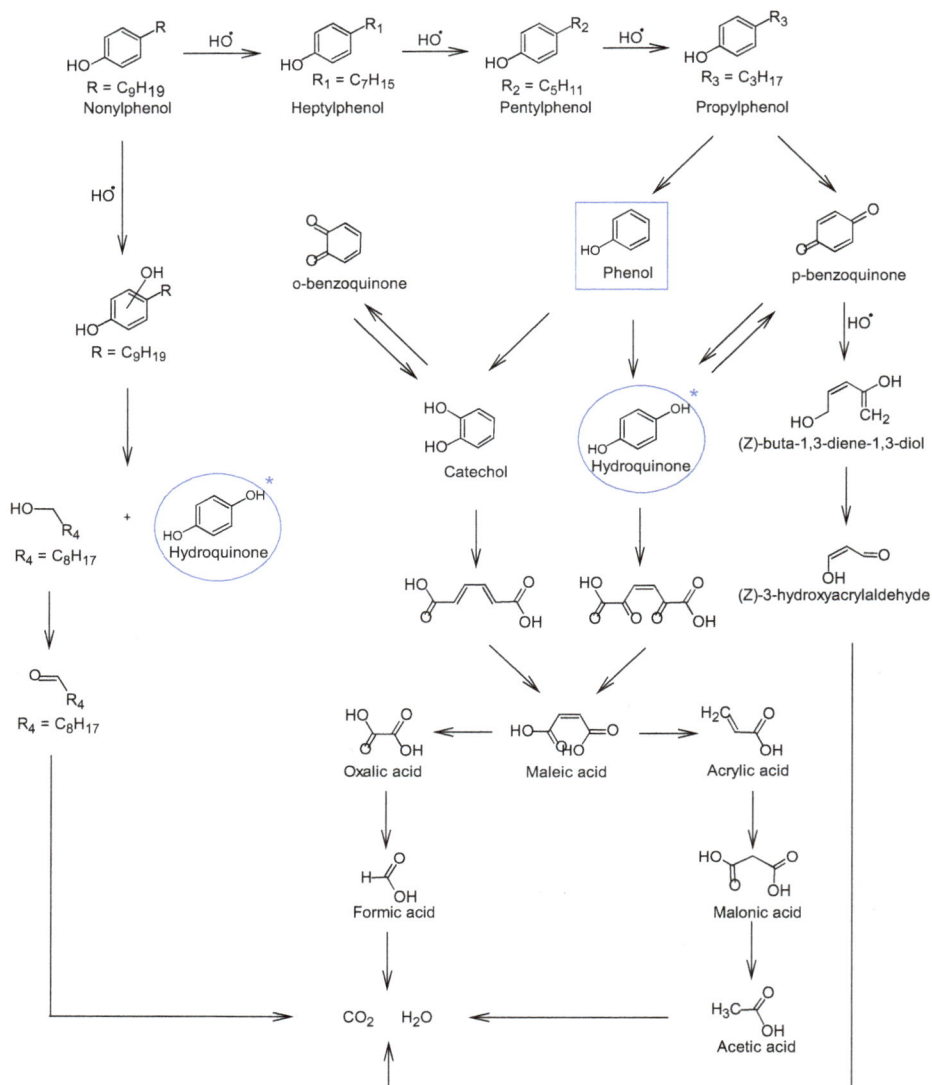

Figure 8.10: Proposed photocatalytic degradation pathway of 4-nonylphenol [256, 258–260, 262, 263]. * Same intermediate photoproduct: hydroquinone. Blue frames indicate the principal intermediates.

8.4 Comparative Assessment of Pilot Plants, Full Treatment, and Hybrid Processes

Photocatalysis have widely been applied at laboratory scale for the treatment of numerous pesticides, pharmaceuticals, and EDCs; but its application at full scale is still very scarce [264].The removal of these contaminants has widely been reported previous to a biological process in order to increase biodegradability and reduce the toxicity of wastewater; or after the biological treatment intending to remove residual bio-recalcitrant contaminants to fit the quality to discharge or to implement further use. Furthermore, wastewater may have been filtered through membranes; thus, rejected water may be treated to remove these types of contaminants [265].

The application of this type of treatment at full scale may be limited by: (a) a low photoconversion efficiency; (b) the presence of substances in the water matrix that scavenge reactive oxygen species (ROS); (c) the use of supported catalysts; (d) the separation processes designed for suspended photocatalysis; (e) the cost of the catalyst, including materials and preparation; and (f) its low efficiency in comparison to other AOPs [264]. Nevertheless, there are certain advantages that impulse to continue developing this treatment technology, such as the low generation of residual matter, the possibility to reuse the catalyst or use supported ones, the low production of toxic compounds, and the low energy use, overall when solar radiation is used. For example, TiO_2 can also be very effective under sunlight, even more effective than sunlight + H_2O_2 processes for disinfection [264, 266].

In fact, one of the main possibilities to develop this technology at industrial-full scale might be focused on its use as tertiary treatment, using solar or LED radiation, because of the necessity to fulfill more stringent worldwide regulations regarding priority and emerging pollutants, including pesticides, pharmaceuticals, and EDCs [264, 265]. Furthermore, and because of the usual low presence of emerging pollutants in municipal WWTPs [267], photocatalysis may be used for the treatment of these compounds and disinfection at the same time, potentially enabling the reuse of this water for irrigation [265].

In this context, different pilot plants, mainly using compound parabolic concentrators (CPC), have been assessed for the removal of pharmaceuticals, pesticides, and EDCs, mostly for effluents of low flow and low concentration of contaminants, as well as for tertiary treatment aiming to remove specific pollutants. For example, researchers of *Plataforma Solar de Almeria* (Almería, Spain) have developed pilot plants using CPCs that have addressed the feasibility of the photocatalytic treatment of wastewater containing pesticides, pharmaceuticals, and EDCs [207], concluding that this technology could be feasible for the treatment of wastewater containing medium-low bio-recalcitrant contaminants, and that the investment linearly depends on the collector surface area.

Alternatively, Almomani et al. [268] tested the removal of four groups of pharmaceuticals (antibiotics, estrogens, acidic, and neutral) in a solar pilot plant in combination with ozone. They found that these pharmaceuticals could be removed and water toxicity could be reduced by TiO_2-photocatalysis in 98% after 60 min of treatment, and that the degradation kinetics and energy use could be improved combining the photocatalytic treatment with ozone. The total removal of the targeted pharmaceuticals and the 78–97% of mineralization were achieved using 100 mg/L of TiO_2 and an initial ozone dose of 0.6 mg/L.

In another trial, Fenoll et al. [91] tested a CPC solar pilot plant combining TiO_2 (300 mg/L) in tandem with $Na_2S_2O_8$ (300 mg/L) for the treatment of the wastewater of three farms containing 4 neo-nicotinoid insecticides (thiamethoxam, imidacloprid, acetamiprid and thiacloprid). The results addressed that the remaining concentration of pesticides were 0–47% for thiamethoxam, 0.1–33% for imidacloprid, 2–66% for acetamiprid, and 0.02–60% for thiacloprid. The differences in the removal of each one were attributed to the more or less abundant presence of other organic and inorganic compounds in this real wastewater. The final economic assessment reported that this could be a cost-effective treatment, and that energy consumption is relatively low (41 kWh/m^3 at 7.7 €/m^3 of energy cost) in comparison with the total treatment cost.

Talwar et al. [269] reported the treatment of the antibiotic metronidazole (MTZ; 2-methyl-5-nitroimidazole-1-ethanol) with a P25-TiO_2 (Degussa, BASF, Germany) layer supported on the beads of a composite made of sand and Fuller's earth in a CPC reactor. They obtained a 60% degradation of MTZ after 15 min of treatment, and its total degradation into numerous small compounds after 6 h, along with a 45–50% COD reduction, reporting a total treatment cost of ≈ 10 \$/m^3.

Philippe et al. [270] assessed the removal of nine pharmaceuticals, one steroid hormone, and three pesticides in a simulated secondary WWTP effluent in a CPC pilot plant, addressing that, when the concentration of pollutants were similar to the concentration in real wastewater ($C_0 = 1$ µg/L), it was possible to achieve more than an 80% removal of contaminants from these complex water matrices, and the inactivation of 5 logs of bacteria, using 3 kJ/L of total energy.

Cabrera-Reina et al. [271] assessed the removal of the antibiotics imipenem (1.575 µmol/L) and meropenem (0.114 µmol/L) in pilot plants composed of CPCs at the Solar Energy Research Centre (CIESOL, Almería) using P25-TiO_2 from Degussa (BASF, Germany) as the catalyst. They obtained very good removal efficiencies for both antibiotics in deionized and natural water, although a higher TiO_2 concentration (50 mg/L versus 20 mg/L) was required to address the same results in natural water than in deionized water ($\approx$80% removal after 30 min of treatment). This is attributable to the presence of other organic and inorganic substances in the water matrix; but, at the same time, it indicates that photocatalysis can be applied to real water matrices without significant photocatalytic activity loss.

Finally, Bansal et al. [272] assessed the use of Fe-TiO$_2$ beads for the treatment of real pharmaceutical effluents by solar radiation, reporting an about 80% removal of the COD in less than 120 min of treatment, and addressing a cost of \$71.56 m^{-3} after 70 cycles of catalyst reuse.

In short, and besides the above-described experiences, there is still an important work to do to scale up the technology of photocatalysis in an economically feasible way. Optimizing operating conditions, improving long-term reliability, and increasing the "quantum yield" are key issues for its success [207]. In addition, there is an important field of research to develop regarding the application of photocatalysis in combination with other processes and as tertiary treatment applied to the removal of specific highly toxic pollutants (priority and emerging), such as pharmaceuticals, pesticides, and EDCs using the sunlight, high efficiency LED technology sources of radiation, and other renewable energies, aiming to reduce the energy cost of the photocatalytic treatment.

8.5 Conclusion and Outlook

Pesticides, PPCPs, and EDCs are contaminating chemical compounds that are very frequently used and, thus, they are consequently commonly found present in water bodies and the environment. They are generally difficult to degrade by conventional water technologies like the separation and biological processes that are usually implemented in urban wastewater treatment plants, although the need for their efficient removal is nowadays growing in concern because of the serious environmental and health problems they are causing. Photocatalysis, thanks to its high capability to produce hydroxyl radical and other different reactive oxygen species, is a promising alternative of treatment for these contaminants.

As it has been reported in the literature reviewed in this chapter, photocatalysis has widely been applied to the removal of pesticides, PPCPs, and EDCs, and has addressed high treatment efficiency results, mainly for low concentration values of these contaminants, that is, an overall degradation > 80% and mineralization > 50%. TiO$_2$ and its different modifications, such as doping with different element aiming to increase its photo-efficiency, have been the main used photocatalyst in the reported trials. Although a great research effort has already been developed aiming to increase quantum yield and efficiency under visible radiation, there is still room for improvement and innovative development of this technology. The major part of the treatment systems herein described has been developed at laboratory scale in batch mode using artificial light and suspended catalysts in concentrations of 10–1,000 mg/L at mild reaction conditions: pH values close to neutral (5–8) and room pressure and temperature.

Because of the characteristics, and, in many cases, high complexity of these compounds, their photocatalytic degradation routes are quite complex, and reaction by-products could be, in some cases, even more toxic than their parent chemical compound. The main intermediates that are generated are products of hydroxylation, dehalogenation, dealkilation, decarboxylation, isomerization, and cyclization reactions. Aromatic compounds are oxidized through ring opening reactions producing aliphatic acids. Complete mineralization could be achieved in some cases, although some structures, such as the triazine ring, are very difficult to oxidize. Further investigation must be developed to better understand the degradation routes of these toxic compounds and identify the by-products that limit the reaction kinetics and treatment efficiency.

Even though photocatalysis has mainly been applied to the treatment of pesticides, PPCPs, and EDCs at laboratory scale, their application at full scale is still scarce and there is an important research and technical work to develop aiming to spread the application of this technology within actual treatment strategies. Its full-scale application is mainly limited by the following issues: (a) the addressed low photoconversion efficiency; (b) the reduction of treatment efficiency attributed to the presence of scavengers of reactive oxygen species (ROS) in complex water matrices; (c) the cost of synthesizing catalysts; (d) the cost of separating the catalyst after treatment; and (e) the low general efficiency that is achieved in comparison to other AOPs in the treatment of wastewater highly loaded with pollutants.

The need to fulfill an everyday more stringent legislation regarding priority and emerging pollutants (which include pesticides, PPCPs, and EDCs) applying sustainable green technologies, together with the several advantages that are provided by the photocatalysis technology (such as the low generation of treatment residues, the possibility of reusing or supporting the catalyst, the low generation of toxic compounds, and the chance to consume very low energy in treatment designs configuring the use of LED technology or directly the sun as radiation sources), open an important field of research and development in the treatment of these toxic compounds focusing on optimizing quantum yield and operating conditions, as well as reducing catalyst synthesis and energy costs, and improving the long-term reliability of this treatment technology.

References

[1]   Salimi M, Esrafili A, Gholami M, Jonidi Jafari A, Rezaei Kalantary R, Farzadkia M, et al. Contaminants of emerging concern: a review of new approach in AOP technologies, Environmental Monitoring and Assessment, 2017, 189(8), 414, 22 pp.

[2]   Diamanti-Kandarakis E, Bourguignon J-P, Giudice LC, Hauser R, Prins GS, Soto AM, et al. Endocrine-disrupting chemicals: an Endocrine Society scientific statement, Endocrine Reviews, 2009, 30, 293–342.

[3]   Murray KE, Thomas SM, Bodour AA. Prioritizing research for trace pollutants and emerging contaminants in the freshwater environment, Environmental Pollution (Barking, Essex: 1987), 2010, 158, 3462–3471.

[4]   Daughton CG, Ternes TA. Pharmaceuticals and personal care products in the environment: agents of subtle change?, Environmental Health Perspectives, 1999, 107, 907–938.

[5]   Gore AC, Crews D, Doan LL, Merrill M. Introduction to Endocrine Disrupting Chemicals (EDCs). A Guide for Public Interest Organizations and Policy-Makers, Endocrine Society, IPEN, 2014.

[6]   Maletz S, Floehr T, Beier S, Klümper C, Brouwer A, Behnisch P, et al. In vitro characterization of the effectiveness of enhanced sewage treatment processes to eliminate endocrine activity of hospital effluents, Water Research, 2013, 47, 1545–1557.

[7]   Jones OAH, Voulvoulis N, Lester JN. Human pharmaceuticals in the aquatic environment a review, Environmental Technology (United Kingdom), 2001, 22, 1383–1394.

[8]   Lee CM, Palaniandy P, Dahlan I. Pharmaceutical residues in aquatic environment and water remediation by TiO_2 heterogeneous photocatalysis: a review, Environmental Earth Sciences, 2017, 76, 611, 19 pp.

[9]   Tijani JO, Fatoba OO, Babajide OO, Petrik LF. Pharmaceuticals, endocrine disruptors, personal care products, nanomaterials and perfluorinated pollutants: a review, Environmental Chemistry Letters, 2016, 14, 27–49.

[10]   Yang Y, Yong SO, Ki HK, Eilhann EK, Yiu FT. Occurrences and removal of pharmaceuticals and personal care products (PPCPs) in drinking water and water/sewage treatment plants: a review, Science of the Total Environment, 2017, 596–597, 303–320.

[11]   Wang J, Wang S. Removal of pharmaceuticals and personal care products (PPCPs) from wastewater: a review, Journal of Environmental Management, 2016, 182, 620–640.

[12]   Ribeiro AR, Nunes OC, Pereira MFR, Silva AMT. An overview on the advanced oxidation processes applied for the treatment of water pollutants defined in the recently launched Directive 2013/39/EU, Environment International, 2015, 75, 33–51.

[13]   Silva L, Lino C, Pena A Selected Pharmaceuticals in Different Aquatic Compartments: part I – Source, Fate and Occurrence, Molecules, 2020, 25(5), 1026, 33 pp.

[14]   EU Decision 2018/840. Commission Implementing Decision (EU) 2018/840 of 5 June 2018, Official Journal of the European Union, 2018, 141, 9–12.

[15]   Tijani JO, Fatoba OO, Petrik LF. A review of pharmaceuticals and endocrine-disrupting compounds: sources, effects, removal, and detections, Water, Air, and Soil Pollution, 2013, 224,1770, 43 pp.

[16]   Mompelat S, Le Bot B, Thomas O. Occurrence and fate of pharmaceutical products and by-products, from resource to drinking water, Environment International, 2009, 35, 803–814.

[17]   Richardson SD. Water analysis: emerging contaminants and current issues, Analytical Chemistry, 2009, 81, 4645–4677.

[18]   Houtman CJ. Emerging contaminants in surface waters and their relevance for the production of drinking water in Europe, Journal of Integrative Environmental Sciences, 2010, 7, 271–295.

[19]   Quadra GR, Oliveira De Souza H, Costa R, Dos S, Fernandez MA, Dos S. Do pharmaceuticals reach and affect the aquatic ecosystems in Brazil? A critical review of current studies in a developing country, Environmental Science and Pollution Research, 2017, 24, 1200–1218.

[20]   Verlicchi P, Al Aukidy M, Zambello E. Occurrence of pharmaceutical compounds in urban wastewater: removal, mass load and environmental risk after a secondary treatment-A review, Science of the Total Environment, 2012, 429, 123–155.

[21]   Onesios KM, Yu JT, Bouwer EJ. Biodegradation and removal of pharmaceuticals and personal care products in treatment systems: a review, Biodegradation, 2009, 20, 441–466.

[22]   A. Polyzos S, Kountouras J, Deretzi G, Zavos C, S. Mantzoros C. The emerging role of endocrine disruptors in pathogenesis of insulin resistance: a concept implicating nonalcoholic fatty liver disease, Current Molecular Medicine, 2011, 12, 68–82.

[23]   Kabir ER, Rahman MS, Rahman I. A review on endocrine disruptors and their possible impacts on human health, Environmental Toxicology and Pharmacology, 2015, 40, 241–258.

[24]   Gore AC, Crews D, Doan LL, Merrill ML. Introduction to endocrine disrupting chemicals (EDCs). A guide for public interest organizations and policy makers, 2014, 76 pp.

[25] Soon SK, Seung JK, Rhee DL, Kwon JL, Gyu SR, Ji HS, et al. Assessment of estrogenic and androgenic activities of tetramethrin in vitro and in vivo assays, Journal of Toxicology & Environmental Health Part A:, 2005, 68, 2277–2289.

[26] McKinlay R, Plant JA, Bell JNB, Voulvoulis N. Endocrine disrupting pesticides: implications for risk assessment, Environment International, 2008, 34, 168–183.

[27] Fingerhut M, Driscoll T, Nelson DI, Concha-Barrientos M, Punnett L, Pruss-Ustin A, et al. Contribution of occupational risk factors to the global burden of disease – A summary of findings, Scandinavian Journal of Work, Environment & Health, 2005, 58–61.

[28] WHO. State of the Science of Endocrine Chemicals-2012, World Health Organization, 2012.

[29] Leong KH, Benjamin Tan LL, Mustafa AM. Contamination levels of selected organochlorine and organophosphate pesticides in the Selangor River, Malaysia between 2002 and 2003, Chemosphere, 2007, 66, 1153–1159.

[30] Kanakaraju D, Glass BD, Oelgemöller M. Advanced oxidation process-mediated removal of pharmaceuticals from water: a review, Journal of Environmental Management, 2018, 219, 189–207.

[31] Borges ME, García DM, Hernández T, Ruiz-Morales JC, Esparza P. Supported photocatalyst for removal of emerging contaminants from wastewater in a continuous packed-bed photoreactor configuration, Catalysts, 2015, 5, 77–87.

[32] Cai Q, Hu J. Decomposition of sulfamethoxazole and trimethoprim by continuous UVA/LED/ TiO$_2$ photocatalysis: decomposition pathways, residual antibacterial activity and toxicity, Journal of Hazardous Materials, 2017, 323, 527–536.

[33] Carbajo J, Jiménez M, Miralles S, Malato S, Faraldos M, Bahamonde A. Study of application of titania catalysts on solar photocatalysis: influence of type of pollutants and water matrices, Chemical Engineering Journal, 2016, 291, 64–73.

[34] Canle M, Fernández Pérez MI, Santaballa JA. Photocatalyzed degradation/abatement of endocrine disruptors, Current Opinion in Green and Sustainable Chemistry, 2017, 6, 101–138.

[35] Konstantinou IK, Albanis TA. Photocatalytic transformation of pesticides in aqueous titanium dioxide suspensions using artificial and solar light: intermediates and degradation pathways, Applied Catalysis B: Environmental, 2003, 42, 319–335.

[36] Mahalakshmi M, Arabindoo B, Palanichamy M, Murugesan V. Photocatalytic degradation of carbofuran using semiconductor oxides, Journal of Hazardous Materials, 2007, 143, 240–245.

[37] Carbajo J, García-Muñoz P, Tolosana-Moranchel A, Faraldos M, Bahamonde A. Effect of water composition on the photocatalytic removal of pesticides with different TiO$_2$ catalysts, Environmental Science and Pollution Research, 2014, 12233–12240.

[38] Luo Z, Li L, Wei C, Li H, Chen D. Role of active oxidative species on TiO$_2$ photocatalysis of tetracycline and optimization of photocatalytic degradation conditions, Journal Of Environmental Biology / Academy of Environmental Biology, India, 2015, 36, 837–843.

[39] Chatzitakis A, Berberidou C, Paspaltsis I, Kyriakou G, Sklaviadis T, Poulios I. Photocatalytic degradation and drug activity reduction of Chloramphenicol, Water Research, 2008, 42, 386–394.

[40] Doll TE, Frimmel FH. Kinetic study of photocatalytic degradation of carbamazepine, clofibric acid, iomeprol and iopromide assisted by different TiO$_2$ materials – Determination of intermediates and reaction pathways, Water Research, 2004, 38, 955–964.

[41] Vaizoğullar AI. TiO$_2$/ZnO supported on sepiolite: preparation, structural characterization, and photocatalytic degradation of flumequine antibiotic in aqueous solution, Chemical Engineering Communications, 2017, 204, 689–697.

[42]	Tran TTH, Bui TTH, Nguyen TL, Man HN, Tran TKC. Phase-pure brookite TiO_2 as a highly active photocatalyst for the degradation of pharmaceutical pollutants, Journal of Electronic Materials, 2019, 48, 7846–7861.

[43]	Sabouni R, Gomaa H. Photocatalytic degradation of pharmaceutical micro-pollutants using ZnO, Environmental Science and Pollution Research, 2019, 26, 5372–5380.

[44]	Serna-Galvis EA, Silva-Agredo J, Giraldo-Aguirre AL, Flórez-Acosta OA, Torres-Palma RA. High frequency ultrasound as a selective advanced oxidation process to remove penicillinic antibiotics and eliminate its antimicrobial activity from water, Ultrasonics Sonochemistry, 2016, 31, 276–283.

[45]	Lutterbeck CA, Machado ÊL, Kümmerer K. Photodegradation of the antineoplastic cyclophosphamide: a comparative study of the efficiencies of UV/H_2O_2, $UV/Fe^{2+}/H_2O_2$ and UV/TiO_2 processes, Chemosphere, 2015, 120, 538–546.

[46]	Rioja N, Benguria P, Peñas FJ, Zorita S. Competitive removal of pharmaceuticals from environmental waters by adsorption and photocatalytic degradation, Environmental Science and Pollution Research, 2014, 21, 11168–11177.

[47]	Tolosana-Moranchel Á, Manassero A, Satuf ML, Alfano OM, Casas JA. Bahamonde A. TiO_2-rGO photocatalytic degradation of an emerging pollutant: kinetic modelling and determination of intrinsic kinetic parameters, Journal of Environmental Chemical Engineering, 2019, 7, 103406.

[48]	Ni T, Li Q, Yan Y, Wang F, Cui X, Yang Z, et al. N,Fe-doped carbon dot decorated gear-shaped WO_3 for highly efficient UV-Vis-NIR-driven photocatalytic performance, Catalysts, 2020, 416, 1–14.

[49]	Zhang X, Wang J, Dong XX, Lv YK. Functionalized metal-organic frameworks for photocatalytic degradation of organic pollutants in environment, Chemosphere, 2020, 242, 125144.

[50]	Wu Z, Liang Y, Yuan X, Zou D, Fang J, Jiang L, et al. MXene Ti3C2 derived Z–scheme photocatalyst of graphene layers anchored TiO_2/g–C_3N_4 for visible light photocatalytic degradation of refractory organic pollutants, Chemical Engineering Journal, 2020, 394, 124921.

[51]	Tahir MB, Sagir M, Shahzad K. Removal of acetylsalicylate and methyl-theobromine from aqueous environment using nano-photocatalyst WO_3-TiO_2 @g-C3N4 composite, Journal of Hazardous Materials, 2019, 363, 205–213.

[52]	Miranda-García N, Suárez S, Sánchez B, Coronado JM, Malato S, Maldonado MI. Photocatalytic degradation of emerging contaminants in municipal wastewater treatment plant effluents using immobilized TiO_2 in a solar pilot plant, Applied Catalysis B: Environmental, 2011, 103, 294–301.

[53]	Khataee AR, Fathinia M, Joo SW. Simultaneous monitoring of photocatalysis of three pharmaceuticals by immobilized TiO_2 nanoparticles: chemometric assessment, intermediates identification and ecotoxicological evaluation, Spectrochimica Acta Part A: Molecular and Biomolecular Spectroscopy, 2013, 112, 33–45.

[54]	Laoufi NA, Hout S, Tassalit D, Ounnar A, Djouadi A, Chekir N, et al. Removal of a persistent pharmaceutical micropollutant by UV/TiO_2 process using an immobilized titanium dioxide catalyst: parametric study, Chemical Engineering Transactions, 2013, 32, 1951–1956.

[55]	Shankaraiah G, Poodari S, Bhagawan D, Himabindu V, Vidyavathi S. Degradation of antibiotic norfloxacin in aqueous solution using advanced oxidation processes (AOPs) – A comparative study, Desalination and Water Treatment, 2016, 57, 27804–27815.

[56]	Expósito AJ, Patterson DA, Mansor WSW, Monteagudo JM, Emanuelsson E, Sanmartín I, et al. Antipyrine removal by TiO_2 photocatalysis based on spinning disc reactor technology, Journal of Environmental Management, 2017, 187, 504–512.

[57] Manassero A, Satuf ML, Alfano OM. Photocatalytic reactors with suspended and immobilized TiO$_2$: comparative efficiency evaluation, Chemical Engineering Journal, 2017, 326, 29–36.

[58] Rimoldi L, Meroni D, Falletta E, Pifferi V, Falciola L, Cappelletti G, et al. Emerging pollutant mixture mineralization by TiO$_2$ photocatalysts, The Role of the Water Medium. Photochemistry & Photobiological Sciences, 2017, 16, 60–66.

[59] Rimoldi L, Meroni D, Cappelletti G, Ardizzone S. Green and low cost tetracycline degradation processes by nanometric and immobilized TiO$_2$ systems, Catalysis Today, 2017, 281, 38–44.

[60] Jiménez-Tototzintle M, Ferreira IJ, Da Silva Duque S, Guimarães Barrocas PR, Saggioro EM. Removal of contaminants of emerging concern (CECs) and antibiotic resistant bacteria in urban wastewater using UVA/TiO$_2$/H$_2$O$_2$ photocatalysis, Chemosphere, 2018, 210, 449–457.

[61] Cerrato G, Bianchi CL, Galli F, Pirola C, Morandi S, Capucci V. Micro-TiO$_2$ coated glass surfaces safely abate drugs in surface water, Journal of Hazardous Materials, 2019, 363, 328–334.

[62] Briones AA, Guevara IC, Mena D, Espinoza I, Sandoval-Pauker C, Guerrero LR, et al. Degradation of meropenem by heterogeneous photocatalysis using TiO$_2$ /fiberglass substrates, Catalysts, 2020, 10, 1–4.

[63] Martín-Sómer M, Pablos C, De Diego A, Van Grieken R, Encinas Á, Monsalvo VM, et al. Novel macroporous 3D photocatalytic foams for simultaneous wastewater disinfection and removal of contaminants of emerging concern, Chemical Engineering Journal, 2019, 366, 449–459.

[64] Lin L, Wang H, Jiang W, Mkaouar AR, Xu P. Comparison study on photocatalytic oxidation of pharmaceuticals by TiO$_2$-Fe and TiO$_2$-reduced graphene oxide nanocomposites immobilized on optical fibers, Journal of Hazardous Materials, 2017, 333, 162–168.

[65] Xing X, Du Z, Zhuang J, Wang D. Removal of ciprofloxacin from water by nitrogen doped TiO$_2$ immobilized on glass spheres: rapid screening of degradation products, Journal of Photochemistry and Photobiology. A, Chemistry, 2018, 359, 23–32.

[66] Moreira NFF, Sampaio MJ, Ribeiro AR, Silva CG, Faria JL, Silva AMT. Metal-free g-C$_3$N$_4$ photocatalysis of organic micropollutants in urban wastewater under visible light, Applied Catalysis B: Environmental, 2019, 248, 184–192.

[67] Li Puma G, Puddu V, Tsang HK, Gora A, Toepfer B. Photocatalytic oxidation of multicomponent mixtures of estrogens (estrone (E1), 17β-estradiol (E2), 17α-ethynylestradiol (EE2) and estriol (E3)) under UVA and UVC radiation: photon absorption, quantum yields and rate constants independent of photon absorp, Applied Catalysis B: Environmental, 2010, 99, 388–397.

[68] Davididou K, Nelson R, Monteagudo JM, Durán A, Expósito AJ, Chatzisymeon E. Photocatalytic degradation of bisphenol-A under UV-LED, blacklight and solar irradiation, Journal of Cleaner Production, 2018, 203, 13–21.

[69] Zúñiga-Benítez H, Aristizábal-Ciro C, Peñuela GA. Heterogeneous photocatalytic degradation of the endocrine-disrupting chemical Benzophenone-3: parameters optimization and by-products identification, Journal of Environmental Management, 2016, 167, 246–258.

[70] Narváez JF, Grant H, Gil VC, Porras J, Bueno Sanchez JC, Ocampo Duque LF, et al. Assessment of endocrine disruptor effects of levonorgestrel and its photoproducts: environmental implications of released fractions after their photocatalytic removal, Journal of Hazardous Materials, 2019, 371, 273–279.

[71] Kaplan R, Erjavec B, Pintar A. Enhanced photocatalytic activity of single-phase, nanocomposite and physically mixed TiO$_2$ polymorphs, Applied Catalysis. A, General, 2015, 489, 51–60.

[72] Samsudin EM, Hamid SBA, Juan JC, Basirun WJ, Centi G. Enhancement of the intrinsic photocatalytic activity of TiO$_2$ in the degradation of 1,3,5-triazine herbicides by doping with N,F, Chemical Engineering Journal, 2015, 280, 330–343.

[73] Gomes JF, Leal I, Bednarczyk K, Gmurek M, Stelmachowski M, Zaleska-Medynska A, et al. Detoxification of parabens using UV-A enhanced by noble metals – TiO$_2$ supported catalysts, Journal of Environmental Chemical Engineering, 2017, 5, 3065–3074.

[74] Meenakshi G, Sivasamy A. Nanorod ZnO/SiC nanocomposite: an efficient catalyst for the degradation of an endocrine disruptor under UV and visible light irradiations, Journal of Environmental Chemical Engineering, 2018, 6(3), 3757–3769.

[75] Petala A, Bontemps R, Spartatouille A, Frontistis Z, Antonopoulou M, Konstantinou I, et al. Solar light-induced degradation of ethyl paraben with CuOx/BiVO$_4$: statistical evaluation of operating factors and transformation by-products, Catalysis Today, 2017, 280, 122–131.

[76] Arthur RB, Ahern JC, Patterson HH. Application of biox photocatalysts in remediation of persistent organic pollutants, Catalysts, 2018, 8(12), 604, 25 pp.

[77] De Liz MV, De Lima RM, Do Amaral B, Marinho BA, Schneider JT, Nagata N, et al. Suspended and immobilized TiO$_2$ photocatalytic degradation of estrogens: potential for application in wastewater treatment processes, Journal of the Brazilian Chemical Society, 2018, 29, 380–389.

[78] Zatloukalová K, Obalová L, Kočí K, Čapek L, Matěj Z, Šnajdhaufová H, et al. Photocatalytic degradation of endocrine disruptor compounds in water over immobilized TiO$_2$ photocatalysts, Iranian Journal of Chemistry & Chemical Engineering, 2017, 36(2), 29–38.

[79] Sacco O, Vaiano V, Rizzo L, Sannino D. Intensification of ceftriaxone degradation under UV and solar light irradiation in presence of phosphors based structured catalyst, Chemical Engineering and Processing: Process Intensification, 2019, 137, 12–21.

[80] Sornalingam K, McDonagh A, Canning J, Cook K, Johir MAH, Zhou JL, et al Photocatalysis of 17α-ethynylestradiol and estriol in water using engineered immersible optical fibres and light emitting diodes, Journal of Water Process Engineering, 2020, 33, 101075.

[81] Chen JQ, Wang D, Zhu MX, Gao CJ. Photocatalytic degradation of dimethoate using nanosized TiO$_2$ powder, Desalination, 2007, 207, 87–94.

[82] Tolosana-Moranchel A, Carbajo J, Faraldos M, Bahamonde A. Solar-assisted photodegradation of isoproturon over easily recoverable titania catalysts, Environmental Science and Pollution Research, 2017, 24, 7821–7828.

[83] Farkas J, Náfrádi M, Hlogyik T, Cora Pravda B, Schrantz K, Hernádi K, et al. Comparison of advanced oxidation processes in the decomposition of diuron and monuron-efficiency, intermediates, electrical energy per order and the effect of various matrices, Environmental Science: Water Research & Technology, 2018, 4, 1345–1360.

[84] Hincapié Pérez M, Vega LP, Zúñiga-Benítez H, Peñuela GA. Comparative Degradation of Alachlor Using Photocatalysis and Photo-Fenton, Water, Air, and Soil Pollution, 2018, 229, 1–12.

[85] Senthilnathan J, Philip L. Photocatalytic degradation of lindane under UV and visible light using N-doped TiO$_2$, Chemical Engineering Journal, 2010, 161, 83–92.

[86] Tang J, Huang X, Huang X, Xiang L, Wang Q. Photocatalytic degradation of imidacloprid in aqueous suspension of TiO$_2$ supported on H-ZSM-5, Environmental Earth Sciences, 2012, 66, 441–445.

[87] Yari K, Seidmohammadi A, Khazaei M, Bhatnagar A, Leili M. A comparative study for the removal of imidacloprid insecticide from water by chemical-less UVC, UVC/TiO$_2$ and UVC/ZnO processes, Journal of Environmental Health Science and Engineering, 2019, 17, 337–351.

[88] Fenoll J, Garrido I, Hellín P, Flores P, Navarro S. Photodegradation of neonicotinoid insecticides in water by semiconductor oxides, Environmental Science and Pollution Research, 2015, 22, 15055–15066.

[89]   Moussavi G, Hossaini H, Jafari SJ, Farokhi M. Comparing the efficacy of UVC, UVC/ZnO and VUV processes for oxidation of organophosphate pesticides in water, Journal of Photochemistry and Photobiology. A, Chemistry, 2014, 290, 86–93.

[90]   Cruz M, Gomez C, Duran-Valle CJ, Pastrana-Martínez LM, Faria JL, Silva AMT, et al. Bare TiO_2 and graphene oxide TiO_2 photocatalysts on the degradation of selected pesticides and influence of the water matrix, Applied Surface Science, 2017, 416, 1013–1021.

[91]   Vela N, Calín M, Yáñez-Gascón MJ, El Aatik A, Garrido I, Pérez-Lucas G, et al Removal of pesticides with endocrine disruptor activity in wastewater effluent by solar heterogeneous photocatalysis using $ZnO/Na_2S_2O_8$, Water, Air, and Soil Pollution, 2019, 230, 1–11.

[92]   Mbiri A, Wittstock G, Taffa DH, Gatebe E, Baya J, Wark M. Photocatalytic degradation of the herbicide chloridazon on mesoporous titania/zirconia nanopowders, Environmental Science and Pollution Research, 2018, 25, 34873–34883.

[93]   Zhang L, Yue X, Liu J, Feng J, Zhang X, Zhang C, et al. Facile synthesis of Bi5O7Br/BiOBr 2D/3D heterojunction as efficient visible-light-driven photocatalyst for pharmaceutical organic degradation, Separation and Purification Technology, 2020, 231, 115917.

[94]   Gupta VK, Fakhri A, Tahami S, Agarwal S. Zn doped CdO nanoparticles: structural, morphological, optical, photocatalytic and anti-bacterial properties, Journal of Colloid and Interface Science, 2017, 504, 164–170.

[95]   Khavar AHC, Moussavi G, Mahjoub AR, Satari M, Abdolmaleki P. Synthesis and visible-light photocatalytic activity of In,S-TiO_2@rGO nanocomposite for degradation and detoxification of pesticide atrazine in water, Chemical Engineering Journal, 2018, 345, 300–311.

[96]   Huang S, Chen C, Tsai H, Shaya J, Lu C. Photocatalytic degradation of thiobencarb by a visible light-driven MoS2 photocatalyst, Separation and Purification Technology, 2018, 197, 147–155.

[97]   Le Cunff J, Tomašić V, Wittine O. Photocatalytic degradation of the herbicide terbuthylazine: preparation, characterization and photoactivity of the immobilized thin layer of TiO2/chitosan, Journal of Photochemistry and Photobiology. A, Chemistry, 2015, 309, 22–29.

[98]   Cunff JL, Tomašić V, Gomzi Z. Photocatalytic degradation of terbuthylazine: modelling of a batch recirculating device, Journal of Photochemistry and Photobiology. A, Chemistry, 2018, 353, 159–170.

[99]   Elatmani K, Oujji NB, Plantara G, Goetz V, Ichou IA. 3D Photocatalytic media for decontamination of water from pesticides, Materials Research Bulletin, 2018, 101, 6–11.

[100] M'Bra IC, García-Muñoz P, Drogui P, Keller N, Trokourey A, Robert D. Heterogeneous photodegradation of Pyrimethanil and its commercial formulation with TiO_2 immobilized on SiC foams, Journal of Photochemistry and Photobiology. A, Chemistry, 2019, 368, 1–6.

[101] Solano Pizarro RA, Herrera Barros AP. Cypermethrin elimination using Fe-TiO_2 nanoparticles supported on coconut palm spathe in a solar flat plate photoreactor, Advanced Composites Letters, 2020, 29, 1–13.

[102] Sabri NA, Nawi MA, Abu Bakar NHH. Recyclable immobilized carbon coated nitrogen doped TiO_2 for photocatalytic degradation of quinclorac under UV–Vis and visible light, Journal of Environmental Chemical Engineering, 2018, 6, 898–905.

[103] Rani M, Shanker U. Degradation of traditional and new emerging pesticides in water by nanomaterials: recent trends and future recommendations, International Journal of Environmental Science and Technology, 2018, 15, 1347–1380.

[104] Tasca AL, Puccini M, Fletcher A. Terbuthylazine and desethylterbuthylazine: recent occurrence, mobility and removal techniques, Chemosphere, 2018, 202, 94–104.

[105] Wang C, Liu H, Qu Y. TiO_2-based photocatalytic process for purification of polluted water: bridging fundamentals to applications, Journal of Nanomaterials, 2013, 319637, 14 pp.

[106] Horikoshi S, Serpone N. Can the photocatalyst TiO_2 be incorporated into a wastewater treatment method? Background and prospects, Catalysis Today, 2020, 340, 334–346.

[107] Kanan S, Moyet MA, Arthur RB, Patterson HH. Recent advances on TiO 2 -based photocatalysts toward the degradation of pesticides and major organic pollutants from water bodies, Catalysis Reviews – Science and Engineering, 2019, 0, 1–65.

[108] Yap HC, Pang YL, Lim S, Abdullah AZ, Ong HC, Wu CH. A comprehensive review on state-of-the-art photo-, sono-, and sonophotocatalytic treatments to degrade emerging contaminants, International Journal of Environmental Science and Technology, 2019, 16, 601–628.

[109] Byrne C, Subramanian G, Pillai SC. Recent advances in photocatalysis for environmental applications, Journal of Environmental Chemical Engineering, 2018, 6, 3531–3555.

[110] Diebold U. The surface science of titanium dioxide, Surface Science Reports, 2002, 48, 53–229.

[111] Kumar SG, Rao KSRK. Polymorphic phase transition among the titania crystal structures using a solution-based approach: from precursor chemistry to nucleation process, Nanoscale, 2014, 6, 11574–11632.

[112] Hanaor DAH, Sorrell CC. Review of the anatase to rutile phase transformation, Journal of Materials Science, 2011, 46, 855–874.

[113] Pillai SC, Periyat P, George R, McCormack DE, Seery MK, Hayden H, et al. Synthesis of high-temperature stable anatase TiO2 photocatalyst, The Journal of Physical Chemistry C, 2007, 111, 1605–1611.

[114] Nolan NT, Seery MK, Hinder SJ, Healy LF, Pillai SC. A systematic study of the effect of silver on the chelation of formic acid to a titanium precursor and the resulting effect on the anatase to rutile transformation of TiO_2, The Journal of Physical Chemistry C, 2010, 114, 13026–13034.

[115] Pelaez M, Nolan NT, Pillai SC, Seery MK, Falaras P, Kontos AG, et al. A review on the visible light active titanium dioxide photocatalysts for environmental applications, Applied Catalysis B: Environmental, 2012, 125, 331–349.

[116] Bacsa RR, Kiwi J. Effect of rutile phase on the photocatalytic properties of nanocrystalline titania during the degradation of p-coumaric acid, Applied Catalysis B: Environmental, 1998, 16, 19–29.

[117] Zhang Q, Gao L, Guo J. Effects of calcination on the photocatalytic properties of nanosized TiO_2 powders prepared by $TiCl_4$ hydrolysis, Applied Catalysis B: Environmental, 2000, 26, 207–215.

[118] Fagan R, McCormack DE, Dionysiou DD, Pillai SC. A review of solar and visible light active TiO_2 photocatalysis for treating bacteria, cyanotoxins and contaminants of emerging concern, Materials Science in Semiconductor Processing, 2016, 42, 2–14.

[119] Czech B, Buda W. Photocatalytic treatment of pharmaceutical wastewater using new multiwall-carbon nanotubes/TiO_2/SiO_2 nanocomposites, Environmental Research, 2015, 137, 176–184.

[120] Peñas-Garzón M, Gómez-Avilés A, Belver C, Rodriguez JJ, Bedia J. Degradation pathways of emerging contaminants using TiO_2-activated carbon heterostructures in aqueous solution under simulated solar light, Chemical Engineering Journal, 2020, 392, 124867.

[121] Yu H, Chen F, Ye L, Zhou H, Zhao T. Enhanced photocatalytic degradation of norfloxacin under visible light by immobilized and modified In_2O_3/TiO_2 photocatalyst facilely synthesized by a novel polymeric precursor method, Journal of Materials Science, 2019, 54, 10191–10203.

[122] Diaz-Angulo J, Lara-Ramos J, Mueses M, Hernández-Ramírez A, Li Puma G, Machuca-Martínez F. Enhancement of the oxidative removal of diclofenac and of the TiO_2 rate of photon absorption in dye-sensitized solar pilot scale CPC photocatalytic reactors, Chemical Engineering Journal, 2020, 381, 122520.

[123] Chen J, Luo H, Shi H, Li G, An T. Anatase TiO_2 nanoparticles-carbon nanotubes composite: optimization synthesis and the relationship of photocatalytic degradation activity of acyclovir in water, Applied Catalysis. A, General, 2014, 485, 188–195.

[124] Xu F, Chen J, Kalytchuk S, Chu L, Shao Y, Kong D, et al. Supported gold clusters as effective and reusable photocatalysts for the abatement of endocrine-disrupting chemicals under visible light, Journal of Catalysis, 2017, 354, 1–12.

[125] Tan TL, Lai CW, Hong SL, Rashid SA. New insights into the photocatalytic endocrine disruptors dimethyl phthalate esters degradation by UV/MWCNTs-TiO_2 nanocomposites, Journal of Photochemistry and Photobiology. A, Chemistry, 2018, 364, 177–189.

[126] Pan Z, Stemmler EA, Cho HJ, Fan W, LeBlanc LA, Patterson HH, et al. Photocatalytic degradation of 17α-ethinylestradiol (EE2) in the presence of TiO_2-doped zeolite, Journal of Hazardous Materials, 2014, 279, 17–25.

[127] Wang R, Ma X, Liu T, Li Y, Song L, Tjong SC, et al. Degradation aspects of endocrine disrupting chemicals: a review on photocatalytic processes and photocatalysts, Applied Catalysis. A, General, 2020, 597, 117547.

[128] Barakat NAM, Nassar MM, Farrag TE, Mahmoud MS. Effective photodegradation of methomyl pesticide in concentrated solutions by novel enhancement of the photocatalytic activity of TiO_2 using $CdSO_4$ nanoparticles, Environmental Science and Pollution Research, 2014, 21, 1425–1435.

[129] Vaiano V, Sacco O, Sannino D, Ciambelli P. Photocatalytic removal of spiramycin from wastewater under visible light with N-doped TiO_2 photocatalysts, Chemical Engineering Journal, 2015, 261, 3–8.

[130] Yola ML, Eren T, Atar N. A novel efficient photocatalyst based on TiO_2 nanoparticles involved boron enrichment waste for photocatalytic degradation of atrazine, Chemical Engineering Journal, 2014, 250, 288–294.

[131] Cavalcante RP, Dantas RF, Bayarri B, González O, Giménez J, Esplugas S, et al. Synthesis and characterization of B-doped TiO_2 and their performance for the degradation of metoprolol, Catalysis Today, 2015, 252, 27–34.

[132] Isari AA, Hayati F, Kakavandi B, Rostami M, Motevassel M, Dehghanifard EN. Cu co-doped TiO_2@functionalized SWCNT photocatalyst coupled with ultrasound and visible-light: an effective sono-photocatalysis process for pharmaceutical wastewaters treatment, Chemical Engineering Journal, 2020, 392, 123685.

[133] Faraldos M, Bahamonde A. Environmental applications of titania-graphene photocatalysts, Catalysis Today, 2017, 285, 13–28.

[134] Luna-Sanguino G, Tolosana-Moranchel A, Duran-Valle C, Faraldos M, Bahamonde A. Optimizing P25-rGO compsites for pesticides degradation: elucidation of photo-mechanism, 2019, 172-7.

[135] Zhang S, Li B, Wang X, Zhao G, Hu B, Lu Z, et al. Recent developments of two-dimensional graphene-based composites in visible-light photocatalysis for eliminating persistent organic pollutants from wastewater, Chemical Engineering Journal, 2020, 390, 124642.

[136] Wang J, Chen H, Tang L, Zeng G, Liu Y, Yan M, et al. Antibiotic removal from water: a highly efficient silver phosphate-based Z-scheme photocatalytic system under natural solar light, Science of the Total Environment, 2018, 639, 1462–1470.

[137] Hernández-Ramírez A, Medina-Ramírez I. Photocatalytic semiconductors: synthesis, characterization, and environmental applications, Photocatalytic Semiconductors. Synthesis, Characterization, and Environmental Applications, 2015, 1–289.

[138] Di G, Zhu Z, Huang Q, Zhang H, Zhu J, Qiu Y, et al. Targeted modulation of g-C3N4 photocatalytic performance for pharmaceutical pollutants in water using ZnFe-LDH derived

mixed metal oxides: structure-activity and mechanism, Science of the Total Environment, 2019, 650, 1112–1121.

[139] Ji B, Zhang J, Zhang C, Li N, Zhao T, Chen F, et al. Vertically aligned ZnO@ZnS nanorod chip with improved photocatalytic activity for antibiotics degradation, ACS Applied Nano Materials, 2018, 1, 793–799.

[140] Al Abri R, Al Marzouqi F, Kuvarega AT, Meetani MA, Al Kindy SMZ, Karthikeyan S, et al. Nanostructured cerium-doped ZnO for photocatalytic degradation of pharmaceuticals in aqueous solution, Journal of Photochemistry and Photobiology. A, Chemistry, 2019, 384, 112065.

[141] Radhika S, Thomas J. Solar light driven photocatalytic degradation of organic pollutants using ZnO nanorods coupled with photosensitive molecules, Journal of Environmental Chemical Engineering, 2017, 5, 4239–4250.

[142] Tobajas M, Belver C, Rodriguez JJ. Degradation of emerging pollutants in water under solar irradiation using novel TiO_2-ZnO/clay nanoarchitectures, Chemical Engineering Journal, 2017, 309, 596–606.

[143] Akkari M, Aranda P, Belver C, Bedia J, Ben Haj Amara A, Ruiz-Hitzky E. Reprint of ZnO/sepiolite heterostructured materials for solar photocatalytic degradation of pharmaceuticals in wastewater, Applied Clay Science, 2018, 160, 3–8.

[144] Tang M, Ao Y, Wang P, Wang C. All-solid-state Z-scheme WO3 nanorod/ZnIn2S4 composite photocatalysts for the effective degradation of nitenpyram under visible light irradiation, Journal of Hazardous Materials, 2020, 387, 121713.

[145] Zhu W, Sun F, Goei R, Zhou Y. Construction of WO_3-g-C_3N_4 composites as efficient photocatalysts for pharmaceutical degradation under visible light, Catalysis Science & Technology, 2017, 7, 2591–2600.

[146] Soares Filho AF, Cruz Filho JF, Lima MS, Carvalho LM, Silva LKR, Costa JS, et al. Photodegradation of 17A-Ethynylstradiol (EE2) on nanostructured material of type WO3-SBA -15, Water, Air, and Soil Pollution, 2018, 229, 268, 16 pp.

[147] Chen W, He ZC, Huang GB, Wu CL, Chen WF, Liu XH. Direct Z-scheme 2D/2D $MnIn_2S_4$/g-C_3N_4 architectures with highly efficient photocatalytic activities towards treatment of pharmaceutical wastewater and hydrogen evolution, Chemical Engineering Journal, 2019, 359, 244–253.

[148] Vadivel S, Hariganesh S, Paul B, Mamba G, Puviarasu P. Highly active novel CeTi2O6/g-C3N5 photocatalyst with extended spectral response towards removal of endocrine disruptor 2,4-dichlorophenol in aqueous medium, Colloids and Surfaces A: Physicochemical and Engineering Aspects, 2020, 592, 124583.

[149] Liu W, Zhou J, Hu Z. Nano-sized g-C3N4 thin layer @ CeO_2 sphere core-shell photocatalyst combined with H_2O_2 to degrade doxycycline in water under visible light irradiation, Separation and Purification Technology, 2019, 227, 115665.

[150] Aanchal, Barman S, Basu S. Complete removal of endocrine disrupting compound and toxic dye by visible light active porous g-C_3N_4/H-ZSM-5 nanocomposite, Chemosphere, 2020, 241, 124981.

[151] Li Y, Fang Y, Cao Z, Li N, Chen D, Xu Q, et al. Construction of g-C3N4/PDI@MOF heterojunctions for the highly efficient visible light-driven degradation of pharmaceutical and phenolic micropollutants, Applied Catalysis B: Environmental, 2019, 250, 150–162.

[152] Hernández-Uresti DB, Vázquez A, Sanchez-Martinez D, Obregón S. Performance of the polymeric g-C_3N_4 photocatalyst through the degradation of pharmaceutical pollutants under UV-vis irradiation, Journal of Photochemistry and Photobiology. A, Chemistry, 2016, 324, 47–52.

[153] Smýkalová A, Sokolová B, Foniok K, Matejka V, Praus P. Photocatalytic degradation of selected pharmaceuticals using g-C$_3$N$_4$ and TiO$_2$ nanomaterials, Nanomaterials, 2019, 9, 1194, 16 pp.

[154] Kumar A, Rana A, Sharma G, Naushad M, Al-Muhtaseb AH, Guo C, et al. High-performance photocatalytic hydrogen production and degradation of levofloxacin by wide spectrum-responsive Ag/Fe$_3$O$_4$ bridged SrTiO$_3$/g-C$_3$N$_4$ plasmonic nanojunctions: joint effect of Ag and Fe$_3$O$_4$, ACS Applied Materials & Interfaces, 2018, 10, 40474–40490.

[155] Zheng L, Pi F, Wang Y, Xu H, Zhang Y, Sun X. Photocatalytic degradation of acephate, omethoate, and methyl parathion by Fe$_3$O$_4$@SiO2@mTiO$_2$ nanomicrospheres, Journal of Hazardous Materials, 2016, 315, 11–22.

[156] Boruah PK, Das MR. Dual responsive magnetic Fe$_3$O$_4$-TiO$_2$/graphene nanocomposite as an artificial nanozyme for the colorimetric detection and photodegradation of pesticide in an aqueous medium, Journal of Hazardous Materials, 2020, 385, 121516.

[157] Mrotek E, Dudziak S, Malinowska I, Pelczarski D, Ryżyńska Z, Zielińska-Jurek A. Improved degradation of etodolac in the presence of core-shell ZnFe$_2$O$_4$/SiO$_2$/TiO$_2$ magnetic photocatalyst, Science of the Total Environment, 2020, 724, 138167, 12 pp.

[158] Wang J, Zhang Q, Deng F, Luo X, Dionysiou DD. Rapid toxicity elimination of organic pollutants by the photocatalysis of environment-friendly and magnetically recoverable step-scheme SnFe$_2$O$_4$/ZnFe$_2$O$_4$ nano-heterojunctions, Chemical Engineering Journal, 2020, 379, 122264.

[159] Yu J, Wang T, Rtimi S. Magnetically separable TiO$_2$/FeOx/POM accelerating the photocatalytic removal of the emerging endocrine disruptor: 2,4-dichlorophenol, Applied Catalysis B: Environmental, 2019, 254, 66–75.

[160] Liu J, Lin H, He Y, Dong Y, Rose E, Menzembere GY. Novel CoS$_2$/MoS$_2$@Zeolite with excellent adsorption and photocatalytic performance for tetracycline removal in simulated wastewater, Journal of Cleaner Production, 2020, 260, 121047.

[161] Hojamberdiev M, Czech B, Göktaş AC, Yubuta K, Kadirova ZC. SnO$_2$@ZnS photocatalyst with enhanced photocatalytic activity for the degradation of selected pharmaceuticals and personal care products in model wastewater, Journal of Alloys and Compounds, 2020, 827, 154339, 13 pp.

[162] Vadaei S, Faghihian H. Enhanced visible light photodegradation of pharmaceutical pollutant, warfarin by nano-sized SnTe, effect of supporting, catalyst dose, and scavengers, Environmental Toxicology and Pharmacology, 2018, 58, 45–53.

[163] Chang Y-K, Wu Y-S, Lu C-S, Lin P-F, Wu T-Y. Photodegradation of alachlor using BiVO$_4$ photocatalyst under visible light irradiation, Water, Air, Soil Pollution, 2015, 226, 194–205.

[164] Bhoi YP, Mishra BG. Photocatalytic degradation of alachlor using type-II CuS/BiFeO$_3$ heterojunctions as novel photocatalyst under visible light irradiation, Chemical Engineering Journal, 2018, 344, 391–401.

[165] Zhang L, Wang W, Sun S, Sun Y, Gao E, Zhang Z. Elimination of BPA endocrine disruptor by magnetic BiOBr@SiO$_2$@Fe$_3$O$_4$ photocatalyst, Applied Catalysis B: Environmental, 2014, 148–149, 164–169.

[166] Zhang F, Zou S, Wang T, Shi Y, Liu P. CeO2/Bi2WO6 heterostructured microsphere with excellent visible-light-driven photocatalytic performance for degradation of tetracycline hydrochloride, Photochemistry and Photobiology, 2017, 93, 1154–1164.

[167] Lu X, Che W, Hu X, Wang Y, Zhang A, Deng F, et al. The facile fabrication of novel visible-light-driven Z-scheme CuInS2/Bi2WO6 heterojunction with intimate interface contact by in situ hydrothermal growth strategy for extraordinary photocatalytic performance, Chemical Engineering Journal, 2019, 356, 819–829.

[168] Liu W, Zhou J, Zhou J. Facile fabrication of multi-walled carbon nanotubes (MWCNTs)/α-Bi$_2$O$_3$ nanosheets composite with enhanced photocatalytic activity for doxycycline degradation under visible light irradiation, Journal of Materials Science, 2019, 54, 3294–3308.

[169] Jiang Y, Guan S, Zhang Y, Liu G, Zheng X, Gao J. Cascade degradation of organophosphorus pollutant by photoenzymatic integrated nanocatalyst, Journal of Chemical Technology and Biotechnology, 2020, 95(9), 2463–2472.

[170] Rasheed T, Adeel M, Nabeel F, Bilal M, Iqbal HMN. TiO$_2$/SiO$_2$ decorated carbon nanostructured materials as a multifunctional platform for emerging pollutants removal, Science of the Total Environment, 2019, 688, 299–311.

[171] Shan AY, Ghazi TIM, Rashid SA. Immobilisation of titanium dioxide onto supporting materials in heterogeneous photocatalysis: a review, Applied Catalysis. A, General, 2010, 389, 1–8.

[172] Morales-Mejía JC, Almanza R, Gutiérrez F. Solar photocatalytic oxidation of hydroxy phenols in a CPC reactor with thick TiO$_2$ films, Energy Procedia, 2014, 57, 597–606.

[173] Jiménez-Tototzintle M, Oller I, Hernández-Ramírez A, Malato S, Maldonado MI. Remediation of agro-food industry effluents by biotreatment combined with supported TiO$_2$/H$_2$O$_2$ solar photocatalysis, Chemical Engineering Journal, 2015, 273, 205–213.

[174] Sivagami K, Ravi Krishna R, Swaminathan T. Optimization studies on degradation of monocrotophos in an immobilized bead photo reactor using design of experiment, Desalination and Water Treatment, 2016, 57, 28822–28830.

[175] Verma A, Toor AP, Bansal P, Sangal V, Sobti A. TiO$_2$-assisted photocatalytic degradation of herbicide 4-chlorophenoxyacetic acid: Slurry and fixed-bed approach, Agnihotri A, Reddy K, Bansal A, ed, Sustainable Engineering. Lecture Notes in Civil Engineering, Vol. 30, Springer, Singapore, 2019.

[176] Perisic DJ, Belet A, Kusic H, Stangar UL, Bozic AL. Comparative study on photocatalytic treatment of diclofenac: Slurry vs, Immobilized Processes. Desalination and Water Treatment, 2017, 81, 170–185.

[177] Švagelj Z, Mandić V, Ćurković L, Biošić M, Žmak I, Gaborardi M. Titania-coated alumina foam photocatalyst for memantine degradation derived by replica method and sol-gel reaction, Materials (Basel), 2020, 13(1), 227, 17 pp.

[178] Santana DR, Espino-Estévez MR, Santiago DE, Méndez JAO, González-Díaz O, Doña-Rodríguez JM. Treatment of aquaculture wastewater contaminated with metronidazole by advanced oxidation techniques, Environmental Nanotechnology, Monitoring & Management, 2017, 8, 11–24.

[179] Zammit I, Vaiano V, Ribeiro AR, Silva AMT, Manaia CM, Rizzo L. Immobilised cerium-doped zinc oxide as a photocatalyst for the degradation of antibiotics and the inactivation of antibiotic-resistant bacteria, Catalysts, 2019, 9, 222, 17 pp.

[180] Byrne C, Nolan M, Banerjee S, John H, Jose S, Periyat P, et al. Advances in the development of novel photocatalysts for detoxification, Visible-light Photocatalysts, 2018, 283–327.

[181] Spasiano D, Marotta R, Malato S, Fernandez-Ibañez P, Di Somma I. Solar photocatalysis: materials, reactors, some commercial, and pre-industrialized applications, A Comprehensive Approach. Applied Catalysis B: Environmental, 2015, 170–171, 90–123.

[182] Ganguly P, Byrne C, Breen A, Pillai SC. Antimicrobial activity of photocatalysts: fundamentals, mechanisms, kinetics and recent advances, Applied Catalysis B: Environmental, 2018, 225, 51–75.

[183] Banerjee S, Pillai SC, Falaras P, O'shea KE, Byrne JA, Dionysiou DD. New insights into the mechanism of visible light photocatalysis, The Journal of Physical Chemistry Letters, 2014, 5, 2543–2554.

[184] Padmanabhan SC, Pillai SC, Colreavy J, Balakrishnan S, McCormack DE, Perova TS, et al. A simple sol – gel processing for the development of high-temperature stable photoactive

anatase titania, Chemistry of Materials: A Publication of the American Chemical Society, 2007, 19, 4474–4481.

[185] Ahmed S, Rasul MG, Brown R, Hashib MA. Influence of parameters on the heterogeneous photocatalytic degradation of pesticides and phenolic contaminants in wastewater: a short review, Journal of Environmental Management, 2011, 92, 311–330.

[186] Cassano AE, Alfano OM. Reaction engineering of suspended solid heterogeneous photocatalytic reactors, Catalysis Today, 2000, 58, 167–197.

[187] Ollis DF, Pelizzetti E, Serpone N. Destruction of water contaminants, Environmental Science & Technology, 1991, 25, 1522–1529.

[188] Doná G, Dagostin JLA, Takashina TA, De Castilhos F, Igarashi-Mafra L. A comparative approach of methylparaben photocatalytic degradation assisted by UV-C, UV-A and Vis radiations, Environmental Technology (United Kingdom), 2018, 39, 1238–1249.

[189] Adesina AA. Industrial exploitation of photocatalysis: progress, perspectives and prospects, Catalysis Surveys from Asia, 2004, 8, 265–273.

[190] Daneshvar N, Rabbani M, Modirshahla N, Behnajady MA. Kinetic modeling of photocatalytic degradation of Acid Red 27 in UV/TiO$_2$ process, Journal of Photochemistry and Photobiology. A, Chemistry, 2004, 168, 39–45.

[191] Saien J, Khezrianjoo S. Degradation of the fungicide carbendazim in aqueous solutions with UV/TiO$_2$ process: optimization, kinetics and toxicity studies, Journal of Hazardous Materials, 2008, 157, 269–276.

[192] Sarkar S, Das R, Choi H, Bhattacharjee C. Involvement of process parameters and various modes of application of TiO$_2$ nanoparticles in heterogeneous photocatalysis of pharmaceutical wastes – A short review, RSC Advances, 2014, 4, 57250–57266.

[193] Hansen AM, Treviño-Quintanilla LG, Márquez-Pacheco H, Villada-Canela M, González-Márquez LC, Guillén-Garcés RA, et al. Atrazine: a controversial herbicide, Rev Int Contam Ambient, 2013, 29, 65–84.

[194] Kar P, Sathiyan G, Gupta RK. Reaction Intermediates During the Photocatalytic Degradation of Emerging Contaminants under Visible or Solar Light, Elsevier Elsevier, Science and Engineering, 2020, 163–193.

[195] Gmurek M, Olak-Kucharczyk M, Ledakowicz S. Photochemical decomposition of endocrine disrupting compounds – A review, Chemical Engineering Journal, 2017, 310, 437–456.

[196] Sacco O, Vaiano V, Han C, Sannino D, Dionysiou DD. Photocatalytic removal of atrazine using N-doped TiO$_2$ supported on phosphors, Applied Catalysis B: Environmental, 2015, 164, 462–474.

[197] Niu J, Dai Y, Yin L, Shang J, Crittenden JC. Photocatalytic reduction of triclosan on Au–Cu$_2$O nanowire arrays as plasmonic photocatalysts under visible light irradiation, Physical Chemistry Chemical Physics: PCCP, 2015, 17, 17421–17428.

[198] Woo OT, Chung WK, Wong KH, Chow AT, Wong PK. Photocatalytic oxidation of polycyclic aromatic hydrocarbons: intermediates identification and toxicity testing, Journal of Hazardous Materials, 2009, 168, 1192–1199.

[199] Zhou Y, Gu X, Zhang R, Lu J. Influences of various cyclodextrins on the photodegradation of phenol and bisphenol A under UV light, Industrial & Engineering Chemistry Research, 2015, 54, 426–433.

[200] Lacson CFZ, De Luna MDG, Dong C, Garcia-Segura S, Lu MC. Fluidized-bed Fenton treatment of imidacloprid: optimization and degradation pathway, Sustainable Environment Research, 2018, 28, 309–314.

[201] Evgenidou E, Konstantinou I, Fytianos K, Albanis T. Study of the removal of dichlorvos and dimethoate in a titanium dioxide mediated photocatalytic process through the examination

of intermediates and the reaction mechanism, Journal of Hazardous Materials, 2006, 137, 1056–1064.

[202] Low GKC, McEvoy SR, Matthews RW. Formation of nitrate and ammonium ions in titanium dioxide mediated photocatalytic degradation of organic compounds containing nitrogen atoms, Environmental Science & Technology, 1991, 25, 460–467.

[203] Sirtori C, Zapata A, Malato S, Agüera A. Formation of chlorinated by-products during photo-Fenton degradation of pyrimethanil under saline conditions. Influence on toxicity and biodegradability, Journal of Hazardous Materials, 2012, 217–218, 217–223.

[204] Tamimi M, Qourzal S, Assabbane A, Chovelon JM, Ferronato C, Ait-Ichou Y. Photocatalytic degradation of pesticide methomyl: determination of the reaction pathway and identification of intermediate products, Photochemical and Photobiological Sciences, 2006, 5, 477–482.

[205] Tomasevic A, Marinkovic A, Mijin D, Radisic M, Porobic S, Prlainovic N, et al. A study of photocatalytic degradation of methomyl and its commercial product Lannate-90, Chemical Industry & Chemical Engineering Quarterly, 2020, 381, 2–2.

[206] Ishag AESA, Abdelbagi AO, Hammad AMA, Elsheikh EAE, Hur JH. Photodegradation of chlorpyrifos, malathion, and dimethoate by sunlight in the Sudan, Environmental Earth Sciences, 2019, 78, 1–14.

[207] Malato S, Maldonado MI, Fernández-Ibáñez P, Oller I, Polo I, Sánchez-Moreno R. Decontamination and disinfection of water by solar photocatalysis: the pilot plants of the Plataforma Solar de Almeria, Materials Science in Semiconductor Processing, 2016, 42, 15–23.

[208] WHO. Atrazine in Drinking-Water, Vol. 2, World Health Organization, 1996.

[209] Graymore M, Stagnitti F, Allinson G. Impacts of atrazine in aquatic ecosystems, Environment International, 2001, 26, 483–495.

[210] Liu Y, Zhu K, Su M, Zhu H, Lu J, Wang Y, et al. Influence of solution pH on degradation of atrazine during UV and UV/H_2O_2 oxidation: kinetics, mechanism, and degradation pathways, RSC Advances, 2019, 9, 35847–35861.

[211] Xu L, Zang H, Zhang Q, Chen Y, Wei Y, Yan J, et al. Photocatalytic degradation of atrazine by $H_3PW_{12}O_{40}$/Ag-TiO_2: kinetics, mechanism and degradation pathways, Chemical Engineering Journal, 2013, 232, 174–182.

[212] Bianchi CL, Pirola C, Ragaini V, Selli E. Mechanism and efficiency of atrazine degradation under combined oxidation processes, Applied Catalysis B: Environmental, 2006, 64, 131–138.

[213] Pellzzetti E, Maurino V, Minero C, Carlin V, Pramauro E, Zerbinati O, et al. Photocatalytic degradation of atrazine and other s-triazine herbicides, Environmental Science & Technology, 1990, 24, 1559–1565.

[214] Zhanqi G, Shaogui Y, Na T, Cheng S. Microwave assisted rapid and complete degradation of atrazine using TiO_2 nanotube photocatalyst suspensions, Journal of Hazardous Materials, 2007, 145, 424–430.

[215] Santacruz-chávez JA, Oros-ruiz S, Prado B, Zanella R. Journal of Environmental Chemical Engineering Photocatalytic Degradation of Atrazine Using TiO_2 Superficially Modified with Metallic Nanoparticles, 2015, 3, 1–7.

[216] Konstantinou K, Albanis T. Degradation pathways and intermediates of photocatalytic transformation of major pesticide groups in aqeous TiO_2 suspensions using artificial and solar light: a review, Applied Catalysis B: Environmental, 2003, 42, 319–335.

[217] Tomašević A, Daja J, Petrović S, Kiss EE, Mijin D. A study of the photocatalytic degradation of methomyl by UV light, Chemical Industry & Chemical Engineering Quarterly, 2009, 15, 17–19.

[218] Tomašević A, Mijin D, Gašic S, Kiss E. The influence of polychromatic light on methomyl degradation in TiO2 and ZnO aqueous suspension, Desalination and Water Treatment, 2014, 52, 4342–4349.

[219] Tomašević A, Kiss E, Petrović S, Mijin D. Study on the photocatalytic degradation of insecticide methomyl in water, Desalination, 2010, 262, 228–234.

[220] Strathmann TJ, Stone AT. Reduction of the pesticides oxamyl and methomyl by FeII: effect of pH and inorganic ligands, Environmental Science & Technology, 2002, 36, 653–661.

[221] Strathmann TJ, Stone AT. Reduction of oxamyl and related pesticides by FeII: influence of organic ligands and natural organic matter, Environmental Science & Technology, 2002, 36, 5172–5183.

[222] Luan J, Ma K, Wang S, Hu Z, Li Y, Pan B. Research on photocatalytic degradation pathway and degradation mechanisms of organics, Current Organic Chemistry, 2010, 14, 645–682.

[223] Loos R, Locoro G, Comero S, Contini S, Schwesig D, Werres F, et al. Pan-European survey on the occurrence of selected polar organic persistent pollutants in ground water, Water Research, 2010, 44, 4115–4126.

[224] Tan C, Gao N, Deng Y, Zhang Y, Sui M, Deng J, et al. Degradation of antipyrine by UV, UV/H$_2$O$_2$ and UV/PS, Journal of Hazardous Materials, 2013, 260, 1008–1016.

[225] Gong H, Chu W, Chen M, Wang Q. A systematic study on photocatalysis of antipyrine: catalyst characterization, parameter optimization, reaction mechanism a toxicity evolution to plankton, Water Research, 2017, 112, 167–175.

[226] Hermosilla D, Han C, Nadagouda M, Machala L, Gascó A, Campo P, et al. Environmentally friendly synthesized and magnetically recoverable designed ferrite photo-catalysts for wastewater treatment applications, Journal of Hazardous Materials, 2020, 381, 121200, 14 pp.

[227] Li S, Hu J. Transformation products formation of ciprofloxacin in UVA/LED and UVA/LED/TiO$_2$ systems: impact of natural organic matter characteristics, Water Research, 2018, 132, 320–330.

[228] Salma A, Thoröe-Boveleth S, Schmidt TC, Tuerk J. Dependence of transformation product formation on pH during photolytic and photocatalytic degradation of ciprofloxacin, Journal of Hazardous Materials, 2016, 313, 49–59.

[229] Zhang Y, Zhou JL, Ning B. Photodegradation of estrone and 17β-estradiol in water, Water Research, 2007, 41, 19–26.

[230] Du P, Chang J, Zhao H, Liu W, Dang C, Tong M, et al. Sea-Buckthorn-Like MnO$_2$ decorated titanate nanotubes with oxidation property and photocatalytic activity for enhanced degradation of 17β-Estradiol under solar light, ACS Applied Energy Materials, 2018, 1, 2123–2133.

[231] Mboula VM, Héquet V, Andrès Y, Gru Y, Colin R, Doña-Rodríguez JM, et al. Photocatalytic degradation of estradiol under simulated solar light and assessment of estrogenic activity, Applied Catalysis B: Environmental, 2015, 162, 437–444.

[232] Souissi Y, Bourcier S, Bouchonnet S, Genty C, Sablier M. Estrone direct photolysis: by-product identification using LC-Q-TOF, Chemosphere, 2012, 87, 185–193.

[233] Perondi T, Michelon W, Junior PR, Knoblauch PM, Chiareloto M, Moreira RDFPM, et al Advanced oxidative processes in the degradation of 17β-estradiol present on surface waters: kinetics, byproducts and ecotoxicity, Environmental Science and Pollution Research, 2020, 27(17), 21032–21039

[234] Mai J, Sun W, Xiong L, Liu Y, Ni J. Titanium dioxide mediated photocatalytic degradation of 17β-estradiol in aqueous solution, Chemosphere, 2008, 73, 600–606.

[235] Jiang L, Huang C, Chen J, Chen X. Oxidative transformation of 17β-estradiol by MnO$_2$ in aqueous solution, Archives of Environmental Contamination and Toxicology, 2009, 57, 221–229.

[236] Caupos E, Mazellier P, Croue JP. Photodegradation of estrone enhanced by dissolved organic matter under simulated sunlight, Water Research, 2011, 45, 3341–3350.

[237] Wang S, Wang X, Li C, Xu X, Wei Z, Wang Z, et al. Photodegradation of 17B-estradiol on silica gel and natural soil by UV treatment, Environmental Pollution (Barking, Essex: 1987), 2018, 242, 1236–1244.

[238] Chowdhury RR, Charpentier P, Ray MB. Photodegradation of estrone in solar irradiation, Industrial & Engineering Chemistry Research, 2010, 49, 6923–6930.

[239] Ohko Y, Iuchi KI, Niwa C, Tatsuma T, Nakashima T, Iguchi T, et al. 17β-estradiol degradation by TiO_2 photocatalysis as a means of reducing estrogenic activity, Environmental Science & Technology, 2002, 36, 4175–4181.

[240] Almeida S, Raposo A, Almeida-González M, Carrascosa C. Bisphenol A: food exposure and impact on human health, Comprehensive Reviews in Food Science and Food Safety, 2018, 17, 1503–1517.

[241] Rochester JR. Bisphenol A and human health: a review of the literature, Reproductive Toxicology (Elmsford, N.Y.), 2013, 42, 132–155.

[242] Kondrakov AO, Ignatev AN, Frimmel FH, Bräse S, Horn H, Revelsky AI. Formation of genotoxic quinones during bisphenol A degradation by TiO_2 photocatalysis and UV photolysis: a comparative study, Applied Catalysis B: Environmental, 2014, 160–161, 106–114.

[243] Sharma J, Mishra IM, Kumar V. Mechanistic study of photo-oxidation of Bisphenol-A (BPA) with hydrogen peroxide (H_2O_2) and sodium persulfate (SPS), Journal of Environmental Management, 2016, 166, 12–22.

[244] Guo CS, Ge M, Liu L, Gao G, Feng Y, Wang Y. Directed synthesis of mesoporous TiO_2 microspheres: catalysts and their photocatalysis for bisphenol A degradation, Environmental Science & Technology, 2010, 44, 419–425.

[245] Barbieri Y, Massad WA, Díaz DJ, Sanz J, Amat-Guerri F, García NA. Photodegradation of bisphenol A and related compounds under natural-like conditions in the presence of riboflavin: kinetics, mechanism and photoproducts, Chemosphere, 2008, 73, 564–571.

[246] Kaneco S, Rahman MA, Suzuki T, Katsumata H, Ohta K. Optimization of solar photocatalytic degradation conditions of bisphenol A in water using titanium dioxide, Journal of Photochemistry and Photobiology. A, Chemistry, 2004, 163, 419–424.

[247] Tsai WT, Lee MK, Su TY, Chang YM. Photodegradation of bisphenol-A in a batch TiO_2 suspension reactor, Journal of Hazardous Materials, 2009, 168, 269–275.

[248] Watanabe N, Horikoshi S, Kawabe H, Sugie Y, Zhao J, Hidaka H. Photodegradation mechanism for bisphenol A at the TiO_2/H_2O interfaces, Chemosphere, 2003, 52, 851–859.

[249] Saggioro EM, Oliveira AS, Pavesi T, Tototzintle MJ, Maldonado MI, Correia FV, et al. Solar CPC pilot plant photocatalytic degradation of bisphenol A in waters and wastewaters using suspended and supported-TiO_2. Influence of photogenerated species, Environmental Science and Pollution Research, 2014, 12112–21.

[250] Katsumata H, Kawabe S, Kaneco S, Suzuki T, Ohta K. Degradation of bisphenol A in water by the photo-Fenton reaction, Journal of Photochemistry and Photobiology A: Chemistry, 2004, 162, 297–305.

[251] Diao ZH, Wei-Qian, Guo PR, Kong LJ, Pu SY. Photo-assisted degradation of bisphenol A by a novel $FeS_2@SiO_2$ microspheres activated persulphate process: synergistic effect, pathway and mechanism, Chemical Engineering Journal, 2018, 349, 683–693.

[252] Ohko Y, Ando I, Niwa C, Tatsuma T, Yamamura T, Nakashima T, et al. Degradation of bisphenol A in water by TiO_2 photocatalyst, Environmental Science & Technology, 2001, 35, 2365–2368.

[253] Gültekin I, Ince NH. Synthetic endocrine disruptors in the environment and water remediation by advanced oxidation processes, Journal of Environmental Management, 2007, 85, 816–832.

[254] Priac A, Morin-Crini N, Druart C, Gavoille S, Bradu C, Lagarrigue C, et al. Alkylphenol and alkylphenol polyethoxylates in water and wastewater: a review of options for their elimination, Arabian Journal of Chemistry, 2017, 10, S3749–73.

[255] Guenther K, Kleist E, Thiele B. Estrogen-active nonylphenols from an isomer-specific viewpoint: a systematic numbering system and future trends, Analytical and Bioanalytical Chemistry, 2006, 384, 542–546.

[256] Noorimotlagh Z, Kazeminezhad I, Jaafarzadeh N, Ahmadi M, Ramezani Z, Silva Martinez S. The visible-light photodegradation of nonylphenol in the presence of carbon-doped TiO_2 with rutile/anatase ratio coated on GAC: effect of parameters and degradation mechanism, Journal of Hazardous Materials, 2018, 350, 108–120.

[257] Noorimotlagh Z, Kazeminezhad I, Jaafarzadeh N, Ahmadi M, Ramezani Z. Improved performance of immobilized TiO_2 under visible light for the commercial surfactant degradation: role of carbon doped TiO_2 and anatase/rutile ratio, Catalysis Today, 2019, 348, 277–289.

[258] Neamțu M, Frimmel FH. Photodegradation of endocrine disrupting chemical nonylphenol by simulated solar UV-irradiation, Science of the Total Environment, 2006, 369, 295–306.

[259] Bechambi O, Najjar W, Sayadi S. The nonylphenol degradation under UV irradiation in the presence of Ag-ZnO nanorods: effect of parameters and degradation pathway, Journal of the Taiwan Institute of Chemical Engineers, 2016, 60, 496–501.

[260] Santos A, Yustos P, Quintanilla A, Rodríguez S, García-Ochoa F. Route of the catalytic oxidation of phenol in aqueous phase, Applied Catalysis B: Environmental, 2002, 39, 97–113.

[261] Mazellier P, Leverd J. Transformation of 4-tert-octylphenol by UV irradiation and by an H_2O_2/UV process in aqueous solution, Photochemical and Photobiological Sciences, 2003, 2, 946–953.

[262] Rachna, Rani M, Shanker U. Sunlight active ZnO@FeHCF nanocomposite for the degradation of bisphenol A and nonylphenol, Journal of Environmental Chemical Engineering, 2019, 7, 103153.

[263] Wei T, Fan Z, Zhao G. Enhanced adsorption and degradation of nonylphenol on electron-deficient centers of photocatalytic surfaces, Chemical Engineering Journal, 2020, 388, 124168.

[264] Iervolino G, Zammit I, Vaiano V, Rizzo L. Limitations and Prospects for Wastewater Treatment by UV and Visible-Light-Active Heterogeneous Photocatalysis: A Critical Review, 2020, 378, 7, 40 pp.

[265] Rueda-Marquez JJ, Levchuk I, Fernández Ibañez P, Sillanpää M. A critical review on application of photocatalysis for toxicity reduction of real wastewaters, Journal of Cleaner Production, 2020, 258, 7, 40 pp.

[266] Fiorentino A, Ferro G, Castro M, Polo-López MI, Fernández-ibañez P, Rizzo L. Journal of Photochemistry and Photobiology B: Biology Inactivation and Regrowth of Multidrug Resistant Bacteria in Urban Wastewater after Disinfection by Solar-Driven and Chlorination Processes, 2015, 148, 43–50.

[267] Malchi T, Maor Y, Tadmor G, Shenker M, Chefetz B. Irrigation of root vegetables with treated wastewater: evaluating uptake of pharmaceuticals and the associated human health risks, Environmental Science & Technology, 2014, 48, 9325–9333.

[268] Almomani F, Bhosale R, Kumar A, Khraisheh M. Potential use of solar photocatalytic oxidation in removing emerging pharmaceuticals from wastewater: a pilot plant study, Solar Energy, 2018, 172, 128–140.

[269] Talwar S, Verma AK, Sangal VK, Štangar UL. Once through continuous flow removal of metronidazole by dual effect of photo-Fenton and photocatalysis in a compound parabolic concentrator at pilot plant scale, Chemical Engineering Journal, 2020, 388, 124184, 13 pp.

[270] Philippe KK, Timmers R, Van Grieken R, Marugan J. Photocatalytic disinfection and removal of emerging pollutants from effluents of biological wastewater treatments, using a newly developed large-scale solar simulator, Industrial & Engineering Chemistry Research, 2016, 55, 2952–2958.

[271] Cabrera-Reina A, Martínez-Piernas AB, Bertakis Y, Xekoukoulotakis NP, Agüera A, Sánchez Pérez JA. TiO$_2$ photocatalysis under natural solar radiation for the degradation of the carbapenem antibiotics imipenem and meropenem in aqueous solutions at pilot plant scale, Water Research, 2019, 166, 115037, 10 pp.

[272] Bansal P, Verma A. Detoxi fi cation of real pharmaceutical wastewater by integrating photocatalysis and photo-Fenton in fixed-mode, Chemical Engineering Journal, 2018, 349, 838–848.

Jesna Louis, Nisha T. Padmanabhan, Honey John

Chapter 9
Commercial Applications and Future Trends of Photocatalytic Materials

9.1 Introduction

One of the greatest challenges in the twenty-first century is raising the living standards along with the increasing global demand for energy and environmental pollution due to the ever-increasing population and industrialization. To avoid the effects of pollutant gases and the greenhouse effect and to secure energy supply and preserve fossil fuels for the coming generations, the use of clean and bountiful solar energy has become inevitable. The concept of photocatalysis was introduced by the pioneering work of Fujishima and Honda in 1972 [1], which attracted immediate research interests. Their demonstration of the splitting of water into H_2 and O_2 using TiO_2 under UV illumination is considered as the genesis of current research and environmentally benign technologies based on photocatalysis. This discovery soon led to the perception of photoelectrochemical (PEC) cells composed of semiconductors when different sites of one particle can act as anodes and cathodes [2]. Thereafter with the growth of nanotechnology, the economical and technological importance of photocatalytic materials considerably increased by tremendously enhancing their catalytic efficiency. Diversity in the application has been discovered and demonstrated using photocatalytic materials ranging from photooxidation/photodegradation of organic pollutants, photocatalytic gas-phase oxidation, antimicrobial and antifogging self-cleaning surfaces, air purification, and wastewater treatment, to photocatalytic water splitting for the production of hydrogen with laterally commercializing many methods and paying their way to different commercial products.

Due to facile product separability, higher stability, and easier recyclability of the photocatalytic materials, it is usually pointed out that heterogeneous photocatalysis is preferable to homogeneous photocatalysis [3]. Heterogeneous photocatalysis is typically the generation of photoinduced charge generations and chemical reactions on the surface of semiconductors upon its exposure to photons of suitable

Jesna Louis, Department of Polymer Science and Rubber Technology, Cochin University of Science and Technology, Kerala, India; Inter University Centre for Nanomaterials and Devices, Cochin University of Science and Technology, Kerala, India
Nisha T. Padmanabhan, Department of Polymer Science and Rubber Technology, Cochin University of Science and Technology, Kerala, India
Honey John, Department of Polymer Science and Rubber Technology, Cochin University of Science and Technology, Kerala, India; Inter University Centre for Nanomaterials and Devices, Cochin University of Science and Technology, Kerala, India, e-mail: honey@cusat.ac.in

https://doi.org/10.1515/9783110668483-009

frequency. The wideband semiconductor TiO_2 stood as the main facet in the earlier stages of photocatalytic research, which had low utilization of solar energy with its characteristic ultraviolet (UV) absorption, which constituted only <5% of solar spectrum energy. Moreover, it was believed in the beginning that the photocatalytic process as a whole is controlled by the kinetic and thermodynamic components, but as the fundamental research progressed it was later known that the factors such as separation of photogenerated charge carriers, positioning of band edge potentials, charge recombination, migration toward reactive sites, surface reactions, and adsorption of molecules at the surface of photocatalysts govern the performance of a photocatalyst [4–6]. Such insights led to the modifications on photocatalytic materials such as metal and nonmetal doping [7, 8], cocatalyst loading [9], plasmonic metal incorporation [10], dye sensitization [11] and various designs like heterojunctions [12], and Z-scheme [13] configurations in a vision to enhance the specific properties for utilizing solar energy for energy harvesting and environmental remediation. Nevertheless, the fundamental research is extensively progressing on optimizing the green technology of photocatalysis by expanding the continuum of their potential day-to-day life applications.

This chapter gives a thorough review of the commercial and industrial applications of photocatalytic materials with special reference to water treatment, air purification, green energy production, transparent conducting oxides and photovoltaics (PV), self-cleaning surfaces, and photocatalytic water splitting. The chapter also gives a short account of the future trends in photocatalytic research and technologies as a reality to practical life.

9.2 Commercial and Industrial Applications of Photocatalytic Materials

The journey from TiO_2 to the present third-generation photocatalysts and its applications had always been a continuous exponential growth. TiO_2 is still the most commercially used photocatalytic semiconductor, especially the benchmark material Degussa P25 (rutile/anatase is 85/15). In the earlier stages of photocatalytic research, TiO_2 along with other binary oxides like ZnO, WO_3, ZrO_2, and CeO_2, and binary chalcogenides such as ZnS, CdS, CdSe, and CuS were the most investigated material photocatalysts. The exploration of binary nitrides emerged in the last two decades, with GaN, Ta_3N_5, and graphitic carbon nitride (g-C_3N_4) being the most prominent ones. GaN is widely used in sensing fields due to its high chemical stability and mechanical strength, while Ta_3N_5 is an auspicious photocatalyst for visible light water splitting due to its narrow direct bandgap (2.1 eV). g-C_3N_4 is a visible light-responsive direct bandgap semiconductor (2.7 eV), which is of n-type and is a nonmetallic conjugated polymer. g-C_3N_4 is now abundantly studied in various fields of photocatalytic research. Thereafter, ternary oxides/chalcogenide semiconductors developed, which can promote various chemistries as compared to binary semiconductors. Ternary oxide photocatalysts mainly include perovskite photocatalysts such as $SrTiO_3$, $LiTaO_3$, $NaTaO_3$, $KTaO_3$, and $AgNbO_3$. $CaIn_2O_4$ and $AgMO_2$ (M = Al, Ga, In) are the other types belonging to ternary oxide photocatalysts. Whereas ternary chalcogenide photocatalysts include mainly the type AB_xC_y (where A = Zn, Ag, Cu, or Cd; B = Ga or In; C = S, Se, or Te). Besides the above-mentioned photocatalysts, aurivillius compounds – $Bi_2A_{n-1}B_nO_{3n+3}$ (A = Na, K, Ca, Ba, Sr, Pb; B = Ti, Mo, Fe, Nb, Ta, W) are also under investigation. Bi_2WO_6 is the simplest example of this group of aurivillius oxides. Such compounds have unique layer structures composed of $(Bi_2O_2)^{2+}$ layers are interleaved with perovskite-like $(A_{n-1}B_nO_{3n+1})$ units [4].

As aforementioned, photocatalytic materials amalgamating multidisciplinary fields of science have spurred the production of countless products finding various applications including water treatment and H_2 production. This part of the chapter includes only such applications that were successful in commercializing giving practical implementations from laboratory to industry. Most of the results are discussed mostly between the years 2018 and 2020 to give a depiction of recent updates in the field.

9.2.1 Air Purification

Air pollution can cause severe health problems like accelerated aging of lungs, decrease in the lung capacity, and occurrence of diseases like asthma, emphysema, bronchitis, and sometimes cancer. Particulate matters and gaseous pollutants are the two air pollutant categories of main concern. Gas-phase photocatalysis can be used for tackling gaseous pollutants in the air, and it is very unlikely to deal with particulate matters. It is identified that gaseous pollutants like NO_x (mainly NO_2), SO_x (mainly SO_2), CO, ozone (O_3), volatile organic compounds (VOCs – aromatics, aldehydes, and halocarbons), and odorous compounds (H_2S and S containing compounds) can be removed effectively using photocatalytic materials. The gas-phase photocatalysis is the synergistic interaction of semiconductor photocatalyst with the incident light and an oxidizing agent (adsorbed O_2 or H_2O molecules) to produce reactive radicals that degrade the gaseous pollutants finally to CO_2 and H_2O of minimal disposal issue.

Despite enormous research on manipulations and modifications of TiO_2 and ZnO for shape control, size dependence, porosity, and preferential facet exposure for efficient air treatment, novel photocatalysts were also explored. Hematite α-Fe_2O_3, iron oxide of n-type semiconductor properties, can be used for the photocatalytic removal of NO_x [14], and bimetal oxides like Bi_2SiO_4, when coupled with AgI, can be used for the removal of formaldehyde [15]. Metal-free photocatalyst g-C_3N_4 when made a synergy with metal oxides like WO_3, the so-formed g-C_3N_4/WO_3 nanocomposite can be used for the visible-light-driven photodegradation of acetaldehyde pollution [16]. Moreover, photocatalytic nanocomposites like $BiVO_4$/g-C_3N_4 [17], Bi/g-C_3N_4 [18],

BiOI/Al_2O_3 [19], and Ag/$BaAl_2O_4$ [20] are also demonstrated for the visible light photodegradation of pollutants such as NO_x, SO_x, and toluene. Newer photocatalysts are far yet to be discovered for photocatalytic removal of gaseous pollutants.

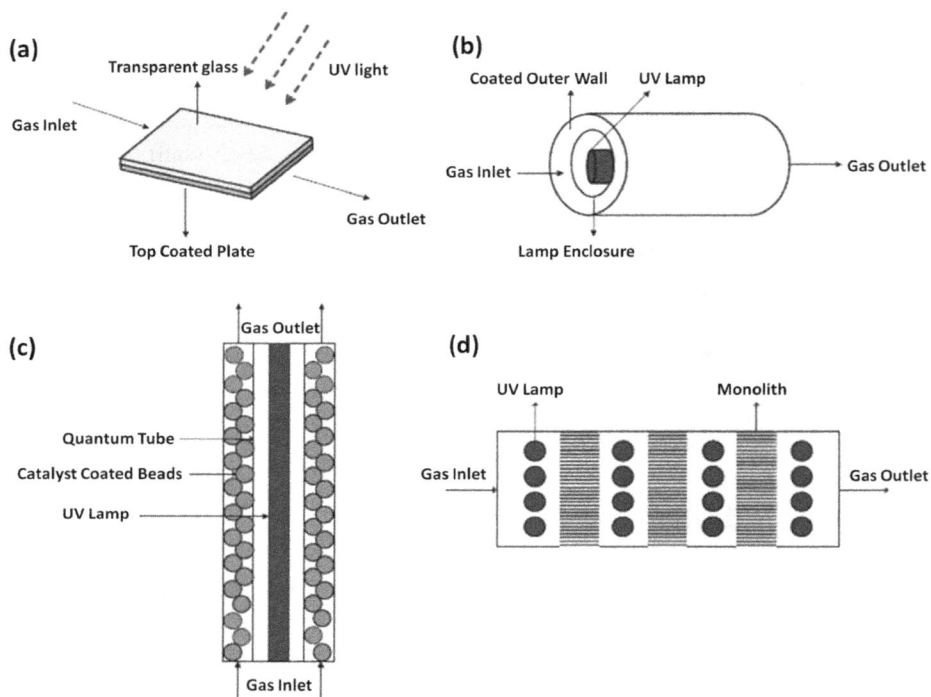

Figure 9.1: Representation models of some gas-phase photocatalytic reactor types: (a) coated plate, (b) annular, (c) packed bed, and (d) monolith (top view) [21].

Typically, in photocatalytic reactors for air treatment, the catalysts are immobilized onto a surface. Each catalyst design has its advantages and limitations. Laboratory-scale photoreactors are usually flat-plate reactor [22] or annular reactor type [23] in which the catalyst is fixed to the reactor wall and designed for high air throughout but can only be commercialized to indoor purification in offices and houses. The reactors in the industrial scale are usually packed bed [24], fluidized bed [25], or monoliths [26] ensuring a large surface for the immobilized catalyst for the intimate interaction of the pollutants with the catalyst and irradiation. While packed-bed reactors are simple to construct and have high efficiency in conversion of VOCs per unit mass of catalyst, but less efficient catalyst utilization and high maintenance are the major hurdles. Whereas monoliths can easily integrate with the traditional heating, ventilation, and air conditioning equipment, and give high throughput but have very low photon transfer, that is, light intensity declines considerably with the depth of monoliths. Alternatively, bubbling fluidized-bed reactors facilitate the deep penetration of light and at the same time ensure continuous catalyst renewal at the illuminated surfaces due to the bubble phase of the fluidized-bed reactor [25]. Hitherto, more advances in research with the further development and designing of photoreactors are required, considering high throughput, low catalyst wastage, effective use of incident light, compact size, low pressure drop, and easy maintenance [21].

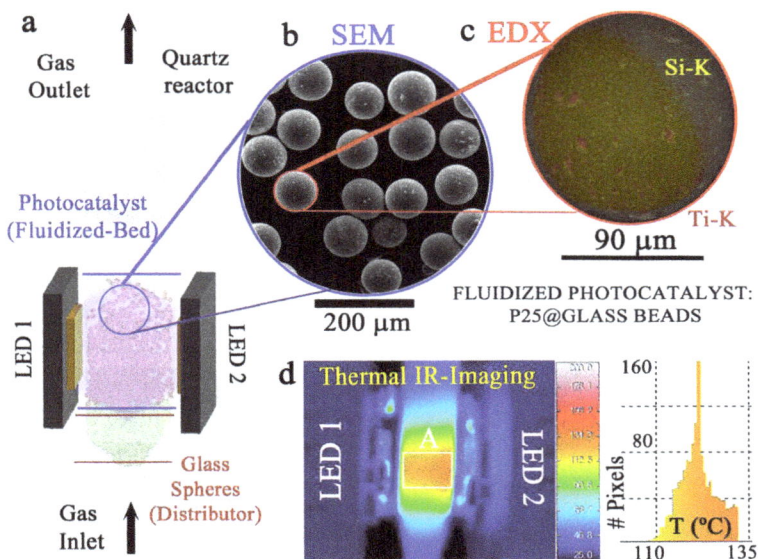

Figure 9.2: (a) Different parts of a fluidized-bed photoreactor; (b) SEM image of P25 (standard TiO₂ nanopowder, rutile:anatase/85:15) impregnated glass beads as the fluidizable photocatalyst; (c) EDX mapping of a glass bead with red color showing decorated P25; (d) infrared camera acquired thermal image of the LED-irradiated reactor wall, histogram showing the thermal distribution for the labeled area A [25].

Commercial development of photocatalytic materials for air purification includes combination of photocatalysts along with other filtering technologies to get synergetic effect of removal of both dirt particulates and harmful gases in air. Today, various TiO₂-based photocatalysts' liquid spray is available commercially for air purification. IN 2004, Daikin Industries Ltd. Japan patented an air purifier based on titanium apatite photocatalyst (Figure 9.3a). It consists of a coarse prefilter, a High efficiency particulate air filter, an ionizer, and a photocatalytic section with UV light. The coarse filter is made of polypropylene netting coated with catechin which removes large household dirt particles and pet hair. HEPA filter is used to remove spores, bacteria, and airborne viruses. Then the plasma ionizer provides a positive charge to dirt particles so that the dirt particle gets attracted to a negatively charged metal plate and the photocatalytic section removes the volatile organic pollutants. A US-based company Abundant Earth patented

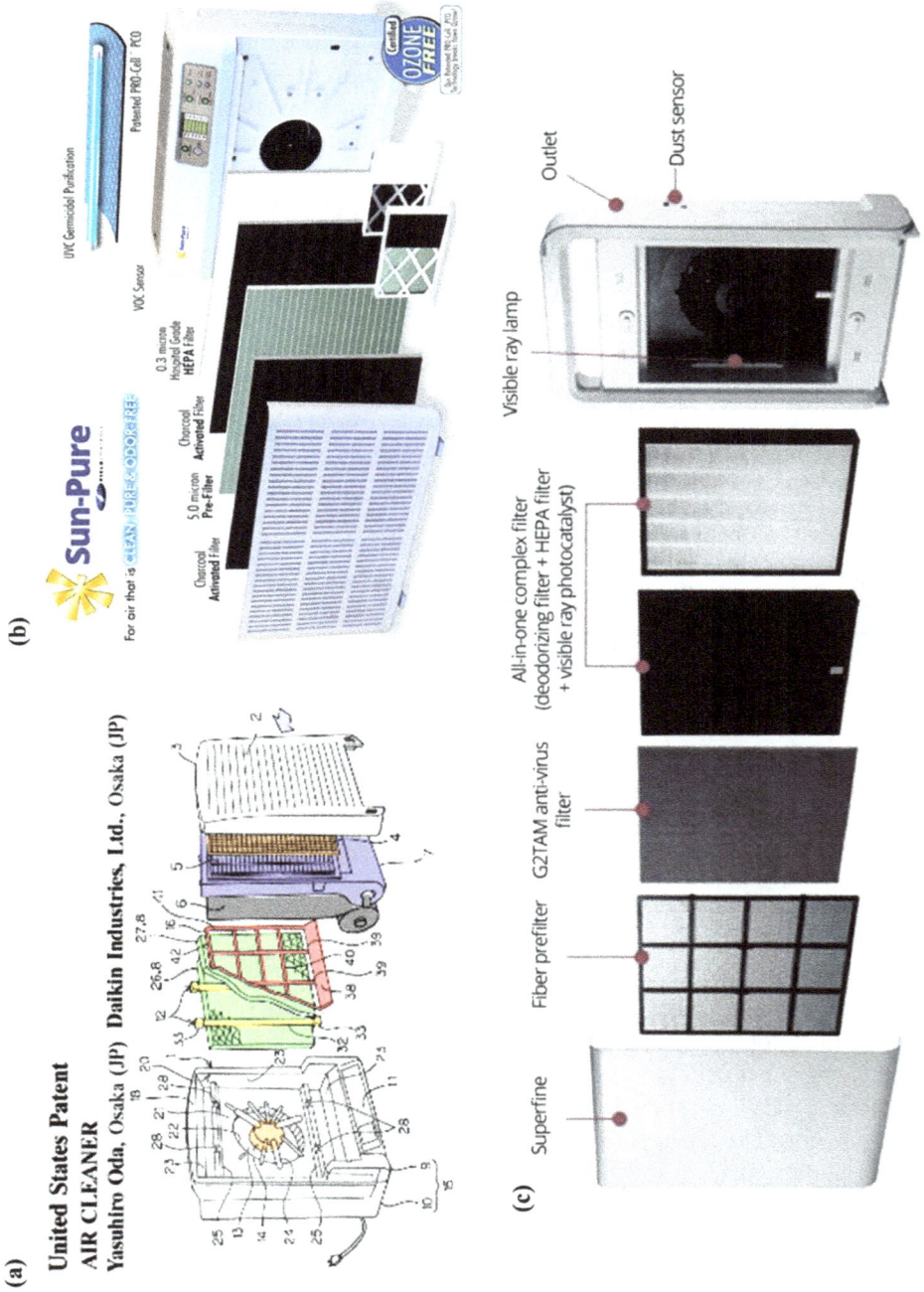

Figure 9.3: (a) An exploded view of air purification system developed by Daikin Industries Ltd. (b) Sun-Pure SP20C PRO-Cell Photocatalytic Air Purifier, and (c) Antivirus L'TOP air purifier [27].

maintenance-free Sun-Pure SP20C PRO-Cell Photocatalytic Air Purifier for advanced full-spectrum indoor air purification (Figure 9.3b). They claimed that the purifier does not produce ozone. The technology is 98.2% efficient toward degradation of VOCs. An antivirus air purifier using visible light is developed by PREXCO, South Korea, which works with indoor lighting (Figure 9.3c). It removes foul odors and prevents sick house syndrome.

9.2.2 Water Purification

According to reports, about 844 million people in the world suffer from inadequacy of drinking water purification system [28]. Advanced oxidation process (AOP) is one of the widely investigated methods for the development of cost-effective technologies. AOP uses diverse methods such as photocatalysis, UV/H_2O_2, ozonation, sonolysis, Fenton-like oxidation, and electrochemical oxidation to generate highly reactive species [29]. Over the past few decades, intense research has been carried out to develop diverse photocatalytic materials for reclamation of water. The production of these photocatalytic materials requires a thorough understanding of the energetics of redox potentials of the materials. Among a wide range of photocatalytic nanomaterials, TiO_2, ZnO, Fe_2O_3, CeO_2, WO_3, and $g-C_3N_4$ have been widely explored for water treatment. With the characteristic nature of the photocatalyst used, water can be purified by organic matter removal, ozonolysis, antimicrobial disinfection, oil–water separation, and heavy-metal removal. Pedrosa et al. synthesized graphene-derived TiO_2 composites for highly active photocatalysis. They have done the ozonation using graphene oxide (GO) and reduced GO (rGO) and introduced TiO_2 nanoparticles to this system as own catalytic phase and to protect the carbon phase from ozone's erosive effect [30]. Recently, TiO_2/PPy composite on melamine sponge by vapor deposition of PPy on in situ–grown TiO_2 is demonstrated to exhibit superhydrophilicity and underwater superhydrophobicity for oil–water separation and thus can be practically applied to oil spills in rivers and oceans [31]. Zhang et al. developed metal-free heterojunction oxygen-doped $g-C_3N_4$ microspheres with hydrothermal carbonation carbon. The photocatalyst exhibits good photocatalytic degradation toward high risk causing waterborne pathogen adenovirus. They concluded that superior virus inactivation efficiency of the photocatalyst is due to effective charge separation by the formation of heterojunction [32]. Ag-doped urea-derived $g-C_3N_4$ was synthesized by in situ photodeposition technique which disinfected *E. coli* in water [33].

Despite the huge number of photocatalysts for enhanced water treatment, the connection between laboratory photocatalytic water treatment and its real-world application is still having gaps. The major hurdle for this disconnection is the separation of photocatalyst material from purified water. In a laboratory scale, this can be easily achieved by centrifugation and filtration process [34]. On large-scale water purification, the separation of these photocatalysts is unrealistic. So the development of simple, cost-effective, and reusable materials that can be separated effortlessly is inevitable for the successful implementation of water purification technology. The use of magnetic photocatalytic materials (magnetophoresis) is a more promising way to separate the photocatalyst from the treated water with minimum contact. The complete recovery of nanomaterials with very low cost for the design of the purification method can be achieved by this method. Magnetic elements such as iron [35], nickel [36], and cobalt [37] and their derivatives are explored for efficient magnetic separation of photocatalysts. The most explored materials are iron-based ferrites and magnetites. A thorough investigation reveals that the photocatalytic efficiency of iron-based composites is reduced due to the transfer of an electron from photocatalytic material to inert magnetic material [38]. This can be avoided by introducing an inert interlayer between the magnetic material and photocatalytic material so that there is no transfer of electrons from the photocatalytic material, and magnetic material occurs [39]. Immobilization of nanomaterials on glass or metals is another way to simplify the separation of photocatalysts. Yaparatne et al. immobilized TiO_2–SiO_2 photocatalyst on a glass slide for the degradation of 2-methyl isoborneol and Geosmin in water. They concluded that the incorporation of SiO_2 could not only provide robustness to the film but also increase photocatalytic efficiency [40].

On the commercial and industrial aspect, photocatalytic water purification using microfluidics chamber-type reactors has earned much attention due to its enhanced mass transport. An important problem overlooked for microfluidics chamber reactor is photocatalyst leaching out due to photodissolution [41]. A seven-layered sandwiched microchamber reactor fabricated using ZnO nanorods had a better performance than the batch reactor [42]. The photocatalytic membrane reactor is another engineering approach that has gained huge attention recently for water purification systems. Reverse osmosis, microfiltration, ultrafiltration, and nanofiltration are the commonly used membrane technologies for water purification. The technology helps to keep the photocatalyst confined to the reactor system providing improved stability and long-term use. A membrane reactor by growing Fe^{2+}-doped ZnO on polyester fabric showed better photocatalytic efficiency along with good cyclability [43]. There is only short time contact between the photocatalyst and pollutant is present in the membrane reactors. So multiple filtration processes are required to achieve successful purification. The degradation efficiency of rhodamine B was found to be 18% by a g-C_3N_4/carbon fiber cloth on the first cycle which increased to 92% on the seventh cycle of filtration [44]. Compound parabolic collector reactor is another emerging technique providing small-scale water purification in single houses, and small and rural areas. A parabolic collector reactor packed with g-C_3N_4–chitosan hydrogel beads was robust for water purification using natural water constituents and was suggested for remote areas where limited access to drinking water service [45]. A reusable and durable BiOBr@TiO_2/carbon hybrid framework containing many microchannels between their building blocks acted

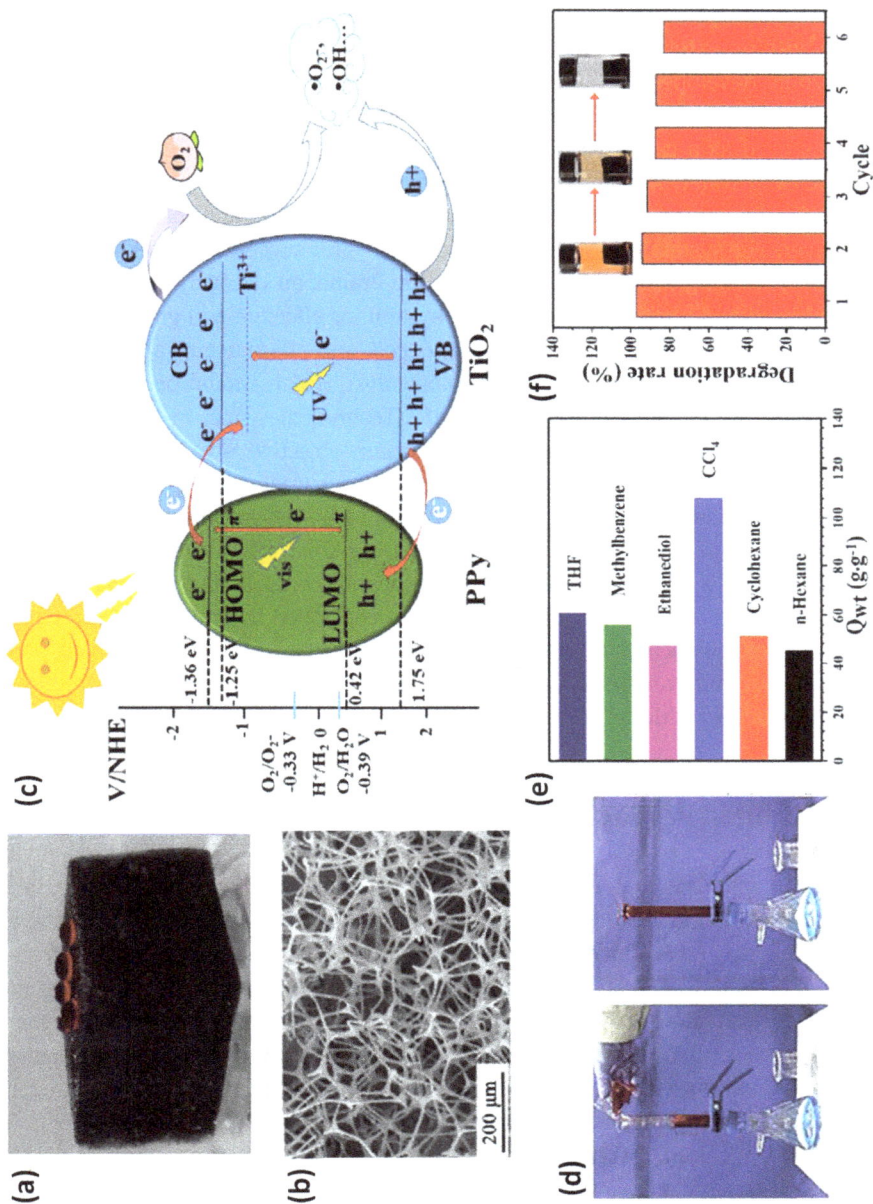

Figure 9.4: TiO$_2$@polypyrrole nanocomposite immobilized on melamine sponge (MS) with oil–water separation and visible light photocatalytic activity for water purification. (a) Image of oil drops on the surface of MS@TiO$_2$@PPy underwater, (b) SEM, (c) the scheme of photocatalytic reaction mechanism, (d) oil–water mixture separation (water and *n*-hexane are dyed with MB and oil red, respectively), (e) adsorption capacity toward different oils, and (f) photocatalytic recycle experiment of MS@TiO$_2$@PPy [31].

as a pathway for water transport by capillary action and can be used for water purification with no external electric input [46].

Panasonic developed a water purification system based on photocatalysis using UV rays from sunlight. They introduced this technology at the Tokyo event 2014. TiO_2–zeolite was used as a catalyst to avoid the quick dispersion of finely powdered TiO_2 nanoparticles in water. A German company Raywox developed a solar photocatalytic water purification system in which the glass tubes present on the device act as a solar light receiver. The system is examined with nano-TiO_2 for successful water purification and currently working with iron salt. Dr. Jonathon Brame, an US Army engineer, developed a fluidized-bed photoreactor treatment system for effective and efficient removal of contaminants such as bacteria, pesticides, *E. coli*, and pharmaceuticals from Swaziland's reservoir. They selected TiO_2 as a photocatalyst, which should be changed after 6 months as a maintenance procedure. TitanPE Technologies, Inc. in China provides nanowater purification systems globally. Their main objective is to provide drinking water disinfection and purification, solar/UV industrial wastewater purification, and water treatment without power supplement. Hirschmann Laborgeräte GmbH & Co. KG and German Aerospace Center (DLR) have designed a solar water treatment system in which a glass tube is used as receiver of solar light and contaminated water along with photocatalyst flows through the glass tubes with a controlled flow rate. Theralux, Australia, developed photocatalytic water purification technology for the swimming pool industry without compromise to health. The system converts sun tan oil and urea to harmless products in addition to instant destruction of bacteria and pathogens.

9.2.3 Hydrogen Production

Hydrogen, a non-carbon fuel, is considered as a potential energy carrier. Commercialization of photocatalytic materials for hydrogen production is at its developing stage. Even though hydrogen is abundantly found in nature (i.e., in biomass, water resources, and fossil fuels), and its production with zero environmental impact is often exigent. The extraction of H_2 from fossil fuels and by gasification of biomass leads to the emission of greenhouse gases like CO_2 and other pollutants like nitrogen oxides and sulfur oxides to the environment. The current commercialization technologies of hydrogen production include natural gas reforming, bioderived liquids reforming, water electrolysis, thermochemical, PEC, and biological processes for which great challenges include high capital and maintenance costs, carbon capture and sequestration, costly reactors, and low-system efficiency [47]. Among the above mentioned, water electrolysis is considered to be the basic industrial process of nearly pure hydrogen production but still requires enormous electrical energy. All these show that the most environmentally benign method is photonic energy-based hydrogen production (i.e., by PEC method and photocatalysis).

Water splitting is an energetically unfavored process in the thermodynamic point of view. It requires a Gibbs free energy change of 237 kJ/mol to uphill the reaction of transforming H_2O to H_2 and O_2 [48]. The prerequisite of a photocatalytic material to perform hydrogen and oxygen evolution half-reaction is the appropriate positioning of band edge positions with both the oxidation and reduction potentials of water (+1.23 and +0 V vs. NHE, respectively) lying within the bandgap of the photocatalyst. Z-scheme photocatalytic water splitting consists of two semiconductors connected in series with redox shuttles [49]. In PEC water splitting, energy transfer takes place between semiconductor electrode (as photoanode or photocathode) and the electrolyte in which it is immersed, such that the Fermi level of the semiconductor is equalized with the redox potential of the electrolyte. PEC reactions are driven by minority carriers on photoexcitation of photoelectrodes. Even if the Fermi level is not desirably placed, PEC redox reactions can be run by applying an external voltage across the

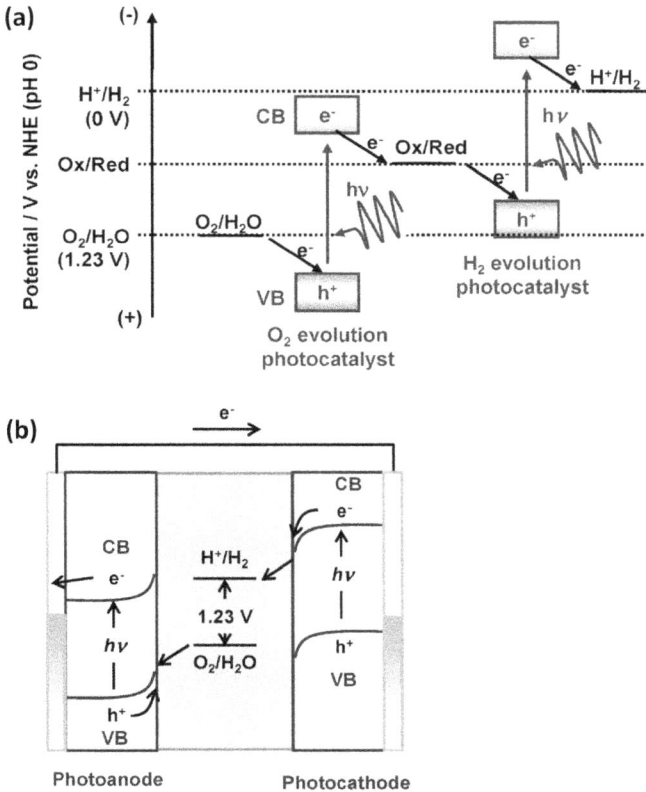

Figure 9.5: (a) Energy diagrams of (a) Z-scheme photocatalytic water splitting, and (b) Z-scheme PEC water splitting [50].

photoelectrode and a counter electrode. Otherwise, it is adopting a Z-scheme PEC water splitting in which a photoelectrode and a photoanode of different semiconductors are tandem connected to each other [50].

Multitudinous photocatalysts have been prepared and proposed for hydrogen evolution reaction (HER), and it is one of the hottest fields of photocatalytic research nowadays. $SrTiO_3$ [51] and $La:NaTaO_3$ [52] along with TiO_2 were the firstly studied photocatalysts for hydrogen production with the major disadvantage of absorbing UV region of solar light. Other UV-reactive systems with appreciable activity are Nb- and Ta-based oxides. In the last years, relevant research was demonstrated on development of visible light water splitting by structural modifications in TiO_2, facet engineering, the combination of TiO_2 with a wide variety of cocatalysts, and also forming heterojunctions of TiO_2 with other semiconductors like ZnO, CdS, SnO_2, Bi_2O_3, Fe_2O_3, MoO_3, ZrO_2, MoS_2, $g-C_3N_4$, or graphene; and their ternary heterojunctions as well with the loading of metal nanoparticles like Au, Ag, Pt, and Pd [53–59]. Besides traditional ones, new photocatalysts have been proposed, of which important ones being $g-C_3N_4$, chalcogenides, oxynitrides, and metal sulfides; and novel photocatalysts such as Sr_2CoWO_6 [60], $MgIn_2S_4$ [61], $AgBiS_2$ [62], $ErVO_4$ [63], NiS [64], SiP_2, and $SiAs_2$[65] are demonstrated in recent years. Lately, 0D($MoS2$)/2D($g-C_3N_4$) heterojunctions in Z scheme were proposed both for photocatalytic and PEC hydrogen evolution, with the introduction of MoS_2 drastically increasing electron transfer and lowering HER potential barriers with enhanced electrical conductivity [66]. Very recently, a 2D–2D $p-MoS_2/n-MgIn_2S_4$ marigold flower-like heterojunction composite is demonstrated with abundant exposed active sites via S–S linkage, capturing electrons from CB of $MgIn_2S_4$ for reducing H_2O to H_2 [61]. Hierarchical structures such as core–shell structures, shelter heterostructures, hollow structures, and raspberry models [67, 68] with enhanced photoactivity are also of particular interest to researchers. But still,

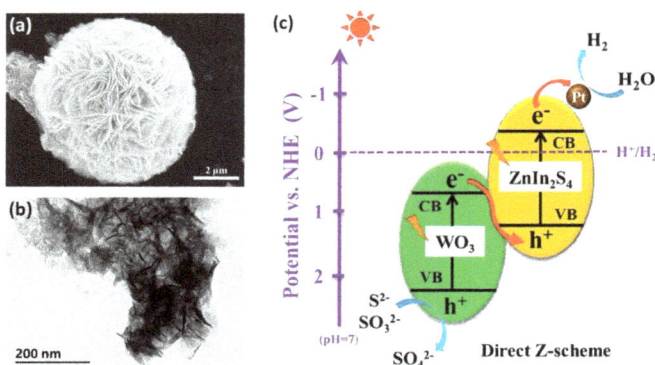

Figure 9.6: An artificially constructed direct Z-scheme $WO_3/ZnIn_2S_4$ heterojunction photocatalyst for efficient hydrogen production. (a) SEM, (b) TEM, and (c) Z-scheme representation on the separation and migration process of photogenerated electrons and holes in the $WO_3/ZnIn_2S_4$ composite [70].

photocatalytic approach of hydrogen fuel generation has a long way to go to deploy these industrially and commercially. The yield per unit mass of the catalyst and efficiency of solar light utilization are the major hurdles faced. Introducing a built-in electric field to enhance the solar efficiency, band structure engineering, inhibiting recombination by reducing the dimension of materials thus shortening the migrating pathway of photogenerated charge carriers, development of different kinds of heterojunctions are the major focuses of today's HER research [69].

9.2.4 Photocatalytic CO_2 Reduction

Anthropogenic CO_2 generated by enormous combustion of fossil fuels brought out global warming and a serious energy crisis. A systematic and effective design of CO_2 capturing techniques is required to address these problems. Photocatalysis mainly involves three principal steps: solar light harvesting, charge separation, and transportation and surface reactions. The photoreduction of CO_2 shows poor efficiency due to some significant issues associated with the aforementioned steps. About 53% of total solar light is in the visible region and only 4% account for UV light [50]. Hence, the development of visible-light-driven photocatalysts improves the photocatalytic efficiency. The photogenerated electrons transfer to the photocatalytic surface and involve in the reduction of the adsorbed CO_2 molecule. Therefore, charge separation efficiency and fast charge transfer can effectively enhance the photoreduction of CO_2. This can be done by successful electron trapping and formation of heterojunctions [71]. The presence of more active sites on the surface of the photocatalyst increases the adsorption of CO_2 molecules and hence facilitates the photoreduction process. Kar et al. synthesized flame-annealed nanotubularTiO$_2$ with a square-shaped cross section [72]. High-temperature annealing leads to phase transformation from anatase to rutile which modifies the nanotube morphology with the circular cross section to the square. This morphology helps to extend the absorption edge to the visible region and boosts the visible light absorption. Adsorption of CO_2 to the photocatalytic surface remains a critical challenge to improve the photocatalytic CO_2 reduction. This can be tackled by creating defect-level vacancies. For example, Gao et al. synthesized vanadium defect-rich o-BiVO$_4$ by hydrothermal method and compared the efficiency of vanadium defect-poor o-BiVO$_4$. The vanadium vacancy-rich o-BiVO$_4$ is more efficient toward CO_2 photoreduction to methanol due to its increased photoabsorption, superior electronic conductivity, and high charge carrier separation [73].

Selectivity of photocatalyst toward CO_2 is another major concern. In the presence of water, since H^+ reduction is more kinetically favorable, HER often competes with CO_2 photoreduction. To tackle this problem, we need to synthesize photocatalysts having a low conduction band edge [75, 76]. Sun et al. synthesized quantum-sized WO$_3$ nanotubes with a large amount of oxygen vacancies, which showed a high efficiency of CO_2 photoreduction to methanol [77]. As we discussed earlier, the adsorption of CO_2 to

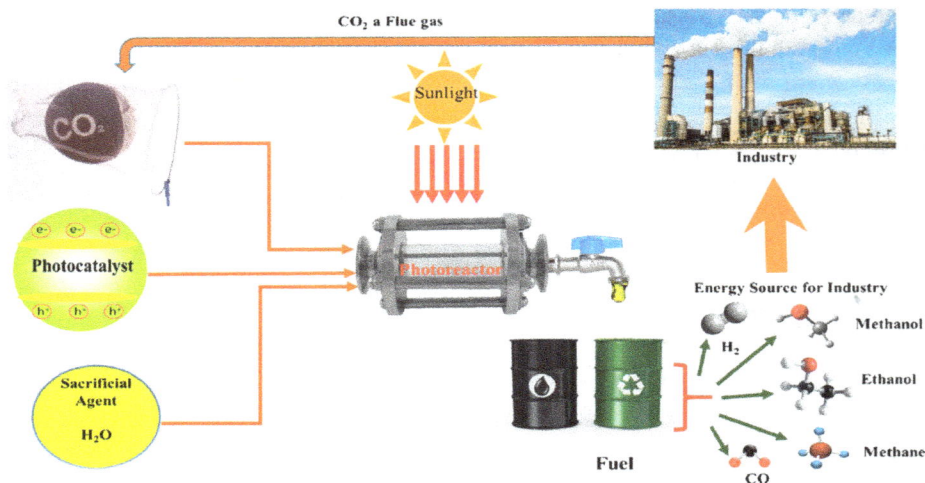

Figure 9.7: Photocatalytic conversion of CO_2 to fuels (mimicking photosynthesis) [74].

catalytic surfaces boosts the CO_2 conversion process. In 2018, Hao et al. developed layered Bi-based photocatalyst ($Bi_2O_2(OH)(NO_3$-BON) for CO_2 conversion. BON contains layered $[Bi_2O_2(OH)]^+$ layers with $[NO_3]^-$ ions aligning along z-axis [78, 79]. This kind of alternate positive- and negative-layered arrangement causes an electric field generation and well-aligned polar $[NO_3]^-$ layer develops polarization field across the BON. Both the electric field and polarization field generated assist the bulk charge separation in the material [80]. Later, in 2019, they halogenated the material to improve the separation efficiency and to boost the CO_2 adsorption to the surface. The halogen ions present on the surface activates the OH group and thereby enhances the CO_2 adsorption process [81]. The fabrication of thin-layered sheets is one of the methods to increase the charge carrier transport as well as their separation [82]. By reducing the thickness of photocatalysts, one can reduce the charge diffusion distance to the surface, thereby enhancing the charge transport process. Besides, the synthesis of ultrathin-layered materials results in the formation of various exposed crystal facets – another powerful factor directing efficient charge separation. Chen et al. developed facet junction controlled thin-layered bismuth-based semiconductor ($BiOIO_3$) by a simple hydrothermal method. They controlled the ratio of exposed {010} top facet to {100} lateral facet by adjusting the acidity of the solution in the hydrothermal process [83]. A better photocatalyst should have more active sites on the surface where the redox reactions occur. Wang et al. in 2018 developed sandwich like ZnI_2S_4–In_2O_3 heterostructure with In_2O_3 microtubes with ZnI_2S_4 sheets aligned in outer and inner surfaces. This hollow heteroarchitecture accelerates the electron–hole pair separation as well as provides large surface area for surface catalysis [84].

9.2.5 Self-Cleaning Surfaces and Coatings

Photocatalytic semiconductor metal oxides find a wider application in self-cleaning coatings including those with antireflective, antimicrobial, antifogging, and anticorrosive properties. Two kinds of self-cleaning methods are available: hydrophilic, which is photoinduced generally exhibited by metal oxides, and the other is hydrophobic. The basic principle underlying superhydrophilicity for such surfaces is that the photogenerated holes diffuse to the surface of the metal oxide, get trapped at the lattice oxygen sites, and finally lead to the formation of hydroxyl linkages with the adsorbed water molecules [85], which in turn increases the affinity of the surface to water. The wettability of the surface to water is often determined by studying the water contact angle (WCA). If the surface is of high surface energy having WCA < 10°, then it can be referred to as a superhydrophilic surface. However, wetting property is dependent on certain other factors like topography, microstructure, and chemical composition of the surface. Various methods such as spray coating, electrospinning, dip coating, layer-by-layer self-assembly, electrochemical method, and plasma treatment are reported to fabricate superhydrophilic self-cleaning surfaces even at large scales for industrial applications.

Transparent self-cleaning coatings with superhydrophilicity, antisoiling, and antireflective nature have attracted them toward PV applications, especially for building-impregnated PV. The sheeting of water like a film by the superhydrophilic coating over the surface which it is coated allows the light ways to pass through rather than being reflected. At the same time, the photocatalytic coating provokes self-cleaning to the panel surface, and hence increases the PV efficiency for a longer period. Due to the low refractive index of SiO_2, introducing photocatalytic materials to such materials is a promising way to develop antireflective films with high mechanical robustness and functional durability. But either antireflectance property or the self-cleaning ability is compromised with the thickness of the coating. However, when a macro-mesoporous structure of TiO_2/SiO_2 is derived using templates of polystyrene nanospheres and Cetyl Trimethyl Ammonium Bromide (CTAB) micelles, the antireflective property significantly enhanced over the range of 350–1,200 nm, with photocatalytic self-cleaning and superhydrophilicity, and highly withstanding to pencil hardness test and sandpaper abrasion test [86]. Recently, morphologically varied ZnO nanostructures grown over Indium tin oxide (ITO) surface were demonstrated with hydrophilic and superhydrophilic self-cleaning coating for PV applications [87]. A commercialized form of antireflective coating is known under the brand name SurfaShield® produced by NanoPhos S. A., which is a water-based suspension of nanocrystalline TiO_2 (anatase/rutile) chemically anchored to SiO_2. The mixture can be air-sprayed to PV modules or glass substrates without any thermal treatment to obtain good adhesion.

Membrane technology is a major attraction in oily wastewater treatment, gas separations, drug removal, humic acid removal, dairy wastewater treatment, etc., due to its eco-friendliness, cost-effectiveness, and easy upscaling. Fouling, requiring expensive and repeating regenerations, is the major drawback faced by the membrane technology. Increasing the hydrophilic nature of membranes is one of the effective techniques. In this way, superhydrophilicity and self-cleaning property of photocatalysts find an important position in antifouling membranes. Recently, Singh et al. demonstrated the fabrication of polysulfone ultrafiltration membrane modified with Cu_2O, showing hydrophilicity and antifouling nature with the capability of visible light–driven photocatalytic pharmaceutical removal [88]. A novel g-C_3N_4/TiO_2/PAA (polyacrylic acid)/polytetrafluoroethylene (PTFE) ultrafiltration membrane is recently fabricated by the surface modification of porous PTFE membranes by g-C_3N_4/TiO_2 via a plasma-enhanced surface graft of PAA before coating g-C_3N_4/TiO_2 layer [89].

Numerous self-cleaning construction materials are produced by depositing photocatalytic nanomaterials into the substrates. Many self-cleaning products are currently available in the market, such as Activ™ self-cleaning glass produced by Pilkington, Active™ Floor and Wall Ceramics, H&C Tiles by Grespania Ceramica, and PARNASOS self-cleaning coatings by COLOROBBIA. Famous spots of application of such construction materials include the Jubilee Church Dives in Misericordia in Rome, Marunouchi Building skyscraper in Tokyo, and the Hospital Manuel Gea Gonzalez in Mexico City [90]. It is also known that photocatalytic TiO_2 is incorporated into various construction materials like glasses, asphalt mixtures, limestone, tiles, cement, and paints for the lifetime maintenance of their aesthetic appearance. As per n-tech Research reports in 2015, the market of self-cleaning research will reach US $3.3 billion in 2020.

9.2.6 Textile Industry

Textiles, often regarded as our second skin, finds great advances with the advent of nanotechnology in terms of enhancing its attributes like water repellence, breathability, softness, durability, antimicrobial property, and fire retardancy. Incorporating photocatalytic materials can impart self-cleaning nature to the textiles with the ability to keep itself tidy and odorless, along with environmentally friendly characteristics like reduced use of detergents, energy-saving, and water-saving. Self-cleaning clothes are actually on the horizons of commercialization. The self-cleaning fabric developed by RMIT in Australia [91], having silver and copper nanoparticles with the renowned ability to absorb visible light and break down organic matter or odor, have already grasped the markets. The case of superhydrophobic fabric is a different approach to attain strong repellence of the fabric to oil/water/dirt. Self-cleaning function by photocatalytic degradation of organic dirt is often driven by the use of photocatalysts, especially UV-driven TiO_2, and ZnO, as well as with visible light-responsive g-C_3N_4 and Bi_2WO_4. Industrially, dip-pad-dry-cure method and dip-coating method are usually employed to construct bonds between the anchored nanoparticles and the fabric. SiO_2 is seen used as a binder on the fabric along with the semiconductor material as it prevents the fabric from photocatalytic degradation [92]. Recently, TiO_2–SiO_2 Janus particle was demonstrated with one half of the

SiO_2 particle covered with TiO_2 nanoparticles and the rest half is available for the attachment with suitable substrates [93], which prevents photodegradation of fabric. Self-cleaning fabric production by coaxial electrospinning is far preferable over non-electrospun fabric, which is due to the high surface to volume ratio of the nanofibrous structures playing a crucial role in enhancing the photocatalytic activity. Lately, a novel electrospun poly(1,4-cyclohexane dimethylene isosorbide terephthalate) nano-fibers embedded with ZnO were fabricated with good self-cleaning performance [94]. Plasma pretreatment of cotton followed by the anchoring of photocatalysts is an effective method to enhance the superhydrophilicity with enhanced coating stability and tensile strength withstanding 50 laundering cycles and 25,000 abrasion cycles [95].

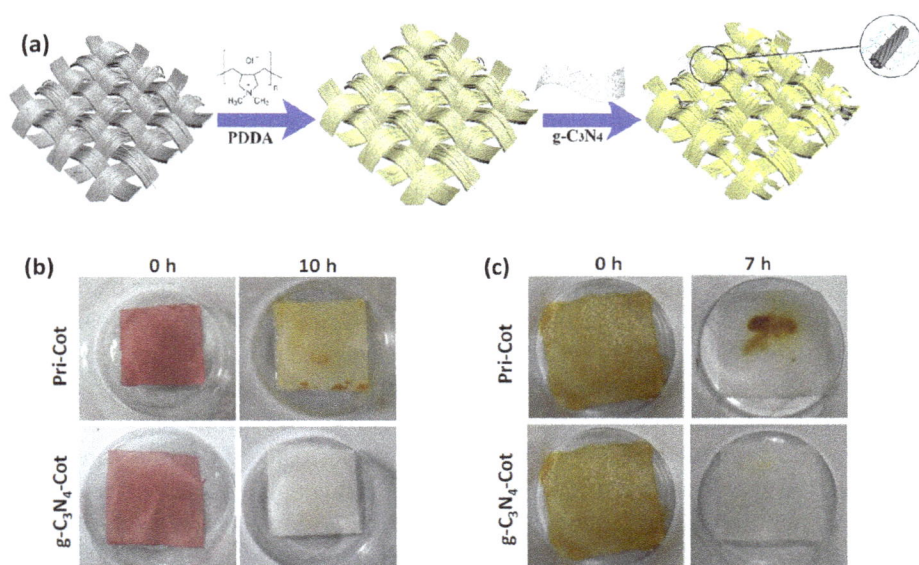

Figure 9.8: Illustration of the process of g-C_3N_4 assembly onto cotton fabrics modified with poly (diallyldimethylammonium chloride) (PDDA) by electrostatic interaction (a). Degradation of fabricated white cotton samples stained with wine red (b) and with coffee (c) after 10 h and 7 h of light irradiation under xenon lamp, respectively [96].

Certain fabrics with superhydrophilicity or underwater superoleophobicity or simulta-neous superhydrophilicity–superoleophobicity have been also demonstrated for oil–water separation. Underwater superoleophobicity is very common in aquatic creatures. For example, shark skin and fish scales have very high oil contact angle underwa-ter and hence have a strong antifouling nature to oil, which helps them to swim in oil–pollutant water without any oil getting adhered to its body [97]. Such biomi-micking finds potential application of oil/water separation in water purification. For oil/water separation, other porous materials like metal meshes and foams are also used in addition to fibers/textiles, to which ceramic and polymers are administered. Li et al. prepared a superhydrophobic fabric exhibiting high oil–water separation efficacy by modifying the surface of the fabric with MnO_2 and stearic acid [98]. Bano et al. showed the fabrication of superhydrophobic meshes coated with TiO_2 and polydimethylsiloxane (PDMS) which is capable of separating four kinds of oil–water mixtures under harsh conditions [99]. But a major difficulty is the clogging of pores by oil droplets leading to the decline of water permeation. Spray-coating a layer of alkylammonium functional silsesquioxane/phytic acid complex followed by a layer of hierarchical structured TiO_2–PDMS composite, imparted cotton fabric superhydrophobicity, and antifouling nature advantageous to oil–water separation, along with flame retardancy, good abrasion resistance, and laundering durability [100].

Multifunctional textiles with the combined antibacterial, UV/NIR (near-infrared) blocking, and self-cleaning properties all springing up in one fabric is another major attraction. TiO_2 photocatalysts along with microcapsules of phase change materials, when applied onto cotton fabric, is reported to have thermal regulation in addition to self-cleaning characteristics [101]. Coating a paint-like slurry of 1:6 mixture of thermochromic core–shell VO_2–TiO_2 and P25 in ethanol to textile can achieve with multiple functions of self-cleaning, thermoregulation, and water-proofing [102]. Recently, dual functions of UV/NIR shielding and full-spectrum-responsive self-cleaning on the cotton fabric have been simultaneously obtained by a facile electrostatic self-assembly of Cs_xWO_3 nanosheets on cotton fabric assisted by a surface modifier of poly(diallyldimethylammonium chloride) [103]. Environmentally benign high-performing breathable membranes with self-cleaning performance are the recent approach of textile research. ZnO introduced to the surface of electrospun polyacrylonitrile nanofibers along with a water-based finishing agent (PM3635) imparted integrated properties with prominent hydrostatic pressure, high water vapor transmission rate, excellent tensile strength, along with significant chemical degradation capability applicable as wearable protective garments in health care and military uniform systems [104].

The currently available photocatalytic self-cleaning textiles in the market includes SELFCLEAR® – TiO_2-incorporated self-cleaning fiber developed and marketed by Japan Exlan Co. Ltd. with the fiber having antibacterial and deodorizing properties. ARCI, India, has developed self-cleaning garments along with UV protection under the tag name "Sun Wash" using suspension of titania microspheres surface engineered with ultrafine silver nanoparticles. The commercialization of self-cleaning textiles with photocatalysts is still on its developing stages. The other self-cleaning textiles available in the market are based on superhydrophobic coatings mostly marketed by Luna, Sharklet Technologies Inc., etc., which are micropatterned surfaces resulting with water repellency and resistant to bioadhesion.

9.2.7 Antimicrobial Disinfection

The photogenerated reactive oxygen species (ROS) produced by the photocatalytic materials are capable of causing cell death in microorganisms by first attacking their cell membrane leading to lipid peroxidation followed by the oxidative attack of the internal components [105], or by the oxidation of coenzyme A that precedes the inhibition of respiration [106], or by impairing their biological functioning [107]. This imparts antimicrobial/antibacterial properties to most photocatalytic materials. When such materials are used on contact surfaces of hospital rooms and washrooms tackles hospital-acquired infections. Moreover, the use of such photocatalysts is also an effective method for the disinfection of highly pathogenic microbes and toxins in the water and is a substitute for chemical disinfectants like chlorine, ozone, and chloramines. Matsunaga et al. were the pioneers to exploit TiO_2 photocatalysts for the disinfection of *Escherichia coli*, *Lactobacillus acidophilus*, and *Saccharomyces cerevisiae* [106]. Since then, several photocatalytic materials have been demonstrated for their antimicrobial property against bacteria, protozoa, fungi, algae, and viruses [108]. Photocatalytic Fe_3O_4, TiO_2, Au, ZnO, and CuO are reported to inhibit the harmful biofilm formation which gives resistivity to drugs in some dangerous superbacteria like *S. aureus* [109]. Composites with g-C_3N_4, fullerene, graphene, Ag_3PO_4, and certain dopants like Cu, Ni, Ce, and Au have imparted visible light antimicrobial activity in TiO_2. $BiVO_4$, g-C_3N_4, MoS_2, Bi_2WO_6, $NiFe_2O_4$, Bi_2MoO_6, $ZnFe_2O_4$, $NiMoO_4$, etc. are low bandgap candidates potential for visible light disinfection process. Decorating these photocatalysts with Au, Ag, and AgX (X = halide) can enhance the bactericidal activity considerably due to the surface plasmon resonance, while Ag^+ ions (in the form of AgX) with intrinsic antibacterial activity also play a crucial role in imparting disinfection property [109]. Still, several issues persist concerning the entry of photocatalysts into the structural integrities of microbes. Understanding the precise mechanism of disinfection could help in the development of efficient photocatalysts that could be fed into some reactors and utilization for the large-scale disinfection of pathogenic microbes/toxins in water effluents from hospitals, industries, and biomedical labs.

Very recently, Kastus® Technologies, an Irish multinational nanotechnology company, developed photocatalytic antimicrobial and antiviral coating that can be applied from ceramic facades to wearable devices and touchscreens, and have proven to block human coronavirus and $99.99\%^2$ of surface bacteria such as *S. aureus* and *E. coli* [111]. PURETi and Nanoland Global are other startups using TiO_2 as the antimicrobial surface cleaner and air purifier that are active with indoor visible light or special UV-A lighting systems making safer environments.

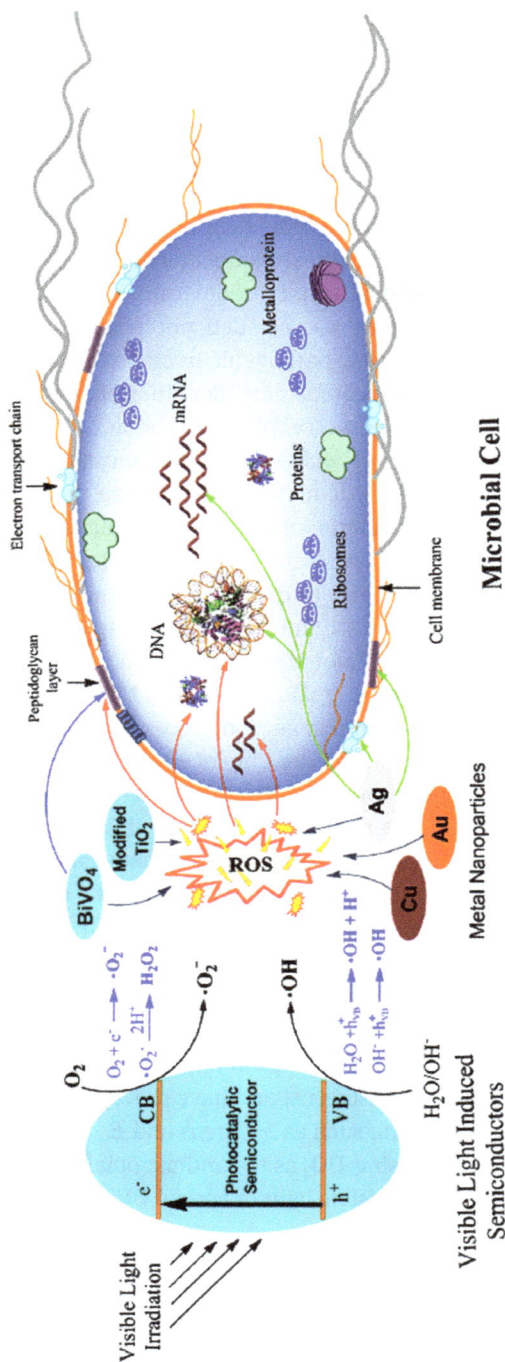

Figure 9.9: The possible antimicrobial mechanism portrayed for different photocatalysts. Photoexcitation of the semiconductor is shown on the left with the generation of reactive oxygen species (ROS), and the red arrows point the targets of ROS. Blue arrow represents the target of $BiVO_4$. Ag, Au, and Cu also generate ROS, with green arrows showing the targets of Ag nanoparticles [110].

9.2.8 Photodetectors

In the last decades, photodetector devices have been explored greatly by researchers for their efficient construction of sensing systems for the wide range of applications including communication, imaging, and environmental monitoring. Photodetectors convert the optical signal to an electrical signal. Upon light irradiation, electron–hole pairs are generated and separated by an in-built electric field in the material or external electric field supplied. Then these photogenerated charge carriers are transported to an external circuit to generate photocurrent. A wide variety of materials like perovskite, semiconductor oxide materials, 2D materials, etc. have been inspected thoroughly [112–114]. Among these, semiconductor nanomaterials are most investigated and are useful building blocks employing photocatalysis mechanism for efficient photodetector applications [115]. The surface effect in the presence and absence of light has much impact on the performance of photodetectors [116]. Besides, fast electron–hole pair recombination reduces the efficiency of photodetectors [117]. The surface effect and fast electron–hole pair recombination led to poor and slow photoresponse for finer devices. Researchers demonstrated a great many strategies like heterojunction formation [118], defect engineering [119], morphology, and surface modification [120] to rectify the limitations of photodetectors. The development of heterojunctions in the photodetector material enhances the electron–hole pair separation by transferring them to different electrodes. Lee et al. synthesized 1D ZnO–2D WSe_2 nanomaterial for van der Waals heterojunction device. The nanomaterial consists of 1D n-type ZnO nanowire and p-type 2D WSe_2 nanosheet. When irradiated with visible and NIR light source, electron–hole pairs are generated only in WSe_2 nanosheet with lower bandgap energy, $E_g = 1.3$ eV. The holes then move to the conduction band of the WSe_2 valence band and electrons move toward the conduction band of ZnO. This results in the generation of photocurrent. They also suggested that the device shows a better performance of photoswitching behavior [121]. $WSe_2/SnSe_2$ van de Waals heterostructures demonstrated to exhibit superior photoresponse properties due to their tailorable band engineering characteristic. The formation of heterojunction reduces the interface barrier of charge carriers by moving photogenerated electron toward $SnSe_2$ and hole toward WSe_2 [122]. The photoresponse of the detectors can be enhanced by increasing the concentration of photogenerated charge carriers. The improvement in the light-harvesting property of materials paves the way to generate a high concentration of electron–hole pairs. Shan et al. reported the fabrication of ZnO nanowire supported by ZnO film shows a good light trapping performance due to large surface area and long optical path [123]. ZnO nanowires act as light trappers and transporter of charge carriers toward ZnO film where the charge carriers are collected by suitable bias electrodes. Two-dimensional-layered MoS_2 has been explored for photodetector device applications but the main drawback is its low visible absorption range. One way to tackle the problem is by incorporating plasmonic nanoparticles with MoS_2. Gold nanoparticles are integrated into the MoS_2 layer by magnetron sputtering technique resulting in enhanced interaction of light with MoS_2 by surface plasmon effect [124].

Photodetector performance also relies upon different nanostructures of the material. The photogenerated charge carriers can transport to electrodes by adopting different dimensions based on the structure of a device material: 0D, 1D, 2D, composites, and arrays are the various nanostructures developed by researchers for photodetector applications. Zero-dimensional carbon-doped ZnO quantum dots were developed by femtosecond laser ablation in liquid technique, which exhibited a good deep UV photodetection with a wavelength less than 280 nm. A stable and flexible device characteristic is achieved by 0D structural property of the material [125]. Hammed et al. synthesized vertically aligned ZnO nanorod by sonochemical method. They first deposited Ti/Zn and Ti/ZnO/Zn on glass substrate by RF sputtering technique. ZnO nanorods grown on Ti/ZnO/Zn seed layer shows rapid response and recovery time [126]. Jiang et al. fabricated photodetectors of selenium quantum dots by facile liquid-phase exfoliation technique, which showed excellent photoresponse behavior resulted from the unique chain-like structure and structure-dependent photocarrier pathways [127]. Self-powered photodetectors are nowadays gaining attention due to its potential application in visible light communication with the unique advantages of energy-saving and cost-effectiveness. An array of ZnO NWs/Sb_2Se_3 heterojunction developed as a self-powered photodetector shows dual-polarity response due to competition between PV and photo-thermoelectric effect, and the response can be increased by incorporation of n-type Sb_2Se_3 films to the system [128].

9.2.9 Gas Sensing

Very intensive research has been carried out for the detection of hazardous gases such as CO, NO_2, NH_3, and CO_2. The detection of gas is done by observing the change in resistance due to charge transfer by the interaction between gas and sensor material and band bending induced by charged molecules. Metal oxide semiconductors such as ZnO, SnO_2, Fe_2O_3, In_2O_3, and WO_3 are the most investigated materials for gas sensing applications [129–131]. At high temperatures or on photo-excitations, oxygen molecules are adsorbed on the surface of n-type metal oxide semiconductors and trap electrons at the surface from the conduction band and leads to formation of electron depletion region. In the presence of reducing gases, the oxygen molecule reacts with these gases and increases electron density and thereby reduces the resistivity of the material. For p-type semiconductors, upon gas exposure the resistivity increases (electron density is reduced).

Conventional gas sensors use high temperature (above 200 °C) for activation since the adsorption and generation of surface oxygen ions occurs only at temperature of 300–450 °C. This high-temperature activation adversely affects the sensing of flammable gases and stability of the sensor. These limitations place photoinduced

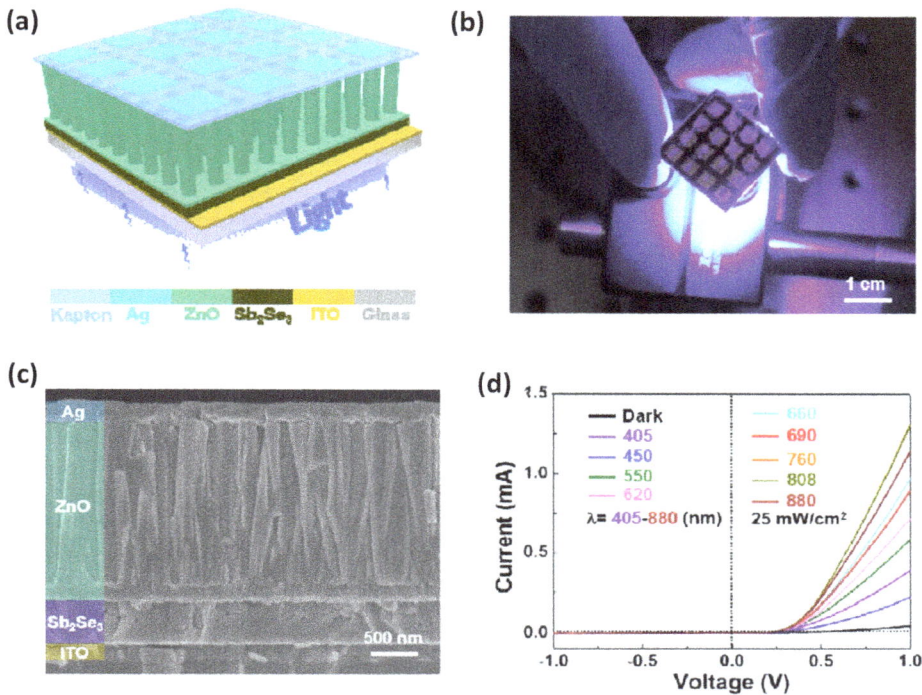

Figure 9.10: Schematic diagram (a) and optical image (b) of self-powered ZnO NWs/Sb_2Se_3 heterojunction photodetector array. Cross-sectional SEM image (c) of the fabricated device. *I–V* characteristics (d) of the photodetector under dark and different wavelengths of light illumination (25 mW/cm^2) [128].

room temperature gas sensors in high demand for commercial applications. Cui et al. synthesized CeO_2–ZnO by simple hydrothermal method for low-temperature acetone sensing. The sample shows 68% greater response in the presence of visible light than that without visible light at 75 °C [132]. Room-temperature NH_3 gas sensing upon UV irradiation is carried out using PANI/TiO_2 core–shell nanofiber which was suggested good for breath analysis due to its high sensitivity and low detection limit or various VOCs [133]. The morphology of nanostructures also plays a significant role in the detection of gaseous molecules. ZnO nanorod/Au hybrid showed visible light–induced NO_2 and NH_3 detection at room temperature with better reversibility and selectivity [134]. While polystyrene sulfonate functionalized ZnO nanowires showed fast response toward ppb-level NO_2 sensing upon UV light irradiation [135]. Careful fabrication of device is another requirement for proper working of gas sensors. Typically, the materials were drop-casted on interdigitated electrodes of silicon substrates as illuminated using a light source [130]. The concentration of target gas is controlled by mixing the gas with dry gas using a mass flow controller. RH/temperature testing device is incorporated into machine to understand the working temperature and relative humidity of the sensor. The output resistor is measured using Keithley Instruments attached to the device which in turn is connected to computer for analysis.

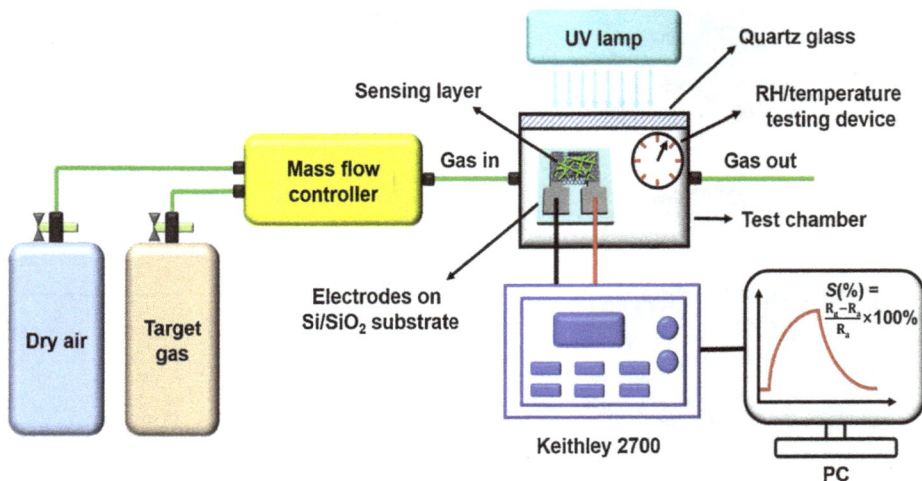

Figure 9.11: Schematic illustration of a typical gas-sensing setup [130].

9.2.10 Ion Photoreduction

Today, high-valent carcinogenic ions in wastewater are important pollutants that have become a severe threat to human health. Various research works suggest that the photocatalytic reduction process could be one of the effective pathways to reduce these carcinogenic high-valent ions to noncarcinogenic low valent ions. High valent toxic and nonbiodegradable chromium, Cr(VI), is a by-product of multiple industrial processes such as paint making, leather tanning, and electroplating [136]. So removal of Cr(VI) from wastewater is indispensable for environmental and human health. Khare et al. developed red-emitting high quantum yield ZnO-doped carbon dots for the aqueous-phase photoreduction of carcinogenic hexavalent Cr(VI) to less harmful low-valent Cr(III) [137]. Bhati et al. synthesized nitrogen –phosphorus-doped fluorescent carbon dots by microwave synthesizer. The material reduces Cr(VI) to Cr(III) using natural sunlight. The reduced form Cr(III) can be readily precipitated out and easily removed [138].

Uranyl ion (U(VI)) accumulation in groundwater and surface water systems by the improper release of nuclear waste to the environment from nuclear power plants is another serious issue [139, 140]. According to the studies, in nuclear industrial wastewater, the concentration of U(VI) reached 28.76 ppm, 0.3–1442.9 ppb in groundwater, and 0.04–9.9 ppb in drinking water [141, 142]. Uranium is highly chemotoxic as well as radiotoxic. Arsenium is another toxic element present in groundwater and industrial water that causes kidney and bladder cancer [143]. Jiang et al. synthesized g-C_3N_4/TiO_2 heterostructures for the simultaneous removal of U(VI) and As(III). They coupled photoreduction of U(VI) to U(IV) and photooxidation of As (III) to As(V). As the concentration of As(III) increases, the photoreduction of U(VI) decreases and photooxidation of As(III) increases [144]. This work shows a vast prospective future of the photoreduction process in the removal of toxic heavy metals from industrial wastewater. In the upper ocean, photoreduction plays an important role to deliver bioavailable Fe(II), essential for marine and freshwater organisms such as nitrate-reducing, phototrophic or microaerophilic Fe(II)-oxidizing bacteria. Most recently, Lueder and coworkers have shown that Fe(III) photoreduction has a high impact on maintaining the Fe cycle in the ocean and they provided a quantitative analysis on photochemically produced Fe(II) in connection with the dissolved organic content [145].

9.2.11 Photovoltaics: Dye-Sensitized Solar Cells

A large variety of solar cells are available in the market, of which dye-sensitized solar cells (DSSC) belonging to thin-film solar cells are attracting both scientific and technological interest and are an alternative to p–n junction PV devices. DSSCs differ from other conventional solar cells in the manner that the functions of light absorption are separated from transport of charge carriers. Initially, visible light photoexcitation occurs in the dye hybridized with semiconductors that not only act as dye supporter but as an electron acceptor (usually from the LUMO of the dye molecule to the conduction band of the semiconductor) and electronic conductor. This results in the current flow across the semiconductor to the charge-collecting electrode and then to the external circuit. Semiconductor metal oxides such as TiO_2, Nb_2O_5, SnO_2, and ZnO, find wide applications in DSSCs and belong to third-generation PV materials with cost-effectiveness and easy processability [146].

DSSCs generally consist of a photoanode which is made up of mesoporous semiconductors, typically like TiO_2, deposited onto transparent conducting substrates of glass (ITO or Fluorine-doped tin oxide (FTO)). A monolayer of dye sensitizer bonded chemically to the surface of the mesoporous semiconductor absorbs light energy, while an organic solvent as electrolyte collects electrons at the counter electrode (usually platinum-coated conductive glass). These electrons then transported through external load to produce electric power. Apart from TiO_2, other semiconductors such as WO_3, ZnO, Nb_2O_5, and Fe_2O_3 have been investigated for DSSC applications by varying morphologies like nanorods, nanobelts, aerogels, nanotubes, and hierarchical structures [147–150]. Considering the dye molecules, ruthenium, porphyrin, natural, and metal-free organic dyes attract much interest in DSSC applications. Kumar et al. synthesized rGO–TiO_2 hybrid samples and co-sensitized with ruthenium-based dyes. They compared the external quantum efficiency (EQE) of devices co-sensitized with metal-based and metal-free organic dyes. Ruthenium metal-based complexes show wide absorption region and higher EQE [151]. Efficient DSSC devices were fabricated based on TiO_2 using Zn–porphyrin complex as a sensitizer. The good spectral and electrochemical characteristic of the designed sensitizer helps to achieve 5.33% efficiency for DSSC [152]. Researchers have extracted dyes from natural sources such as roots, leaves, flowers, and stems as sensitizers in DSSC. A maximum efficiency of only 2% is achieved with these sensitizers due to their limited absorption region. Maurya et al. extracted dye from *Cassia fistula* flower, showing broad absorption between 400 and 500 nm. The absorption region is redshifted in dye-loaded TiO_2 thin film due to modified Highest energy Occupied Molecular Orbital-Lowest energy Unoccupied Molecular Orbital energy level of TiO_2–dye complex and they correlated the results with density functional calculations [153]. Four different aromatic metal-free electron donor molecules were developed for sensitizers and found that triphenylene-based dye molecule in DSSC shows excellent photoconversion efficiency [154]. Commonly used counter electrodes for DSSC application is platinum and carbon-based materials due to good electrocatalytic activity and conductivity. Li et al. developed TiO_2-based DSSC at low temperature by atomic-layer deposition using platinum as counter electrode [155]. High cost of platinum, low abundance, and formation of platinum (IV) iodide limits its use in DSSC. Murugadoss and colleagues developed platinum-free cobalt nickel selenide/graphene nanohybrid by simple hydrothermal method to use as counter electrode in DSSC. The synthesized low-cost nanohybrid material shows better photocatalytic activity and low charge resistance compared to platinum electrode [156].

Cost-effective, nonhazardous, and environmentally friendly nature of DSSCs make them to use as sustainable energy source and widely accepted for commercial applications. G24 innovations developed inexpensive, silicon- and cadmium-free first commercial DSSCs with less than 1 mm thickness, which can be used in cameras, portable electronics, and powering cell phones and can be embedded into a tent for providing light for camping events. Dai et al. from the Institute of Plasma Physics have designed a 500 W DSSC primary power station on roof top as an initial step toward outdoor commercial applications. A Japanese company developed solid-state DSSCs which can be incorporated into consumer electronics and sensors. They demanded that the DSSCs can operate under indoor light also. They built DSSCs into office desk that can produce electricity with indoor LED lighting. It supplies continuous electricity for recharge portable devices. The company is currently working to develop transparent cell for mobile usage.

9.3 Future Perspectives

Despite certain commercialized applications of photocatalytic materials especially self-cleaning windows to some portable depollution systems are already available in the market, the fundamental research in photocatalysis will continue to be a hot topic in the near future. TiO_2, as a cost-effective and eco-friendly semiconductor, will continue to dominate its market at least for the short term, particularly for massive uses like in water treatment. Among other commercially introduced semiconductors, ZnO can be applied in some areas, but its photostability is a major concern [157]. A constant investigation of novel phases as potentially improved photocatalysts is most likely leading to exciting results shortly. The integration of theoretical simulations and advanced techniques with no doubt can lead to enhanced performance of materials in the visible region of the solar spectrum. Still, under the industrial point of view, the development of viable photocatalysts is crucial which is profitable but at the same time visible light–responsive and eco-friendly.

   H_2 production and CO_2 photoreduction are the two major challenging artificial photosynthesis reactions that produce chemical energy from solar energy. To date, considerable advances in both fields through in-depth mechanism investigation and controlled construction and modifications of the catalyst surface have been carried out. But the scientific and technological abilities still have a long way to go to make these artificial photosynthesis processes for practical use. In comparison to photocatalytic water splitting for H_2 production, CO_2 photoreduction is still in its infancy, mainly in terms of activating and absorbing CO_2 molecules to the photocatalyst surface and also competing simultaneous reactions. Some effective strategies applied for H_2 production, such as heterojunctions, interfacial layers, surface coating layers, and photocatalytic sheets, can address the challenges of CO_2 reduction as well [158]. Hybrid systems consisting of artificial photocatalysts with some natural photocatalytic enzymes (e.g., photosystems I and II) can be considered, which clubs the merits of both synthetic and natural photosynthetic systems [158]. Properties like wettability of the catalyst surface, interactions, performances under acidic and basic conditions, and the internal electric field created need to be considered and rationalized for the effective catalyst design.

   The design of photoreactors is an essential factor in increasing the commercial and economic viability of the photocatalytic materials. In fact, the accurate engineering approach of the reactor enhances the efficiency of the photocatalytic process. In a typical laboratory-scale photoreactor, all principal steps of photocatalysis take place within the reactor. The reactor designing should be done very carefully by evaluating all the possible factors that influence the performance. The light source, the geometrical configuration of the reactor, percentage of light that reaches the photocatalytic material, materials used for reactor designing, and removal of heat generated as a result of light irradiation are the main key factors that must get much attention during the construction of the reactor. However, still no clear information about the reactor designs, especially for industrial H_2 production and CO_2 photoreduction, is known to date. All studies so far were done at the laboratory scale with personalized microreactors using small volumes of precursor mixtures. A long way has to be gone investigating such large-scale reactors that are capable to work under solar light, and such methodologies have to be commercialized. Of course, the technology to simultaneously separate and store the products until usage is also required.

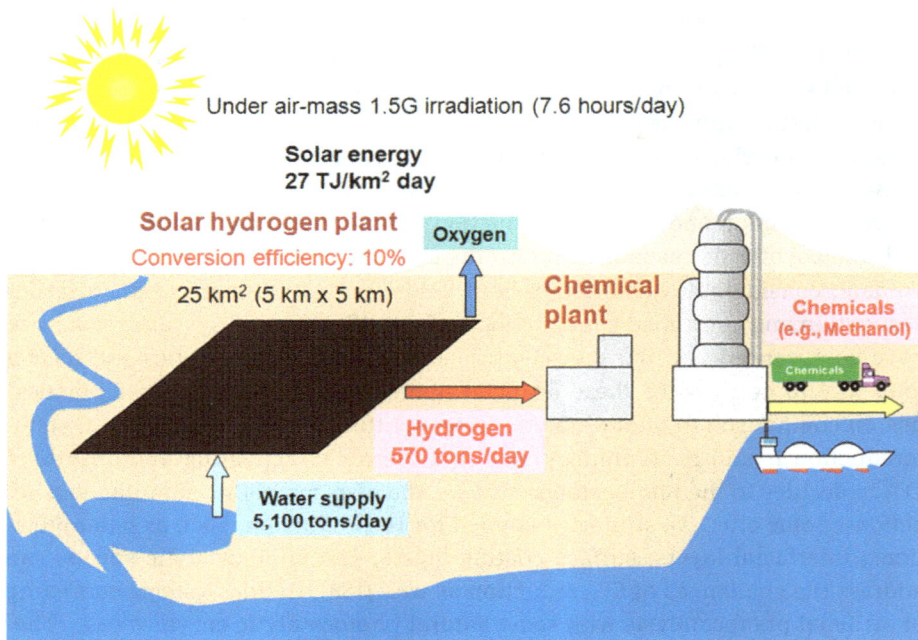

Figure 9.12: Possible scheme for large-scale H_2 production via solar water splitting [159].

The production and application of photocatalytic materials as in many products and technologies are increasing rapidly but a limited number of studies have gone through the potential risk of these materials both for the environment and for humans. Toxic mechanisms associated with a photocatalytic semiconductor is mainly associated with (a) the nanoparticle size which is detrimental to biological molecules, (b) solubility properties with the release of metal ions to the solution, and (c) the most important being the production of ROS during the photoexcitation of the semiconductors that leads to oxidation of protein or an oxidative stress to cellular organisms resulting in cellular death. As the shreds of evidence for toxic effects of some nanoengineered materials are emerging, intense research and assessment on the correlation of physicochemical properties of each kind of developed semiconductor to their toxicity is required. The toxicity both on photoilluminated and nonilluminated conditions should be considered for the photocatalytic materials. Thus, it is clear that intense interdisciplinary research is urged in developing green technology for photocatalysts at the industrial level accompanied by environmental safety to palliate unforeseen consequences and thereby contribute materials for sustainable energy production and environmental remediation.

9.4 Conclusions

This chapter displayed the applications of both UV and visible light–responsive photocatalytic materials on the industrial point of view. Certain photocatalyst-based commercialized products are also presented. Still we note that the field is having a longer pathway in perfectly exploring such photocatalytic materials as final products in the market. The photocatalytic research and industry faces low efficiency of photocatalysts in solar-to-chemical energy conversion. Therefore, for the broader and scaled-up commercialization of photocatalytic materials, development and optimization of promising visible light–harvesting candidates with improved performance along with high physical and chemical stability is essential. Moreover, a good understanding about the toxic effects of these photocatalytic products on both flora and fauna is crucial before any such commercialization for the day-to-day use.

References

[1]   Fujishima A, Honda K. Electrochemical photolysis of water at a semiconductor electrode, Nature, 1972, 238, 37–38.
[2]   Bard AJ. Photoelectrochemistry and heterogeneous photo-catalysis at semiconductors, Journal of Photochemistry, 1979, 10, 59–75.
[3]   Herrmann J-M. Heterogeneous photocatalysis: fundamentals and applications to the removal of various types of aqueous pollutants, Catalysis Today, 1999, 53, 115–129.
[4]   Xu C, Ravi Anusuyadevi P, Aymonier C, Luque R, Marre S. Nanostructured materials for photocatalysis, Chemical Society Reviews, 2019, 48, 3868–3902.
[5]   Ashwin Kishore MR, Ravindran P. Tailoring the electronic band gap and band edge positions in the C2N monolayer by P and as substitution for photocatalytic water splitting, The Journal of Physical Chemistry C, 2017, 121, 22216–22224.
[6]   Primc D, Zeng G, Leute R, Walter M, Mayrhofer L, Niederberger M. Chemical substitution – Alignment of the surface potentials for efficient charge transport in nanocrystalline TiO2 photocatalysts, Chemistry of Materials, 2016, 28, 4223–4230.
[7]   Kumaravel V, Mathew S, Bartlett J, Pillai SC. Photocatalytic hydrogen production using metal doped TiO2: a review of recent advances, Applied Catalysis B: Environmental, 2019, 244, 1021–1064.
[8]   Xing W, Chen G, Li C, Han Z, Hu Y, Meng Q. Doping effect of non-metal group in porous ultrathin g-C3N4 nanosheets towards synergistically improved photocatalytic hydrogen evolution, Nanoscale, 2018, 10, 5239–5245.
[9]   Meng A, Zhang L, Cheng B, Yu J. Dual cocatalysts in TiO2 photocatalysis, Advanced Materials, 2019, 31, 1807660.
[10] Liang X, Wang P, Li M, et al. Adsorption of gaseous ethylene via induced polarization on plasmonic photocatalyst Ag/AgCl/TiO2 and subsequent photodegradation, Applied Catalysis B: Environmental, 2018, 220, 356–361.
[11] Youssef Z, Colombeau L, Yesmurzayeva N, et al. Dye-sensitized nanoparticles for heterogeneous photocatalysis: cases studies with TiO2, ZnO, fullerene and graphene for water purification, Dyes and Pigments, 2018, 159, 49–71.

[12]   Lu H, Hao Q, Chen T, et al. A high-performance Bi2O3/Bi2SiO5 p-n heterojunction
        photocatalyst induced by phase transition of Bi2O3, Applied Catalysis B: Environmental,
        2018, 237, 59–67.
[13]   Li B, Lai C, Zeng G, et al. Facile hydrothermal synthesis of Z-Scheme Bi2Fe4O9/Bi2WO6
        heterojunction photocatalyst with enhanced visible light photocatalytic activity, ACS Applied
        Materials & Interfaces, 2018, 10, 18824–18836.
[14]   Balbuena J, Cruz-Yusta M, Cuevas AL, et al. Hematite porous architectures as enhanced air
        purification photocatalyst, Journal of Alloys and Compounds, 2019, 797, 166–173.
[15]   Wan Z, Zhang G. Synthesis and facet-dependent enhanced photocatalytic activity of Bi2SiO5/
        AgI nanoplate photocatalysts, Journal of Materials Chemistry A, 2015, 3, 16737–16745.
[16]   Katsumata K-I, Motoyoshi R, Matsushita N, Okada K. Preparation of graphitic carbon nitride
        (g-C3N4)/WO3 composites and enhanced visible-light-driven photodegradation of
        acetaldehyde gas, Journal of Hazardous Materials, 2013, 260, 475–482.
[17]   Sun R, Shi Q, Zhang M, et al. Enhanced photocatalytic oxidation of toluene with a coral-like
        direct Z-scheme BiVO4/g-C3N4 photocatalyst, Journal of Alloys and Compounds, 2017, 714,
        619–626.
[18]   Jiang G, Li X, Lan M, et al. Monodisperse bismuth nanoparticles decorated graphitic carbon
        nitride: enhanced visible-light-response photocatalytic NO removal and reaction pathway,
        Applied Catalysis B: Environmental, 2017, 205, 532–540.
[19]   Xia D, Hu L, He C, et al. Simultaneous photocatalytic elimination of gaseous NO and SO2 in a
        BiOI/Al2O3-padded trickling scrubber under visible light, Chemical Engineering Journal,
        2015, 279, 929–938.
[20]   Zhu Z, Liu F, Zhang W. Fabricate and characterization of Ag/BaAl2O4 and its photocatalytic
        performance towards oxidation of gaseous toluene studied by FTIR spectroscopy, Materials
        Research Bulletin, 2015, 64, 68–75.
[21]   Boyjoo Y, Sun H, Liu J, Pareek VK, Wang S. A review on photocatalysis for air treatment: from
        catalyst development to reactor design, Chemical Engineering Journal, 2017, 310, 537–559.
[22]   Passalia C, Alfano OM, Brandi RJ. Integral design methodology of photocatalytic reactors for
        air pollution remediation, Molecules, 2017, 22.
[23]   Qin Y, Wang Z, Jiang J, Xing L, Wu K. One-step fabrication of TiO2/Ti foil annular photoreactor
        for photocatalytic degradation of formaldehyde, Chemical Engineering Journal, 2020, 394,
        124917.
[24]   Claes T, Dilissen A, Leblebici ME, Van Gerven T. Translucent packed bed structures for high
        throughput photocatalytic reactors, Chemical Engineering Journal, 2019, 361, 725–735.
[25]   Bueno-Alejo CJ, Hueso JL, Mallada R, Julian I, Santamaria J. High-radiance LED-driven
        fluidized bed photoreactor for the complete oxidation of n-hexane in air, Chemical
        Engineering Journal, 2019, 358, 1363–1370.
[26]   Hosseini S, Moghaddas H, Masoudi Soltani S, Kheawhom S. Technological applications of
        honeycomb monoliths in environmental processes: a review, Process Safety and
        Environmental Protection, 2020, 133, 286–300.
[27]   https://patentsgooglecom/patent/US6761859B1/en, https://wwwexplainthatstuffcom/
        how-photocatalytic-air-purifiers-workhtml, http://wwwabundantearthcom/store/
        SP20CAirPurifierhtml
[28]   World Health O, United Nations Children's Fund. Progress on Drinking Water, Sanitation and
        Hygiene: 2017 update and SDG baselines, World Health Organization, Geneva, 2017.
[29]   Ribeiro AR, Nunes OC, Pereira MFR, Silva AMT. An overview on the advanced oxidation
        processes applied for the treatment of water pollutants defined in the recently launched
        Directive 2013/39/EU, Environment International, 2015, 75, 33–51.

[30] Pedrosa M, Pastrana-Martínez LM, Pereira MFR, Faria JL, Figueiredo JL, Silva AMT. N/S-doped graphene derivatives and TiO2 for catalytic ozonation and photocatalysis of water pollutants, Chemical Engineering Journal, 2018, 348, 888–897.

[31] Yan S, Li Y, Xie F, et al. Environmentally safe and porous MS@TiO2@PPy monoliths with superior visible-light photocatalytic properties for rapid oil–water separation and water purification, ACS Sustainable Chemistry & Engineering, 2020, 8, 5347–5359.

[32] Zhang C, Zhang M, Li Y, Shuai D. Visible-light-driven photocatalytic disinfection of human adenovirus by a novel heterostructure of oxygen-doped graphitic carbon nitride and hydrothermal carbonation carbon, Applied Catalysis B: Environmental, 2019, 248, 11–21.

[33] Wei F, Li J, Dong C, Bi Y, Han X. Plasmonic Ag decorated graphitic carbon nitride sheets with enhanced visible-light response for photocatalytic water disinfection and organic pollutant removal, Chemosphere, 2020, 242, 125201.

[34] Gadisa BT, Appiah-Ntiamoah R, Kim H. In-situ derived hierarchical ZnO/Zn-C nanofiber with high photocatalytic activity and recyclability under solar light, Applied Surface Science, 2019, 491, 350–359.

[35] Mishra P, Patnaik S, Parida K. An overview of recent progress on noble metal modified magnetic Fe3O4 for photocatalytic pollutant degradation and H2 evolution, Catalysis Science & Technology, 2019, 9, 916–941.

[36] Kong L, Dong Y, Jiang P, Wang G, Zhang H, Zhao N. Light-assisted rapid preparation of a Ni/g-C3N4 magnetic composite for robust photocatalytic H2 evolution from water, Journal of Materials Chemistry A, 2016, 4, 9998–10007.

[37] Zheng J, Zhang L. Designing 3D magnetic peony flower-like cobalt oxides/g-C3N4 dual Z-scheme photocatalyst for remarkably enhanced sunlight driven photocatalytic redox activity, Chemical Engineering Journal, 2019, 369, 947–956.

[38] Shi Z, Xiang Y, Zhang X, Yao S. Photocatalytic activity of Ho-doped anatase titanium dioxide coated magnetite, Photochemistry and Photobiology, 2011, 87, 626–631.

[39] Yao H, Fan M, Wang Y, Luo G, Fei W. Magnetic titanium dioxide based nanomaterials: synthesis, characteristics, and photocatalytic application in pollutant degradation, Journal of Materials Chemistry A, 2015, 3, 17511–17524.

[40] Yaparatne S, Tripp CP, Amirbahman A. Photodegradation of taste and odor compounds in water in the presence of immobilized TiO2-SiO2 photocatalysts, Journal of Hazardous Materials, 2018, 346, 208–217.

[41] İkizler B, Peker SM. Synthesis of TiO2 coated ZnO nanorod arrays and their stability in photocatalytic flow reactors, Thin Solid Films, 2016, 605, 232–242.

[42] Zhao P, Qin N, Wen JZ, Ren CL. Photocatalytic performances of ZnO nanoparticle film and vertically aligned nanorods in chamber-based microfluidic reactors: reaction kinetics and flow effects, Applied Catalysis B: Environmental, 2017, 209, 468–475.

[43] Ashar A, Bhatti IA, Ashraf M, et al. Fe3+ @ ZnO/polyester based solar photocatalytic membrane reactor for abatement of RB5 dye, Journal of Cleaner Production, 2020, 246, 119010.

[44] Shen X, Zhang T, Xu P, Zhang L, Liu J, Chen Z. Growth of C3N4 nanosheets on carbon-fiber cloth as flexible and macroscale filter-membrane-shaped photocatalyst for degrading the flowing wastewater, Applied Catalysis B: Environmental, 2017, 219, 425–431.

[45] Zheng Q, Aiello A, Choi YS, et al. 3D printed photoreactor with immobilized graphitic carbon nitride: a sustainable platform for solar water purification, Journal of Hazardous Materials, 2020, 399, 123097.

[46] Mei Y, Su Y, Li Z, et al. BiOBr nanoplates@TiO2 nanowires/carbon fiber cloth as a functional water transport network for continuous flow water purification, Dalton Transactions, 2017, 46, 347–354.

[47] Fajrina N, Tahir M. A critical review in strategies to improve photocatalytic water splitting towards hydrogen production, International Journal of Hydrogen Energy, 2019, 44, 540–577.

[48] Colón G. Towards the hydrogen production by photocatalysis, Applied Catalysis. A, General, 2016, 518, 48–59.

[49] Maeda K. Z-Scheme water splitting using two different semiconductor photocatalysts, ACS Catalysis, 2013, 3, 1486–1503.

[50] Hisatomi T, Kubota J, Domen K. Recent advances in semiconductors for photocatalytic and photoelectrochemical water splitting, Chemical Society Reviews, 2014, 43, 7520–7535.

[51] Carr RG, Somorjai GA. Hydrogen production from photolysis of steam adsorbed onto platinized SrTiO3, Nature, 1981, 290, 576–577.

[52] Kato H, Kudo A. New tantalate photocatalysts for water decomposition into H2 and O2, Chemical Physics Letters, 1998, 295, 487–492.

[53] Sun B, Zhou W, Li H, et al. Synthesis of particulate hierarchical tandem heterojunctions toward optimized photocatalytic hydrogen production, Advanced Materials, 2018, 30, 1804282.

[54] Zhao T, Xing Z, Xiu Z, et al. CdS quantum dots/Ti3+-TiO2 nanobelts heterojunctions as efficient visible-light-driven photocatalysts, Materials Research Bulletin, 2018, 103, 114–121.

[55] Pan J, Dong Z, Wang B, et al. The enhancement of photocatalytic hydrogen production via Ti3+ self-doping black TiO2/g-C3N4 hollow core-shell nano-heterojunction, Applied Catalysis B: Environmental, 2019, 242, 92–99.

[56] Khan I, Qurashi A. Sonochemical-assisted in situ electrochemical synthesis of Ag/α-Fe2O3/ TiO2 nanoarrays to harness energy from photoelectrochemical water splitting, ACS Sustainable Chemistry & Engineering, 2018, 6, 11235–11245.

[57] Police AKR, Vattikuti SVP, Mandari KK, et al. Bismuth oxide cocatalyst and copper oxide sensitizer in Cu2O/TiO2/Bi2O3 ternary photocatalyst for efficient hydrogen production under solar light irradiation, Ceramics International, 2018, 44, 11783–11791.

[58] Shang J, Xu X, Liu K, Bao Y, Yangyang, He M. LSPR-driven upconversion enhancement and photocatalytic H2 evolution for Er-Yb: TiO2/MoO3-xnano-semiconductor heterostructure, Ceramics International, 2019, 45, 16625–16630.

[59] Hafeez HY, Lakhera SK, Bellamkonda S, et al. Construction of ternary hybrid layered reduced graphene oxide supported g-C3N4-TiO2 nanocomposite and its photocatalytic hydrogen production activity, International Journal of Hydrogen Energy, 2018, 43, 3892–3904.

[60] Idris AM, Liu T, Shah JH, et al. A novel double perovskite oxide semiconductor Sr2CoWO6 as bifunctional photocatalyst for photocatalytic oxygen and hydrogen evolution reactions from water under visible light irradiation, Solar RRL, 2020, 4, 1900456.

[61] Swain G, Sultana S, Parida K. Constructing a novel surfactant-free MoS2 nanosheet modified MgIn2S4 marigold microflower: an efficient visible-light driven 2D-2D p-n heterojunction photocatalyst toward HER and pH regulated NRR, ACS Sustainable Chemistry & Engineering, 2020, 8, 4848–4862.

[62] Ganguly P, Mathew S, Clarizia L, et al. Theoretical and experimental investigation of visible light responsive AgBiS2-TiO2 heterojunctions for enhanced photocatalytic applications, Applied Catalysis B: Environmental, 2019, 253, 401–418.

[63] Obregón S, Vázquez A, Hernández-Uresti DB. Nanocrystalline ErVO4 as a novel photocatalyst for degradation of organic compounds and solar fuels production, Journal of Materials Science: Materials in Electronics, 2018, 29, 3967–3972.

[64] Zhang Y, Peng Z, Guan S, Fu X. Novel β-NiS film modified CdS nanoflowers heterostructure nanocomposite: extraordinarily highly efficient photocatalysts for hydrogen evolution, Applied Catalysis B: Environmental, 2018, 224, 1000–1008.

[65] Shojaei F, Mortazavi B, Zhuang X, Azizi M. Silicon diphosphide (SiP2) and silicon diarsenide (SiAs2): novel stable 2D semiconductors with high carrier mobilities, promising for water splitting photocatalysts, Materials Today Energy, 2020, 16, 100377.

[66] Liu Y, Zhang H, Ke J, et al. 0D (MoS2)/2D (g-C3N4) heterojunctions in Z-scheme for enhanced photocatalytic and electrochemical hydrogen evolution, Applied Catalysis B: Environmental, 2018, 228, 64–74.

[67] Yang L, Zhou W, Lu J, et al. Hierarchical spheres constructed by defect-rich MoS 2 /carbon nanosheets for efficient electrocatalytic hydrogen evolution, Nano Energy, 2016, 22, 490–498.

[68] Lu Y, Cheng X, Tian G, et al. Hierarchical CdS/m-TiO2/G ternary photocatalyst for highly active visible light-induced hydrogen production from water splitting with high stability, Nano Energy, 2018, 47, 8–17.

[69] Fu C-F, Wu X, Yang J. Material design for photocatalytic water splitting from a theoretical perspective, Advanced Materials, 2018, 30, 1802106.

[70] Wang Y, Chen D, Hu Y, et al. An artificially constructed direct Z-scheme heterojunction: WO3 nanoparticle decorated ZnIn2S4 for efficient photocatalytic hydrogen production, Sustainable Energy & Fuels, 2020, 4, 1681–1692.

[71] White JL, Baruch MF, Pander JE, et al. Light-Driven heterogeneous reduction of carbon dioxide: photocatalysts and photoelectrodes, Chemical Reviews, 2015, 115, 12888–12935.

[72] Kar P, Zeng S, Zhang Y, et al. High rate CO2 photoreduction using flame annealed TiO2 nanotubes, Applied Catalysis B: Environmental, 2019, 243, 522–536.

[73] Gao S, Gu B, Jiao X, et al. Highly efficient and exceptionally durable CO2 photoreduction to methanol over freestanding defective single-unit-cell bismuth vanadate layers, Journal of the American Chemical Society, 2017, 139, 3438–3445.

[74] Khan AA, Tahir M. Recent advancements in engineering approach towards design of photo-reactors for selective photocatalytic CO2 reduction to renewable fuels, Journal of CO2 Utilization, 2019, 29, 205–239.

[75] Klepser BM, Bartlett BM. Anchoring a molecular iron catalyst to solar-responsive WO3 Improves the rate and selectivity of photoelectrochemical water oxidation, Journal of the American Chemical Society, 2014, 136, 1694–1697.

[76] Zhang N, Li X, Ye H, et al. Oxide defect engineering enables to couple solar energy into oxygen activation, Journal of the American Chemical Society, 2016, 138, 8928–8935.

[77] Sun S, Watanabe M, Wu J, An Q, Ishihara T. Ultrathin WO3·0.33H2O nanotubes for CO2 photoreduction to acetate with high selectivity, Journal of the American Chemical Society, 2018, 140, 6474–6482.

[78] Hao L, Huang H, Guo Y, Zhang Y. Multifunctional Bi2O2(OH)(NO3) nanosheets with {001} active exposing facets: efficient photocatalysis, dye-sensitization, and piezoelectric-catalysis, ACS Sustainable Chemistry & Engineering, 2018, 6, 1848–1862.

[79] Huang H, He Y, Li X, et al. Bi2O2(OH)(NO3) as a desirable [Bi2O2]2+ layered photocatalyst: strong intrinsic polarity, rational band structure and {001} active facets co-beneficial for robust photooxidation capability, Journal of Materials Chemistry A, 2015, 3, 24547–24556.

[80] Chen F, Huang H, Guo L, Zhang Y, Ma T. The role of polarization in photocatalysis, Angewandte Chemie International Edition, 2019, 58, 10061–10073.

[81] Hao L, Kang L, Huang H, et al. Surface-halogenation-induced atomic-site activation and local charge separation for superb CO2 photoreduction, Advanced Materials (Deerfield Beach, Fla.), 2019, 31, 1900546.

[82] Liu G, Yu JC, Lu GQ, Cheng H-M. Crystal facet engineering of semiconductor photocatalysts: motivations, advances and unique properties, Chemical Communications, 2011, 47, 6763–6783.

[83] Chen F, Huang H, Ye L, et al. Thickness-dependent facet junction control of layered BiOIO3 single crystals for highly efficient CO2 photoreduction, Advanced Functional Materials, 2018, 28, 1804284.

[84] Wang S, Guan BY, Lou XWD. Construction of ZnIn2S4–In2O3 hierarchical tubular heterostructures for efficient CO2 photoreduction, Journal of the American Chemical Society, 2018, 140, 5037–5040.

[85] Banerjee S, Dionysiou DD, Pillai SC. Self-cleaning applications of TiO2 by photo-induced hydrophilicity and photocatalysis, Applied Catalysis B: Environmental, 2015, 176–177, 396–428.

[86] Jin B, He J, Yao L, Zhang Y, Li J. Rational design and construction of well-organized macro-mesoporous SiO2/TiO2 nanostructure toward robust high-performance self-cleaning antireflective thin films, ACS Applied Materials & Interfaces, 2017, 9, 17466–17475.

[87] Nundy S, Ghosh A, Mallick TK. Hydrophilic and superhydrophilic self-cleaning coatings by morphologically varying ZnO microstructures for photovoltaic and glazing applications, ACS Omega, 2020, 5, 1033–1039.

[88] Singh R, Yadav VSK, Purkait MK. Cu2O photocatalyst modified antifouling polysulfone mixed matrix membrane for ultrafiltration of protein and visible light driven photocatalytic pharmaceutical removal, Separation and Purification Technology, 2019, 212, 191–204.

[89] Chi L, Qian Y, Guo J, Wang X, Arandiyan H, Jiang Z. Novel g-C3N4/TiO2/PAA/PTFE ultrafiltration membrane enabling enhanced antifouling and exceptional visible-light photocatalytic self-cleaning, Catalysis Today, 2019, 335, 527–537.

[90] Cedillo-González EI, Hernández-López JM, Ruiz-Valdés JJ, Barbieri V, Siligardi C. Self-cleaning TiO2 coatings for building materials: the influence of morphology and humidity in the stain removal performance, Construction and Building Materials, 2020, 237, 117692.

[91] Anderson SR, Mohammadtaheri M, Kumar D, et al. Surface plasmon resonance: robust nanostructured silver and copper fabrics with localized surface plasmon resonance property for effective visible light induced reductive catalysis (Adv. Mater. Interfaces 6/2016), Advanced Materials Interfaces, 2016, 3.

[92] Saad SR, Mahmed N, Abdullah MMAB, Sandu AV. Self-cleaning technology in fabric: a review, IOP Conference Series: Materials Science and Engineering, 2016, 133, 012028.

[93] Panwar K, Jassal M, Agrawal AK. TiO2–SiO2 Janus particles for photocatalytic self-cleaning of cotton fabric, Cellulose, 2018, 25, 2711–2720.

[94] Khan MQ, Lee H, Koo JM, et al. Self-cleaning effect of electrospun poly (1,4-cyclohexanedimethylene isosorbide terephthalate) nanofibers embedded with zinc oxide nanoparticles, Textile Research Journal, 2017, 88, 2493–2498.

[95] Hu J, Gao Q, Xu L, et al. Functionalization of cotton fabrics with highly durable polysiloxane–TiO2 hybrid layers: potential applications for photo-induced water–oil separation, UV shielding, and self-cleaning, Journal of Materials Chemistry A, 2018, 6, 6085–6095.

[96] Fan Y, Zhou J, Zhang J, et al. Photocatalysis and self-cleaning from g-C3N4 coated cotton fabrics under sunlight irradiation, Chemical Physics Letters, 2018, 699, 146–154.

[97] Bixler GD, Bhushan B. Bioinspired rice leaf and butterfly wing surface structures combining shark skin and lotus effects, Soft Matter, 2012, 8, 11271–11284.

[98] Li D, Guo Z. Stable and self-healing superhydrophobic MnO2@fabrics: applications in self-cleaning, oil/water separation and wear resistance, Journal of Colloid and Interface Science, 2017, 503, 124–130.

[99] Bano S, Zulfiqar U, Zaheer U, Awais M, Ahmad I, Subhani T. Durable and recyclable superhydrophobic fabric and mesh for oil–water separation, Advanced Engineering Materials, 2018, 20, 1700460.

[100] Guo W, Wang X, Huang J, et al. Construction of durable flame-retardant and robust superhydrophobic coatings on cotton fabrics for water-oil separation application, Chemical Engineering Journal, 2020, 398, 125661.

[101] Scacchetti FAP, Pinto E, Soares G. A multifunctional cotton fabric using TiO2 and PCMs: Introducing thermal comfort and self-cleaning properties, IOP Conference Series: Materials Science and Engineering, 2017, 254, 122011.

[102] Lu X, Sun Y, Chen Z, Gao Y. A multi-functional textile that combines self-cleaning, water-proofing and VO2-based temperature-responsive thermoregulating, Solar Energy Materials and Solar Cells, 2017, 159, 102–111.

[103] Peng L, Chen W, Su B, Yu A, Jiang X. CsxWO3 nanosheet-coated cotton fabric with multiple functions: UV/NIR shielding and full-spectrum-responsive self-cleaning, Applied Surface Science, 2019, 475, 325–333.

[104] Zhao J, Zhu W, Wang X, Liu L, Yu J, Ding B. Environmentally benign modification of breathable nanofibrous membranes exhibiting superior waterproof and photocatalytic self-cleaning properties, Nanoscale Horizons, 2019, 4, 867–873.

[105] Saito T, Iwase T, Horie J, Morioka T. Mode of photocatalytic bactericidal action of powdered semiconductor TiO2 on mutans streptococci, Journal of Photochemistry and Photobiology. B, Biology, 1992, 14, 369–379.

[106] Matsunaga T, Tomoda R, Nakajima T, Wake H. Photoelectrochemical sterilization of microbial cells by semiconductor powders, FEMS Microbiology Letters, 1985, 29, 211–214.

[107] Oguma K, Katayama H, Ohgaki S. Photoreactivation of Escherichia coli after low- or medium-pressure UV disinfection determined by an endonuclease sensitive site assay, Applied and Environmental Microbiology, 2002, 68, 6029–6035.

[108] Liou J-W, Chang -H-H. Bactericidal effects and mechanisms of visible light-responsive titanium dioxide photocatalysts on pathogenic bacteria, Archivum Immunologiae et Therapiae Experimentalis, 2012, 60, 267–275.

[109] You J, Guo Y, Guo R, Liu X. A review of visible light-active photocatalysts for water disinfection: features and prospects, Chemical Engineering Journal, 2019, 373, 624–641.

[110] Regmi C, Joshi B, Ray SK, Gyawali G, Pandey RP. Understanding mechanism of photocatalytic microbial decontamination of environmental wastewater, Frontiers in Chemistry, 2018, 6.

[111] https://www.retail-focus.co.uk/kastus-launches-a-new-generation-of-antimicrobial-and-antiviral-screen-protectors/

[112] Xu Z, Yu Y, Arya S, et al. Frequency- and power-dependent photoresponse of a perovskite photodetector down to the single-photon level, Nano Letters, 2020, 20, 2144–2151.

[113] Li M-Q, Yang N, Wang -G-G, Zhang H-Y, Han J-C. Highly preferred orientation of Ga2O3 films sputtered on SiC substrates for deep UV photodetector application, Applied Surface Science, 2019, 471, 694–702.

[114] Muench JE, Ruocco A, Giambra MA, et al. Waveguide-integrated, plasmonic enhanced graphene photodetectors, Nano Letters, 2019, 19, 7632–7644.

[115] Pargoletti E, Hossain UH, Di Bernardo I, et al. Room-temperature photodetectors and VOC sensors based on graphene oxide–ZnO nano-heterojunctions, Nanoscale, 2019, 11, 22932–22945.

[116] Liu K, Sakurai M, Aono M. ZnO-Based Ultraviolet Photodetectors, Sensors (Basel, Switzerland), 2010, 10, 8604–8634.

[117] Hamdaoui N, Ben Elkamel I, Mezni A, Ajjel R, Beji L. Highly efficient, low cost, and stable self–powered UV photodetector based on Co2+: ZnO/Sn diluted magnetic semiconductor nanoparticles, Ceramics International, 2019, 45, 17729–17736.

[118] Jia C, Wu D, Wu E, et al. A self-powered high-performance photodetector based on a MoS2/ GaAs heterojunction with high polarization sensitivity, Journal of Materials Chemistry C, 2019, 7, 3817–3821.

[119] Xie Y, Liang F, Chi S, et al. Defect engineering of MoS2 for room-temperature terahertz photodetection, ACS Applied Materials & Interfaces, 2020, 12, 7351–7357.

[120] Huang -J-J, Lin C-H, Ho Y-R, Chang Y-H. Aluminium oxide passivation films by liquid phase deposition for TiO2 ultraviolet solid–liquid heterojunction photodetectors, Surface & Coatings Technology, 2020, 391, 125684.

[121] Lee YT, Jeon PJ, Han JH, et al. Mixed-dimensional 1D ZnO–2D WSe2 van der Waals heterojunction device for photosensors, Advanced Functional Materials, 2017, 27, 1703822.

[122] Xue H, Dai Y, Kim W, et al. High photoresponsivity and broadband photodetection with a band-engineered WSe2/SnSe2 heterostructure, Nanoscale, 2019, 11, 3240–3247.

[123] Shan C, Zhao M, Jiang D, et al. Improved responsivity performance of ZnO film ultraviolet photodetectors by vertical arrays ZnO nanowires with light trapping effect, Nanotechnology, 2019, 30, 305703.

[124] Guo J, Li S, He Z, et al. Near-infrared photodetector based on few-layer MoS2 with sensitivity enhanced by localized surface plasmon resonance, Applied Surface Science, 2019, 483, 1037–1043.

[125] Mitra S, Aravindh A, Das G, et al. High-performance solar-blind flexible deep-UV photodetectors based on quantum dots synthesized by femtosecond-laser ablation, Nano Energy, 2018, 48, 551–559.

[126] Hammed NA, Aziz AA, Usman AI, Qaeed MA. The sonochemical synthesis of vertically aligned ZnO nanorods and their UV photodetection properties: effect of ZnO buffer layer, Ultrasonics Sonochemistry, 2019, 50, 172–181.

[127] Jiang X, Huang W, Wang R, et al. Photocarrier relaxation pathways in selenium quantum dots and their application in UV-Vis photodetection, Nanoscale, 2020, 12, 11232–11241.

[128] Ouyang B, Zhao H, Wang ZL, Yang Y. Dual-polarity response in self-powered ZnO NWs/ Sb2Se3 film heterojunction photodetector array for optical communication, Nano Energy, 2020, 68, 104312.

[129] Yu L, Li C, Ma S, et al. Optoelectronic gas sensor sensitized by hierarchically structured ZnO nanorods/Ag nanofibers via on-chip fabrication, Materials Letters, 2019, 242, 71–74.

[130] Li W, Guo J, Cai L, et al. UV light irradiation enhanced gas sensor selectivity of NO2 and SO2 using rGO functionalized with hollow SnO2 nanofibers, Sensors and Actuators. B, Chemical, 2019, 290, 443–452.

[131] Yu H-L, Wang J, Zheng B, et al. Fabrication of single crystalline WO3 nano-belts based photoelectric gas sensor for detection of high concentration ethanol gas at room temperature, Sensors and Actuators. A, Physical, 2020, 303, 111865.

[132] Cui J, Pan G, Yang X, Zhu M, Huang C, Qi J. Enhanced acetone sensing performance of CeO2-ZnO at low temperature and its photo-excitation effect, Materials Science in Semiconductor Processing, 2020, 118, 105221.

[133] Seif AM, Nikfarjam A, Hajghassem H. UV enhanced ammonia gas sensing properties of PANI/ TiO2 core-shell nanofibers, Sensors and Actuators. B, Chemical, 2019, 298, 126906.

[134] Wang J, Fan S, Xia Y, Yang C, Komarneni S. Room-temperature gas sensors based on ZnO nanorod/Au hybrids: visible-light-modulated dual selectivity to NO2 and NH3, Journal of Hazardous Materials, 2020, 381, 120919.

[135] Wang J, Yu M, Li X, Xia Y. UV-enhanced NO2 gas sensing properties of polystyrene sulfonate functionalized ZnO nanowires at room temperature, Inorganic Chemistry Frontiers, 2019, 6, 176–183.

[136] Yang Y, Wang G, Deng Q, Ng DHL, Zhao H. Microwave-assisted fabrication of nanoparticulate TiO2 microspheres for synergistic photocatalytic removal of Cr(VI) and methyl orange, ACS Applied Materials & Interfaces, 2014, 6, 3008–3015.
[137] Khare P, Bhati A, Anand SR, Gunture, Sk S. Brightly fluorescent zinc-doped red-emitting carbon dots for the sunlight-induced photoreduction of Cr(VI) to Cr(III), ACS Omega, 2018, 3, 5187–5194.
[138] Bhati A, Anand SR, Saini D, Gunture, Sonkar SK. Sunlight-induced photoreduction of Cr(VI) to Cr(III) in wastewater by nitrogen-phosphorus-doped carbon dots, npj Clean Water, 2019, 2, 12.
[139] Manos MJ, Kanatzidis MG. Layered metal sulfides capture uranium from seawater, Journal of the American Chemical Society, 2012, 134, 16441–16446.
[140] Bryant PA. Chemical toxicity and radiological health detriment associated with the inhalation of various enrichments of uranium, Journal of Radiological Protection, 2014, 34, N1–N6.
[141] Babu MNS, Somashekar RK, Kumar SA, Shivanna K, Krishnamurthy V, Eappen KP. Concentration of uranium levels in groundwater, International Journal of Environmental Science & Technology, 2008, 5, 263–266.
[142] Keshtkar AR, Mohammadi M, Moosavian MA. Equilibrium biosorption studies of wastewater U(VI), Cu(II) and Ni(II) by the brown alga Cystoseira indica in single, binary and ternary metal systems, Journal of Radioanalytical and Nuclear Chemistry, 2015, 303, 363–376.
[143] Bundschuh J, Litter M, Ciminelli VST, et al. Emerging mitigation needs and sustainable options for solving the arsenic problems of rural and isolated urban areas in Latin America – A critical analysis, Water Research, 2010, 44, 5828–5845.
[144] Jiang X-H, Xing Q-J, Luo X-B, et al. Simultaneous photoreduction of Uranium(VI) and photooxidation of Arsenic(III) in aqueous solution over g-C3N4/TiO2 heterostructured catalysts under simulated sunlight irradiation, Applied Catalysis B: Environmental, 2018, 228, 29–38.
[145] Lueder U, Jørgensen BB, Kappler A, Schmidt C. Fe(III) photoreduction producing Feaq2+ in oxic freshwater sediment, Environmental Science & Technology, 2020, 54, 862–869.
[146] Grätzel M. Dye-sensitized solar cells, Journal of Photochemistry and Photobiology C: Photochemistry Reviews, 2003, 4, 145–153.
[147] Liu Q, Sun Y, Yao M, et al. Au@Ag@Ag2S heterogeneous plasmonic nanorods for enhanced dye-sensitized solar cell performance, Solar Energy, 2019, 185, 290–297.
[148] Wang C-T, Lin H-S, Wang W-P. Hydrothermal synthesis of Fe and Nb-doped titania nanobelts and their tunable electronic structure toward photovoltaic application, Materials Science in Semiconductor Processing, 2019, 99, 85–91.
[149] Ramasubbu V, Kumar PR, Mothi EM, et al. Highly interconnected porous TiO2-Ni-MOF composite aerogel photoanodes for high power conversion efficiency in quasi-solid dye-sensitized solar cells, Applied Surface Science, 2019, 496, 143646.
[150] Fu N, Jiang X, Chen D, et al. Au/TiO2 nanotube array based multi-hierarchical architecture for highly efficient dye-sensitized solar cells, Journal of Power Sources, 2019, 439, 227076.
[151] Kumar KA, Subalakshmi K, Senthilselvan J. Effect of co-sensitization in solar exfoliated TiO2 functionalized rGO photoanode for dye-sensitized solar cell applications, Materials Science in Semiconductor Processing, 2019, 96, 104–115.
[152] Nasirian A, Mirkhani V, Moghadam M, Tangestaninejad S, Mohammadpour Baltork I. Efficient dye-sensitized solar cell based on a new porphyrin complex as an inorganic photosensitizer, Journal of Chemical Sciences, 2020, 132, 75.
[153] Chandra Maurya I, Singh S, Srivastava P, Maiti B, Bahadur L. Natural dye extract from Cassia fistula and its application in dye-sensitized solar cell: experimental and density functional theory studies, Optical Materials, 2019, 90, 273–280.

[154] Zhang H, Chen Z-E, Tian H-R. Molecular engineering of metal-free organic sensitizers with polycyclic benzenoid hydrocarbon donor for DSSC applications: the effect of the conjugate mode, Solar Energy, 2020, 198, 239–246.

[155] Li Y, Ma L, Yoo Y, Wang G, Zhang X, Ko MJ. Atomic layer deposition: a versatile method to enhance TiO2 nanoparticles interconnection of dye-sensitized solar cell at low temperature, Journal of Industrial and Engineering Chemistry, 2019, 73, 351–356.

[156] Murugadoss V, Panneerselvam P, Yan C, Guo Z, Angaiah S. A simple one-step hydrothermal synthesis of cobalt nickel selenide/graphene nanohybrid as an advanced platinum free counter electrode for dye sensitized solar cell, Electrochimica Acta, 2019, 312, 157–167.

[157] Fresno F, Portela R, Suárez S, Coronado JM. Photocatalytic materials: recent achievements and near future trends, Journal of Materials Chemistry A, 2014, 2, 2863–2884.

[158] Chen S, Qi Y, Li C, Domen K, Zhang F. Surface strategies for particulate photocatalysts toward artificial photosynthesis, Joule, 2018, 2, 2260–2288.

[159] Maeda K, Domen K. Photocatalytic water splitting: recent progress and future challenges, The Journal of Physical Chemistry Letters, 2010, 1, 2655–2661.

Index

https://doi.org/10.1515/9783110668483-010

www.ingramcontent.com/pod-product-compliance
Lightning Source LLC
Chambersburg PA
CBHW080703220326
41598CB00033B/5291